Experientia Supplementum

Volume 115

Experientia Supplementum (EXS) is a multidisciplinary book series originally created as a supplement to the journal *Experientia* which appears now under the cover of *Cellular and Molecular Life Sciences*.

***

The edited volumes focus on selected topics of biological or biomedical research, discussing current methodologies, technological innovations, novel tools and applications, new developments and recent findings.

The series is a valuable source of information not only for scientists and graduate students in medical, pharmacological and biological research, but also for physicians as well as practitioners in industry.

***

EXS is indexed in Medline and Chemical Abstract Service (CAS).

Mohammad Fahad Ullah • Aamir Ahmad
Editors

Non-coding RNAs

Insights on the Regulatory Genome

Springer

Editors
Mohammad Fahad Ullah
Medical Laboratory Technology,
Faculty of Applied Medical Science
University of Tabuk
Tabuk, Saudi Arabia

Aamir Ahmad
Translational Research Institute
Hamad Medical Corporation
Doha, Qatar

ISSN 1664-431X ISSN 2504-3692 (electronic)
Experientia Supplementum
ISBN 978-3-032-06947-4 ISBN 978-3-032-06948-1 (eBook)
https://doi.org/10.1007/978-3-032-06948-1

This Springer imprint is published by the registered company Springer Nature Switzerland AG
The registered company address is: Gewerbestrasse 11, 6330 Cham, Switzerland

Foreword

Research on non-coding RNAs has come a long way since the realization that these RNA molecules hold a lot of promise, particularly in the way of diagnosis, prognosis, and perhaps treatment of several human diseases. A number of non-coding RNAs, ranging from linear microRNAs and long-non-coding RNAs to circular RNAs, and several others are now under investigation. The book *Non-coding RNAs: Insights on the Regulatory Genome* is a timely contribution to the understanding of the role of ncRNAs in the setting and progression of human diseases and cancer. It is recognized that ncRNAs are deregulated in various types of cancer, and can provide a signature for the typing of cancers, and are of prognostic value for classification, such as metastatic potential, response to chemotherapeutic drugs, and cancer cell metabolism. This book provides a comprehensive exploration of the world of ncRNAs, unraveling their functional potential and shedding light on their significance in the regulatory genome. One of the relatively more explored functions of lncRNAs is their ability to sponge and regulate the expression of microRNAs (miRNAs), the other ncRNAs that are much shorter in length. The regulation through miRNAs is not the only mode of lncRNAs' functionality. There is a role of ncRNAs in stoichiometric saturation of a target sequence, such as protection of mRNA untranslated regions (UTRs) from miRNA binding: ncRNAs act as decoys by protecting target sequences in mRNA and proteins from undesired interactions. Novel performers have attracted the attention of ncRNA research, such as circular RNAs that have a transcriptional role in drug resistance through miRNAs/mRNAs axis, and tRNA-derived miRNAs. These ncRNA families are involved in the response to cellular stress, including ER stress, metabolic stress (serum starvation) or heat shock, and in the silencing of the immune response. Non-coding RNAs have been implicated in metabolic reprogramming, such as mitochondrial miRNAs as key players in tumor survival. Various aspects have been explored relating the expression of ncRNAs with development of neurological disorders and cardiovascular diseases, blood disorders, carcinogenesis, and metastasis. The deregulated expression of ncRNAs can provide cell-specific biomarkers in cancers, useful for the diagnosis and prognosis of the disease and its metastatic potential.

This book is a timely contribution for RNA world scientists, in the prospect of disease treatment, addressing the present challenges and future perspectives in ncRNA in relation to human diseases and therapy. The editors, Drs. Mohammad Fahad Ullah and Aamir Ahmad, are renowned scientists studying ncRNAs in various human diseases settings. Dr. Ullah has found miRNA polymorphisms linked to susceptibility to diseases such as PCOS and cancer, while Dr. Ahmad, in addition to studying ncRNAs in cancer metastasis and therapy resistance, has been editor of special issues for several reputed journals, and is Editor-in-Chief for the journal *Non-coding RNA Research.*

Agrofood Department
Institute of Sciences of Food Productions (CNR-ISPA)
Lecce, Italy
National Research Council of Italy
Rome, Italy

Palmiro Poltronieri

Preface

The *Non-coding RNAs: Insights on the Regulatory Genome* has been a satisfying experience, the timing of which could not have been more pertinent provided the fact that the world recognized the significance of non-coding RNAs with the Nobel Prize 2024 in Physiology or Medicine awarded to Victor Ambros and Gary Ruvkun for their discovery of microRNAs. These RNA molecules, which are encoded by a large span of DNA and were previously considered to be a junk due to the absence of any translation protein product, are evident and critical regulators of gene expression and various biological processes. It is increasingly being realized that the complex interplay of RNA diversity, rather than just the number of proteins, significantly influences the developmental complexity of organisms from a worm like *C. elegans* to humans. This provides an in-depth exploration of various categories of ncRNAs, their biogenesis, mechanisms of action, and implications in health and disease. The book is structured into 14 chapters that navigate the reader through the diverse roles of ncRNAs, from their involvement in neurological and cardiovascular diseases to their implications in cancer and metabolic disorders.

Chapter Overviews

Chapter 1, "Non-coding RNAs: The Not So Junk DNA Is the Blueprint of the Transcriptional Landscape of the RNA World," introduces the subject of the book highlighting various categories of ncRNAs, their biogenesis, mechanisms of action, their critical roles in regulating gene expression, cellular processes, and maintaining genomic stability. Chapter 2, "Roles of ncRNAs in Neurological Disorders and Cardiovascular Diseases," explores the crucial roles of ncRNAs such as long non-coding RNAs (lncRNAs) and microRNAs (miRNAs) in neurological disorders and cardiovascular diseases. The text discusses how ncRNAs orchestrate molecular regulatory networks in these diseases. Chapter 3, "Role of SLC Transporters and Non-coding RNA in Regulating the Genetic Landscape of the Blood Disorders," addresses the complex interplay between solute carrier (SLC) transporters and ncRNAs in regulating genetic and metabolic pathways in hematological disorders. It highlights the multifaceted roles of SLC transporters and ncRNAs in blood cell regulation, providing evidence of their impact on hematopoiesis and leukemogenesis. Further, it also explores how SLC transporters mediate nutrient uptake, drug resistance, and metabolic adaptations in leukemic cells, and examines the involvement of ncRNAs in modulating transporter activity. Chapter 4, "Reconnaissance of the Good, the Bad and the Ugly Role of tRNAs in Carcinogenesis and Metastasis," describes the multiple roles of transfer RNAs (tRNAs) in cancer development and metastasis. The chapter provides an overview of the role of tRNAs, aminoacyl-tRNA synthetases, and tRNA-derived small RNAs in the onset and progression of cancer. Chapter 5, "The Potential of Non-coding RNAs: siRNAs in Cancer Research and Therapy—Challenges and Solutions," provides a comprehensive review of the biogenesis, functions, and roles of small interfering RNAs (siRNAs) in cancer research and therapy. The chapter discusses how siRNAs function primarily by silencing specific genes through RNA interference (RNAi). It also addresses the challenges and potential of siRNAs as a powerful tool for cancer treatment, offering novel strategies for targeted therapies and personalized medicine. Chapter 6, "Exploring the Role of Long Non-coding RNAs as Biomarkers in Cancers," explores the potential of lncRNAs as biomarkers in cancer. It discusses how dysregulated lncRNAs are associated with various disorders, including cancer, and can be biomarkers for diagnosis and prognosis. It further highlights how the aberrant expression of lncRNAs regulates tumorigenesis, proliferation, metastasis, cancer stage, tumor size, and overall survival rates. Chapter 7, "MicroRNAs as Biomarkers for the Diagnosis and Prognosis of Different Diseases and Cancer," discusses the potential of miRNAs as reliable biomarkers for disease diagnosis, prognosis, and response prediction to treatment. The chapter emphasizes the identification of miRNAs in bodily fluids as a compelling possibility for the creation of noninvasive biomarkers. It also connects the deregulation of miRNA expression to the emergence of metabolic, cardiovascular, neurodegenerative, and cancerous disorders. Chapter 8, "The Transcriptional Role of Circular RNAs in Drug Resistance Through miRNAs/mRNAs Axis," explores the transcriptional role of circular RNAs (circRNAs) in drug resistance through the miRNA/mRNA axis. It highlights the connection between cancers and non-coding RNAs, focusing on how circRNAs regulate miRNAs to control mRNA degradation in the context of drug resistance. Chapter 9, "Silhouette of Probiotics in the Regulation of miRNAs and lncRNAs in Carcinogenesis and Metastasis," summarizes seminal studies associated with the anticancer and anti-metastatic role of probiotics in animal models and critically examines how probiotics modulate non-coding RNAs to prevent/inhibit cancer progression. Chapter 10, "Non-coding RNAs in Metabolic Reprogramming of Breast Cancer: Dietary Opportunities and Translational Implications," discusses the intricate relationship between ncRNAs and metabolic reprogramming in breast cancer. It focuses on innovative techniques to target ncRNAs and the potential of dietary strategies to influence these regulatory pathways, highlighting the roles of microRNAs, long non-coding RNAs, and circular RNAs in coordinating the complex regulatory networks underlying metabolic reprogramming. Chapter 11, "MitomiRs as

Key Players in Metabolic Reprogramming of Blood Cell Development," discusses the role of mitochondrial microRNAs (MitomiRs) in metabolic reprogramming during blood cell development. The chapter explores the biosynthesis, function, and possible therapeutic implications of MitomiRs, emphasizing their influence on mitochondrial gene expression and cellular functions like oxidative stress responses, energy production, and apoptosis. Chapter 12, "Non-Coding RNA in Clinical Trials: Diagnostic and Therapeutical Prospects," is an interesting write-up that explores the clinical potential of ncRNAs, focusing on their diagnostic and therapeutic prospects. The chapter discusses the use of ncRNAs in clinical applications, highlighting their roles in regulating cellular processes and their detection in various body fluids. It also addresses the challenges associated with ncRNAs, such as off-target effects and limitations in delivery systems, while emphasizing the need for broader research into investigating other types of ncRNAs. Chapter 13, "Bioinformatics Approaches in Non-coding RNAs Research," explores the recent advancements in ncRNA databases and the innovative in silico strategies that enable the prediction and analysis of biological interactions, particularly for miRNAs and lncRNAs. The chapter discusses the structural properties, classification, and functions of various types of ncRNAs, highlighting their crucial roles in cellular processes. Chapter 14, "Challenges and Perspectives in ncRNA Research," concludes the book with a discussion on the challenges and perspectives in ncRNA research. The chapter focuses on the promises that ncRNAs hold and the associated challenges that are impeding their utilization in clinics, particularly in the context of diagnosis, prognosis, and treatment of several human diseases.

We express our gratitude to all the authors for the valuable contributions. It is their willingness to share their expert opinions that has empowered us to bring forth this piece of scientific literature with an impactful content. We appreciate the support of Dr. Tanja Weyandt (Editor: Springer-Nature) for working out the procedural framework of our book proposal. Further, we also acknowledge the support of Dr. Ramya Ravichandran (Production Editor: Springer-Nature) for her meaningful insight in developing this project. We are indeed honored to have Prof. Palmiro Poltronieri (National Research Council of Italy), introducing the substance of the book in the foreword. Lastly, we wish that *Non-coding RNAs: Insights on the Regulatory Genome* serves as a valuable resource for those seeking to understand the complex roles of ncRNAs in biology and medicine.

Tabuk, Saudi Arabia — Mohammad Fahad Ullah
Doha, Qatar — Aamir Ahmad

About the Book

Non-coding RNAs (ncRNAs) have emerged as significant regulators of gene expression and biological processes in recent years. Once considered "junk" or functionless molecules, ncRNAs have captured the attention of researchers worldwide due to their diverse roles in gene regulation and their implications in various diseases. This transformation reflects the relentless pursuit of scientific inquiry, driven by technological advancements and a curiosity for unraveling the complexities of the genome. These versatile molecules orchestrate a symphony of regulatory mechanisms that govern critical phases of cellular behavior from development to disease. Their significance lies in their ability to fine-tune gene expression, influence cellular pathways, and maintain genomic stability, thereby shaping the very essence of life. This book, *Non-coding RNAs: Insights on the Regulatory Genome*, provides a comprehensive exploration of ncRNAs, unraveling their functional potential and shedding light on their significance in the regulatory genome. Specific chapters examine the roles of microRNAs (miRNAs), long non-coding RNAs (lncRNAs), circular RNAs (circRNAs), and other ncRNA classes in various biological processes, such as development, disease pathogenesis, and cellular homeostasis that would find resonance with scientific curiosity of the audience while engaging with a meaningful content.

Contents

About the Editors

Mohammad Fahad Ullah, PhD is Professor of Biochemistry in the Department of Medical Laboratory Technology, Faculty of Applied Medical Sciences, University of Tabuk, Saudi Arabia. He received his academic degrees along with a gold medal in MSc (Biochemistry) from Aligarh Muslim University, India. Further, he worked as a Research Associate at Experimental Oncology Laboratory, Department of Biomedical & Diagnostic Sciences, University of Tennessee, USA. His research interests include studies on small molecules (low molecular weight compounds) in the context of their mechanistic actions and therapeutic potential against chronic diseases including cancer and diabetes. His work has also demonstrated the role of polymorphic variations in certain microRNAs in increasing the risk and susceptibility to diseases such as PCOS and cancer. He is a member of AACR (USA)/Royal Society of Chemistry (UK) and member of the editorial/reviewer board of a number of scientific journals. He has more than 14 years of experience in teaching Biochemistry to the students of biological and health sciences. His academic works include 75 publications in reputed journals (WOS/h-index: 30) and 4 books entitled *Critical Dietary Factors in Cancer Chemoprevention* (Springer, Switzerland), *Illustrated Notes on Biomolecules* (Partridge, Singapore), *Nutraceuticals and Natural Product Derivatives: Disease Prevention & Drug Discovery* (John Wiley & Sons, UK), and *Biocatalysis: Enzymatic Basics and Applications* (Springer, Switzerland). He has been ranked among 2% scientists globally for high citation impact published by Stanford University in 2023 and 2025.

Aamir Ahmad, PhD is a Principal Investigator and Senior Scientist at Hamad Medical Corporation, Doha, Qatar. He received his academic degrees from Aligarh Muslim University, Aligarh, India, with gold medals for highest marks in Department (Biochemistry) as well as Faculties of Science and Life Sciences combined. He completed his post-doctoral training at National Cancer Institute, National Institutes of Health, Bethesda, MD, USA. He has been a member of leading cancer organizations such as the American Association for Cancer Research, Union for International Cancer Control, American Society of Clinical Oncology, and Cancer Epigenetics

Society, and has authored more than 250 peer-reviewed publications (h-index: 79) and 30 book chapters. He serves as Editor-in-Chief of the journal *Non-coding RNA Research* in addition to serving as editorial board member of several journals. He has edited 8 books and was ranked among the top 2% scientists globally for career-long citation impact, in the list published by Stanford University in 2020 and also in the follow-up lists 2021 through 2024, based on data from Mendeley and Scopus.

Chapter 1
Noncoding RNAs: The Not So "Junk" DNA Is the Blueprint of the Transcriptional Landscape of the RNA World

Mohammad Fahad Ullah

Contents

Abstract In the past, the vast genomic DNA, encoding transcripts now known as noncoding RNAs were considered as "junk DNA." However, over the years, studies have demonstrated the critical roles of these noncoding RNAs in various biological

---

M. F. Ullah (✉)
Department of Medical Laboratory Technology, Faculty of Applied Medical Science, University of Tabuk, Tabuk, Saudi Arabia
e-mail: m.ullah@ut.edu.sa

M. F. Ullah, A. Ahmad (eds.), *Non-coding RNAs*, Experientia Supplementum 115,
https://doi.org/10.1007/978-3-032-06948-1_1

functions, leading to a major transformation in our understanding of molecular biology. Noncoding RNAs (ncRNAs) have emerged as critical players in the regulation of gene expression, cellular processes, and the maintenance of genomic stability. It is the complex interplay of the RNA diversity rather than the number of proteins that seems to have a significant role in the developmental complexity of organisms. This chapter provides an in-depth exploration of various categories of ncRNAs, their biogenesis, mechanisms of action, and implications in health and disease. With technological advancements in genomic sequencing and bioinformatics, our understanding of ncRNAs has expanded significantly, revealing their vast potential as therapeutic targets and biomarkers.

Keywords Noncoding RNAs · Gene regulation · RNA biogenesis · Nucleases · Spliceosome

Abbreviation

circRNAs	Circular RNAs
endo-siRNAs	Endogenous small interfering RNAs
exRNAs	Extracellular RNAs
lncRNAs	Long noncoding RNAs
miRNAs	microRNAs
moRNAs	miRNA-offset-RNAs
piRNAs	Piwi-interacting RNAs
rRNAs	Ribosomal RNAs
scaRNAs	Small cajal body RNAs
sdRNAs	sno-derived RNAs
siRNAs	Small interfering RNAs
snoRNAs	Small nucleolar RNAs
snRNAs	Small nuclear RNAs
tiRNAs	Transcription initiation RNAs
tRFs	tRNA-derived RNA fragments
tRNAs	Transfer RNAs
t-UCRs	Transcribed-ultraconserved regions
vtRNAs	Vault RNAs

1.1 Introduction

Unlike messenger RNAs (mRNAs) that encode proteins, noncoding RNAs (ncRNAs) do not serve this classical function. However, ncRNA performs important functions in regulating gene expression, with many cellular functions attributed to them as important regulators of an array of biological processes. These include a cluster of molecular functions such as regulating gene expression, cell cycle regulation, regulating apoptosis or programmed cell death, regulating cell differentiation, chromatin remodeling, epigenetic modifications, and mediating homeostasis (Mangiavacchi et al. 2023). Previously, large genomic DNA regions encoding transcripts referred to as noncoding RNAs were perceived as "junk DNA." However, extensive research has elucidated noncoding RNA functions in a variety of biological functions, which has led to a paradigm shift in our comprehension of molecular biology. Within the human transcriptome, protein coding genes comprise a fraction of the whole, while approximately 66% of the human transcriptome incorporates various noncoding RNA species (Hubbard et al. 2002). Think of our DNA as a massive library, where surprisingly, only a tiny fraction, less than 2% genes, contains the blueprints for building proteins (The ENCODE Project Consortium 2004). Noncoding RNAs, transcribed from the majority of DNA, play vital roles in cellular processes, particularly in regulation, despite not encoding proteins. Based on their length, ncRNAs are broadly classified into two categories: small ncRNAs (sncRNAs), which are shorter than 200 nucleotides and long noncoding RNAs (lncRNAs), exceeding 200 nucleotides in length (Mattick et al. 2023; Smal et al. 2024). It is interesting to note that human beings, with all their complexities, share a similar number of protein-coding genes with a tiny and much lesser complex organism like *Caenorhabditis elegans*. There exist about roughly 20,000 genes in both cases, despite the massive difference in the number of cells—around 30 trillion in humans versus a mere 1000 in the worm (Hillier et al. 2005; Ezkurdia et al. 2014). This suggests that it is the complex interplay of the RNA diversity rather than the number of proteins that seems to have a significant role in the developmental complexity of organisms. All noncoding RNAs are generated from longer precursors through the action of different RNases, which include both exo- and endonucleases targeting single and double-stranded RNA. Mature noncoding RNAs evade enzymatic degradation through various protective mechanisms (Falaleeva and Stamm 2013). A variety of RNases, including exo- and endonucleases that target single- and double-stranded RNA, work to produce all noncoding RNAs from longer precursors. Using a variety of defense mechanisms, mature noncoding RNAs avoid enzymatic destruction (Falaleeva and Stamm 2013). In order to protect them against exonucleases that target single-stranded RNAs, double-stranded RNA structures are formed at their ends. Moreover, base modifications such as pseudouridylation and 2′-O-methylation reduces RNase sensitivity, and end capping, adenylation, or uridylation provide additional defense against exonucleases. It has been observed that these RNAs form ribonuclear protein complexes with proteins, which provide an additional protection from the nuclease activity. Recent technological advancements in genomic sequencing and bioinformatics have enhanced our understanding of ncRNAs significantly, expanding their functional importance in cellular pathways (Micheel et al. 2021). In the sections ahead, a brief account of the different classes of noncoding RNAs has been provided to create an enhanced understanding of their diverse structures and functions, mostly focused on humans.

1.2 Transfer RNA (tRNA)

Living organisms require transfer RNA (tRNA) as an essential element of translational mechanism, which delivers precise amino acids to ribosomes for protein synthesis. The cellular RNA composition contains tRNAs as its most plentiful elements, which make up nearly 15% of all cellular RNA (Su et al. 2020). Transfer RNA functions as the adapter between genetic instructions and messenger RNA (mRNA) to generate proteins. A single tRNA transports its particular amino acid to ribosomes, which pair it with specific mRNA codons. The complex synthesis pathway accomplishes precise protein synthesis because accurate protein sequences form a critical base for biological functions. Scientific research has established that one mutant tRNA in human cells can result in major mistranslation events (Lant et al. 2018). Studies of tRNA along with its newly discovered functions in cellular processes show that this molecule plays a fundamental role in gene expression regulation and cellular metabolic processes despite its initial role in protein synthesis.

Every species preserves its fundamental tRNA structural features at a high degree of conservation. A three-dimensional L-shape configuration combines with the cloverleaf structure of this molecule. A tRNA molecule comprises 76–90 nucleotides, which exhibit several identifiable structural components starting from the (i) Amino Acid Arm at one end, (ii) Anticodon Arm: This arm comprises a triplet of nucleotides that, during translation, base pair with the complementary codon on the mRNA, (iii) D-Loop (Dihydrouridine Loop): This loop is distinguished by the presence of dihydrouridine residues and is involved in the interaction with aminoacyl-tRNA synthetases, which are the enzymes that charge tRNAs with their corresponding amino acids. (iv) TΨC Loop: This loop's pseudouridine residue is essential for preserving the tRNA's structural stability and promoting interactions with the ribosome. The three-dimensional organizational pattern of tRNA serves dual functions to detect codons and guarantee proficient and precise protein synthesis.

The formation of tRNA requires multiple essential steps beginning with transcription and ending with post-transcriptional modifications. The transcription process uses the following steps: tRNA gene transcription occurs through RNA polymerase in all organisms and by RNA polymerase III in eukaryotic cells. During transcription, RNA polymerase III generates pre-tRNA, which contains intron sequences together with 5′ and 3′ end extra nucleotides. Mature tRNA formation begins with pre-tRNA processing into mature tRNA following three main steps that include RNase P-directed 5′ leader sequence removal and 3′ end tRNA CCA sequence production through trimming and intron splicing, which employs a spliceosome or tRNA-specific mechanism to eliminate introns while keeping the exons intact. Mature tRNA receives essential post-transcriptional modifications through nucleotide methylation and uridine modification into pseudouridine and atypical base queuosine installation, which stabilize and direct tRNA structure for binding amino acids and ribosomes. The tRNA achieves maturity through aminoacylation where aminoacyl-tRNA synthetases join tRNA with its corresponding amino acid. Aminoacylation works in a precise manner as enzymes verify tRNA and distinct amino acids to create complexes between the activated amino acid and tRNA. As mentioned, the core function of tRNA is to act as an adaptor molecule during protein synthesis. However, beyond its classical role in translation, tRNA has also been implicated in various cellular processes such as gene regulation, cellular signaling, and defense mechanisms. Thus, transfer RNA is not merely a conduit in the flow of genetic information but is a multifunctional entity that plays indispensable roles in cellular metabolism, gene expression, and adaptation. Cells adopt two primary strategic responses to counter stressful conditions through the transient Integrated Stress Response (ISR) pathway, which allows them either to recover homeostasis or initiate programmed cell death. The protein kinase General Control Nondepressible 2 (GCN2) functions as the central element of ISR by detecting nutrients depletion. The pathway relies on deacylated tRNA to detect amino acid deficiency because the direct binding between deacylated tRNA and GCN2 removes self-inhibitory mechanisms leading to pathway activation (Masson 2019). A role of initiator-tRNA has also been suggested in pre-mRNA splicing, which has a critical significance in the regulation of gene expression and protein synthesis (Kamhi et al. 2010). The reverse transcription phase of retroviral and retrotransposon life cycles generates double-stranded DNA (dsDNA) from genomic RNA that subsequently integrates into the host genome. During their replication cycle, these take over essential components of host cell translational mechanisms to use cellular transfer RNAs as reverse transcription primers (Jin and Musier-Forsyth 2019). Bacterial RNA is a potent trigger of the mammalian innate immune response via Toll-like receptors. To evade this response, 2′-O-methylation—an abundant modification found in mammalian cells—on guanosine 18 (Gm18) in bacterial tRNA has been shown to inhibit immune activation. This indicates that bacteria may utilize tRNA modifications to diminish the host defense mechanisms (Keller et al. 2018). Moreover, tRNA can prevent cytochrome c (cyt c)-mediated apoptosis through a specific interaction with cyt c, with the reduction of cyt c potentially serving as a mechanism for cells to avoid apoptosis during conditions of high tRNA expression, such as during growth. In contrast, the degradation of tRNAs increases apoptosis (Mei et al. 2010).

1.3 Transfer RNA-Derived Small RNAs (tsRNAs)

During the process of biogenesis, transfer RNAs (tRNAs) can be selectively cleaved into a range of small fragments by various enzymes, resulting in tRNA-derived small RNAs (tsRNAs), which usually range from 15 to 40 nucleotides in length (Kim et al. 2020). The generation of tsRNAs does not occur through random tRNA degradation because these small fragments result from directed enzymatic cleavage of either pre-tRNA precursors or mature tRNA molecules at particular sites (Wen et al. 2021). Endonucleases like angiogenin (ANG), RNase Z, and Dicer create different categories of tsRNAs by targeting specific regions of pre-tRNAs or mature tRNAs through proper enzymatic cleavage. Different types of tsRNAs have been identified depending on their cleavage sites with tRFs and tiRNAs.

The tRNA halves also known as tRNA-derived stress-inducible RNAs (tiRNAs) are generated when angiogenin (ANG) cleaves the middle of the anticodon loop of mature tRNAs in response to stress stimuli such as heat shock, cold shock, hypoxia, starvation, viral infections, and oxidative stress. Depending on whether the cleavage site includes the 5′ or 3′ sequence of the anticodon loop, tiRNAs (31–40 nucleotides in length) can be categorized as either 5′-tiRNA or 3′-tiRNA. Alternatively, tRNA-derived fragments (tRFs), which range in length from 14 to 30 nucleotides can also arise from either precursor tRNAs or mature tRNAs. These fragments are classified into different types such as tRF-1, tRF-3, tRF-5, tRF-2, and i-tRF based on their specific sites of origin within the tRNA molecules. Among these, tRF-1 (14–30 nucleotides) is created when RNase Z/ELAC2 cleaves the 3′ end of pre-tRNAs. It is characterized by a poly(U) tail, earning it the designation 3′UtRF. The tRF-3 fragment is produced by Dicer and ANG cleaving the TψC loop of mature tRNAs and can be further divided into tRF-3a (18 nucleotides) and tRF-3b (22 nucleotides). The tRF-5 fragment is formed through Dicer's cleavage of the D-loop, D stem, or the 5′ half of the anticodon stem of mature tRNAs. Based on the different cleavage sites, tRF-5 can also be further classified into three subtypes that include tRF-5a (14–16 nucleotide), tRF-5b (22–24 nucleotide), and tRF-5c (28–30 nucleotide). Moreover, tRF-2 is generated from the anticodon loop under hypoxic stress, while i-tRF primarily originates from the internal region of mature tRNAs. Additionally, i-tRF can be further categorized into A-tRF, D-tRF, and V-tRF based on different starting cleavage sites (Chen et al. 2024).

It is understood that tsRNAs are involved in several regulatory mechanisms that encompass epigenetic regulation, transcriptional and post-transcriptional regulation, as well as translation regulation (Li et al. 2024). The tsRNAs have been identified in a range of physiological conditions, notably being present in high quantities in serum, mammalian sperm, various bodily fluids, and extracellular vesicles (Chen and Zhou 2023). Previous reports have indicated that tRNA decomposition products were found in the urine of 27 cancer patients with 13 distinct types of malignancies, suggesting that the levels of these decomposition products may be related to the severity of the disease (Speer et al. 1979). tsRNAs can interact with the endonuclease Dicer and associate with AGO protein, leading to the formation of the RNA-induced silencing complex (RISC) and consequently regulating gene expression at both transcriptional and post-transcriptional levels, such as tRF-3 causing sequence-specific repression of mRNA transcription in B-cell lymphoma (Zhou et al. 2024). These have also shown to possess regulatory activities in translation since tiRNAs feature a segment of the terminal oligoguanine (TOG) motif at their 5′ end (Kim 2019), which grants them translational repression capabilities by facilitating the formation of tetramer-containing G-quadruplex (G4) structures.

1.4 Ribosomal RNAs (rRNAs)

Ribosomes function as macromolecular production assemblies, which rapidly and precisely turn amino acids into proteins mainly guided by ribosomal RNA (rRNA). The rRNA serves as a crucial component in the ribosome because it maintains both structural integrity and functional catalytic activity (Sloan et al. 2017). It has been established that both the ribosomes and tRNAs are highly conserved in the evolutionary process that shows a relatively high percentage of similarity in all living forms. Dynamic connections between rRNA, tRNA, and peripheral proteins guide the tRNA molecules through the ribosome's center, positioning the subunits and aiding their movement during peptide elongation (Noller et al. 2017). The human 80S ribosome consists of 80 ribosomal proteins and 4 noncoding RNAs. In contrast, prokaryotic ribosomes 70S are made up of three rRNAs: 23S and 5S in the large 50S subunit, along with 16S in the small 30S subunit. Mammalian ribosomes (80S) contain four rRNAs: 28S (5050 nucleotides), 5.8S (150 nucleotides), and 5S (120 nucleotides) rRNAs in the large 60S subunit, and 18S (2100 nucleotides) rRNA in the small 40S subunit. The 18S rRNA plays a critical role in mRNA binding and initiation of translation. The 5.8S and 28S rRNAs are involved in peptide bond formation and the overall structure of the ribosome, whereas the 5S rRNA through unique interactions helps stabilize the ribosomal structure. Secondary structures have been characterized for each of these rRNA types, essential for their function in ribosome assembly and catalysis. For example, the 18S rRNA consists of eight helices that form a compact, L-shaped structure, which is critical for its interaction with tRNA, base-pairing with mRNA, and molding with other ribosomal components (Martin et al. 2016).

DNA that codes for rRNA (rDNA) exists in multiple tandem repeats and is located in specific regions of the nuclear genome, typically on several chromosomes in the short arms of the five acrocentric chromosomes: 13, 14, 15, 21, and 22 (Hori et al. 2021). Each eukaryotic nuclear rDNA copy is organized into three genes—18S, 5.8S, and 28S—interspersed by two internal transcribed spacer regions, known as ITS-1 and ITS-2. Surrounding these tandem copies are two additional regions: an external transcribed spacer (ETS) and a non-transcribed spacer (NTS). The repeated copies of the nuclear rDNA are arranged in a tandem fashion, incorporating all three coding regions alongside both intergenic and intragenic noncoding regions. Additionally, transcripts for the 5S rRNAs are also organized in tandem, with most located on chromosome 1qter. Cellular transcription is a crucial process necessary for protein synthesis, particularly in actively dividing and metabolically active cells that need a diverse protein inventory for cell structure and function. The transcription of rRNA genes makes up a significant portion of all cellular transcription, essential for cell growth and maintaining cellular protein levels. Four types of rRNA are incorporated into the small and large ribosomal subunits within the nucleus before being transported to the cytoplasm, where translation occurs.

During the initiation of translation, the tRNA and mRNA complex first interacts with the decoding center, also known as the peptidyl transferase center, before moving to the catalytic site (Tirumalai et al. 2021). Notably, the functional core of the PTC is composed of rRNA, which contributes significantly to the ribosome's ability to accurately decode genetic information and catalyze peptide bond formation. The rRNA within the DC inspects tRNA molecules for proper pairing with mRNA codons through conformational changes that allow for proofreading. The specificity of tRNA recognition is enhanced by rRNA's ability to interact with not only the anticodon region but also other elements of the tRNA structure, ensuring that only correctly matched tRNA is allowed to participate in peptide synthesis. Additionally, the PTC, primarily composed of 23S rRNA in prokaryotes and 28S rRNA in eukaryotes, houses the active site that facilitates peptide bond formation between amino acids. The rRNA not only positions the substrates within the ribosome but also stabilizes the transition state during the chemical reaction, showcasing ribosomal RNA's unique enzymatic properties. This catalysis is often referred to as ribozymatic activity, demonstrating that RNA can function as a biocatalyst (Steitz and Moore 2003).

1.5 Small Nucleolar RNAs (snoRNAs)

Small nucleolar RNAs (snoRNAs) are a heterogeneous group of noncoding RNAs that originate in the nucleolus and are involved in the maturation of rRNA and the modification of RNA molecules (Webster and Ghalei 2023). Discovered in the early 1990s, snoRNAs have gained attention due to their critical functions in ribosome biogenesis and cellular homeostasis. SnoRNAs are typically 60–300 nucleotides in length and possess a conserved secondary structure (Kufel and Grzechnik 2019). SnoRNAs can be classified based on their structural features and functional roles. The two primary classes are C/D box snoRNAs (70–120 nucleotides) and H/ACA box snoRNAs (60–75 nucleotides). C/D box snoRNAs typically contain two conserved sequence motifs, the C box (RUGAUGA) and the D box (CUGA), which are responsible for guiding the modification of target RNAs. The C boxes are located in the 5′ region, while the D boxes are found in the 3′ region. The classic structure consists of two stem-loop structures, where the first stem contains the C box, and the second stem contains the D box. This arrangement allows C/D snoRNAs to carry out 2′-O-methylation of specific ribose moieties in rRNA. H/ACA box snoRNAs, in contrast, contain two H boxes (ANANNA) and two ACA motifs, typically positioned within the terminal region. Like C/D snoRNAs, H/ACA box snoRNAs also exhibit a stem-loop structure; however, they involve the pseudouridylation of target RNAs. This type of snoRNA is required for the conversion of specific uridine residues in rRNA to pseudouridine, a modification that stabilizes RNA secondary structures and enhances ribosome function. Additionally, studies have suggested other functions of SnoRNAs that include RNA modification with N4-acetylcytidine (ac4C), miRNA-like effects, and regulation of alternative splicing (Huang et al. 2022).

SnoRNAs are encoded in the genome and are usually part of larger polycistronic transcription units, meaning they can be co-transcribed with other RNA species. They are primarily transcribed by RNA polymerase II, though some snoRNAs can be transcribed by RNA polymerase III. Once transcribed, the primary snoRNA transcripts undergo a series of maturation steps including *Capping and Splicing*: The snoRNA transcripts often undergo capping and, in the case of polycistronic units, require splicing to generate mature snoRNAs; *Export*: Mature snoRNAs are then exported from the nucleus to the nucleolus, their primary site of action; *Assembly with Proteins:* In the nucleolus, snoRNAs associate with specific proteins, forming small nucleolar ribonucleoproteins (snoRNPs). C/D box snoRNAs associate with four key proteins—Nop1p, Nop56p, Nop58p, and Snu13p—creating functional small nucleolar ribonucleoproteins (snoRNPs). Similarly, H/ACA box snoRNAs produce functional snoRNPs by interacting with Cbf5p, Gar1p, Nhp2p, and Nop10p. This assembly is crucial for their stability and functional activity. The maturation of snoRNPs involves the interaction of snoRNAs with specific proteins that help in the recognition and modification of target substrates. The assembly of snoRNPs is a finely tuned process that dictates the biological availability and activity of snoRNAs within the nucleolus (Massenet et al. 2017).

1.6 Small Cajal Body RNAs (scaRNAs)

Cajal (coiled) bodies are conserved subnuclear organelles located within the nucleoplasm. They are among the most notable and widely researched nucleoplasmic organelles, second only to the nucleolus (Trinkle-Mulcahy and Sleeman 2017). Small Cajal body-associated RNAs (scaRNAs) are a subclass of the snoRNA family that accumulate in these bodies (Darzacq et al. 2002). These RNAs have several important functions, particularly in facilitating spliceosome formation by modifying small nuclear RNAs (snRNAs), promoting ribosome maturation via ribosomal RNAs (rRNAs), and assisting in telomere elongation. Additionally, human telomerase RNA (hTR) features a box H/ACA RNA domain that shares a common localization element specific to Cajal bodies with a certain subclass of box H/ACA RNAs (Jády et al. 2006). Complementary elements in scaRNAs allow them to hybridize with their target sequences, which in turn promotes either site-specific pseudouridylation or methylation of targets by associated proteins, akin to snoRNA-derived RNAs. The interaction between scaRNAs and these proteins is shepherd by specific sequence motifs that are present in the scaRNAs: either the box H/ACA domain, creating a hairpin-hinge-hairpin-tail structure that engages with the pseudouridylation complex (including dyskerin, NOP10, NHP2, and GAR1), or the box C/D domain, which takes on a kink-turn-like structure that interacts with the methylation complex (composed of fibrillarin, 15.5 kDa/NHP2L1, NOP56, and NOP58) (Bergstrand et al. 2022). It has been suggested that both box C/D and H/ACA snoRNPs gather in Cajal bodies before advancing to the nucleoli; however, box H/ACA snoRNAs transit through the Cajal body more swiftly than box C/D snoRNAs. The temporary accumulation of box C/D and possibly box H/ACA snoRNPs in Cajal bodies strongly implies that certain phases of snoRNP biogenesis take place in this nuclear organelle. It has been demonstrated that the Cajal body harbor fibrillarin, dyskerin, and Gar1 snoRNP proteins, along with the putative snoRNP assembly and processing factor, Nopp140, suggestive of its function in snoRNP biogenesis (Kiss et al. 2006).

1.7 Small Nucleolar RNAs-Derived RNAs (sdRNAs)

Mature and functional snoRNAs are processed into smaller fragments referred to as snoRNA-derived RNAs (sdRNAs) (Wajahat et al. 2021). Studies have indicated that these sdRNAs are not merely random degradation products, rather their expression is frequently conserved across different species (Chauhan et al. 2024). Certain sdRNAs that are highly abundant can originate from snoRNAs that are expressed at lower levels. Additionally, the presence of thermodynamically and metabolically stable fragments derived from snoRNAs suggests a specific stabilization mechanism for these sdRNAs. sdRNAs can be categorized into different groups based on their origin and size: H/ACA box snoRNAs typically produce fragments ranging from 20 to 24 nucleotides, primarily derived from the 3′ end. In contrast, C/D box snoRNAs are known to generate two distinct types of fragments: one that is greater than 27 nucleotides in length and another that spans 17 to 19 nucleotides, predominantly originating from the 5′ end. These fragments mostly contain C and D boxes indicative of their parental snoRNA origin. Biogenesis of H/ACA box snoRNAs-derived fragments demonstrated the requirement of microprocessor complex (DICER-DGCR8) similar to the canonical miRNA biogenesis whereas smaller fragments from C/D box snoRNAs are believed to be generated by a pathway independent of DICER. It has been hypothesized that sdRNAs have certain regulatory roles in translation as the production of sdRNAs increases during stress conditions since it is known that protein synthesis usually decreases under such conditions (Wajahat et al. 2021). These sdRNA can bind to Argonaute-2 to create the RNA-Induced Silencing Complex (RISC) in which mature sdRNA acts as a molecular guide, steering RISC toward a specific target mRNA. When the sdRNA perfectly pairs with the target mRNA, RISC cleaves the mRNA, leading to its degradation or alternatively the binding of the complex to target mRNA itself impedes the translation (Coley et al. 2022).

1.8 Endogenous Small Interfering RNAs (Endo-siRNAs)

Endogenous small interfering RNAs (endo-siRNAs), which range from 18 to 30 nucleotides in length are a class of sncRNAs that bind to particular Argonaute (Ago) proteins forming a functional ribonucleoprotein complex called the RISC. This complex directed by its siRNA component can target complementary RNAs, thereby influencing gene expression at both transcriptional and post-transcriptional stages through a mechanism known as RNA interference (RNAi) (Isenmann et al. 2023). They influence diverse biological processes acting through short antisense RNAs that act in the cytoplasm, where they interfere with gene expression by inhibiting translation by degrading cytoplasmic mRNA. Initially, the process of RNAi involves cleavage of long dsRNA into siRNAs by Dicer. Some very interesting observations were made in earlier studies on plants demonstrating that when researchers added extra copies of a gene to enhance a particular trait it resulted in diminishing the trait further, a case of co-suppression. A study aimed at creating more vividly colored flowers by adding extra gene copies involved in the pigmentation pathway led instead to a rise in variegated and white flower colors, accompanied by a decrease in the steady-state levels of mRNAs for the targeted gene (Napoli et al. 1990). It was discovered that these additional gene copies produced structured (hairpin) RNA, which could generate small RNAs that silenced gene expression (Hamilton and Baulcombe 1999). Over the years, such works contributed to the characterization of these siRNAs and identified double-stranded RNA (dsRNA) as the trigger for gene silencing. In invertebrates, RNA interference (RNAi) serves as a mechanism of innate immunity, as double-stranded RNA (dsRNA) is often generated during viral infections. RNA viruses create dsRNA during their replication process within host cells, while DNA viruses typically produce complementary sense and antisense RNA transcripts. Moreover, in humans, their major role has been associated with gene regulation and genomic stability. Retrotransposons, such as long interspersed nuclear element-1 (LINE-1 or L1), are mutagens that are capable of generating deleterious mutations by inserting themselves into genes and affecting gene function in the human genome. Studies have shown the role of siRNAs in functional silencing of such retrotransposons to maintain genomic stability (Chen et al. 2012).

1.9 Small Interfering RNAs (siRNAs)

Small interfering RNAs (siRNAs) are artificially developed ds RNA molecules that are 19–23 nucleotides in length. They are commonly utilized in molecular biology experiments for the transient silencing of a specific gene (Dana et al. 2017). Upon binding to their target transcript through sequence complementarity, they trigger an RNAi response. Duplex RNA oligonucleotides promote the cleavage of matching target transcripts, facilitating post-transcriptional silencing of target genes of interest in a sequence-specific manner. To achieve specific gene silencing, RNA interference can be initiated through chemically or enzymatically synthesized ds siRNA, or by the intracellular production of siRNA from vector-driven precursor small hairpin (sh) RNAs (Kang et al. 2023). In the vector-driven method, an oligonucleotide that includes the siRNA sequence, followed by a ~9 nucleotide loop and the reverse complement of the siRNA sequence, is inserted into plasmids or viral vectors to express shRNA endogenously. This shRNA is then exported from the nucleus and processed in the cytoplasm by Dicer into siRNA (Sheng et al. 2020). Because shRNA is continuously synthesized within the cell, the resulting gene silencing can stay longer, lasting for several weeks. In contrast, synthetic siRNA effects are typically short-lived, averaging only about 3–5 days, due to intracellular degradation (Subramanya et al. 2010).

1.10 Transcription Initiation RNAs (tiRNAs)

Transcription initiation RNAs (tiRNAs) are a class of promoter-associated small noncoding RNAs, which are about 18 nucleotides in length and are derived from regions proximal to transcription start sites-TSSs (mapped within −60 to +120 nucleotide of TSSs) (Yu et al. 2018). These tiRNAs are enriched at GC-rich promoters and regions of active transcription, implicating them in transcriptional regulation. These have a peak density at ~15 nucleotides downstream of RNA polymerase II (RNAPII) transcription in humans. The site of their activity is thus close to their biogenesis from the sequences upstream of the first nucleosome. Some interesting studies have demonstrated their role in transcription regulation due to their association with genomic CCCTC-binding factor (CTCF) and RNAPII occupancy sites as CTCF sites that co-localize with RNAPII are highly enriched for tiRNAs (Taft et al. 2011). Moreover, they are also known to be related to epigenetic regulation due to their association with chromatin activation mark H3K4me3 since these loci usually contain modified histones associated with transcription initiation, including histone H3 trimethylated at lysine 4 (H3K4me3) (Hobbs et al. 2018).

1.11 Small Nuclear RNAs (snRNAs)

Small nuclear RNAs (snRNAs) are a class of small, noncoding RNAs from ≈60 to 300 nucleotides in length that are predominantly located in the nucleus of eukaryotic cells. Alongside the major small nuclear RNAs, which include U1, U2, U4, U5, and U6 (Séraphin 1995), there exists a distinct group of minor spliceosomal snRNAs, comprising U11, U12, U4atac, and U6atac (Su et al. 2024). These minor snRNAs play a vital role in the splicing of a specific category of introns, often referred to as "minor class" introns. Both sets of snRNAs are transcribed from DNA by either DNA-dependent RNA polymerase II (Pol II) or DNA-dependent RNA polymerase III (Pol III), depending on their specific sequences and functions (Dergai et al. 2018). After their transcription, these snRNAs undergo critical processing steps, including modifications and cleavage at the 3′ end, which are necessary for their maturation. Despite their small size, these molecules are integral to various cellular processes. snRNAs are primarily known for their role in the spliceosome, a dynamic molecular complex responsible for the removal of introns from pre-mRNA transcripts. Pre-mRNA splicing refers to a critical step in the process of gene expression, occurring either concurrently with transcription (co-transcriptional) or shortly after it has taken place (post-transcriptional). During this process, noncoding sequences known as introns are eliminated from the initial RNA transcript, known as pre-mRNA. Simultaneously, the coding sequences called exons are precisely linked together. This intricate operation is essential for generating mature mRNAs that can be translated into proteins, thus ensuring that the final RNA product contains only the necessary genetic information required for proper cellular function.

The snRNAs are associated with various proteins to create RNA-protein complexes known as small nuclear ribonucleoproteins (snRNPs) within the cell nucleus (Morais et al. 2021). These snRNPs play a crucial role in pre-mRNA splicing by identifying essential sequence elements found in introns, which leads to the formation of active spliceosomes. This recognition occurs mainly through base-pairing interactions between the snRNAs and the pre-mRNA. The intricate orchestration and versatility of snRNAs indicate their importance in post-transcriptional regulation and cellular homeostasis.

1.12 Vault RNAs (vtRNAs)

Vault RNAs are a highly conserved subgroup of noncoding RNAs, named after their initial identification as part of the largest known hollow barrel-shaped ribonucleoprotein complexes known as "vaults" (Tanaka and Tsukihara 2012). These vaults are primarily found in the cytoplasm close to the nuclear pore complexes. Notably, 95% of vtRNAs are uniformly dispersed throughout the cytoplasm, while only 5% are directly associated with vault particles, which have a highly conserved structure in eukaryotes. Vault RNAs, which range in length from 80 to 140 nucleotides, are longer than miRNAs, but are still classified as part of the small noncoding RNA family (Gallo et al. 2022; Aghajani 2023). In humans, four vault RNAs are encoded on chromosome 5q31 at two loci: the VTRNA-1 locus, which contains the genetic sequences for vault RNA1-1 (98 nucleotides), vault RNA1-2 (88 nucleotides), and vault RNA1-3 (88 nucleotides), and the VTRNA-2 locus, which encodes vault RNA2-1 (100 nucleotides). All vault RNA genes are regulated by a type 2 polymerase III promoter and feature box A and box B motifs typically seen in tRNA genes. These motifs are essential for the binding of transcription factors TFIIIC and TFIIIB, which assist in the recruitment of polymerase III to the transcription start site. The functional significance of vault RNAs (vtRNAs) is thought to encompass a diverse range of roles within cellular processes (Taube et al. 2024). These include their involvement in innate immunity, particularly in the context of interferon (IFN)-associated antiviral defenses, which help protect cells from viral infections. Additionally, vtRNAs are implicated in regulating cell proliferation, suggesting a role in cell growth and division. Moreover, vtRNAs facilitate nucleocytoplasmic transport, aiding the movement of molecules between the nucleus and the cytoplasm. They are also believed to contribute to intracellular detoxification processes, helping to eliminate harmful substances within the cell. Another critical function of vtRNAs lies in their association with multidrug resistance, where they may play a role in protecting cells from the effects of chemotherapeutic agents. Further, vtRNAs have been associated with apoptosis, particularly highlighting the role of the noncoding vtRNA1-1 in inhibiting both extrinsic and intrinsic apoptotic pathways through an NF-κB-mediated mechanism. Lastly, vtRNAs are associated with p62-dependent autophagy, a process essential for cellular homeostasis and the degradation of damaged organelles and proteins (Johansen 2019). Once transcribed by

RNA polymerase III, vtRNAs can facilitate post-transcriptional gene silencing by DICER-dependent pathway that is distinct from the conventional microRNA pathway and operates independently of Drosha. This unique process is illustrated by vtRNA 1-2 and vtRNA 2-1. After processing by DICER, vtRNAs are fragmented into small vault RNAs (svtRNAs) that are around 22–24 nucleotides long and are capable of binding to the Argonaute RISC Catalytic Component 2 (AGO2) protein to execute regulate post-transcriptional gene expression regulation.

1.13 Circular RNAs (CircRNAs)

Circular RNAs are noncoding RNA molecules characterized by their closed-loop structure, formed by a covalent bond between the 5′ and 3′ ends. Unlike linear RNAs, which have distinct start and end points, CircRNAs create a continuous loop that renders them resistant to exonuclease degradation, thereby increasing their stability and longevity in cells (Greene et al. 2017). The distinct structure of circRNAs grants them a longer half-life and greater resistance to RNase R compared to linear RNAs. Most circRNAs are derived from well-characterized protein-coding genes and can comprise a single exon or multiple exons and range between 100 nucleotides and over 4 kb (Lasda and Parker 2014). These can be characterized as exonic circular RNAs, which consist solely of exons and account for approximately 85% of the total circular RNA population, intronic circular RNAs, which are formed from intronic lariats that arise during RNA splicing, exon-intron circular RNAs including both introns and exons with localization in the nucleus, intergenic circular RNAs, which are located in the regions between genes and do not correspond to any specific coding sequences, and antisense circular RNAs transcribed from the opposite strand of known genes, enabling potential interactions with their complementary sense transcripts. As naturally occurring biomolecules, circular RNAs are produced from precursor messenger RNAs through the backsplicing of exons in eukaryotic cells. The primary mechanism by which circRNAs are formed is back-splicing, wherein a downstream splice donor site connects with an upstream splice acceptor site, resulting in the creation of a covalently closed molecule. This process typically occurs within pre-mRNA transcripts and may involve the inclusion of exons, introns, or both. Certain splicing factors have been identified as facilitators of back-splicing, including proteins such as muscleblind-like (MBL) and members of the heterogeneous nuclear ribonucleoprotein (hnRNP) family, which have been shown to enhance the circularization process. These proteins promote interactions between splice sites in the pre-mRNA, thereby facilitating the back-splicing event. The efficiency of circRNA synthesis is also impacted by specific sequence motifs present within the pre-mRNA. Notably, the presence of complementary sequences, termed intronic repeats, near the splice sites, can aid in the circularization. These regions help stabilize the loop structure, thereby increasing the probability of back-splicing occurring. A prevalent function of circRNAs is their utilization as molecular sponges for microRNAs through sequestration thereby upregulating or downregulating miRNA target gene expression (Dong et al. 2020). CircRNAs have also been identified as protein sponges for various RNA-binding proteins. By sequestering RBPs, circRNAs may either inhibit or enhance the functions of these proteins, thereby participating in the regulation of mRNA splicing, translation, or stability. Additionally, circRNAs also function as protein scaffolds and translatable templates.

1.14 Extracellular Noncoding RNAs (exRNAs)

Extracellular noncoding RNAs have emerged as key players in cell communication and intercellular signaling (Tosar et al. 2021). Extracellular RNAs are heterogeneous population of different ribonucleic acids present in a variety of biological fluids, including serum, plasma, urine, and saliva (Sohail et al. 2022). These exRNAs encompass a rich and diverse array of RNA subpopulations that are typically shielded from degradation (Galvanin et al. 2019). This protection is primarily achieved through two mechanisms: the encapsulation of exRNAs within membranous vesicles (extracellular vesicles) and their association with protein or lipids such as Argonaute 2 (Ago2) and high-density lipoproteins (HDLs). The variety and quantity of extracellular RNAs are significantly influenced by their sources, including intracellular compartments, as well as blood and various tissues. Additionally, studies have demonstrated that the profiles of extracellular RNAs (exRNAs) vary significantly across different cellular compartments. For example, while extracellular vesicles (EVs) are primarily enriched in microRNAs, they also contain a range of other RNA species. These include small nucleolar RNAs, Piwi-interacting RNAs, long noncoding RNAs, transfer RNAs and their fragments, Y RNAs, ribosomal RNAs, mitochondrial RNAs, and protein-coding RNAs. In contrast, the composition of high-density lipoprotein (HDL): RNA complexes predominantly consist of noncoding RNAs, such as miRNAs, tRNA fragments, ribosomal RNAs, snoRNAs, and lncRNAs, with no detectable protein-coding RNAs present (Sadik et al. 2018). Their ability to transfer biological signals between cells underlines their potential in various processes such as cell differentiation, apoptosis, and cell-to-cell communication.

1.15 Y RNAs

Y RNAs are an emerging class of highly conserved, small ncRNAs, which are localized in nucleus and cytoplasm and are considered to be bioactive molecules with diverse biological significance. In humans, there are four Y RNAs, which are expressed from the genes Y1, Y3, Y4, and Y5 located in close proximity and transcribed in nucleus by RNA polymerase III (Valkov and Das 2020). These Y RNAs vary in size from 84 to 113 nucleotides. Y RNAs adopt unique hairpin-shaped structures, created by the base-pairing of their 3′ and 5′ ends in the mature form. These structures feature at least two primary stems, separated by a prominent single-stranded loop rich in pyrimidines. Notably, the main stem at the 5′/3′ end contains a highly conserved cytosine bulge, which serves as the high-affinity binding site for the Ro60 protein, a key partner of Y RNAs. The binding of Ro60 to Y RNA is necessary for Y RNA nuclear export and the stabilization of these RNA molecules in the cytoplasm. There is evidence to suggest that Y RNAs have a functional role in RNA stability and quality control and are essential factors for the initiation step of chromosomal DNA replication in human cell nuclei (Kowalski and Krude 2015).

1.16 MicroRNAs (miRNAs)

MicroRNAs are a class of small, noncoding RNAs, which function as vital regulatory molecules that influence a wide range of biological processes through their ability to modulate gene expression by either degrading protein-coding transcripts or inhibit their translation into proteins (Alpuche-Lazcano et al. 2024). These are 18–25 nucleotides in length that are encoded either within introns, exons, or in intergenic regions (Ratti et al. 2020). MicroRNAs undergo a series of steps, beginning with transcription and primary processing in the nucleus, followed by precursor processing and their mature function in the cytoplasm (O'Brien et al. 2018). After being transcribed by RNA Polymerase II, the initial microRNA transcript, known as the primary microRNA (pri-miRNA), is produced within the nucleus. This transcript features a characteristic stem-loop structure, with the microRNA itself localized within the stem portion of this structure. The pri-miRNA undergoes processing by an RNAse III enzyme called Drosha, which cleaves the transcript to form shorter precursor microRNA (pre-miRNA). This pre-miRNA is then exported as Exportin-5 complex from the nucleus into the cytoplasm. In the cytoplasmic environment, the pre-miRNA is subjected to further cleavage by another RNAse III enzyme, Dicer, which processes the pre-miRNA into a mature microRNA duplex. This duplex typically consists of two strands: the guide strand, which will be incorporated into the RNA-induced silencing complex (RISC), and the passenger strand, which is usually degraded. After processing, the mature microRNA strand is loaded onto an Argonaute protein to form the RISC, a critical complex involved in gene silencing and the regulation of target messenger RNAs (mRNAs). Argonaute (Ago) serves as the central component of the microRNA-induced silencing complex (miRISC), where it is directed by the microRNA to bind to complementary sequences located in the 3′ untranslated region (UTR) of the target mRNA. It is believed that the expansion of metazoan miRNAs likely played a significant role in the evolution of complexity observed in the organisms (Dexheimer and Cochella 2020). Since their discovery (lin-4) in 1993, research has shed light on the critical role of miRNAs in human pathophysiology. Numerous studies have established links between their regulatory functions and different stages of growth and development, for instance, miR-1 and miR-23a are involved in the regulation of developmental pathways (Safa et al. 2020; Qian et al. 2024); miR-19a and miR-101 are engaged in the regulation of cell differentiation and proliferation (Wang et al. 2025; Liu et al. 2022).

1.17 MicroRNA-Offset-RNAs (moRNAs)

MicroRNA-offset RNAs (moRNAs) are small RNAs that are approximately 20 nucleotides in length and resemble microRNAs (Gaffo et al. 2020). They are produced through an alternative processing pathway involving the Drosha/DGCR8 complex, which acts on microRNA precursors. Specifically, moRNAs are generated from the pre-miRNA hairpins during this processing, resulting in a distinct class of small RNAs that may have unique functions in gene regulation and cellular processes (Bortoluzzi et al. 2011). This alternative pathway highlights the complexity of small RNA biogenesis and suggests that moRNAs could play important roles in various biological contexts. MoRNAs exhibit a remarkable degree of sequence conservation across different species, suggesting their fundamental importance in biological functions. Moreover, the precursors of microRNAs that give rise to moRNAs are typically ancient from an evolutionary standpoint, implying that they have been maintained over long periods of evolutionary history (Langenberger et al. 2009). Such longevity may reflect their vital contributions to the regulatory networks within cells, underscoring their significance in the evolution of complex biological systems. Research has shown that nearly half of the known microRNA genes can produce multiple stable small RNA molecules, specifically up to four distinct species: two miRNAs and two microRNA-offset RNAs (Agnelli et al. 2019). This generation of different small RNA types is believed to occur via RNAse III-like processing of pre-miRNA hairpins. The moRNAs are derived from sequences that are situated directly adjacent to the mature miRNAs, denoted as the −5p and −3p strands (Shi et al. 2009). Studies have confirmed that moRNAs are not merely remnants of RNA processing but are indeed functional molecules that may exert regulatory effects similar to those of traditional microRNAs. For instance, moR-21 has been identified as having a significant role in inhibiting cell proliferation and regulating gene expression in vascular smooth muscle cells (Zhao et al. 2016). Moreover, the study also documented that the protein argonaute 2 (Ago2) is essential for the gene repression activities mediated by moR-21. This dependency on Ago2 indicates that moRNAs, much like conventional miRNAs, likely utilize RNA-induced silencing complexes to exert their effects on target genes.

1.18 Piwi-Interacting RNAs (piRNAs)

P-element-induced wimpy testis (PIWI)-interacting RNAs (piRNAs) are a class of noncoding RNAs that stand out for their unique features and functions, particularly in germline development and genome integrity (Sarkar et al. 2017). Studies have shown that piRNAs are also expressed in a variety of human somatic tissues (Qian et al. 2021). Unlike microRNAs (miRNAs) and small interfering RNAs (siRNAs), which are derived from double-stranded precursor RNA, piRNAs are produced from long single-stranded precursors and interact with Piwi proteins with its functions primarily localized in the nucleus (Sun et al. 2022). The precursors of piRNA are subsequently processed involving two distinct mechanisms to form mature piRNAs: the primary synthesis of piRNAs and the "ping-pong" amplification. The mature piRNAs are typically 24–32 nucleotides in length, making them one of the largest small noncoding RNA species. They possess a distinctive feature: a 2′-O-methyl modification at the 3′ end, which is thought to contribute to their stability and interaction with Piwi proteins. Maintaining the integrity of the germline genome is essential for ensuring the continuity and survival of a species through successive generations. Germline cells, which include sperm and egg cells, are the sole carriers of genetic material that will be passed down to offspring. Any alterations or damages to the genome within these cells can have profound implications, leading to the potential inheritance of mutations that may adversely affect the health and viability of future generations. One of the most thoroughly studied functions of piwi-interacting RNAs (piRNAs) is their role in silencing transposable elements, which are sequences of DNA that can change positions within the genome (Sarkar et al. 2017). In the germline cells, piRNAs are adept at recognizing and binding to RNA transcripts that originate from these transposable elements. Through this binding, piRNAs facilitate the degradation of the transposon-derived RNA, effectively preventing its mobilization and replication throughout the genome. To achieve the silencing of transposable elements at both post-transcriptional and transcriptional levels, mature piwi-interacting RNAs (piRNAs) are incorporated into germline-specific Argonaute proteins, also known as PIWI proteins. This assembly forms a sophisticated molecular complex known as the piRNA-induced silencing complex (piRISC) (Haase 2022). Once loaded onto PIWI proteins, the piRNAs serve as guides that direct the piRISC to target RNA molecules derived from transposable elements. Through base pairing with these target RNAs, the piRISC can either promote the degradation of the RNA, effectively preventing it from being translated into functional proteins, or inhibit the transcription process, limiting the initial production of the TE-derived RNA. This silencing mechanism is crucial for preserving genomic integrity, as unchecked activity of transposable elements can have severe consequences. When transposons are allowed to move freely within the genome, they can insert themselves into and disrupt essential genes, leading to mutations that can impair cellular function. Additionally, their activity can result in chromosomal instability, a condition in which chromosomes undergo abnormal changes such as breakage or fusion, potentially leading to various diseases.

1.19 Long Noncoding RNAs (lncRNAs)

Long noncoding RNAs (lncRNAs) are a category of noncoding RNAs that are typically larger than 200 nucleotides and play crucial regulatory roles in cellular processes (Zhang et al. 2019). These molecules are primarily found in the nucleus and associated with chromatin. Notably, lncRNAs represent the most significant portion of noncoding transcripts in mammals, making up approximately 80% of all noncoding RNA species, primarily consisting of lncRNAs that are transcribed by RNA polymerase II (Pol II), although other RNA polymerases also contribute to their transcription (Mattick et al. 2023). lncRNAs frequently feature a 7-methyl guanosine (m7G) cap at their 5′ ends and possess polyadenylation at their 3′ ends, undergoing splicing processes akin to those of messenger RNAs (mRNAs) (Tsagakis et al. 2020). lncRNAs exhibit lower conservation across different species compared to more established RNA types, such as mRNAs, miRNAs, and snoRNAs (Kashi et al. 2016). The lack of interspecies conservation may be suggestive of their nonfunctional presence or alternatively, that they carry species-specific regulatory roles. LncRNAs are not as diverse in structure, specificity and functions as sncRNAs and accordingly their classification is limited compared to sncRNAs which have many classes as described in the preceding sections. lncRNAs are found in various genomic locations and can be categorized based on their transcriptional origins. Those that are transcribed from intergenic regions that do not overlap with protein-coding genes are referred to as intergenic lncRNAs (such as HOTAIR, MALAT1, and GAPLINC). In contrast, lncRNAs that are transcribed entirely from the introns of protein-coding genes are known as intronic lncRNAs (like ncC11orf49 and ncHDAC5) (Ma et al. 2013). It has been shown that lincRNAs are transcriptionally activated in a manner similar to mRNAs; they are generally more conserved than introns and antisense transcripts, exhibit more tissue-specific expression than protein-coding genes, and tend to be more stable than intronic lncRNAs. Consequently, there exists a substantial variety of long intergenic non-coding RNAs (lincRNAs) that operate through various mechanisms, including *cis* or *trans* transcriptional regulation, translational control, splicing regulation, and chromatin modulations. The "K4-K36" domain, characterized by trimethylation of histone H3 at lysine 4 (H3K4) at the 5′ end and trimethylation at lysine 36 (H3K36) within the gene, serves as a marker of active transcription in protein-coding genes. Interestingly, this domain is commonly found in transcriptionally active lincRNAs (Ørom et al. 2010). Long noncoding RNAs can also be categorized as either sense lncRNAs (e.g., DAPALR) or antisense lncRNAs (e.g., CDKN2B-AS1), depending on whether they are transcribed from the sense strand or the antisense strand of protein-coding genes, respectively (Mercer et al. 2009). Furthermore, they can be classified as *cis*-lncRNAs (*cis*-acting lncRNAs, such as Xist/Air) that regulate genes located nearby, or *trans*-lncRNAs (*trans*-acting lncRNAs, such as HOTAIR) that influence the expression of distant genes (Taylor et al. 2015). Numerous lncRNAs regulate the expression of neighboring genes by influencing their transcription, and they also impact other aspects of chromatin biology, including DNA replication and the cellular response to DNA damage and repair. The negatively charged nature of RNA has the ability to neutralize the positively charged tails of histone proteins, which results in the loosening of chromatin structure, a process known as chromatin decompaction. When chromatin is decompacted, it becomes more accessible for the transcription machinery, facilitating gene expression. This dynamic RNA-mediated mechanism for the opening and closing of chromatin functions as a rapid regulatory switch, allowing cells to promptly adjust their gene expression profiles in response to various signals or environmental changes. The IncRNAs have the ability to engage with chromatin-modifying complexes and guide them to specific gene promoters, thereby influencing the transcriptional activity of those genes, either by activating or repressing their expression. From a mechanistic standpoint, both *cis*-acting and *trans*-acting lncRNAs play pivotal roles in shaping the chromatin landscape. These lncRNAs can modify the chromatin environment through multiple avenues: some achieve this indirectly by binding to proteins that have the capacity to interact with both RNA and DNA, thus facilitating a change in chromatin structure through these protein interactions. In other instances, lncRNAs might bind directly to DNA in a sequence-specific manner, repositioning and modifying chromatin to enhance or hinder the transcription of target genes. Enhancers are noncoding regions of the genome that play a critical role in regulating the spatial and temporal expression of genes throughout development. When enhancers are active, they can produce two primary types of long noncoding RNAs: enhancer RNAs (eRNAs), such as E2–3, and enhancer-associated long noncoding RNAs (elncRNAs), such as the splicing variant SE-lncRNA HCCL5 (Kim et al. 2015). The key difference between these two types of transcripts lies in their characteristics. eRNAs are typically short, bidirectional transcripts that have a 5′ cap, but they are usually unspliced, lack a polyadenylated tail, and are transient. In contrast, elncRNAs tend to be unidirectional, possess a polyadenylated tail, and undergo splicing. Some of these have been classified as super enhancers with distinctive roles in disease pathology (Song et al. 2024). This distinction in structural features influences their stability and functional roles within the cell. Genome-wide and global nuclear run-on sequencing techniques have unveiled a distinct population of low-copy small RNAs that originate from the promoter regions within the mammalian genome. These small RNAs, each measuring less than 200 nucleotides, are transcribed from specialized transcription sites that operate in a bidirectional manner, meaning they generate RNA molecules in both directions but from nonoverlapping regions. This category of long noncoding RNAs is known as promoter-associated noncoding RNAs (pancRNAs), which have been identified across a wide range of eukaryotic organisms, from yeast to humans (Chellini et al. 2020). PancRNAs are characterized as transcripts that are generated within a few hundred base pairs of the transcription start sites (TSSs) of either protein-coding or noncoding genes. These transcripts function as *cis*-acting elements, playing a key role in the transcriptional regulation of adjacent genes. pancRNAs are believed to function as key mediators of sequence-specific epigenetic alterations. One proposed molecular mechanism for this involves the regulation of CpG methylation and demethylation. A notable example of this is seen in the Sphk1 gene and its associated pancRNA variants,

Khps1, which demonstrate how pancRNAs can influence methylation patterns (Mercatelli et al. 2020). Additionally, pancRNAs may serve a broader role by enabling swift activation or repression of adjacent genes (Kurokawa 2011; Uesaka et al. 2017). They could accomplish this by modifying chromatin structure or by recruiting transcriptional regulators that facilitate these changes.

1.20 Transcribed-Ultraconserved Regions (t-UCRs)

Transcribed-ultraconserved regions (t-UCRs) represent a distinctive category of long noncoding RNAs that originate from ultraconserved regions (UCRs) within the genome (Wang et al. 2023). These regions are characterized by their remarkable stability across species, with 481 identified transcripts being at least 200 nucleotides in length, demonstrating complete conservation in humans and other organisms, such as mice (Zambalde et al. 2022). The extraordinary level of conservation that t-UCRs exhibit is particularly significant, especially when compared to the broader class of long noncoding RNAs (lncRNAs), which typically show a lower conservation rate of around 30–40%. This high degree of evolutionary conservation is often interpreted as an indicator of biological relevance, suggesting that t-UCRs may play vital roles in various cellular processes and physiology (Zhou et al. 2017). These are located on all chromosomes except chromosome 21 and Y. Most of these segments do not overlap with any coding region, mostly appearing in clusters near coding regions for transcription factors and molecules involved in developmental processes. The genomic landscape of UCRs indicates that while some are found overlapping with coding exons, most are situated within introns or intergenic regions. Research has revealed that exonic UCRs are particularly abundant in genes related to RNA binding and splicing regulation, suggesting that they may play a vital role in sustaining the expression of splicing factors (Carvalho de Oliveira 2023). Non-exonic UCRs often cluster around transcription factors and developmental genes, where they are believed to play a role in transcriptional regulation by serving as enhancers for nearby distal genes. These findings support the idea that t-UCRs are potential regulatory elements that are significant in the early phases of vertebrate development, differentiation, and intercellular coordination (Fabris and Calin 2017).

1.21 Conclusion

The increasing recognition of noncoding RNAs (ncRNAs) has revolutionized our understanding of the transcriptome and its role in cellular biology. While once dismissed as "junk DNA," its transcriptional product ncRNAs have emerged as pivotal regulators of gene expression and key players in a multitude of essential biological processes, including cell differentiation, apoptosis, and chromatin remodeling. With approximately 66% of the human transcriptome composed of these noncoding molecules, their influence extends far beyond that of protein-coding genes, which constitute less than 2% of the genome. This demonstrates the importance of RNA diversity—not merely protein quantity—as a driving force behind the developmental and cognitive complexity seen across organisms. As research continues to unveil the sophisticated mechanisms by which noncoding RNAs exert their effects, including their resistance to degradation through various protective strategies, it is clear that ncRNAs are indispensable for maintaining cellular homeostasis and regulating genetic mechanisms. Technological advancements in genomic sequencing and bioinformatics are believed to unfold further discoveries in this ever-changing field, paving the way for novel insights into the functions of noncoding RNAs and their potential implications in health and disease. Table 1.1 summarizes different classes of ncRNAs discussed within this chapter.

Table 1.1 Diversity of noncoding RNAs

S. no.	Noncoding RNA	Length (approximate)	Specific attribute	Structure[a]
1.	Transfer RNAs (tRNAs)	~ 76–90 nucleotides	Translation of genetic code Protein synthesis	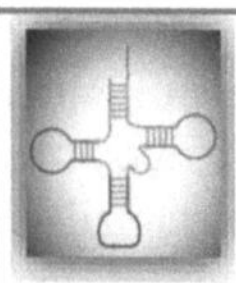
2.	tRNA-derived RNA fragments (tRFs)	~ 15–40 nucleotides	Response to stress stimuli Epigenetic regulation Translation regulation	
3.	Ribosomal RNAs (rRNAs)	~ 28S (5050 nucleotides) 5.8S (150 nucleotides) 5S (120 nucleotides) 18S (2100 nucleotides)	mRNA binding Initiation of translation Peptide bond formation Stabilize ribosomal structure	
4.	Small nucleolar RNAs (snoRNAs)	~ 60–300 nucleotides	Maturation of rRNA Ribosome biogenesis	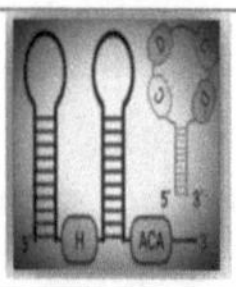
5.	Small cajal body RNAs (scaRNAs)	~ 400 nucleotides	Facilitate spliceosome formation Promote ribosome maturation Assist in telomere elongation	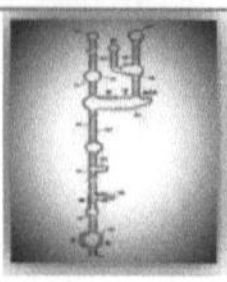
6.	sno-derived RNAs (sdRNAs)	~ 20–24 nucleotides (H/ACA box) ~ 17–19 nucleotides/>27 (C/D box)	Translation regulation	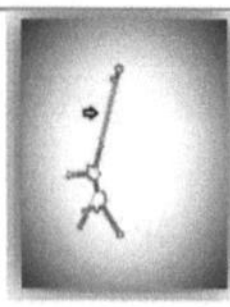

(continued)

Table 1.1 (continued)

S. no.	Noncoding RNA	Length (approximate)	Specific attribute	Structure[a]
7.	Endogenous small interfering RNAs (endo-siRNAs)	~ 18–30 nucleotides	Transcriptional, post-transcriptional and translation regulation Maintain genomic stability	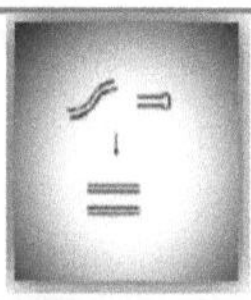
8.	Small interfering RNAs (siRNAs)	~ 19–23 nucleotides	Artificially developed Gene silencing	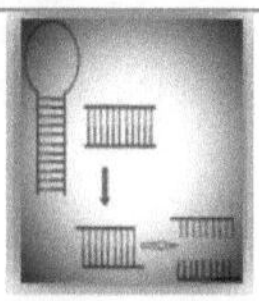
9.	Transcription initiation RNAs (tiRNAs)	~ 18 nucleotides	Transcriptional regulation Epigenetic regulation	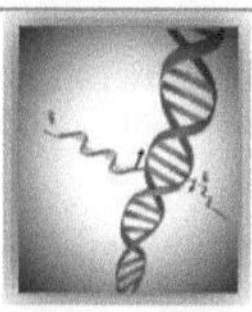
10.	Small nuclear RNAs (snRNAs)	~ 60–300 nucleotides	Role in the spliceosome	
11.	Vault RNAs (vtRNAs)	~ 80–140 nucleotides	Innate immunity Nucleocytoplasmic transport	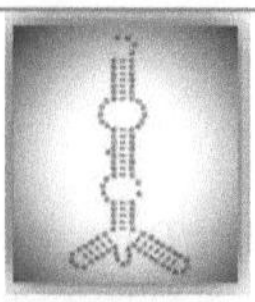
12.	Circular RNAs (circRNAs)	~ 100 nucleotides—4 kb (majority <1000 nucleotide)	A longer half-life and greater resistance to nucleases Protein scaffolds and translatable templates	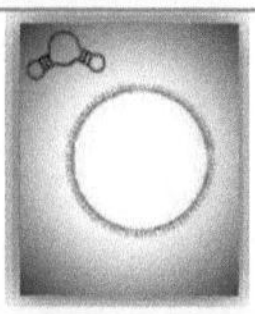
13.	Extracellular RNAs (exRNAs)	Varied size depending on the diversity of noncoding RNA	Biological signals Cell-to-cell communication	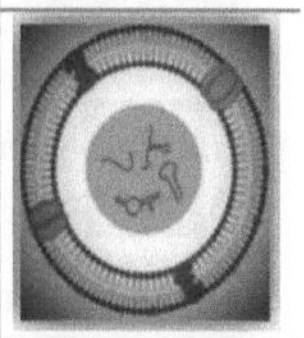
14.	Y RNAs	~ 84–113 nucleotides	Functional role in RNA stability Chromosomal DNA replication	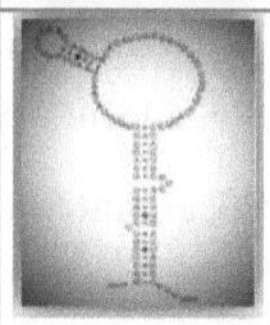

(continued)

Table 1.1 (continued)

S. no.	Noncoding RNA	Length (approximate)	Specific attribute	Structure[a]
15.	MicroRNAs (miRNAs)	~ 18–24 nucleotides	Regulation of gene expression Significance in growth and development Cell proliferation and cell differentiation	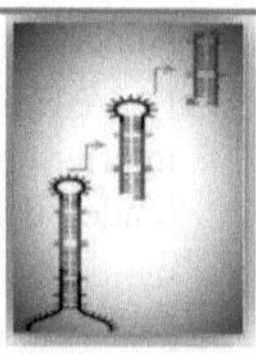
16.	miRNA-offset-RNAs (moRNAs)	~ 20 nucleotides	Effects on target genes Evolutionary significance	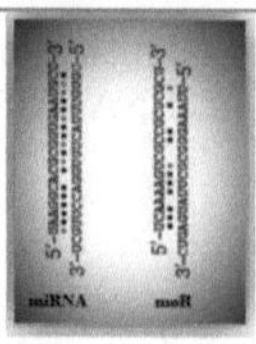
17.	Piwi-interacting RNAs (piRNAs)	~ 24–32 nucleotides	Germline development Genome integrity Silencing transposable elements	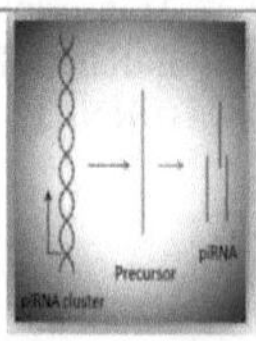
18.	Long noncoding RNAs (lncRNAs)	> 200 nucleotides	Approximately 80% of all noncoding RNA species Species-specific regulatory roles Transcription control Chromatin decompaction.	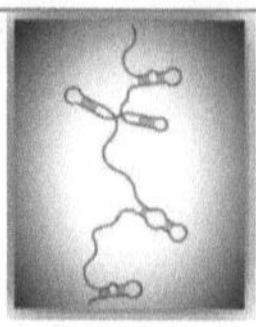
19.	Transcribed-ultraconserved regions (t-UCRs)	~ 200 nucleotides	Expression of splicing factors Transcriptional regulation	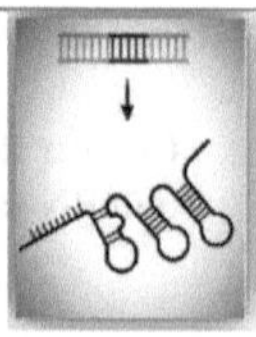

[a]These are schematic representation. For detailed structure, refer to textbooks and biological molecules databases.

Acknowledgments The author is alone responsible for the content and writing of the manuscript. An AI-assisted copy editing was used to improve spelling, punctuation, grammar, and style.

Conflict of Interest The author has no conflict of interest to disclose.

Data Availability Statement There is no associated data and all content has been described within the manuscript.

Ethical Statement The manuscript is a review and this has not used any human subject or animal model.

Funding No external funding has been received.

References

Aghajani MM (2023) Vault RNAs (vtRNAs): rediscovered non-coding RNAs with diverse physiological and pathological activities. Genes Dis 11(2):772–787

Agnelli L, Bisognin A, Todoerti K, Manzoni M, Taiana E, Galletti S, Cutrona G, Gaffo E, Bortoluzzi S, Neri A (2019) Expanding the repertoire of miRNAs and miRNA-offset RNAs expressed in multiple myeloma by small RNA deep sequencing. Blood Cancer J 9(3):21

Alpuche-Lazcano SP, Scarborough RJ, Gatignol A (2024) MicroRNAs and long non-coding RNAs during transcriptional regulation and latency of HIV and HTLV. Retrovirology 21(1):5

Bergstrand S, O'Brien EM, Coucoravas C, Hrossova D, Peirasmaki D, Schmidli S, Dhanjal S, Pederiva C, Siggens L, Mortusewicz O, O'Rourke JJ, Farnebo M (2022) Small Cajal body-associated RNA 2 (scaRNA2) regulates DNA repair pathway choice by inhibiting DNA-PK. Nat Commun 13(1):1015

Bortoluzzi S, Biasiolo M, Bisognin A (2011) MicroRNA-offset RNAs (moRNAs): by-product spectators or functional players? Trends Mol Med 17(9):473–474

Carvalho de Oliveira J (2023) Transcribed ultraconserved regions: new regulators in cancer signaling and potential biomarkers. Genet Mol Biol 46:e20220125

Chauhan W, Sudharshan S, Kafle S, Zennadi R (2024) SnoRNAs: exploring their implication in human diseases. Int J Mol Sci 25(13):7202

Chellini L, Frezza V, Paronetto MP (2020) Dissecting the transcriptional regulatory networks of promoter-associated noncoding RNAs in development and cancer. J Exp Clin Cancer Res 39(1):51

Chen Q, Zhou T (2023) Emerging functional principles of tRNA-derived small RNAs and other regulatory small RNAs. J Biol Chem 299(10):105225

Chen L, Dahlstrom JE, Lee SH, Rangasamy D (2012) Naturally occurring endo-siRNA silences LINE-1 retrotransposons in human cells through DNA methylation. Epigenetics 7(7):758–771

Chen Q, Li D, Jiang L, Wu Y, Yuan H, Shi G, Liu F, Wu P, Jiang K (2024) Biological functions and clinical significance of tRNA-derived small fragment (tsRNA) in tumors: current state and future perspectives. Cancer Lett 587:216701

Coley AB, DeMeis JD, Chaudhary NY, Borchert GM (2022) Small nucleolar derived RNAs as regulators of human cancer. Biomedicine 10(8):1819

Dana H, Chalbatani GM, Mahmoodzadeh H, Karimloo R, Rezaiean O, Moradzadeh A, Mehmandoost N, Moazzen F, Mazraeh A, Marmari V, Ebrahimi M, Rashno MM, Abadi SJ, Gharagouzlo E (2017) Molecular mechanisms and biological functions of siRNA. Int J Biomed Sci 13(2):48–57

Darzacq X, Jády BE, Verheggen C, Kiss AM, Bertrand E, Kiss T (2002) Cajal body-specific small nuclear RNAs: a novel class of 2′-O-methylation and pseudouridylation guide RNAs. EMBO J 21(11):2746–2756

Dergai O, Cousin P, Gouge J, Satia K, Praz V, Kuhlman T, Lhôte P, Vannini A, Hernandez N (2018) Mechanism of selective recruitment of RNA polymerases II and III to snRNA gene promoters. Genes Dev 32(9–10):711–722

Dexheimer PJ, Cochella L (2020) MicroRNAs: from mechanism to organism. Front Cell Dev Biol 8:409

Dong P, Xu D, Xiong Y, Yue J, Ihira K, Konno Y, Watari H (2020) The expression, functions and mechanisms of circular RNAs in gynecological cancers. Cancers (Basel) 12(6):1472

Ezkurdia I, Juan D, Rodriguez JM, Frankish A, Diekhans M, Harrow J, Vazquez J, Valencia A, Tress ML (2014) Multiple evidence strands suggest that there may be as few as 19,000 human protein-coding genes. Hum Mol Genet 23(22):5866–5878

Fabris L, Calin GA (2017) Understanding the genomic ultraconservations: T-UCRs and cancer. Int Rev Cell Mol Biol 333:159–172

Falaleeva M, Stamm S (2013) Processing of snoRNAs as a new source of regulatory non-coding RNAs: snoRNA fragments form a new class of functional RNAs. BioEssays 35(1):46–54

Gaffo E, Bortolomeazzi M, Bisognin A, Di Battista P, Lovisa F, Mussolin L, Bortoluzzi S (2020) MiR&moRe2: a bioinformatics tool to characterize microRNAs and microRNA-offset RNAs from small RNA-Seq data. Int J Mol Sci 21(5):1754

Gallo S, Kong E, Ferro I, Polacek N (2022) Small but powerful: the human vault RNAs as multifaceted modulators of pro-survival characteristics and tumorigenesis. Cancers (Basel) 14(11):2787

Galvanin A, Dostert G, Ayadi L, Marchand V, Velot É, Motorin Y (2019) Diversity and heterogeneity of extracellular RNA in human plasma. Biochimie 164:22–36

Greene J, Baird AM, Brady L, Lim M, Gray SG, McDermott R, Finn SP (2017) Circular RNAs: biogenesis, function and role in human diseases. Front Mol Biosci 4:38

Haase AD (2022) An introduction to PIWI-interacting RNAs (piRNAs) in the context of metazoan small RNA silencing pathways. RNA Biol 19(1):1094–1102

Hamilton AJ, Baulcombe DC (1999) A species of small antisense RNA in posttranscriptional gene silencing in plants. Science 286(5441):950–952

Hillier LW, Coulson A, Murray JI, Bao Z, Sulston JE, Waterston RH (2005) Genomics in C. elegans: so many genes, such a little worm. Genome Res 15(12):1651–1660

Hobbs M, Ender C, Baillie GJ, Crawford J, Ru K, Taft RJ, Mattick JS (2018) Transcription initiation RNAs are associated with chromatin activation mark H3K4me3. bioRxiv:265827

Hori Y, Shimamoto A, Kobayashi T (2021) The human ribosomal DNA array is composed of highly homogenized tandem clusters. Genome Res 31(11):1971–1982

Huang ZH, Du YP, Wen JT, Lu BF, Zhao Y (2022) snoRNAs: functions and mechanisms in biological processes, and roles in tumor pathophysiology. Cell Death Discov 8(1):259

Hubbard T, Barker D, Birney E, Cameron G, Chen Y, Clark L, Cox T, Cuff J, Curwen V, Down T, Durbin R, Eyras E, Gilbert J, Hammond M, Huminiecki L, Kasprzyk A, Lehvaslaiho H, Lijnzaad P, Melsopp C, Mongin E, Pettett R, Pocock M, Potter S, Rust A, Schmidt E, Searle S, Slater G, Smith J, Spooner W, Stabenau A, Stalker J, Stupka E, Ureta-Vidal A, Vastrik I, Clamp M (2002) The Ensembl genome database project. Nucleic Acids Res 30(1):38–41

Isenmann M, Stoddart MJ, Schmelzeisen R, Gross C, Della Bella E, Rothweiler RM (2023) Basic principles of RNA interference: nucleic acid types and in vitro intracellular delivery methods. Micromachines 14(7):1321

Jády BE, Richard P, Bertrand E, Kiss T (2006) Cell cycle-dependent recruitment of telomerase RNA and Cajal bodies to human telomeres. Mol Biol Cell 17(2):944–954

Jin D, Musier-Forsyth K (2019) Role of host tRNAs and aminoacyl-tRNA synthetases in retroviral replication. J Biol Chem 294(14):5352–5364

Johansen T (2019) Selective autophagy: RNA comes from the vault to regulate p62/SQSTM1. Curr Biol 29(8):R297–R299

Kamhi E, Raitskin O, Sperling R, Sperling J (2010) A potential role for initiator-tRNA in pre-mRNA splicing regulation. Proc Natl Acad Sci USA 107(25):11319–11324

Kang H, Ga YJ, Kim SH, Cho YH, Kim JW, Kim C, Yeh JY (2023) Small interfering RNA (siRNA)-based therapeutic applications against viruses: principles, potential, and challenges. J Biomed Sci 30(1):88

Kashi K, Henderson L, Bonetti A, Carninci P (2016) Discovery and functional analysis of lncRNAs: methodologies to investigate an uncharacterized transcriptome. Biochim Biophys Acta 1859(1):3–15

Keller P, Freund I, Marchand V, Bec G, Huang R, Motorin Y, Eigenbrod T, Dalpke A, Helm M (2018) Double methylation of tRNA-U54 to 2′-O-methylthymidine (Tm) synergistically decreases immune response by toll-like receptor 7. Nucleic Acids Res 46(18):9764–9775

Kim HK (2019) Transfer RNA-derived small non-coding RNA: dual regulator of protein synthesis. Mol Cells 42:687–692

Kim TK, Hemberg M, Gray JM (2015) Enhancer RNAs: a class of long noncoding RNAs synthesized at enhancers. Cold Spring Harb Perspect Biol 7(1):a018622

Kim HK, Yeom JH, Kay MA (2020) Transfer RNA-derived small RNAs: another layer of gene regulation and novel targets for disease therapeutics. Mol Ther 28(11):2340–2357

Kiss T, Fayet E, Jády BE, Richard P, Weber M (2006) Biogenesis and intranuclear trafficking of human box C/D and H/ACA RNPs. Cold Spring Harb Symp Quant Biol 71:407–417

Kowalski MP, Krude T (2015) Functional roles of non-coding Y RNAs. Int J Biochem Cell Biol 66:20–29

Kufel J, Grzechnik P (2019) Small nucleolar RNAs tell a different tale. Trends Genet 35(2):104–117

Kurokawa R (2011) Promoter-associated long noncoding RNAs repress transcription through a RNA binding protein TLS. Adv Exp Med Biol 722:196–208

Langenberger D, Bermudez-Santana C, Hertel J, Hoffmann S, Khaitovich P, Stadler PF (2009) Evidence for human microRNA-offset RNAs in small RNA sequencing data. Bioinformatics 25(18):2298–2301

Lant JT, Berg MD, Sze DHW, Hoffman KS, Akinpelu IC, Turk MA, Heinemann IU, Duennwald ML, Brandl CJ, O'Donoghue P (2018) Visualizing tRNA-dependent mistranslation in human cells. RNA Biol 15(4–5):567–575

Lasda E, Parker R (2014) Circular RNAs: diversity of form and function. RNA 20(12):1829–1842

Li Y, Yu Z, Jiang W, Lyu X, Guo A, Sun X, Yang Y, Zhang Y (2024) tRNA and tsRNA: from heterogeneity to multifaceted regulators. Biomolecules 14(10):1340

Liu N, Yang C, Gao A, Sun M, Lv D (2022) MiR-101: An Important Regulator of Gene Expression and Tumor Ecosystem.Cancers (Basel). Nov 28;14(23):5861

Ma L, Bajic VB, Zhang Z (2013) On the classification of long non-coding RNAs. RNA Biol 10(6):925–933

Mangiavacchi A, Morelli G, Orlando V (2023) Behind the scenes: how RNA orchestrates the epigenetic regulation of gene expression. Front Cell Dev Biol 11:1123975

Martin F, Ménétret JF, Simonetti A, Myasnikov AG, Vicens Q, Prongidi-Fix L, Natchiar SK, Klaholz BP, Eriani G (2016) Ribosomal 18S rRNA base pairs with mRNA during eukaryotic translation initiation. Nat Commun 7:12622

Massenet S, Bertrand E, Verheggen C (2017) Assembly and trafficking of box C/D and H/ACA snoRNPs. RNA Biol 14(6):680–692

Masson GR (2019) Towards a model of GCN2 activation. Biochem Soc Trans 47(5):1481–1488

Mattick JS, Amaral PP, Carninci P, Carpenter S, Chang HY, Chen LL, Chen R, Dean C, Dinger ME, Fitzgerald KA, Gingeras TR, Guttman M, Hirose T, Huarte M, Johnson R, Kanduri C, Kapranov P, Lawrence JB, Lee JT, Mendell JT, Mercer TR, Moore KJ, Nakagawa S, Rinn JL, Spector DL, Ulitsky I, Wan Y, Wilusz JE, Wu M (2023) Long non-coding RNAs: definitions, functions, challenges and recommendations. Nat Rev Mol Cell Biol 24(6):430–447

Mei Y, Yong J, Liu H, Shi Y, Meinkoth J et al (2010) tRNA binds to cytochrome c and inhibits caspase activation. Mol Cell 37:668–678

Mercatelli N, Palombo R, Paronetto MP (2020) Emerging contribution of PancRNAs in cancer. Cancer 12(8):2035

Mercer TR, Dinger ME, Mattick JS (2009) Long non-coding RNAs: insights into functions. Nat Rev Genet 10:155–159

Micheel J, Safrastyan A, Wollny D (2021) Advances in non-coding RNA sequencing. Noncoding RNA 7(4):70

Morais P, Adachi H, Yu YT (2021) Spliceosomal snRNA epitranscriptomics. Front Genet 12:652129

Napoli C, Lemieux C, Jorgensen R (1990) Introduction of a chimeric chalcone synthase gene into petunia results in reversible co-suppression of homologous genes in trans. Plant Cell 2(4):279–289

Noller HF, Lancaster L, Zhou J, Mohan S (2017) The ribosome moves: RNA mechanics and translocation. Nat Struct Mol Biol 24(12):1021–1027

O'Brien J, Hayder H, Zayed Y, Peng C (2018) Overview of microRNA biogenesis, mechanisms of actions, and circulation. Front Endocrinol (Lausanne) 9:402

Ørom UA, Derrien T, Beringer M, Gumireddy K, Gardini A, Bussotti G et al (2010) Long noncoding RNAs with enhancer-like function in human cells. Cell 143(1):46–58

Qian L, Xie H, Zhang L, Zhao Q, Lü J, Yu Z (2021) Piwi-interacting RNAs: a new class of regulator in human breast cancer. Front Oncol 11:695077
Qian X, Jiang Y, Yang Y, Zhang Y, Xu N, Xu B, Pei K, Yu Z, Wu W (2024) Recent advances of miR-23 in humandiseases and growth development. Noncoding RNA Res. Dec 30;11:220-233
Ratti M, Lampis A, Ghidini M, Salati M, Mirchev MB, Valeri N, Hahne JC (2020) MicroRNAs (miRNAs) and long non-coding RNAs (lncRNAs) as new tools for cancer therapy: first steps from bench to bedside. Target Oncol 15(3):261–278
Sadik N, Cruz L, Gurtner A, Rodosthenous RS, Dusoswa SA, Ziegler O, Van Solinge TS, Wei Z, Salvador-Garicano AM, Gyorgy B, Broekman M, Balaj L (2018) Extracellular RNAs: a new awareness of old perspectives. Methods Mol Biol 1740:1–15
Safa A, Bahroudi Z, Shoorei H, Majidpoor J, Abak A, Taheri M, Ghafouri-Fard S (2020) miR-1: A comprehensive reviewof its role in normal development and diverse disorders. Biomed Pharmacother. 132:110903
Sarkar A, Volff JN, Vaury C (2017) piRNAs and their diverse roles: a transposable element-driven tactic for gene regulation? FASEB J 31(2):436–446
Séraphin B (1995) Sm and Sm-like proteins belong to a large family: identification of proteins of the U6 as well as the U1, U2, U4 and U5 snRNPs. EMBO J 14(9):2089–2098
Sheng P, Flood KA, Xie M (2020) Short hairpin RNAs for strand-specific small interfering RNA production. Front Bioeng Biotechnol 8:940
Shi W, Hendrix D, Levine M, Haley B (2009) A distinct class of small RNAs arises from pre-miRNA-proximal regions in a simple chordate. Nat Struct Mol Biol 16:183–189
Sloan KE, Warda AS, Sharma S, Entian KD, Lafontaine DLJ, Bohnsack MT (2017) Tuning the ribosome: the influence of rRNA modification on eukaryotic ribosome biogenesis and function. RNA Biol 14(9):1138–1152
Smal M, Memoli D, Alexandrova E, Di Rosa D, D'Agostino Y, Russo F, Giurato G, Nassa G, Tarallo R, Weisz A, Rizzo F (2024) Small non-coding RNA transcriptomic profiling in adult and fetal human brain. Sci Data 11(1):767
Sohail AM, Khawar MB, Afzal A, Hassan A, Shahzaman S, Ali A (2022) Multifaceted roles of extracellular RNAs in different diseases. Mil Med Res 9(1):43
Song P, Han R, Yang F (2024) Super enhancer lncRNAs: a novel hallmark in cancer. Cell Commun Signal 22(1):207
Speer J, Gehrke CW, Kuo KC, Waalkes TP, Borek E (1979) tRNA breakdown products as markers for cancer. Cancer 44(6):2120–2123
Steitz TA, Moore PB (2003) RNA, the first macromolecular catalyst: the ribosome is a ribozyme. Trends Biochem Sci 28(8):411–418
Su Z, Wilson B, Kumar P, Dutta A (2020) Noncanonical roles of tRNAs: tRNA fragments and beyond. Annu Rev Genet 54:47–69
Su Y, Wu J, Chen W, Shan J, Chen D, Zhu G, Ge S, Liu Y (2024) Spliceosomal snRNAs, the essential players in pre-mRNA processing in eukaryotic nucleus: from biogenesis to functions and spatiotemporal characteristics. Adv Biol (Weinh) 8(7):e2400006
Subramanya S, Kim SS, Manjunath N, Shankar P (2010) RNA interference-based therapeutics for human immunodeficiency virus HIV-1 treatment: synthetic siRNA or vector-based shRNA? Expert Opin Biol Ther 10(2):201–213
Sun YH, Lee B, Li XZ (2022) The birth of piRNAs: how mammalian piRNAs are produced, originated, and evolved. Mamm Genome 33(2):293–311
Taft RJ, Hawkins PG, Mattick JS, Morris KV (2011) The relationship between transcription initiation RNAs and CCCTC-binding factor (CTCF) localization. Epigenetics Chromatin 4:13
Tanaka H, Tsukihara T (2012) Structural studies of large nucleoprotein particles, vaults. Proc Jpn Acad Ser B Phys Biol Sci 88(8):416–433
Taube M, Lisiak N, Totoń E, Rubiś B (2024) Human vault RNAs: exploring their potential role in cellular metabolism. Int J Mol Sci 25(7):4072
Taylor DH, Chu ET, Spektor R, Soloway PD (2015) Long non-coding RNA regulation of reproduction and development. Mol Reprod Dev 82(12):932–956

The ENCODE Project Consortium (2004) The ENCODE (ENCyclopedia of DNA elements) project. Science 306:636–640
Tirumalai MR, Rivas M, Tran Q, Fox GE (2021) The peptidyl transferase center: a window to the past. Microbiol Mol Biol Rev 85(4):e0010421
Tosar JP, Witwer K, Cayota A (2021) Revisiting extracellular RNA release, processing, and function. Trends Biochem Sci 46(6):438–445
Trinkle-Mulcahy L, Sleeman JE (2017) The Cajal body and the nucleolus: "in a relationship" or "it's complicated"? RNA Biol 14(6):739–751
Tsagakis I, Douka K, Birds I, Aspden JL (2020) Long non-coding RNAs in development and disease: conservation to mechanisms. J Pathol 250(5):480–495
Uesaka M, Agata K, Oishi T, Nakashima K, Imamura T (2017) Evolutionary acquisition of promoter-associated non-coding RNA (pancRNA) repertoires diversifies species-dependent gene activation mechanisms in mammals. BMC Genomics 18(1):285
Valkov N, Das S (2020) Y RNAs: biogenesis, function and implications for the cardiovascular system. Adv Exp Med Biol 1229:327–342
Wajahat M, Bracken CP, Orang A (2021) Emerging functions for snoRNAs and snoRNA-derived fragments. Int J Mol Sci 22(19):10193
Wang T, Li F, Lu Z (2023) Ultra-conserved RNA: a novel biological tool with diagnostic and therapeutic potential. Discov Oncol 14(1):41
Wang K, Wen J, Liang T, Hu H, Li S, Shen L, Ren T, Yao Y, Xie J, Ding J, Chen J, Tang YD, Zhu Y, Gao C (2025)Enhancing miR-19a/b induced cardiomyocyte proliferation in infarcted hearts by alleviating oxidant stress andcontrolling miR-19 release. Biomaterials. Jan;312:122732.
Webster SF, Ghalei H (2023) Maturation of small nucleolar RNAs: from production to function. RNA Biol 20(1):715–736
Wen JT, Huang ZH, Li QH, Chen X, Qin HL, Zhao Y (2021) Research progress on the tsRNA classification, function, and application in gynecological malignant tumors. Cell Death Discov 7(1):388
Yu D, Ma X, Zuo Z, Wang H, Meng Y (2018) Classification of transcription boundary-associated RNAs (TBARs) in animals and plants. Front Genet 9:168
Zambalde EP, Adamoski D, Gradia DF, Rabinovich I, Rodrigues AC, Ivan C, Ribeiro EMSF, Calin GA, Carvalho de Oliveira J (2022) Transcribed ultraconserved regions are associated with clinicopathological features in breast cancer. Biomolecules 12(2):214
Zhang X, Wang W, Zhu W, Dong J, Cheng Y, Yin Z, Shen F (2019) Mechanisms and functions of long non-coding RNAs at multiple regulatory levels. Int J Mol Sci 20(22):5573
Zhao J, Schnitzler GR, Iyer LK, Aronovitz MJ, Baur WE, Karas RH (2016) MicroRNA-offset RNA alters gene expression and cell proliferation. PLoS One 11(6):e0156772
Zhou J, Wang R, Zhang J, Zhu L, Liu W, Lu S, Chen P, Li H, Yin B, Yuan J, Qiang B, Shu P, Peng X (2017) Conserved expression of ultra-conserved noncoding RNA in mammalian nervous system. Biochim Biophys Acta Gene Regul Mech 1860(12):1159–1168
Zhou M, He X, Zhang J, Mei C, Zhong B, Ou C (2024) tRNA-derived small RNAs in human cancers: roles, mechanisms, and clinical application. Mol Cancer 23(1):76

Chapter 2
Role of ncRNAs in Neurological Disorders and Cardiovascular Diseases

Muhammad Imran Sajid, Fatima Abid Khan, Hadia Mohsin, Muhammad Bilal Maqbool, Fahad Mahmood, Maira Yaseen, Maryam Younas, Aria Naziri, Khawaja Husnain Haider, and Rakesh Kumar Tiwari

Contents

M. I. Sajid
Department of Pharmacy, The University of Lahore, Lahore, Pakistan

F. A. Khan · H. Mohsin · M. B. Maqbool · F. Mahmood · M. Yaseen · M. Younas
Faculty of Pharmacy, University of Central Punjab, Lahore, Pakistan

A. Naziri · R. K. Tiwari (✉)
Department of Biomedical Sciences, College of Osteopathic Medicine of the Pacific–Northwest, Western University of Health Sciences, Lebanon, OR, USA
e-mail: rtiwari@westernu.edu

K. H. Haider
College of Medicine, Sulaiman Al Rajhi University, Al-Bukairiyah, Saudi Arabia

M. F. Ullah, A. Ahmad (eds.), *Non-coding RNAs*, Experientia Supplementum 115,
https://doi.org/10.1007/978-3-032-06948-1_2

Abstract For a long time, noncoding RNAs (ncRNAs) were considered irrelevant fragments of the genome, dismissed as genetic noise. However, recent breakthroughs have unveiled their crucial Role in regulating gene expression, influencing fundamental biological processes such as chromatin remodeling, epigenetic modifications, and cellular communication. Among them, long noncoding RNAs (lncRNAs) and microRNAs (miRNAs) have drawn considerable attention due to their strong association with neurodegenerative disorders and cardiovascular diseases (CVDs). Despite their apparent differences, these conditions share molecular regulatory networks that ncRNAs help orchestrate. LncRNAs, like ANRIL and MEG3, play a Role in vascular integrity and cardiac fibrosis, while MIAT and MALAT1 are implicated in heart failure and ischemic injury. In Alzheimer's disease, BACE1-AS and BC200 contribute to the buildup of amyloid plaques and tau protein tangles, worsening cognitive decline. The ability of ncRNAs to act as molecular sponges—binding to miRNAs and modulating gene expression—demonstrates their intricate Role in disease progression. With advances in sequencing technologies and computational biology, ncRNAs are emerging as promising biomarkers and therapeutic targets. New approaches, including CRISPR-based gene editing and RNA therapeutics, present exciting possibilities for intervention. However, challenges such as stability, precise delivery, and potential side effects must be addressed before these treatments can be translated into clinical practice. This chapter delves into the expanding field of ncRNA research, highlighting its potential to reshape the future of precision medicine and targeted therapies.

Keywords Noncoding RNAs · Neurodegenerative diseases · Cardiovascular diseases · RNA interference · Precision medicine

Abbreviations

51A	51A Long Noncoding RNA
AD	Alzheimer's Disease
ADODs	Alzheimer's Disease and Other Dementias
AKT	Protein Kinase B
ANRIL	Antisense Noncoding RNA in the INK4 Locus
APP	Amyloid Precursor Protein
ATG4a	Autophagy-Related 4A Cysteine Peptidase
Aβ	Amyloid Beta
BACE1	Beta-Secretase 1
BACE1-AS	Beta-Secretase 1 Antisense RNA
BC200	Brain Cytoplasmic 200 RNA
BDNF	Brain-Derived Neurotrophic Factor
BDNF-AS	Brain-Derived Neurotrophic Factor Antisense RNA
CARL	Cardiac Apoptosis-Related Long Noncoding RNA

CRISPR-Cas9	Clustered Regularly Interspaced Short Palindromic Repeats-Associated Protein 9
CVDs	Cardiovascular Diseases
ChIPBase	Chromatin Immunoprecipitation Database
DGCR8	DiGeorge Syndrome Critical Region 8 Protein
DINO	Damage-Induced Noncoding RNA
DNMTs	DNA Methyltransferases
EBF3-AS	Early B-cell Factor 3 Antisense RNA
EIF3J-AS1	Eukaryotic Translation Initiation Factor 3 Subunit J Antisense RNA 1
EVs	Extracellular Vesicles
FZD3	Frizzled Class Receptor 3
GAS5	Growth Arrest Specific 5
GSK3β	Glycogen Synthase Kinase 3 Beta
HFpEF	Heart Failure with Preserved Ejection Fraction
HOTAIR	HOX Transcript Antisense RNA
HULC	Highly Upregulated in Liver Cancer
IRES	Internal Ribosomal Entry Site
LIPCAR	Long Intergenic Noncoding RNA Predicting Cardiac Remodeling
LRRK2	Leucine-Rich Repeat Kinase 2
Linc00507	Long Intergenic Noncoding RNA 507
LncBase	Long Noncoding RNA Database
MALAT1	Metastasis-Associated Lung Adenocarcinoma Transcript 1
MAPT	Microtubule-Associated Protein Tau
MEG3	Maternally Expressed Gene 3
MIAT	Myocardial Infarction-Associated Transcript
MRI	Magnetic Resonance Imaging
Mhrt	Myosin Heavy Chain-Associated RNA Transcript
NDM29	Noncoding RNA Downstream of MYCN 29
NEAT1	Nuclear-Enriched Abundant Transcript 1
NONCODE	A Comprehensive Database of Noncoding RNAs
Nrf2	Nuclear Factor Erythroid 2-Related Factor 2
PI3K	Phosphoinositide 3-Kinase
PINK1	PTEN-Induced Kinase 1
PRC2	Polycomb Repressive Complex 2
R-loops	RNA-DNA Hybrid Structures
RISC	RNA-Induced Silencing Complex
SK3	Small Conductance Calcium-activated Potassium Channel 3
SNHG1	Small Nucleolar RNA Host Gene 1
SNPs	Single Nucleotide Polymorphisms
Slncky	Computational Tool for LncRNA Analysis
Syk	Spleen Tyrosine Kinase
TGF-β	Transforming Growth Factor Beta
TTC3	Tetratricopeptide Repeat Domain 3
Wisper	Wisp2 Super-Enhancer Associated RNA

XIST	X-Inactive Specific Transcript
ZEB1-AS	Zinc Finger E-Box Binding Homeobox 1 Antisense RNA
circRNAs	Circular RNAs
iPSCs	Induced Pluripotent Stem Cells
lncRNAs	Long Noncoding RNAs
miR-21	MicroRNA-21
miRNAs	MicroRNAs
ncRNAs	Noncoding RNAs
piRNAs	Piwi-interacting RNAs
pri-miRNA	Primary microRNA
rRNA	Ribosomal RNA
shRNAs	Short Hairpin RNAs
siRNAs	Small Interference RNAs
snRNAs	Small Nuclear RNAs
snoRNAs	Small Nucleolar RNAs
snoRNPs	Small Nucleolar Ribonucleoproteins

2.1 Introduction

In recent decades, the noncoding RNA (ncRNA) field has undergone a paradigm shift. Once considered the genome's "dark matter," ncRNAs are now recognized as key regulators of genetic and cellular processes (Consortium 2007). Astonishingly, only 1.5% of the human genome encodes proteins, while nearly 80% is transcribed into noncoding RNA molecules, highlighting their profound biological significance (Kornienko et al. 2013).

ncRNAs play a critical Role in shaping molecular pathways, influencing gene expression, and maintaining cellular homeostasis. Dysregulation of these molecules is intricately linked to various diseases, including cancer, cardiovascular diseases (CVDs), and neurodegenerative disorders such as Alzheimer's disease (AD) (Abubakar et al. 2024). These conditions represent some of the most significant global health challenges, with CVDs accounting for approximately 17.9 million deaths annually (31% of global mortality) and over 55 million people living with dementia worldwide—a number projected to triple by 2050 (Montgomery 2024). Likewise, Alzheimer's disease and other dementias (ADODs) represent a significant global health challenge, with 57 million individuals affected worldwide in 2019 and projections indicating this number will rise to 153 million by 2050. The economic impact is equally staggering, with an estimated global cost of $14,513 billion from 2020 to 2050, equivalent to 0.421% of the annual global GDP. This burden is unevenly distributed, affecting lower-middle-income countries disproportionately in terms of informal caregiving, which constitutes up to 85.45% of their economic burden (Chen et al. 2024).

The world of ncRNAs is vast and complex, with significant diversity and complexity. Long noncoding RNAs (lncRNAs) are also highly diverse, with over

270,000 unique transcripts in the human genome, out of which a great number is still uncharacterized (Cao 2014). The ncRNAs are instrumental in RNA maturation, RNA processing, signaling, gene expression, and protein synthesis (Esteller 2011). They have unique biogenesis pathways. For instance, microRNAs (miRNAs) are processed from pri-miRNAs through Drosha/DGCR8 and Dicer. DiGeorge Syndrome Critical Region 8 protein (DGCR8) is an RNA-binding protein that recognizes specific motifs within the pri-miRNA hairpin (O'Brien et al. 2018). DGCR8 recruits Drosha to the pri-miRNA, thereby guiding the cleavage process (O'Brien et al. 2018). The lncRNAs can be transcribed by RNA polymerase II; some are spliced and polyadenylated. Circular RNAs (circRNAs) are formed via back splicing. Small interference RNAs (siRNAs) are processed from double-stranded RNA by Dicer. Short hairpin RNAs (shRNAs) are transcribed in the nucleus and then processed by Drosha/DGCR8, then Dicer (O'Brien et al. 2018). Briefly, ncRNAs act through diverse mechanisms. For instance, miRNAs primarily induce translational repression or mRNA degradation, lncRNAs function as scaffolds, guides, and decoys, circRNAs act as miRNA sponges, and siRNAs and shRNAs induce RNA interference.

This chapter comprehensively explores ncRNAs, beginning with their classification, history, and biogenesis. The mechanisms through which ncRNAs orchestrate molecular and cellular functions are discussed, focusing on their emerging Role in neurological and cardiovascular diseases. The intersection between neurodegenerative and cardiovascular pathologies is highlighted, given the growing evidence that shared molecular pathways contribute to both conditions. The complex involvement of lncRNAs in disease progression, their diagnostic and prognostic potential. Furthermore, the ethical and translational challenges associated with harnessing ncRNAs for clinical applications are explored, emphasizing the need for precision medicine approaches and interdisciplinary research.

The final section of this chapter discusses the prospects of ncRNA research, focusing on integrating bioinformatics, high-throughput sequencing, artificial intelligence, and multi-omics approaches. These technological advancements are expected to drive the discovery of novel ncRNA biomarkers and therapeutic targets, ultimately paving the way for RNA-based precision medicine. As the field continues to evolve, ncRNAs hold immense promise in revolutionizing our understanding of human diseases and developing targeted interventions to improve patient outcomes.

2.1.1 Historical Perspectives of ncRNAs

The study of ncRNAs has progressed through several pivotal discoveries (Kresge et al. 2005; Consortium 2007; Wang et al. 2021). Figure 2.1 illustrates the significant developments in the progress of ncRNAs. Below is a timeline highlighting key milestones.

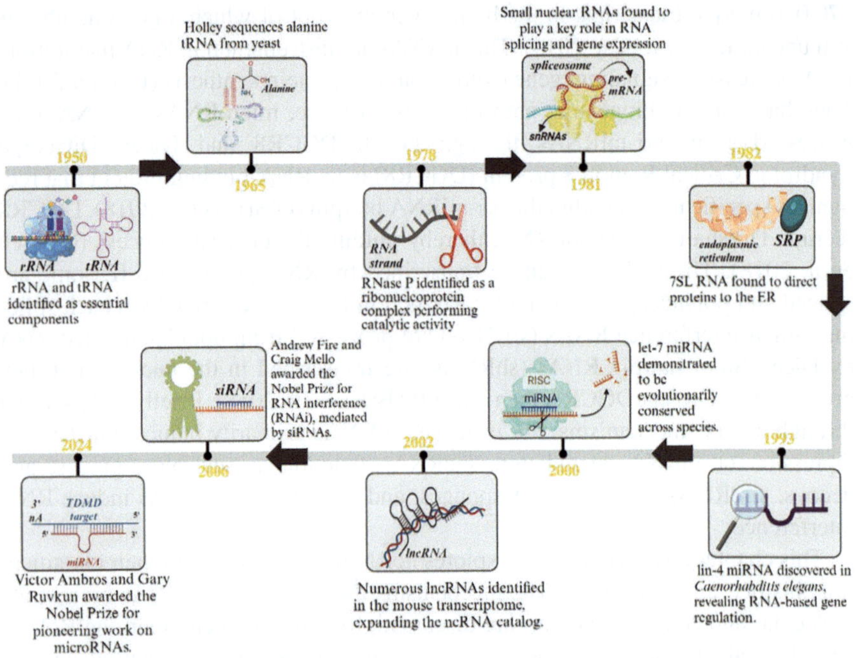

Fig. 2.1 The major milestones in the discovery of lncRNA. (Created in BioRender. Miller, S. (2025) https://BioRender.com/x40a776)

1950s: The ribosomal RNA (rRNA) and transfer RNA (tRNA) were identified as essential components in protein synthesis. These foundational discoveries were instrumental in understanding the Role of noncoding RNAs (rRNA and tRNA) in translation (Palade 1955; Hoagland et al. 1958; Holley et al. 1965), significantly advancing our understanding of the molecular mechanisms underlying protein synthesis.

1965: Robert W. Holley and colleagues characterized the first noncoding RNA by sequencing an alanine tRNA from baker's yeast, marking a significant advancement in molecular biology (Holley et al. 1965).

1978: Sidney Altman and colleagues identified RNase P, a ribonucleoprotein complex involved in tRNA processing, highlighting RNA's catalytic potential (Altman et al. 1980).

1981: Joan A. Steitz and her team discovered small nuclear RNAs (snRNAs), which play crucial Role in RNA splicing. This discovery expanded the understanding of RNA's involvement in gene expression regulation (Reddy and Singh 1991).

1982: Peter Walter and colleagues identified 7SL RNA, the RNA component of the signal recognition particle, essential for directing proteins to the endoplasmic reticulum, underscoring RNA's Role in protein targeting (Andrews et al. 1987).

1993: Victor Ambros and his co-workers Rosalind Lee and Rhonda Feinbaum reported the discovery of lin-4, the first microRNA (miRNA), in *Caenorhabditis elegans*, revealing a novel mechanism of gene regulation (Ambros 2008).

2000: Gary Ruvkun's laboratory characterized let-7 miRNA in *C. elegans* and demonstrated its evolutionary conservation across species, establishing the significance of miRNAs in gene regulation (Ruvkun et al. 1989).

2002: The FANTOM consortium identified numerous long noncoding RNAs (lncRNAs) in the mouse transcriptome, expanding the catalog of ncRNAs and their potential regulatory Role (Abugessaisa et al. 2021).

2006: Andrew Fire and Craig Mello were awarded the Nobel Prize in Physiology or Medicine for their discovery of RNA interference (RNAi), a process mediated by small interfering RNAs (siRNAs), highlighting the therapeutic potential of ncRNAs (Fire et al. 1998).

2024: Victor Ambros and Gary Ruvkun were awarded the Nobel Prize in Physiology or Medicine for their pioneering work on microRNAs and their Role in gene regulation, cementing the importance of ncRNAs in biology and medicine (Kay 2024).

These breakthroughs underscore the shift from a protein-centric view of biology to one where ncRNAs are central players in normal physiology and disease. The diversity of ncRNAs, including small noncoding RNAs, long noncoding RNAs (lncRNAs), microRNAs (miRNAs), and circular RNAs (circRNAs), has opened up exciting avenues in diagnostics, therapeutics, and disease modeling (Ammad et al. 2024).

2.1.2 *Types of ncRNAs*

The ncRNAs encompass a diverse group of RNA molecules that do not encode proteins but perform crucial regulatory functions. They are broadly categorized based on their length and structure into small noncoding RNAs (sncRNAs) (≤200 nucleotides) and long noncoding RNAs (lncRNAs) (>200 nucleotides). sncRNAs include microRNAs (miRNAs), which are Short (~22 nucleotides) RNA molecules that regulate gene expression by binding to target mRNAs. For example, miR-126 modulates vascular integrity (Yu et al. 2020), small Interfering RNAs (siRNAs) (~20–24 nucleotides) that play a Role in RNA interference and antiviral defense [22], and Piwi-interacting RNAs (piRNAs) (~24 to 33 nucleotides). The length of the piRNAs distinguishes them from miRNAs and siRNAs. The specific length of piRNAs is determined by the associated PIWI proteins, which have unique insertion modules (Liu and Lu 2024). This size specification is crucial for the proper functioning of the piRNA-induced silencing complex (pi-RISC) in gene regulation and transposon repression, which is vital for genome stability and transposon silencing (Kang et al. 2023).

The lncRNAs regulate chromatin architecture, transcriptional activity, and post-transcriptional processes. Examples include Metastasis-Associated Lung Adenocarcinoma Transcript 1 (MALAT1), implicated in cancer metastasis, and HOX transcript antisense RNA (HOTAIR), which mediates histone modification (Chen et al. 2020). Circular RNAs exhibit a wide range of lengths, typically spanning from a few hundred to several thousand nucleotides (Liu et al. 2021). circRNAs are generated by back-splicing; these molecules act as miRNA sponges, protein scaffolds, or regulators of transcription (Zhou et al. 2020a, b). Table 2.1 provides a comprehensive comparison of ncRNAs. The detailed mechanism of action of various ncRNAs is illustrated in Figs. 2.2, 2.3, 2.4, 2.5, 2.6, and 2.7 in Sect. 2.1.3.

2.1.3 Mechanism of Actions of Noncoding RNAs

2.2 Biogenesis, Molecular Classification, and Functions of lncRNAs

lncRNAs are transcribed from intergenic regions (or protein-coding regions), introns, or enhancer regions and processed like mRNA, with functionalities conferred by splicing and capping (Ma et al. 2013; Gonzalez et al. 2015; Losko et al. 2016). They are classified based on their genomic location and transcriptional context relative to protein-coding genes (PCGs) (Li et al. 2010; Jeck and Sharpless 2014). Figure 2.8 explains in detail the biogenesis of lncRNAs and their types. LncRNA transcripts are transcribed by RNA polymerase II and play critical Role in gene transcription regulation and modulate essential biological processes, including inflammation, cell proliferation, synaptic plasticity, and neural regeneration (Groszer et al. 2001; Fitch and Silver 2008; Plank and Dean 2014). They also play critical Role in chromatin modification, modulating chromatin states, histone modifications, DNA methylation, gene transcription, epigenetic modifications, mRNA stability, and translation regulation (Groszer et al. 2001; Fitch and Silver 2008; Batista and Chang 2013; Gomes et al. 2013; Mukhopadhyay and Singh 2014; Plank and Dean 2014; Archer et al. 2015; Han and Chang 2015; Hu et al. 2018; Statello et al. 2021). Functionally, lncRNAs act as signals, decoys, guides, and scaffolds (Consortium 2007). Figure 2.9 illustrates the functions of lncRNAs, while Table 2.2 provides a comparative analysis of lncRNAs with other ncRNAs.

Evolutionarily, lncRNAs are conserved in promoter regions (Pang et al. 2006) but show species-specific diversity (Haerty and Ponting 2013), with tools like "slncky" aiding analysis (Chen et al. 2016a, 2016b). They manage critical processes such as X-chromosome inactivation (e.g., Xist) (Yang et al. 2011) and serve as

Table 2.1 A comprehensive comparison of noncoding RNAs (ncRNAs)

Type	microRNA (miRNA)	Long noncoding RNA (lncRNA)	Circular RNA (circRNA)	Small nuclear RNA (snRNA)	Small nucleolar RNA (snoRNA)	Transfer RNA (tRNA)	Ribosomal RNA (rRNA)	Enhancer RNA (eRNA)
Size	Short, typically 21–23 nt or < 200 nt (Correia et al. 2021)	Long, >200 nucleotides (nt) with an average length of ~1000 nt (Ilieva 2024)	It can be a variable similar in size to lncRNAs but also smaller. Derived from back-splicing of precursor mRNA (Ilieva 2024)	Small, ~150 nt, component of spliceosomes (Bhatti et al. 2021)	Small, ~60–300 nt (Bhatti et al. 2021)	Small, ~70–90 nt (Kamps and Robinson 2024)	Larger, various sizes, and essential for ribosome structure (Kamps and Robinson 2024)	Small, smaller than even lncRNA (Mattick et al. 2023)
Biogenesis	Processed from longer RNA precursors (pri-miRNA) by Drosha and Dicer enzymes (Bhatti et al. 2021). The mitochondrial genome can encode it as well (Kuthethur et al. 2022)	Transcribed by RNA polymerase II (Quinn and Chang 2016). It can have 5′ caps, 3′ polyadenylation, and splicing (Quinn and Chang 2016). Unique biogenesis mechanisms linked to subcellular localization (Statello et al. 2021). Some lncRNAs transcribed from the mitochondrial genome (mt-lncRNAs) (Burzio et al. 2009)	Generated from precursor mRNA back-splicing (Ilieva 2024). Can involve RNA-binding proteins, N6-methyladenosine (m6A) modification (Weingarten-Gabbay et al. 2016), or IRES[a] for translation initiation (Fan et al. 2022)	Synthesized by RNA polymerase II or III, depending on the specific snRNA, often associate with proteins to form snRNPs[b] (Bhatti et al. 2021)	Synthesized by RNA polymerase II or III, it often resides in the nucleolus and can also be found in Cajal bodies (Bhatti et al. 2021). Process rRNA maturation (Reddy and Busch 1983)	Transcribed by RNA polymerase III. Undergoes extensive modifications before becoming functional	Transcribed by RNA polymerase I8. Undergoes processing and modifications to mature into its functional form (Bhatti et al. 2021)	Arise from enhancer regions of protein-coding genes (Bhatti et al. 2021); transcribed bidirectionally (Archer et al. 2015)

(continued)

Table 2.1 (continued)

Subcellular localization	Primarily in the cytoplasm (Statello et al. 2021)	Mainly in the nucleus but also found in the cytoplasm, mitochondria, exosomes, and ribosomes—nuclear retention is regulated by sequence motifs (Statello et al. 2021)	Found in both the nucleus and cytoplasm (Statello et al. 2021)	Primarily in the nucleus, often in the nucleoplasm, Cajal bodies, or paraspeckles (Rao et al. 2009)	Primarily in the nucleolus, but can also be in other nuclear compartments (Bhatti et al. 2021)	Primarily in the cytoplasm, present in ribosomes (Bhatti et al. 2021)	Primarily in the cytoplasm as part of ribosomes (Bhatti et al. 2021)	Primarily in the nucleus, associated with chromatin and enhancer regions (Li et al. 2013)
Conservation	Generally well conserved across species (Qu and Adelson 2012)	Poorly conserved between species, which may suggest a Role in increasing the complexity of species (Qu and Adelson 2012)	Less well-studied in terms of conservation (Qu and Adelson 2012)	Highly conserved (Qu and Adelson 2012)	Generally well-conserved (Johnsson et al. 2014)	Generally well-conserved (Johnsson et al. 2014)	Highly conserved across species (Johnsson et al. 2014)	Variable; some can be highly conserved, but others cannot (Johnsson et al. 2014)
Expression	It is generally expressed in all cells and can be dysregulated in various diseases (Soutschek and Schratt 2023)	Tissue-specific, dynamic expression patterns are dysregulated in various diseases, and expression levels vary considerably (Shi et al. 2013)	Ubiquitous expression (Salzman 2016)	It is generally expressed in all cells, as they play a fundamental Role in splicing (Lafontaine 2015)	They are generally expressed in all cells and are essential for RNA processing and maturation(Lafontaine 2015)	It is expressed in all cells, but levels vary depending on the cell type and metabolic needs (Torres et al. 2019)	Expressed in all cells at variable levels, generally the most abundant type of RNA (Torres et al. 2019)	It can be dynamically expressed in response to cellular signals (Arnold et al. 2020)

[a]*IRES* Internal Ribosomal Entry Site, [b]*snRNPs* small nuclear ribonuclear proteins

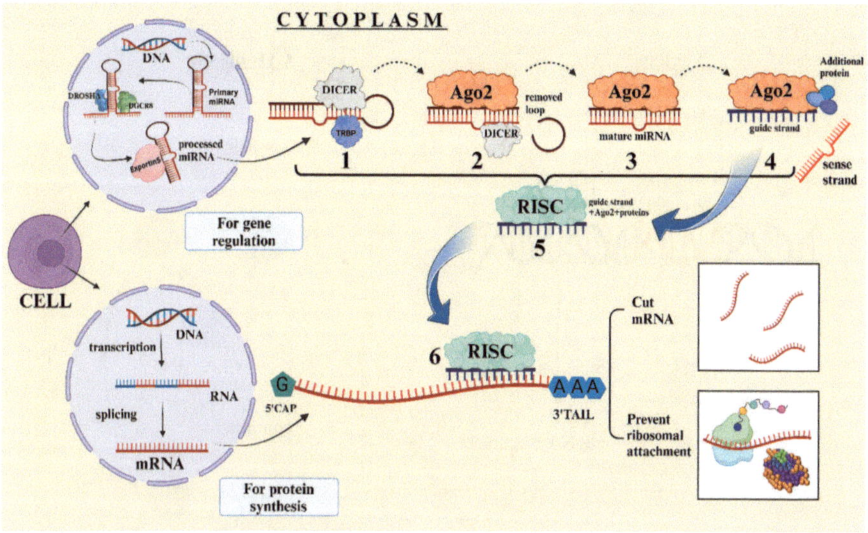

Fig. 2.2 Biogenesis of miRNA and their mechanism of action. In the nucleus, the gene is transcribed into primary miRNA (hairpin loop-shaped), which DGCR8 and DROSHA recognize; they associate together to form a microprocessor complex, which cuts miRNA into smaller miRNA by removing its tails and is exported out to cytoplasm via exportin 5. (1) Inside cytoplasm, RNA proteins, and DICER recognize miRNA and cleaves its loop head. (2) Ago2 interacts with the dicer. (3) Binds the miRNA. (4) miRNA unwound and one strand is released, i.e., sense strand, and the rest is called the guide strand. (5) Guide strand+Ago2 + additional proteins form the RNA-induced silencing complex (RISC), which targets and inactivates multiple genes. (6) Messenger RNA (mRNA) and miRNA have a complementary sequence that results in pairing and mRNA is destroyed. (Created in BioRender. Miller, S. (2025) https://BioRender.com/j48o742)

potential biomarkers (e.g., HULC for liver cancer) (Ayers 2013). Mechanistically, they form RNA–DNA interactions (R-loops, triplexes) (Santos-Pereira and Aguilera 2015), modulate chromatin complexes (e.g., PRC2, DNMTs)(Rinn et al. 2007; Am 2009; Mohammad et al. 2010; Schmitz et al. 2010; Shore et al. 2012; Grote et al. 2013; Marín-Béjar et al. 2013; Kotzin et al. 2016; Wang et al. 2016a, b), influence genome structures (e.g., NEAT1 in nuclear bodies) (Clemson et al. 2009; Sasaki et al. 2009; Sunwoo et al. 2009), and regulate transcription factors and signaling pathways (e.g., GAS5, DINO) (Kino et al. 2010; Schmitt et al. 2016). Emerging evidence highlights the functional relevance of small open reading frames (sORFs) within some lncRNAs, which are translated into micro-peptides and may perform vital biological Role [28].

Dysregulation of lncRNAs is associated with various diseases, including cancer, autoimmune disorders, neurodegenerative disorders, and cardiovascular conditions (Gutschner and Diederichs 2012). Some lncRNAs act as molecular scaffolds or sponges, while others encode functional peptides (Di Gesualdo et al. 2014). Their critical involvement in disease pathogenesis makes them promising targets for therapeutic intervention and valuable biomarkers for diagnosis and prognosis.

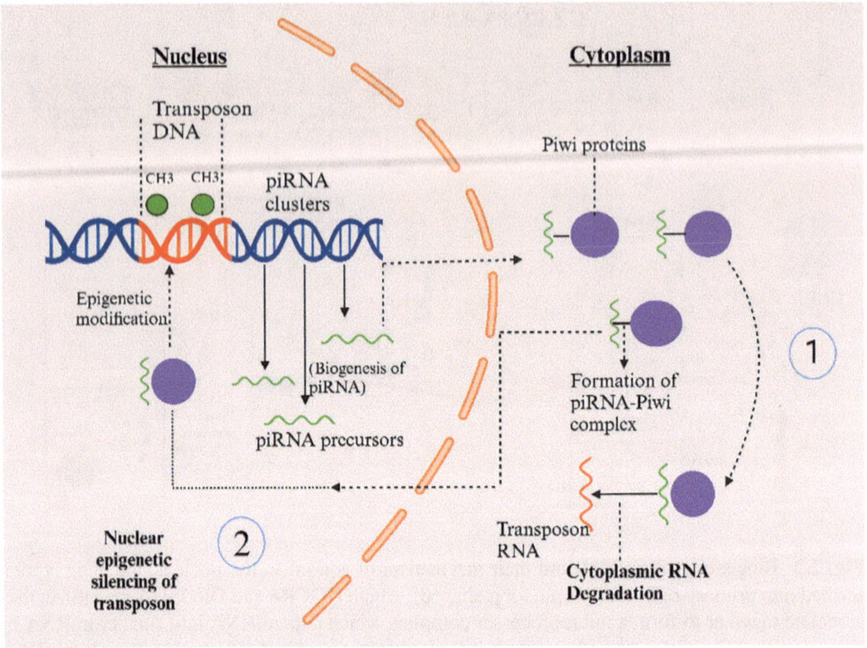

Fig. 2.3 Biogenesis and mechanism of piRNA-mediated transposon silencing. In the nucleus, transposon-derived DNA segments within piRNA clusters are transcribed to produce piRNA precursors. These precursors undergo processing and are subsequently exported to the cytoplasm, where they associate with Piwi proteins to form the piRNA-Piwi complex. (1) This complex identifies and binds to complementary transposon RNAs, leading to their degradation in the cytoplasm, thus silencing transposons at the posttranscriptional level. (2) The piRNA-Piwi complex translocate back to the nucleus, guiding epigenetic modifications, such as DNA methylation (CH_3), on transposon loci. This transcriptional silencing reinforces the suppression of transposon activity, ensuring genomic stability by preventing transposon mobilization. (Created in BioRender. Miller, S. (2025) https://BioRender.com/h90q592)

2.3 Noncoding RNAs in Cardiovascular and Neurodegenerative Disorders

Several studies have explored the Role of ncRNAs in neurodegenerative diseases and CVDs. For instance, in cardiovascular diseases, ncRNAs, such as ANRIL and MEG3, drive cardiac remodeling and fibrosis through the TGF-β and PI3K/AKT pathways (Yeh et al. 2020). ANRIL promotes atherogenesis by enhancing vascular smooth muscle proliferation, while MALAT1 mitigates ischemic damage by reducing oxidative stress. These molecules and MIAT are promising biomarkers for heart failure and coronary artery disease due to their Role in epigenetic regulation (Li et al. 2022). Similarly, miRNAs like miR-21 have been linked to cardiac fibrosis (Vassar and Kandalepas 2011). In other studies, it was revealed that Mhrt and ANRIL influence inflammation, apoptosis, and cardiac hypertrophy (Congrains

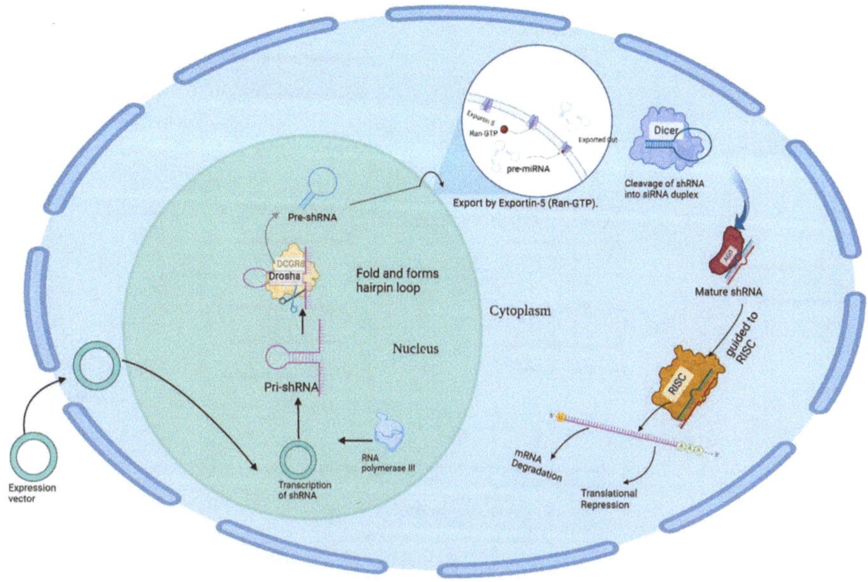

Fig. 2.4 Biogenesis and mechanism of action of shRNA. RNA polymerase II/III transcribes DNA into primary microRNA (pri-miRNA) containing a hairpin structure in the nucleus. The pri-miRNA is processed by the microprocessor complex (Drosha and DGCR8) into precursor microRNA (pre-miRNA), which is exported to the *cytoplasm* via exportin-5 (Ran-GTP-dependent). In the cytoplasm, the pre-miRNA is cleaved by Dicer to generate a mature miRNA duplex. The duplex is loaded into the RNA-induced silencing complex (RISC), where Argonaute (AGO) retains the guide strand and degrades the passenger strand. The RISC-miRNA complex binds to the target mRNA, leading to its *degradation* (perfect complementarity) or *translational repression* (partial complementarity). This process ensures precise regulation of gene expression. (Created in BioRender. Miller, S. (2025) https://BioRender.com/l38b745)

et al. 2012; Han et al. 2014; Wang et al. 2014a, b; Wu and Arora 2015), while in AD, lncRNAs such as BACE1-AS and BC200 are linked to amyloid-beta production, tau hyperphosphorylation, and synaptic dysfunction (Wang et al. 2014a, b).

2.3.1 Role of ncRNAs in CVS Diseases

This section describes the ncRNA's function and how it affects cardiovascular processes. Circulating ncRNAs offer a noninvasive avenue for disease diagnosis and prognosis. For instance, ncRNAs, like MALAT1 and LIPCAR, differentiate between stable and unstable coronary artery disease (Barbalata et al. 2023). Circulating ncRNAs can provide prognostic insights as well. For example, elevated levels of miR-155 and MALAT1 correlate with myocardial infarction and atherosclerosis severity, offering insights into disease progression (Barbalata et al. 2023). Similarly, miR-27a and miR-30d regulate oxidative stress and cardiac fibrosis, while lncRNAs

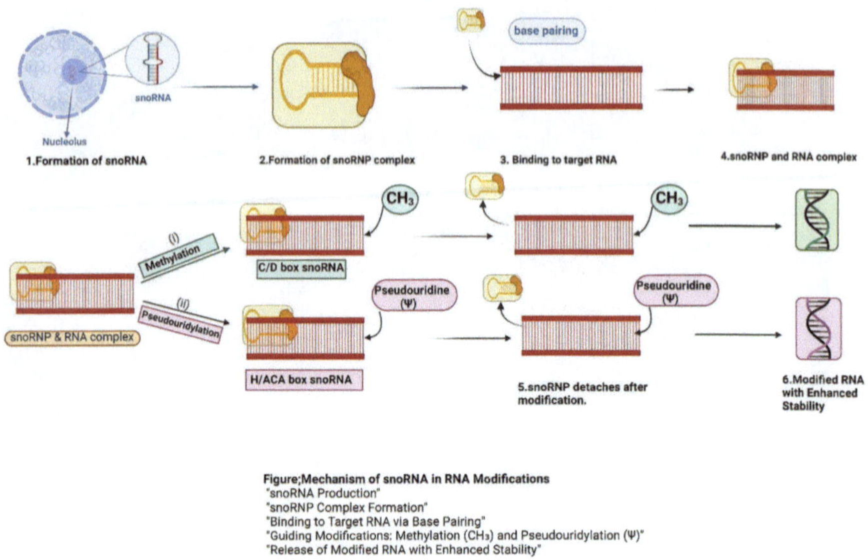

Fig. 2.5 *Biogenesis and mechanism of snoRNA* in RNA modifications. The figure depicts the Role of snoRNA in RNA modifications: (1) snoRNA production: Transcription and processing in the nucleolus. (2) snoRNP formation: Association of snoRNA with proteins to form snoRNP complexes. (3) Base pairing: snoRNP binds target RNA via sequence-specific base pairing. (4) RNA modifications: (i) Methylation (CH_3): Guided by C/D box snoRNA (ii) Pseudouridylation (Ψ): Guided by H/ACA box snoRNA. (5) snoRNP Detachment: Complex dissociates post-modification. (6) Modified RNA: Chemically modified RNA gains enhanced stability and functionality. (Created in BioRender. Miller, S. (2025) https://BioRender.com/q06r120)

influence mitochondrial oxidative homeostasis (Tian et al. 2020, 2022; Chen et al. 2021). Another study revealed that miR-223 delivered via EVs reduces inflammation and cell death in cardiomyocytes, showcasing this mechanism's therapeutic potential (Zhang et al. 2022). Likewise, miR-155 encapsulated in exosomes can be transferred from vascular smooth muscle cells (VSMCs) to endothelial cells (ECs), exacerbating endothelial dysfunction and influencing inflammation and vascular damage (Cánovas-Cervera et al. 2023). For example, MALAT1-loaded exosomes counteract drug-induced cardiotoxicity by inhibiting miR-92a (Xia et al. 2020; Zhang et al. 2023; Climent and García-Giménez 2024).

Several ncRNAs have been implicated in cardiac hypertrophy, such as miR-208a, miR-19a/b, and miR-155, which are upregulated in cardiac hypertrophy and contribute to cardiac remodeling. Conversely, miR-1, miR-101, and miR-185 are downregulated and inhibit hypertrophy signaling. Similarly, miR-499 regulates the SK3 channel and has a potential Role in atrial fibrillation. Several ncRNAs, including circRNAs, have been thoroughly investigated in cardiovascular diseases. The detailed analysis of the reported ncRNAs is described in detail in Table 2.3 with the relevant citations.

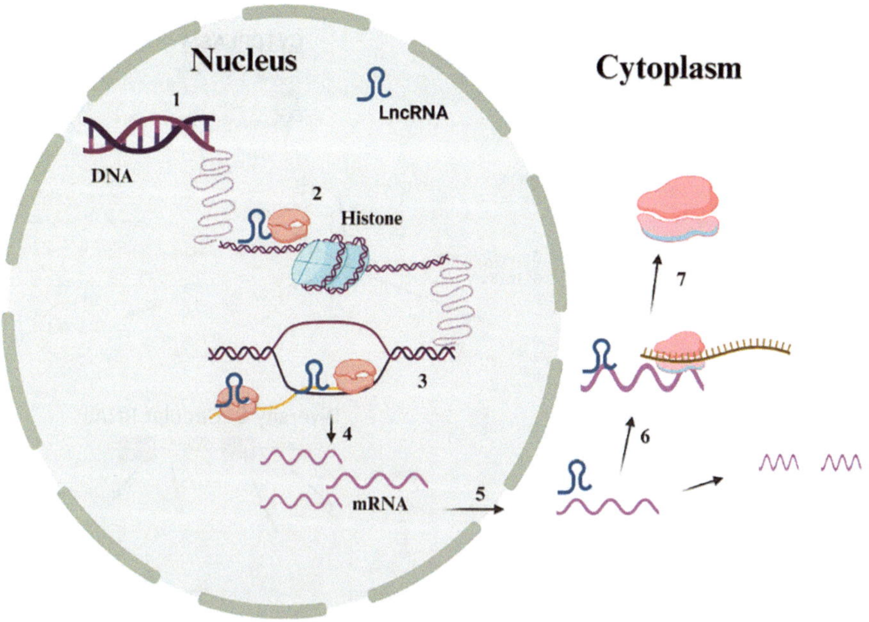

Fig. 2.6 Biogenesis and mechanism of action of lncRNA. (1) DNA, tightly packed into chromatin, is accessible to transcriptional machinery, which determines whether a gene is expressed or silenced. (2) Long noncoding RNA (lncRNA) interacts with histone-modifying proteins, altering chromatin structure. (3) By recruiting histone modifiers, lncRNA plays a crucial Role in regulating gene expression. (4) lncRNAs interact with transcription factors and RNA polymerase to either promote or inhibit gene transcription. Some lncRNAs act as scaffolds, bringing together transcriptional activators or repressors to specific gene loci, thereby influencing gene expression. (5) Once transcription is initiated, an mRNA molecule is synthesized from the DNA template and undergoes processing, including splicing, capping, and polyadenylation. (6) The mature mRNA is transported from the nucleus to the cytoplasm to be translated into a functional protein. (7) In the cytoplasm, lncRNAs bind to mRNA and regulate its stability, degradation, or translation efficiency. Some lncRNAs act as decoys, preventing ribosomes from binding to mRNA, while others stabilize mRNA, ensuring its prolonged presence for translation. (8) Finally, the ribosome binds to mRNA, initiating protein synthesis. Depending on the function of a particular lncRNA, it can either enhance translation by facilitating ribosome binding or inhibit it by blocking translation factors. (Created in BioRender. Miller, S. (2025) https://BioRender.com/c15k224)

2.3.2 *Role of ncRNAs in Neurodegenerative Diseases*

Since the groundbreaking discovery of lncRNAs, as discussed earlier, and their critical importance in enhancing our understanding, diagnosis, and treatment of Alzheimer's disease, researchers have increasingly focused on this promising field. These efforts have yielded remarkable findings, unveiling new pathways and possibilities for exploring the intricate Role of lncRNAs in Alzheimer's pathology. Figure 2.10 illustrates the involvement of lncRNAs in Alzheimer's disease.

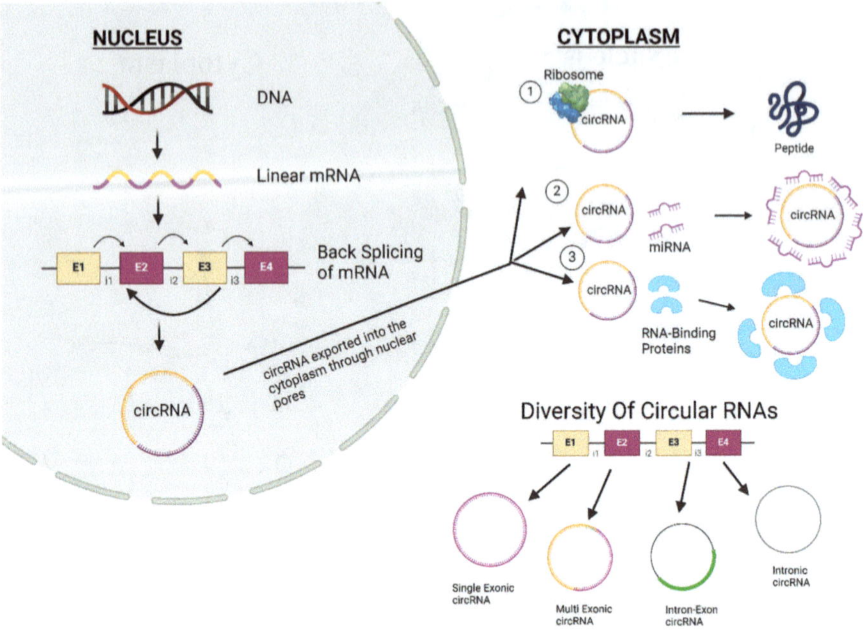

Fig. 2.7 The diagram depicts circRNA formation through back-splicing, producing single-exonic, multi-exonic, intron-exon, and intronic types. CircRNAs are formed via a unique process called *back-splicing*, where parts of a gene (exons or introns) loop back to form a circle. Once formed in the nucleus, circRNAs are transported to the cytoplasm. In the cytoplasm, they perform several important functions. (1) Protein Production**:** Some circRNAs act as templates for making small proteins. (2) miRNA Sponging**:** They bind to and block microRNAs (miRNAs), preventing miRNAs from regulating other genes. (3) Protein Interaction: CircRNAs bind to RNA-binding proteins (RBPs) and regulate their activity. (Created in BioRender. Miller, S. (2025) https://BioRender.com/f90a204)

lncRNAs have a crucial Role in AD pathogenesis, regulating essential processes such as amyloid-β (Aβ) processing, tau hyperphosphorylation, oxidative stress, neuroinflammation, mitochondrial failure, synaptic dysfunction and neuronal death (Lauretti et al. 2021; Lan et al. 2022). Dysregulation of specific lncRNAs, including BACE1-AS, NEAT1, and MALAT1, contributes to disease development by boosting Aβ buildup, affecting tau stability, and causing neuronal death and inflammation (Cao et al. 2019; Ma et al. 2019; Khodayi et al. 2022). Exosomal lncRNAs, such as ENST00000608936 and ENST00000433747, show potential as noninvasive biomarkers for AD detection, with over 95% sensitivity and specificity using advanced machine learning (ML) models (Mosquera-Heredia et al. 2024). However, the precise biological mechanisms and interactions of lncRNAs remain poorly understood, particularly in understudied populations and gender-specific circumstances (Maurano et al. 2012).

For instance, BACE1-AS stabilizes BACE1 mRNA and enhances Aβ synthesis, while 51A and NDM29 increase dangerous Aβ isoforms through faulty APP processing (Gu and Guo 2013; Gu et al. 2018). NEAT1 and Linc00507 induce tau

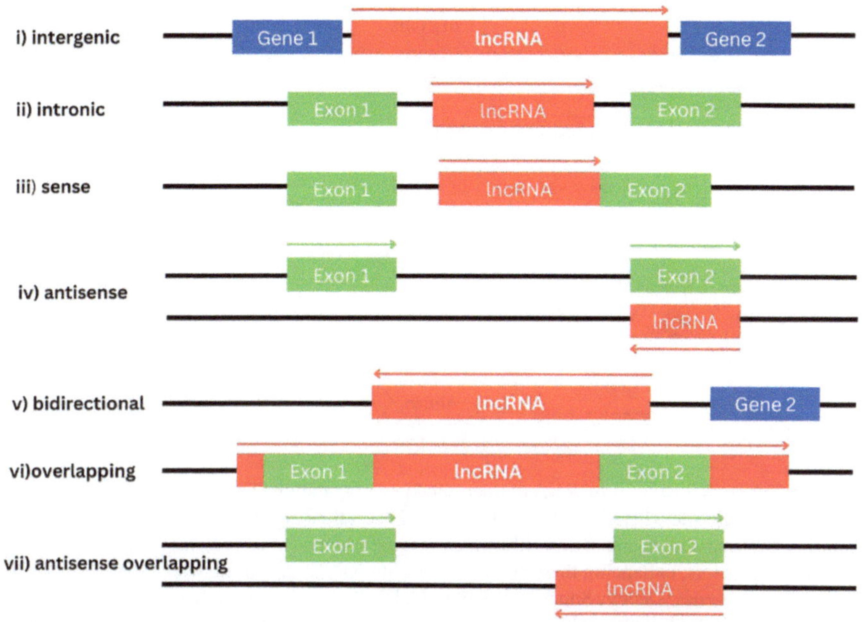

Fig. 2.8 *Classification of long noncoding RNAs (lncRNAs)* based on their genomic location and transcriptional context relative to protein-coding genes (PCGs). Green boxes represent exons of PCGs, blue boxes represent protein-coding genes, and red boxes represent lncRNAs. Arrows indicate the direction of transcription (Ma et al. 2013). (i) Intergenic lncRNA, also known as "lincRNA," is located in the intergenic region at least 1 kb away from neighboring PCGs. It is transcribed from either DNA strand without overlapping PCG sequences, e.g., H19, a well-characterized intergenic lncRNA involved in genomic imprinting and regulation of cell proliferation. (ii) Intronic lncRNA: transcribed solely from the intronic regions of PCGs without overlapping exons. These lncRNAs function independently of the coding capacity of their host genes, e.g., MIR155HG, transcribed from the intron of a PCG and involved in modulating immune and inflammatory responses. (iii) Sense lncRNA: transcribed from the sense strand of PCGs. They contain exons of PCGs, either overlapping with part of the PCG sequence or covering the entire coding sequence through an intron, e.g., MALAT1 (Metastasis-Associated Lung Adenocarcinoma Transcript 1), which overlaps with the sense strand of its host gene and plays a Role in cancer progression and alternative splicing regulation. (iv) Antisense lncRNA: Transcribed from the antisense strand of PCGs. These lncRNAs overlap with exonic or intronic regions or span the entire PCG coding sequence, e.g., ANRIL (Antisense Noncoding RNA in the INK4 Locus) transcribed antisense to the INK4B-ARF-INK4A locus and is associated with cardiovascular diseases and cancer. (v) Bidirectional lncRNA: oriented head-to-head with a PCG, transcribed from the opposite direction, and located within 1 kb of the PCG promoter, e.g., EIF3J-AS1, a bidirectional lncRNA transcribed divergently from EIF3J, and implicated in transcriptional regulation, (vii) Overlapping lncRNA: Contains exonic regions of PCGs, overlapping partially or fully with the PCG exons. These lncRNAs may regulate their overlapping PCG in *cis*, e.g., UCA1 (Urothelial Cancer Associated 1) overlaps partially with its host gene and is implicated in tumor development and progression, and (viii) Antisense Overlapping lncRNA, transcribed from the antisense strand and overlaps with the exonic regions of PCGs, e.g., KCNQ1OT1 overlaps with the KCNQ1 gene and regulates its epigenetic expression

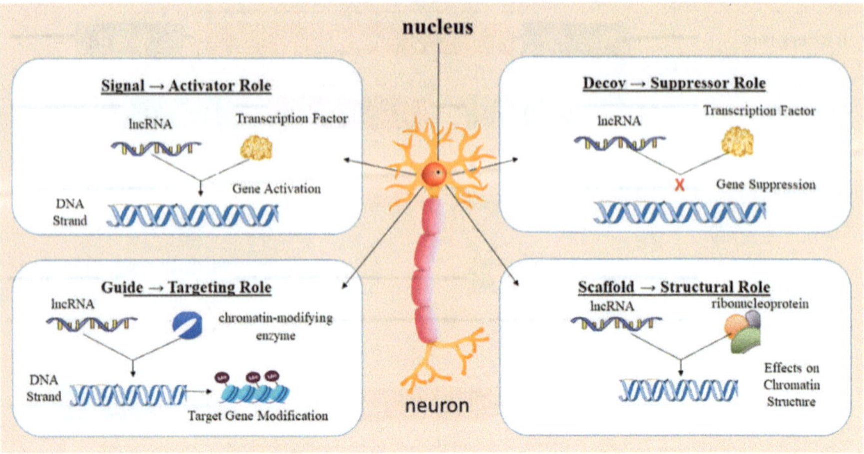

Fig. 2.9 Functional Role of lncRNAs in neuronal gene regulation. The central neuron highlights the nucleus as the site where lncRNAs originate and exert their regulatory effects on gene expression. Four primary mechanisms are illustrated: Signal (Activator Role): lncRNAs interact with transcription factors, enhancing their binding to specific DNA sequences, which leads to the activation of target genes. Decoy (Suppressor Role): lncRNAs sequester transcription factors by competing with DNA for binding, inhibiting gene expression. This mechanism is symbolized by a red cross (X), denoting suppression. Guide (Targeting Role): lncRNAs recruit chromatin-modifying enzymes to specific DNA loci, resulting in epigenetic modifications such as histone methylation (indicated as "Me"). These modifications alter chromatin structure and regulate target gene expression. Scaffold (Structural Role): lncRNAs function as molecular scaffolds, facilitating the assembly of ribonucleoprotein complexes and contributing to chromatin remodeling and DNA structural organization. (Adapted and modified from Mathy and Chen 2017)

Table 2.2 Functional Role of lncRNAs, miRNAs, and other noncoding RNAs in diseases

Feature	lncRNAs	miRNAs	circRNAs	Other ncRNAs (e.g., piRNAs, pseudogenes)
Function	Genomic imprinting, chromatin modification, transcriptional control, and protein translation (Xuan et al. 2023)	Target mRNAs to regulate stability and translation, key Role in immune responses and neuronal development (Xuan et al. 2023)	Act as miRNA sponges, modulate transcription, and serve as biomarkers (Good 2023)	Emerging Role in gene regulation, epigenetics, and RNA stability
Molecular Role	Act as scaffolds, guides, or decoys in transcriptional processes (Xuan et al. 2023)	Regulate mRNA expression through complementary binding (Good 2023)	Interact with miRNAs and RNA-binding proteins; regulate RNA stability(Good 2023)	Limited understanding may include Role in RNA silencing and genome stability

(continued)

Table 2.2 (continued)

Feature	lncRNAs	miRNAs	circRNAs	Other ncRNAs (e.g., piRNAs, pseudogenes)
Formation	Transcribed from genomic DNA, usually long sequences (>200 nucleotides) (Good 2023)	Processed from precursor RNAs (pri-miRNAs) into short (~22 nucleotides) sequences (Good 2023)	Formed via back-splicing of exonic or intronic regions (Xuan et al. 2023)	Diverse origins, including transposons and pseudogenes
Disease association	Implicated in neurodegeneration, cancer, cardiovascular disorders, and diabetes (Xuan et al. 2023)	Dysregulated expression is linked to cancer, immune disorders, and neurodegeneration (Good 2023)	Emerging biomarkers for diseases like cancer and neurological disorders (Good 2023)	Potential links to epigenetic regulation and tumorigenesis
Regulatory interactions	Interact with miRNAs and mRNAs to form regulatory networks in neuronal differentiation and synaptic plasticity (Xuan et al. 2023)	Directly target mRNAs to influence gene expression; can interact with lncRNAs	Sponge miRNAs, influencing downstream mRNA and protein expression	Limited data; likely interact with other ncRNAs and proteins
Challenges	Delivery methods are hindered by the blood–brain barrier and limited functional annotation (Xuan et al. 2023)	Precise targeting remains complex due to off-target effects (Good 2023)	Technological gaps in understanding their full spectrum of functions	Underexplored due to a lack of advanced tools and functional annotations
Research directions	Computational and experimental approaches for refined RNA-disease predictions; lncRNA-miRNA-mRNA network exploration (Xuan et al. 2023)	Advanced bioinformatics to map RNA interactions; therapeutic targeting (Good 2023)	Investigate Role as disease biomarkers and therapeutic agents	Further exploration of biological mechanisms and disease links
Potential applications	Biomarkers and therapeutics for neurodegeneration and other diseases (Xuan et al. 2023)	Biomarkers and therapeutic tools for cancer, immune, and neurological disorders (Good 2023)	Biomarkers for disease diagnosis and progression (Good 2023)	Possible use in targeted therapies and prognostic markers

hyperphosphorylation through the FZD3/GSK3β and miR-181c-5p pathways, respectively (Kickstein et al. 2010). BC200 and BDNF-AS inhibit synaptic plasticity and reduce BDNF expression, hastening cognitive decline (Ji et al. 2010; Sosińska et al. 2015). NEAT1 inhibits mitophagy by interacting with PINK1, which causes mitochondrial dysfunction (Zhou et al. 2018; Lim et al. 2020). SNHG1

Table 2.3 (continued)

ncRNA Type	Specific ncRNA	CVD Type	Role and mechanism	Clinical significance/findings	References
	miR-1, miR-133a, miR-208a, miR-208b, miR-499p	Myocardial infarction, acute coronary syndrome	Regulations identified	Significant in myocardial infarction and acute coronary events	Zhu et al. (2021), Singh et al. (2023)
	miR-320a, miR-200b, miR-622, miR-1228, miR-499, miR-223, miR-1306, miR-18a, miR-26b, miR-27a	Heart failure	Regulations identified	Potential biomarkers and therapeutic targets in heart failure	Zhu et al. (2021), Han et al. (2024), Wei et al. (2024)
	miR-499	Arrhythmia	Regulation of the SK3 channel	Potential Role in atrial fibrillation	Sharma et al. (2023)
	miR-21	Postinfarction heart failure	Role in atrial pro-fibrillatory fibrotic remodeling	Potential therapeutic target to prevent adverse remodeling	Adam et al. (2012)
	miR-206	Pulmonary arterial hypertension	Regulates pulmonary artery smooth muscle cell proliferation and differentiation	Modulation may alleviate pulmonary hypertension	Jalali et al. (2012a, b)
	miR-130a	Arrhythmias	Downregulation of Connexin43 in cardiomyocytes	Leads to cardiac arrhythmias	Osbourne et al. (2014)
	miR-34a	Cardiac aging and function	Regulates cardiac aging and function	Potential therapeutic target for age-related cardiac dysfunction	Boon et al. (2013)
	miR-539	Cardiomyocytes during anoxia	Impairs PHB2 downregulation leading to mitochondrial fission and apoptosis	Inhibition of miR-539 protects cardiomyocytes	Wang et al. (2014a, b)
	miR-489	Cardiac hypertrophy	Targeted by lncRNA CHRF, regulates Myd88 expression	Modulation of miR-489 and CHRF impacts cardiomyocyte hypertrophy	Wang et al. (2014a, b)

	miR-188-3p	Myocardial infarction	Targeted by APF lncRNA, regulates autophagy	APF lncRNA modulates autophagy via miR-188-3p	Wang et al. (2015)
	miR-145	Vascular smooth muscle cells	Regulated by circ_Lrp6, affects vascular smooth muscle cell function	circ_Lrp6 acts as a sponge regulating miRNA-145 function	Hall et al. (2019)
	miR-92a-3p	Doxorubicin-induced cardiac senescence	Sponged by lncRNA MALAT1, affects ATG4a expression	MALAT1 modulates cardiac senescence by targeting miR-92a-3p	Xia et al. (2020)
	miR-15b	Myocardial infarction	Sponged by circular RNA Ttc3, affects cardiac function	Circular RNA Ttc3 regulates cardiac function by sponging miR-15b	Cai et al. (2019)
	Reduced levels of miR-126, members of the miR-17-92 cluster, inflammation-related miR-155, and smooth muscle-enriched miR-145	Coronary Artery Disease (CAD)	Expression differences	Expression differences observed compared with healthy controls	Fichtlscherer et al. (2010)
Long noncoding RNAs	Myocardial infarction-associated transcript (MIAT)	Myocardial infarction	SNPs in MIAT are associated with increased expression and altered protein binding	Reduced expression in patients with ST-segment elevation myocardial infarction	Ishii et al. (2006)
	Myosin heavy chain-associated RNA transcript (MHRT/MyHEART)	Cardiac hypertrophy and heart failure	Interacts with Brg1/Baf chromatin remodeling complex, inhibits Myh6/7 switch	Overexpression protects the heart and reduces pathological switch of Myh6/7 expressions	Han et al. (2014)

(continued)

Table 2.3 A detailed tapestry of noncoding RNAs in cardiovascular diseases

ncRNA Type	Specific ncRNA	CVD Type	Role and mechanism	Clinical significance/findings	References
MicroRNAs	miR-208a	Cardiac hypertrophy (CH)	Upregulation leads to cardiac remodeling	Increased miR-208a levels are associated with cardiac remodeling. Target for therapeutic intervention in cardiac hypertrophy	Huang et al. (2021a, b)
	miR-19a/b	CH	Upregulation induces cardiac remodeling in response to angiotensin II infusion	Potential target for managing cardiac remodeling caused by angiotensin II	Liu et al. (2018a, b)
	miR-155	CH	Upregulation contributes to cardiac remodeling	Inhibition may reduce cardiac remodeling in hypertrophy	Luo et al. (2023)
	miR-199a	CH	Upregulation maintains cell size in cardiomyocytes	Modulation could help regulate cardiomyocyte size during hypertrophy	Song et al. (2010)
	miR-1	CH	Downregulation induces cardiac hypertrophy	Restoring miR-1 levels may inhibit cardiac hypertrophy	Karakikes et al. (2013)
	miR-101	CH	Downregulation inhibits cardiac hypertrophy signaling	Enhancing miR-101 activity may suppress hypertrophy signaling	Wei et al. (2015)
	miR-185	CH	Downregulation inhibits CH hypertrophy signaling	Potential target for enhancing signaling pathways that inhibit hypertrophy	Kim et al. (2015)
	miR-33a, miR-122, miR-126, miR-1, miR-221/222	Arteriosclerosis	Regulations identified	Involved in the pathobiology of arteriosclerosis	Chistiakov et al. (2015), Kumar et al. (2019), Šatrauskienė et al. (2021), Tanashyan et al. (2023)

	Cardiac hypertrophy-related factor (CHRF)	Cardiac hypertrophy	Elevated by angiotensin II, binds to miR-489 to regulate Myd88 expression	Induces cardiomyocyte hypertrophy in vitro and apoptosis in vivo	Wang et al. (2014a, b)
	Long intergenic noncoding RNA predicting cardiac remodeling (LIPCAR)	Heart failure	Circulating lncRNA that predicts survival	High levels associated with poor outcomes in heart failure patients	Kumarswamy et al. (2014)
	Metastasis-associated lung adenocarcinoma transcript 1(MALAT1)	Endothelial cell function and vessel growth	Regulates endothelial cell function and vessel growth	Also implicated in cardioprotection during myocardial ischemia-reperfusion injury	Michalik et al. (2014)
	Wisp2 super-enhancer–associated RNA(Wisper)	Cardiac fibrosis and remodeling	Controls cardiac fibrosis and remodeling	Potential therapeutic target to reduce fibrosis	Micheletti et al. (2017)
	ZFAS1	Myocardial infarction	Induces mitochondria-mediated apoptosis	Potential therapeutic target to reduce apoptosis post-MI7	Jiao et al. (2019)
	H19	Atherosclerosis	Contributes to atherosclerosis and induces ischemic stroke	Upregulates acid phosphatase	Huang et al. (2019)
	NEXN-AS1	Atherosclerosis	Mitigates atherosclerosis by regulating the actin-binding protein NEXN	Potential therapeutic target for atherosclerosis	Hu et al. (2019)
	ANRIL	Atherosclerosis, coronary artery disease	Regulates vascular smooth muscle cell function	Overexpression inhibits phenotypic switching of vascular smooth muscle cells	Hu et al. (2020)

(continued)

Table 2.3 (continued)

ncRNA Type	Specific ncRNA	CVD Type	Role and mechanism	Clinical significance/findings	References
	XR007793	Vascular smooth muscle cell	Regulates proliferation and migration via suppressing miR23b	Potential target to modulate vascular smooth muscle cell behavior	Wu et al. (2018)
	ENST00000609755.1	Coronary heart disease	Knockdown confers protection against early oxLDL-induced disease	Potential therapeutic target for early coronary heart disease	Sun et al. (2021)
	ZNF593-AS	Dilated cardiomyopathy	Alleviates contractile dysfunction	Potential therapeutic target for DCM	Fan et al. (2021a, b)
	CARL	Anoxia-induced mitochondrial fission	Binds to miR-539, blocks its function in mitochondrial fission	Protects cardiomyocytes by derepressing PHB	Hui et al. (2017)
Circular RNAs	circRNA ANRIL	Coronary atherosclerosis	Reduces vascular endothelial injury, oxidative stress, and inflammation	Potential therapeutic target for atherosclerosis	Shi et al. (2020)
	circGNAQ	Atherosclerosis	Inhibits endothelial cell senescence	Potential therapeutic target for preventing atherosclerosis progression	Wu et al. (2021)
	circFndc3b	Myocardial infarction	Modulates cardiac repair via FUS/VEGF-A axis	Potential therapeutic target for cardiac repair post-MI	Garikipati et al. (2019)
	circWDR77	Vascular smooth muscle cells	Targets FGF-2 to regulate proliferation and migration by sponging miR-124	Potential therapeutic target for modulating vascular smooth muscle cell behavior52	Chen et al. (2017)

	circ_RUSC2	Vascular smooth muscle cells	Upregulates SYK expression by targeting miR-661	Potential therapeutic target for modulating vascular smooth muscle cell function52	Sun et al. (2019)
	circRNA 010567	Myocardial infarction	Regulates miRNA-141/ DAPK1 axis	Potential therapeutic target for MI53	Zhao et al. (2021)
	circNfix	Cardiac hypertrophy	Attenuates pressure overload-induced hypertrophy via regulating miR-145-5p/ ATF3 axis	Potential therapeutic target for cardiac hypertrophy	Pan et al. (2021)
	circTtc3	Myocardial infarction	Regulates cardiac function by sponging miR-15b	Potential therapeutic target post-MI	Cai et al. (2019)
	circ_Lrp6	Vascular smooth muscle cells	Acts as a sponge regulating miRNA-145 function6	Influences vascular smooth muscle cell function	Hall et al. (2019)
	circNlgn	Doxorubicin-induced cardiac remodeling and fibrosis	Mediates cardiac remodeling and fibrosis	Potential therapeutic target for cardiac remodeling	Du et al. (2021)
	hsa_circ_0001879	Heart failure with preserved ejection fraction (HFpEF)	Identification	Diagnostic and therapeutic potential	Wang et al. (2019)

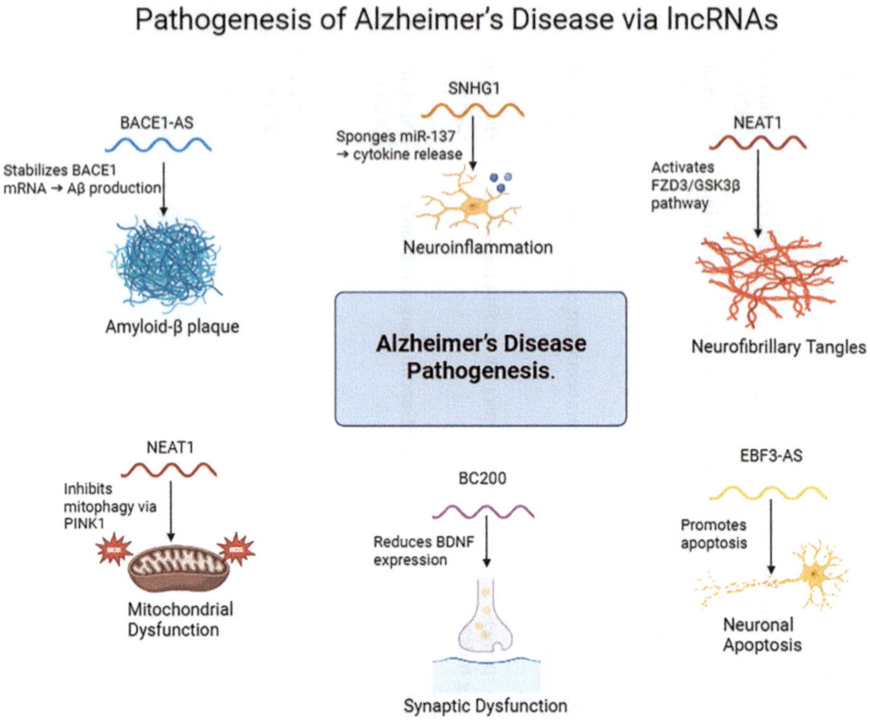

Fig. 2.10 The pathogenesis of Alzheimer's disease (AD) is significantly influenced by long non-coding RNAs (lncRNAs), which regulate various hallmark pathological processes. BACE1-AS stabilizes BACE1 mRNA, leading to increased production of amyloid-β (Aβ) and driving the formation of amyloid plaques, a hallmark feature of AD. SNHG1 acts as a sponge for miR-137, thereby enhancing cytokine release and contributing to neuroinflammation, exacerbating neuronal damage. NEAT1 plays a dual Role in AD pathology: It activates the FZD3/GSK3β signaling pathway, promoting tau hyperphosphorylation, and leading to the formation of neurofibrillary tangles; it also inhibits mitophagy by suppressing PINK1, causing mitochondrial dysfunction that impairs cellular energy regulation and neuronal survival. BC200 reduces brain-derived neurotrophic factor (BDNF) expression, impairing synaptic function and contributing to cognitive decline through synaptic dysfunction. Lastly, EBF3-AS promotes neuronal apoptosis by activating pro-apoptotic signaling pathways, accelerating neurodegeneration and programmed cell death. Together, these lncRNAs orchestrate the molecular mechanisms underlying amyloid plaque deposition, neurofibrillary tangle formation, synaptic loss, mitochondrial dysfunction, neuroinflammation, and neuronal apoptosis, collectively defining AD pathophysiology. (Created in BioRender. Khan, F. (2025) https://BioRender.com/a51x250)

facilitates the inflammatory response by sponging miR-(Cao et al. 2018) while concurrently alleviating neuroinflammation (Cao et al. 2018; Meng et al. 2021). EBF3-AS promotes apoptotic processes, in contrast to MALAT1, which attenuates neuronal apoptosis and inflammatory responses (Zhuang et al. 2020; Lan et al. 2022). Table 2.4 provides a comprehensive overview of key ncRNAs, their dysregulation trends, and their involvement in AD pathogenesis.

Table 2.4 A detailed overview of noncoding RNAs in neurodegenerative diseases

NcRNA name	Type	Disease implication	Expression in disease	Mechanism of action	References
BACE1-AS	lncRNA	Alzheimer's disease (AD)	Upregulated	Promotes Aβ production by increasing BACE1 activity and may also regulate BACE1 mRNA expression. Potential blood-based biomarker	Modarresi et al. (2011), Li et al. (2019)
MALAT1	lncRNA	AD, Parkinson's disease (PD)	Downregulated in AD, upregulated in PD	Neuroprotective in AD by inhibiting neuronal apoptosis and neuroinflammation via miR-30b/CNR1 network and PI3K/AKT activation. Promotes inflammation in PD through epigenetic suppression of Nrf21	Chen et al. (2016a, 2016b), Cai et al. (2020)
NEAT1	lncRNA	AD, PD, brain tumors, stroke	Upregulated	Limits Aβ production in AD, promotes autophagy and apoptosis in PD by impairing miR-374c-5p. May aggravate Aβ-induced neuronal damage via miR-107	He et al. (2016), Ke et al. (2019), Liu et al. (2020)
MAPT-AS1	lncRNA	AD	Downregulated	Inhibits tau translation by pairing with rRNA, silencing increases neuronal tau levels1	Herrera-Rivero et al. (2015), Tauber et al. (2023)
SNHG8	lncRNA	AD	Downregulated	Interacts with TIA1, an RNA-binding protein associated with tau and stress granules	Bhagat et al. (2023)
BC200	lncRNA	AD	Upregulated	Promotes neuronal apoptosis and neuroinflammation, potential blood biomarker for late-onset AD1	Mus et al. (2007), Sosińska et al. (2015), Khodayi et al. (2022)

Table 2.4 (continued)

NcRNA name	Type	Disease implication	Expression in disease	Mechanism of action	References
ANRIL	lncRNA	AD, atherosclerosis	Varies	Associated with atherosclerosis, may be involved in AD. Knockdown suppresses apoptosis and pro-inflammatory cytokines, may regulate ADIPOR1, VAMP3, and C11ORF10	Zhou et al. (2020a, b), Shou et al. (2024)
Long intergenic nonprotein coding RNA 507 (LINC00507)	lncRNA	AD	Upregulated	Part of the LINC00507/miRNA-181c-5p/TTBK1/MAPT axis, regulates tau hyperphosphorylation	Yan et al. (2020)
RPPH1	lncRNA	AD	Upregulated	Protects against amyloid-β induced neuronal injury via the miR-122/Wnt1 axis	Gu et al. (2020, 2021)
SNHG1	lncRNA	AD, PD	Upregulated	Promotes neuroinflammation in both AD and PD. May also promote alpha-synuclein aggregation and toxicity by targeting miR-15b-5p. Regulates miR-7/NLRP3 pathway and the miR-125b-5p/MAPK1 axis	Cao et al. (2018), Chen et al. (2018)
XIST	lncRNA	AD	Downregulated	Silencing attenuates AD-related BACE1 alteration through miR-124	Wang et al. (2021)
TUG1	lncRNA	AD, PD	Upregulated	Knockdown depresses apoptosis of hippocampal neurons in AD, promotes inflammation in PD	Li et al. (2020)

HOTAIR	lncRNA	PD	Upregulated	Promotes PD through upregulating LRRK2, associated with stroke	Lu et al. (2022), Wang et al. (2022a, b)
BDNF-AS	lncRNA	AD	Varies	Regulates the miR-9-5p/BACE1 pathway affecting neurotoxicity	Ding et al. (2022)
SOX21-AS1	lncRNA	AD	Varies	Experimentally validated ceRNA	Zhang et al. (2019a, b)
LINC00094	lncRNA	AD	Varies	Experimentally validated ceRNA	Zhu et al. (2019)
MAGI2-AS3	lncRNA	AD	Varies	Functions as a competing endogenous RNA	Zhang and Wang (2021)
LINC01311	lncRNA	AD	Varies	Experimentally validated ceRNA	Fan et al. (2021a, b)
Abhd11os	lncRNA	Huntington's Disease	Varies	Neuroprotective against an N-terminal fragment of mutant huntingtin	Francelle et al. (2015)
SNHG14	lncRNA	PD	Downregulated	Interfering mitigates dopaminergic neuron injury by regulating miR-133b/α--synuclein pathway	Zhang et al. (2019a, b)
miR-124	miRNA	Neurodegeneration	Varies	Implicated in the regulation of BACE1 expression and neuroinflammation	Zhang and Wang (2021)
miR-137	miRNA	Neurodegeneration	Varies	Role in synaptic plasticity	Cao et al. (2018)
miR-15b-5p	miRNA	Neurodegeneration, PD, AD	Varies	Targets of SNHG1 and other lncRNAs, impact neuroinflammation and alpha-synuclein aggregation	Chen et al. (2018)
miR-7	miRNA	Neuroinflammation, AD	Varies	Regulated by SNHG1, influences NLRP3 pathway	Cao et al. (2018)

Microtubule-Associated Protein Tau Antisense RNA 1: MAPT AS1
Antisense Noncoding RNA in the INK4 Locus), ANRIL, also known as CDKN2B-AS1

2.4 Diagnostic and Therapeutic Potential of ncRNAs

Dysregulated circulating ncRNAs, including BACE1-AS, NEAT1, MALAT1, and GAS5, have surfaced as key factors contributing to the pathogenesis of AD, demonstrating a significant correlation with the severity of the disease (Shobeiri et al. 2023). BACE1-AS promotes the production of amyloid-beta (Aβ) by enhancing the stability of BACE1 mRNA, thereby facilitating the formation of plaques, whereas NEAT1 induces tau hyperphosphorylation and neuronal apoptosis through mechanisms involving miRNA sponging (Faghihi et al. 2008; Ke et al. 2019). In contrast, MALAT1 displays neuroprotective characteristics by reducing inflammation and neuronal cell death, while GAS5 is a highly reliable diagnostic biomarker (Ma et al. 2019; Cao et al. 2021). Sophisticated machine learning algorithms utilizing lncRNA profiles have attained diagnostic accuracy rates of up to 95%, underscoring their potential in the early detection of AD (Shobeiri et al. 2023). Diagnostic applications of lncRNAs (Long and Holtzman 2019) with examples like MALAT1 in metastasis and H19 in imprinting (Jalali et al. 2012a, b; Satpathy and Chang 2015). Recent advancements in deep learning have significantly enhanced lncRNA research, particularly in disease contexts like cardiovascular conditions, by facilitating the identification of lncRNA-disease associations and exploring their genetic, epigenetic, and posttranscriptional regulatory Role (Liu et al. 2015; Guo et al. 2021).

In the intricate world of molecular biology, circulating ncRNAs have emerged as key players in human health, particularly in the context of neurological and cardiovascular diseases (Zhu et al. 2024). These remarkable molecules are characterized by their stability in body fluids, safeguarded by exosomes or protein complexes, which allow them to endure even in harsh extracellular environments. This inherent resilience ensures their reliable blood, plasma, or cerebrospinal fluid detection, pav-machine learning can advance lncRNA discovery and diagnostics. Targeting BACE1-AS and NEAT1 offers therapy while optimizing biomarkers and neuroprotective lncRNAs like MALAT1 may prevent neuronal loss and boost cognitive resilience in Alzheimer's (Bernard et al. 2010; Fotuhi et al. 2019; Khodayi et al. 2022; Black et al. 2024). These RNA molecules modulate gene expression, cellular differentiation, and intercellular communication, often through extracellular vesicles such as exosomes (Valadi et al. 2007). In CVDs, lncRNAs such as MALAT1 and LIPCAR serve as promising biomarkers and therapeutic targets, showing the potential to mitigate cardiac remodeling and enhance mitochondrial metabolism (Zhao et al. 2017; Barbalata et al. 2023; Le and Nhu 2023). Similarly, in AD, circulating lncRNAs like BACE1 and its antisense transcript BACE1-AS function as stable biomarkers for early detection and disease monitoring while amplifying pathological processes, including amyloid beta (Aβ) production, tau hyperphosphorylation, and chronic neuroinflammation (Faghihi et al. 2008; Zhu et al. 2012; Banzhaf-Strathmann et al. 2014; Feng et al. 2018). Therapeutic strategies aimed at these lncRNAs encompass the inhibition of BACE1-AS to mitigate Aβ accumulation, silencing of NEAT1 to restore the balance of miRNAs, and the upregulation of MALAT1 to fortify cognitive resilience (Liu et al. 2014, 2020; Ma et al. 2019).

Beyond their structural advantages, lncRNAs perform critical regulatory Role, influencing gene expression, cellular signaling, and epigenetic mechanisms. In neurological diseases, such as Alzheimer's and Parkinson's, specific circulating lncRNAs—like HOTAIR and BC200—have been linked to disrupted neuronal signaling and epigenetic alterations (Mus et al. 2007; Fan et al. 2019). Similarly, circulating lncRNAs such as ANRIL and MALAT1 regulate endothelial function, inflammation, and plaque stability in cardiovascular health. Their elevated expression in conditions like atherosclerosis and myocardial infarction highlights their significant Role in angiogenesis and cardiac remodeling (Petkovic et al. 2023). These lncRNAs act as molecular mirrors, reflecting the pathological changes within the body, thus cementing their place as potential biomarkers for diseases like stroke and heart failure (Viereck and Thum 2017).

Noncoding RNAs possess a unique ability to interact with DNA, RNA, and proteins, enabling them to orchestrate complex cellular pathways and making them attractive targets for therapeutic intervention (Grillone et al. 2024; Nappi 2024). Emerging approaches such as antisense oligonucleotides, CRISPR-Cas9 gene editing, and small molecule inhibitors offer promising avenues to modulate lncRNA activity (Mahato et al. 2024). While challenges like delivery efficiency and specificity persist, advancements in delivery systems and a deeper understanding of lncRNA biology drive the development of precision therapies. These efforts highlight the transformative potential of lncRNA-targeted strategies in addressing unmet clinical needs.

Noncoding RNAs hold significant therapeutic potential due to their Role in regulating gene expression, acting as miRNA sponges, and influencing transcription and protein expression (Zhang et al. 2021). Effective delivery systems, including viral vectors (e.g., adeno-associated virus), nonviral approaches like nanoparticles and liposomes, and exosome-based carriers, are crucial for addressing degradation, inefficient delivery, biological barriers, and safety concerns. Chemical modifications, such as locked nucleic acids (LNAs), have been proposed to enhance stability and specificity. Key advancements include lncRNA CHROME, which regulates cholesterol metabolism by modulating miR-33, and lncRNA H19, which reduces smooth muscle cell apoptosis in vascular conditions (Huang et al. 2020). While challenges like delivery efficiency and specificity persist, advancements in delivery systems and a deeper understanding of lncRNA biology drive the development of precision therapies. These efforts highlight the transformative potential of lncRNA-targeted strategies in addressing unmet clinical needs.

2.4.1 Evaluating the Diagnostic Accuracy and Prognostic Potential of lncRNAs

lncRNAs are emerging as critical regulators in gene expression and cellular pathways, with significant diagnostic and therapeutic potential in various diseases (Bella et al. 2023; Zhu et al. 2023). In psychiatric and neurodegenerative disorders, lncRNAs such as NEAT1, MALAT1, and MEG3 exhibit disease-specific expression patterns (Li et al. 2018; Cai et al. 2020; Quan et al. 2020), influencing pathways like Tau protein regulation in AD and amyloid β synthesis via BACE1-AS (Faghihi et al. 2008; Simone et al. 2021). Similarly, lncRNAs HCG15 and Morrbid have shown high diagnostic accuracy (AUC > 0.93) and prognostic value in acute myocardial infarction (AMI), linked to inflammatory responses and cardiomyocyte apoptosis (Kotzin et al. 2016; Lin et al. 2021; Sasmita et al. 2022; Yu et al. 2023). Despite promising insights, challenges include the complex interplay of genetic and environmental factors, variability in study designs, and limited large-scale validation (Crump et al. 2013; Wang and Sun 2020). Solutions like high-throughput sequencing, multi-omics integration, and multicenter studies are proposed to enhance understanding and clinical application (Huang et al. 2024; Zhu et al. 2024). Future directions emphasize developing lncRNA-based biomarkers, therapies, and personalized medicine approaches to improve early disease detection, progression monitoring, and tailored interventions (Huang et al. 2024, Zhu et al. 2024).

2.4.2 Challenges in Delivery and Translational Studies

Despite these innovations, poor sequence conservation across species and limited translational studies hinder progress. However, the journey to fully harness the potential of circulating ncRNAs is challenging. Inconsistent methodologies for sampling, RNA isolation, and detection across studies have hampered reproducibility. Additionally, the multifunctionality of ncRNAs complicates the interpretation of their Role in diseases, while transitioning these molecules from research to clinical diagnostics presents regulatory and logistical hurdles (Das et al. 2020). Table 2.5 describes the several detection techniques employed for dectecting the lnRNAs with their key applications and limitations.

To address these obstacles, researchers advocate for developing universal standards for ncRNA analysis. Integrating ncRNA data with proteomic and genomic profiles could provide comprehensive insights into disease mechanisms. At the same time, advancements in high-throughput sequencing and digital PCR technologies promise enhanced sensitivity and specificity in detection (Jun 2019). Newer tools such as engineered heart tissues (EHTs) and living myocardial slices bridge preclinical and clinical gaps, while refined targeting methods like GalNAc conjugation and bioinformatics-based RNA target identification aim to address delivery and specificity hurdles (Huang et al. 2020). Advances in delivery systems, including biocompatible nanocarriers and natural exosomes, enhance the precision of targeting ncRNAs, yet challenges remain in achieving efficient tissue-specific delivery, stability, and immunogenicity minimization (Taghvimi et al. 2024).

Other challenges in ncRNA research include their context-specific Role, functional complexity, limited clinical validation, and delivery mechanisms, which hinder translational research. Highly variable expression across tissues necessitates rigorous validation for clinical use, while delivery systems for ncRNA-targeted

Table 2.5 Noncoding RNA detection techniques (Kotzin et al. 2016, Lin et al. 2021, Sasmita et al. 2022, Yu et al. 2023)

Detection technique	Sensitivity	Specificity	Cost	Time required	Key applications	Limitations
qRT-PCR	High	High	Low	Less	Quantitative detection of known lncRNAs (e.g., HOTAIR, LIPCAR)	Requires prior knowledge of target sequences; limited to known lncRNAs
Microarray	Moderate	Moderate	Moderate	Moderate	High-throughput profiling of multiple lncRNAs (e.g., MALAT1 expression profiling)	Lower sensitivity and specificity compared to qRT-PCR and NGS
Next-generation sequencing (NGS)	Very high	Very high	High	More	Discovery of novel lncRNAs and in-depth profiling (e.g., RNA-seq for H19)	Expensive and time-intensive; requires computational expertise
Northern Blot	Moderate	High	Moderate	Moderate	Validation of specific lncRNA expression levels (e.g., transcript size verification)	Labor-intensive and less suited for high-throughput analysis
Droplet Digital PCR (ddPCR)	Very high	Very high	Moderate	Less	Absolute quantification of low-abundance lncRNAs (e.g., UCA1 in cancer)	Higher cost than qRT-PCR; limited by droplet partitioning efficiency

therapies remain underdeveloped (Archer et al. 2015; Statello et al. 2021). Advanced techniques like single-cell RNA sequencing, CRISPR/Cas9 (Liu et al. 2016), multi-omics approaches, and biomarker validation are vital for leveraging lncRNAs in disease management (Ran et al. 2013). Future efforts should emphasize RNA-targeted therapies, noninvasive biomarkers, and translational studies to advance precision medicine in CVDs and AD (Archer et al. 2015, Statello et al. 2021). Their unique ability to interact with DNA, RNA, and proteins enables them to orchestrate complex cellular pathways, making them attractive targets for therapeutic intervention (Grillone et al. 2024). Novel approaches such as antisense oligonucleotides, CRISPR-Cas9 gene editing, and small molecule inhibitors offer promising avenues to modulate ncRNA activity (Mahato et al. 2024). While challenges like delivery efficiency and specificity persist, advancements in delivery systems and a deeper understanding of lncRNA biology drive the development of precision therapies. These efforts highlight the transformative potential of lncRNA-targeted strategies in addressing unmet clinical needs. Recent advancements in RNA sequencing and deep learning approaches, such as convolutional neural networks (LeCun et al. 1989) and extended short-term memory models (Schmidhuber and Hochreiter 1997), have enhanced our ability to distinguish coding from noncoding transcripts. However, challenges persist, including the incomplete characterization of ncRNAs and their condition-specific expression profiles (Nabi et al. 2023). Addressing these requires interdisciplinary research and integrated data resources to harness lncRNAs as biomarkers or therapeutic targets (Huang et al. 2021). Antisense oligonucleotides and nanoparticle delivery systems offer new avenues for targeting lncRNAs by reversing epigenetic changes mediated by lncRNAs; these strategies pave the way for personalized medicine in cardiovascular and neurodegenerative diseases (Srinivas et al. 2024). Future research should focus on unraveling ncRNA functions in disease mechanisms, including potential AD–CVD interactions, and leveraging their profiles for precision medicine (Greco et al. 2022).

Some other challenges include the intricate interplay between lncRNAs and molecular pathways and the absence of standardized protocols for clinical detection and application (Zhao et al. 2018; Climent and García-Giménez 2024). Future research must integrate advanced transcriptomic and proteomic technologies to overcome these barriers and unravel ncRNA mechanisms (Zhao et al. 2018, Climent and García-Giménez 2024). RNA-targeted therapies, such as inhibitors of BACE1-AS, represent promising therapeutic avenues (Faghihi et al. 2008; Mazdeh et al. 2018). Moreover, developing comprehensive diagnostic panels incorporating lncRNA profiles and exploring early disease stages are critical to advancing precision medicine in cardiovascular and neurodegenerative diseases (Liu et al. 2018a, b). These efforts will help unlock the complete diagnostic and therapeutic potential of lncRNAs in clinical practice (Laura Francés et al. 2023). Proposed solutions encompass robust nanocarriers, biomimetic designs, and RNA modifications to improve stability, specificity, and immune evasion (Zhang et al. 2021). Multi-omics approaches are emphasized for understanding ncRNA mechanisms, and integrating AI for therapeutic design could optimize delivery strategies (Zhang et al. 2021; Taghvimi et al. 2024).

2.4.3 Technological Advances and Future Directions

Future directions include the development of tissue-specific lncRNAs to enhance safety and reduce off-target effects, advancing clinical trials for safety and efficacy, and exploring combination therapies to improve outcomes (Huang et al. 2020; Zhang et al. 2021). Personalized therapies based on ncRNA profiles and CRISPR-based modulation highlight the potential expansion of ncRNAs as diagnostic and prognostic biomarkers (Taghvimi et al. 2024). Standardized protocols for ncRNA validation and safety assessment are essential to bridge the gap between bench and bedside (Harries 2012). Initiatives to integrate multidisciplinary approaches, including bioinformatics, molecular biology, and clinical sciences, can expedite the development of lncRNA-based therapies while ensuring their reliability and efficacy (Jathar et al. 2017). Prospective avenues for research include the advancement of biomarker development, the strategic targeting of ncRNAs through RNA interference or CRISPR methodologies, and the implementation of transcriptomic profiling to reveal innovative therapeutic approaches (Magistri et al. 2015).

Recent innovations in technology have markedly improved our comprehension and management of Alzheimer's and cardiovascular diseases, revealing common pathways and shared risk factors (Chen et al. 2022). Furthermore, advancements in deep learning have significantly enhanced ncRNA research, particularly in disease contexts like cardiovascular conditions, by facilitating the identification of lncRNA-disease associations and exploring their genetic, epigenetic, and posttranscriptional regulatory Role (Liu et al. 2015; Balas and Johnson 2018; Guo et al. 2021). Nevertheless, obstacles such as an inadequate comprehension of ncRNA interactions across diverse populations, the absence of standardized detection methodologies, and fluctuations in expression attributable to gender and environmental influences remain prevalent (Shobeiri et al. 2023). Future initiatives must rigorously validate these biomarkers within heterogeneous cohorts, elucidate gender-specific Role of ncRNAs in AD, and utilize advanced methodologies such as RNA interference and CRISPR-Cas9 for accurate targeting, thereby heralding a novel epoch of ncRNA-oriented diagnostics and therapeutics (Guttman and Rinn 2012; Ransohoff et al. 2018).

The research underscores the utility of polygenic risk scores in forecasting genetic susceptibilities, thus facilitating the early identification of these ailments (He et al. 2024). Exosomal RNA biomarkers obtained from biological fluids present noninvasive diagnostic methodologies that connect neurodegeneration with cardiovascular impairments (Ngowi et al. 2024). Sophisticated imaging modalities, including MRI-based cerebral mapping and three-dimensional cardiac imaging, furnish intricate insights into neurovascular and microvascular dysfunction, highlighting their interrelated characteristics (Carvalhal et al. 2024). Stem cell methodologies, encompassing induced pluripotent stem cells and organoids, adeptly model disease progression with remarkable accuracy, promoting tailored therapeutic interventions (Azam et al. 2024). Prospective endeavors concentrate on integrating multi-omics data, artificial intelligence, and wearable health technologies to formulate precision medicine frameworks and ongoing monitoring mechanisms (farooq Mohi-U-din et al. 2024). This interdisciplinary strategy holds significant potential for revolutionizing patient outcomes and alleviating the worldwide impact of these disorders (farooq Mohi-U-din et al. 2024). Bioinformatics and machine learning can advance lncRNA discovery and diagnostics. Targeting BACE1-AS and NEAT1 offers therapy while optimizing biomarkers and neuroprotective lncRNAs like MALAT1 may prevent neuronal loss and boost cognitive resilience in Alzheimer's (Bernard et al. 2010; Fotuhi et al. 2019; Khodayi et al. 2022; Black et al. 2024). These RNA molecules modulate gene expression, cellular differentiation, and intercellular communication, often through extracellular vesicles such as exosomes (Valadi et al. 2007). In CVDs, ncRNAs such as MALAT1 and LIPCAR serve as promising

Table 2.6 Emerging biomedical technologies and their applications

Technology	Application	Example	Reference
Polygenic risk scores	Predict genetic predisposition for early detection	Identification of genetic variants associated with Alzheimer's and cardiovascular diseases	Leonenko et al. (2021)
Exosomal RNA biomarkers	Noninvasive diagnostics	RNA extracted from biofluids to link neurodegeneration and cardiovascular dysfunction	Wang et al. (2022a, b)
Advanced imaging techniques	Neurovascular and microvascular mapping	MRI-based brain imaging and 3D cardiac imaging for detailed visualization	Kang et al. (2020), Zeng et al. (2022)
Stem Cell Technologies	Disease modeling and personalized therapies	Induced pluripotent stem cells (iPSCs) and organoids for precision medicine	Novelli et al. (2022)

biomarkers and therapeutic targets, showing the potential to mitigate cardiac remodeling and enhance mitochondrial metabolism (Zhao et al. 2017; Barbalata et al. 2023; Le and Nhu 2023). Similarly, in AD, circulating ncRNAs like BACE1 and its antisense transcript BACE1-AS function as stable biomarkers for early detection and disease monitoring while amplifying pathological processes, including amyloid beta (Aβ) production, tau hyperphosphorylation, and chronic neuroinflammation (Faghihi et al. 2008; Zhu et al. 2012; Banzhaf-Strathmann et al. 2014; Feng et al. 2018) (Table 2.6).

Bioinformatics and machine learning can advance lncRNA discovery and diagnostics. Targeting BACE1-AS and NEAT1 offers therapy, while optimizing biomarkers and neuroprotective lncRNAs like MALAT1 may prevent neuronal loss and boost cognitive resilience in Alzheimer's (Bernard et al. 2010, Fotuhi et al. 2019, Khodayi et al. 2022, Black et al. 2024). These RNA molecules modulate gene expression, cellular differentiation, and intercellular communication, often through extracellular vesicles such as exosomes (Valadi et al. 2007). In CVDs, lncRNAs such as MALAT1 and LIPCAR serve as promising biomarkers and therapeutic targets, showing the potential to mitigate cardiac remodeling and enhance mitochondrial metabolism (Zhao et al. 2017, Barbalata et al. 2023, Le and Nhu 2023). Similarly, in AD, circulating lncRNAs like BACE1 and its antisense transcript BACE1-AS function as stable biomarkers for early detection and disease monitoring while amplifying pathological processes, including amyloid beta (Aβ) production, tau hyperphosphorylation, and chronic neuroinflammation (Faghihi et al. 2008, Zhu et al. 2012, Banzhaf-Strathmann et al. 2014, Feng et al. 2018).

2.4.4 *High-Throughput Sequencing, Bioinformatics, and AI in lncRNA Research*

High-throughput sequencing methodologies, coupled with bioinformatics and artificial intelligence (AI), have fundamentally transformed the field of ncRNA research (Fig. 2.11), providing profound insights into the pathophysiology of cardiovascular and AD (Leptidis et al. 2022; Alkhatabi and Pushparaj 2024). Cutting-edge RNA

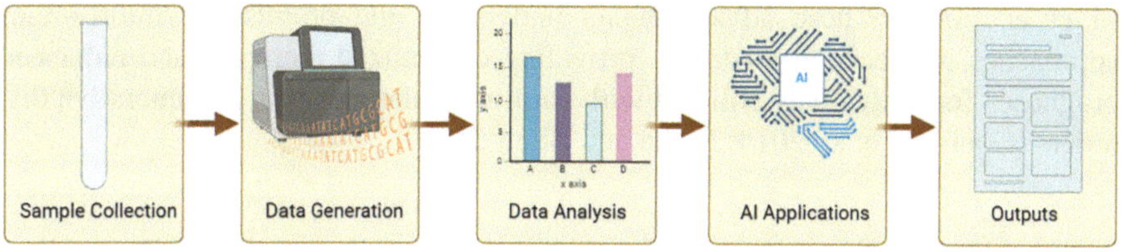

Fig. 2.11 Workflow of technological integration in lncRNA research and precision medicine. The workflow begins with sample collection, where biological specimens such as blood or cerebrospinal fluid are obtained to isolate lncRNAs. This is followed by data generation, which utilizes high-throughput sequencing to produce genomic and transcriptomic datasets. Next, Data Analysis employs bioinformatics and statistical tools to identify disease-specific lncRNA signatures. AI Applications are integrated to enhance biomarker discovery, predictive modeling, and functional annotation. Finally, the process concludes with Outputs, translating these findings into diagnostic tools, therapeutic targets, and precision medicine solutions. (Created in BioRender. Khan, F. (2025) https://BioRender.com/y98y871)

sequencing and single-cell sequencing innovations have facilitated the discovery of disease-specific ncRNAs, such as MALAT1, associated with cardiac pathology and BACE1-AS linked to Alzheimer's, which are potential biomarkers and therapeutic targets (Skroblin and Mayr 2014). Bioinformatics resources, exemplified by NONCODE and LncBase, support functional annotation and pathway analysis, elucidating the regulatory networks that connect lncRNAs to neuroinflammation mechanisms and vascular dysfunction (Fritah et al. 2014; Fang et al. 2018). AI-enhanced models, particularly those employing deep learning frameworks, facilitate the prediction of lncRNA-disease associations and improve diagnostic accuracy through the profiling of exosomal RNA (Tan et al. 2022).

2.4.5 *Multi-Omics Approaches and Personalized Medicine in lncRNA Applications*

Multi-omics methodologies, which incorporate genomics, transcriptomics, proteomics, and epigenomics, have revolutionized research on long noncoding RNAs (lncRNAs) by facilitating a more profound comprehension of their contributions to intricate pathologies such as cardiovascular diseases and AD (Sohag et al. 2021). The integration of epigenomic and proteomic data has elucidated the functions of lncRNAs associated with methylation, exemplified by NEAT1's Role in neuroinflammation and vascular remodeling (Alimohammadi et al. 2022; Adiga et al. 2024). Artificial intelligence (AI)-enabled instruments, encompassing deep learning frameworks such as EVlncRNA-Dpred, augment functional annotation and forecast lncRNA-disease correlations, enhancing diagnostic accuracy and therapeutic advancements (Zhou et al. 2023). Clinical implementations encompass the utilization of exosomal lncRNAs like H19 as noninvasive biomarkers and the engineering of antisense oligonucleotides directed at disease-specific lncRNAs (Rakicevic 2023;

Li et al. 2024). These advancements underscore the promise of multi-omics approaches in precision medicine, providing customized therapies and enhanced prognoses for individuals afflicted with cardiovascular and neurodegenerative disorders (Gourraud et al. 2014; Molla and Bitew 2024).

2.5 Ethical, Regulatory, Translational Considerations and Clinical Applications of lncRNAs

The ncRNAs are transformative molecules with the potential to redefine therapeutic approaches for complex diseases. However, translating their promise into clinical reality is fraught with ethical, regulatory, and translational challenges. Addressing these issues with rigor and innovation is essential for realizing the full potential of lncRNA-based therapies (Fernandes et al. 2019). Challenges include distinguishing functional ncRNAs from transcriptional noise (Niu and Jiang 2013) and limited clinical translation (Wang et al. 2016a, b, Wang et al. 2017a, b). Solutions involve bioinformatics tools (lncRNAdb, ChIPBase) and experimental validation (Chen et al. 2012). Future directions focus on functional studies, therapeutic applications, and systems biology integration to understand and leverage lncRNAs in precision medicine (Dhanoa et al. 2018). Also, the intricate interplay between ncRNAs and molecular pathways and the absence of standardized protocols further complicate their clinical translation and application (Zhao et al. 2018; Climent and García-Giménez 2024). Future research must integrate advanced transcriptomic and proteomic technologies to overcome these barriers and unravel lncRNA mechanisms (Zhao et al. 2018, Climent and García-Giménez 2024). RNA-targeted therapies, such as inhibitors of BACE1-AS, represent promising therapeutic avenues (Faghihi et al. 2008; Mazdeh et al. 2018). Moreover, developing comprehensive diagnostic panels incorporating ncRNA profiles and exploring early disease stages are critical to advancing precision medicine in cardiovascular and neurodegenerative diseases (Liu et al. 2018a, b). These efforts will help unlock the complete diagnostic and therapeutic potential of lncRNAs in clinical practice (Laura Francés et al. 2023).

The novelty and complexity of ncRNAs introduce unique ethical and regulatory challenges. A critical issue is ensuring safety, as the multifunctional nature of ncRNAs increases the potential for off-target effects and unintended molecular interactions (Wilkinson and Campbell 2013). Regulatory agencies demand robust preclinical evidence to address these risks, but the limited availability of long-term safety data remains a significant hurdle (Cech and Steitz 2014). Integrating computational techniques with molecular biology has transformed the understanding of ncRNAs, enabling deeper functional insights. The challenges include the complexity of ncRNA functions, context specificity, and the need for high-quality multi-omics datasets. Addressing these challenges requires sophisticated deep-learning frameworks and comprehensive lncRNA databases (Kim and Lee 2023). Future research should focus on elucidating lncRNA Role in immune modulation and neurodegenerative diseases, leveraging graph-based and network-learning approaches to uncover their broader biological relationships (Cai et al. 2023). The diverse functional repertoire of ncRNAs and identifying reliable therapeutic targets pose an uphill task (Monteiro et al. 2019). Solutions include leveraging subcellular localization studies and advanced technologies like organelle immunoaffinity purification to uncover disease-relevant ncRNA interactions (Bink et al. 2023). Future efforts should focus on integrating ncRNA profiling into precision medicine and developing nucleic acid-based therapies to revolutionize diagnostics and treatment strategies for these conditions (Gluba-Sagr et al. 2024).

Key ncRNAs, such as ANRIL, MALAT1, MIAT, and BACE1-AS, play significant Role in the pathogenesis of coronary artery disease, heart failure, and neurodegenerative conditions like AD, where they regulate amyloid-beta processing and synaptic function (Ishii et al. 2006; Folkersen et al. 2009; Holdt et al. 2010; Harismendy et al. 2011; Watts et al. 2013). Despite their potential, challenges such as limited functional characterization, transcript complexity, and poor sequence conservation across species remain barriers to clinical translation (Uchida and Dimmeler 2015; Ribas and Costa 2017). Addressing these hurdles requires RNA-targeted therapeutics, noninvasive biomarker discovery, and integrative multi-omics approaches, paving the way for innovative diagnostic and therapeutic strategies in precision medicine (Uchida and Dimmeler 2015, Ribas and Costa 2017).

Ethically, the intricate mechanisms of lncRNA therapeutics complicate obtaining truly informed consent, as patients may not fully understand the associated risks. The use of delivery vectors, such as viral systems, raises additional concerns regarding immunogenicity and potential genomic integration (Statello et al. 2021). Conducting exhaustive safety evaluations using advanced in vitro and in vivo models, fostering transparency in data reporting, and developing globally harmonized regulatory guidelines can harness their full potential (Halasz and Carpenter 2022). Ethical oversight committees and international collaborations can ensure that patient safety and equitable access remain central to developing lncRNA therapeutics (Hostiuc et al. 2024).

2.5.1 Case Studies Highlighting Ethical and Regulatory Challenges

Case Study 1 ANRIL and Cardiovascular Diseases

The ncRNA ANRIL, associated with coronary artery disease (CAD) and atherosclerosis, exemplifies ncRNA therapeutics' ethical and regulatory complexities. Genetic screening for ANRIL polymorphisms in clinical studies raised concerns over patient privacy and informed consent. Moreover, regulatory bodies demanded extensive preclinical studies to ensure the safety and specificity of ANRIL-targeted therapies, addressing their multifaceted Role in cellular processes. The success of early studies underscores the importance of transparent communication with patients and robust regulatory frameworks in advancing ncRNA-based treatments for cardiovascular diseases (Holdt et al. 2010).

Case Study 2 BACE1-AS and AD
BACE1-AS stabilizes BACE1 mRNA, a key enzyme in amyloid-beta production linked to Alzheimer's disease. Targeting BACE1-AS presents ethical challenges, particularly in obtaining informed consent for therapies that cross the blood–brain barrier (BBB). Translational hurdles include the limited applicability of animal models for studying BACE1-AS mechanisms. Using human-derived iPSC models and antisense oligonucleotides (ASOs), researchers have demonstrated the feasibility of reducing amyloid-beta levels in preclinical settings. This case highlights the critical need for innovative research models and ethical safeguards in developing lncRNA-based Alzheimer's therapies (Fotuhi et al. 2019).

2.5.2 Bridging Basic Research and Clinical Applications: Future of lncRNA Therapeutics

The translational journey of lncRNA-based therapies presents formidable challenges. Poor conservation of lncRNAs across species limits the utility of traditional animal models. At the same time, their structural complexity demands cutting-edge techniques such as cryo-electron microscopy and small-angle X-ray scattering to elucidate functional mechanisms (Shi et al. 2013; Sanbonmatsu 2022). Despite these challenges, advancements in computational biology and high-throughput screening are accelerating the identification of therapeutic lncRNAs (Abbas et al. 2016). Nanotechnology-based carriers and biodegradable systems have emerged as promising tools for overcoming delivery challenges (Li et al. 2023). Additionally, emphasizing the functional conservation of structural motifs rather than sequence homology may provide deeper insights into their therapeutic potential (Johnsson et al. 2014).

2.5.2.1 Case Studies Demonstrating Translational Successes

Case Study 1 MIAT and Heart Failure
MIAT, a myocardial infarction-associated transcript, illustrates the challenges of translating lncRNA research into clinical applications. Targeting MIAT to reduce cardiac remodeling and prevent heart failure post-myocardial infarction required advanced nanotechnology-based delivery systems. These systems ensured tissue-specific delivery and minimized off-target effects. Preclinical trials demonstrated improved cardiac function in animal models, emphasizing the Role of innovative delivery platforms in bridging basic research and clinical application gaps (Ishii et al. 2006).

Case Study 2 NEAT1 and Alzheimer's Disease
NEAT1 is implicated in neuroinflammation and synaptic dysfunction in AD, presenting a promising target for lncRNA-based interventions. Researchers faced translational challenges in delivering NEAT1-modulating therapies across the

BBB. Developing nanoparticle carriers and viral vectors overcame these barriers, enabling precise delivery to brain tissues. These advancements paved the way for clinical trials, demonstrating the importance of interdisciplinary approaches in accelerating lncRNA-based therapies for neurodegenerative diseases (Wu et al. 2022).

2.6 Conclusion

Noncoding RNAs (ncRNAs) have increasingly been recognized as essential regulators in various biological processes, particularly in neurological and cardiovascular diseases. This chapter has highlighted ncRNAs' significant Role, including lncRNAs and miRNAs, in regulating gene expression, cellular signaling, and disease progression. In conditions like AD and cardiovascular disorders, ncRNAs have been shown to influence critical pathways such as amyloid-beta processing, tau hyperphosphorylation, and cardiac remodeling.

The potential of ncRNAs as diagnostic markers and therapeutic targets is becoming more evident with ongoing research. ncRNAs such as BACE1-AS, MALAT1, and NEAT1 are emerging as promising candidates for early disease detection and monitoring, providing insights into disease severity and progression. Additionally, ncRNAs offer new opportunities for therapeutic intervention, although challenges related to their delivery, stability, and precise regulation still need to be addressed before clinical application.

Despite these challenges, the continued advancements in sequencing technologies, bioinformatics tools, and therapeutic strategies indicate that ncRNAs have the potential to contribute significantly to the development of more personalized and targeted treatment options in the future. Further research and interdisciplinary approaches will be necessary to fully translate ncRNA-based findings into clinical practice and overcome the existing barriers to their application in managing neurological and cardiovascular diseases.

Compliance with Ethical Standards

Conflict of interest: The authors declare no conflict of interest.

Funding: No funding received.

Acknowledgment: All the figures were created with BioRender.com except Figs. 2.8 and 2.9.

Ethical Approval: This article does not contain any studies with human participants or animals performed by any authors.

References

Abbas Q, Raza SM, Biyabani AA et al (2016) A review of computational methods for finding non-coding RNA genes. Genes (Basel) 7(12). https://doi.org/10.3390/genes7120113

Abubakar M, Hajjaj M, Naqvi Z e Z et al (2024) Non-coding RNA-mediated gene regulation in cardiovascular disorders: current insights and future directions. J Cardiovasc Transl Res 17(4):739–767. https://doi.org/10.1007/s12265-023-10469-4

Abugessaisa I, Ramilowski JA, Lizio M et al (2021) FANTOM enters 20th year: expansion of transcriptomic atlases and functional annotation of non-coding RNAs. Nucleic Acids Res 49(D1):D892–D898. https://doi.org/10.1093/nar/gkaa1054

Adam O, Löhfelm B, Thum T et al (2012) Role of miR-21 in the pathogenesis of atrial fibrosis. Basic Res Cardiol 107:1–12. https://doi.org/10.1007/s00395-012-0278-0

Adiga D, Eswaran S, Srinath S et al (2024) Noncoding RNAs in Alzheimer's disease: overview of functional and therapeutic significance. Curr Top Med Chem 24(19):1615–1634. https://doi.org/10.2174/0115680266293212240405042540

Alimohammadi M, Makaremi S, Rahimi A et al (2022) DNA methylation changes and inflammaging in aging-associated diseases. Epigenomics 14(16):965–986. https://doi.org/10.2217/epi-2022-0143

Alkhatabi HA, Pushparaj PN (2024) Untangling the complex mechanisms associated with Alzheimer's disease in elderly patients using high-throughput RNA sequencing data and next-generation knowledge discovery methods: focus on potential gene signatures and drugs for dementia. Heliyon. https://doi.org/10.1016/j.heliyon.2024.e41266

Altman S, Bowman EJ, Garber RL et al (1980) Aspects of RNase P structure and function. Transfer RNA Biol Aspects 2:71–82. https://doi.org/10.1016/0092-8674(88)90527-2

Am K (2009) Many human large intergenic noncoding RNAs associate with chromatin-modifying complexes and affect gene expression. Proc Natl Acad Sci U S A 106:11667–11672. https://doi.org/10.1073/pnas.0904715106

Ambros V (2008) The evolution of our thinking about microRNAs. Nat Med 14(10):1036–1040. https://doi.org/10.1038/nm1008-1036

Ammad M, Javed Z, Sadia H et al (2024) Advancements in long non-coding RNA-based therapies for cancer: targeting, delivery, and clinical implications. Med Oncol 41(11):292. https://doi.org/10.1007/s12032-024-02534-y

Andrews DW, Walter P, Ottensmeyer F (1987) Evidence for an extended 7SL RNA structure in the signal recognition particle. EMBO J 6(11):3471–3477. https://doi.org/10.1002/j.1460-2075.1987.tb02671.x

Archer K, Broskova Z, Bayoumi AS et al (2015) Long non-coding RNAs as master regulators in cardiovascular diseases. Int J Mol Sci 16(10):23651–23667. https://doi.org/10.3390/ijms161023651

Arnold PR, Wells AD, Li XC (2020) Diversity and emerging Role of enhancer RNA in regulation of gene expression and cell fate. Front Cell Dev Biol 7:377. https://doi.org/10.3389/fcell.2019.00377

Ayers D (2013) Long non-coding RNAs: novel emergent biomarkers for cancer diagnostics. J Cancer Res Treat 1(2):31–35. https://doi.org/10.12691/jcrt-1-2-2

Azam F, Saba AR, Pervaiz A et al (2024) Stem cells: the foundation of regenerative medicine. In: Ijaz MU, Ehsan N, Imran M, Yousaf S (eds) Zoology: advancements and research trends. FahumSci, Lahore. https://doi.org/10.61748/zool.2024/37

Balas MM, Johnson AM (2018) Exploring the mechanisms behind long noncoding RNAs and cancer. Noncoding RNA Res 3(3):108–117. https://doi.org/10.1016/j.ncrna.2018.03.001

Banzhaf-Strathmann J, Benito E, May S et al (2014) Micro RNA-125b induces tau hyperphosphorylation and cognitive deficits in Alzheimer's disease. EMBO J 33(15):1667–1680. https://doi.org/10.15252/embj.201387576

Barbalata T, Niculescu LS, Stancu CS et al (2023) Elevated levels of circulating lncRNAs LIPCAR and MALAT1 predict an unfavorable outcome in acute coronary syndrome patients. Int J Mol Sci 24(15):12076. https://doi.org/10.3390/ijms241512076

Batista PJ, Chang HY (2013) Long noncoding RNAs: cellular address codes in development and disease. Cell 152(6):1298–1307. https://doi.org/10.1016/j.cell.2013.02.012

Bella F, Muscatello MRA, D'Ascola A et al (2023) Gene expression analysis of nc-RNAs in bipolar and panic disorders: a pilot study. Genes 14(9):1778. https://doi.org/10.3390/genes14091778

Bernard D, Prasanth KV, Tripathi V et al (2010) A long nuclear-retained non-coding RNA regulates synaptogenesis by modulating gene expression. EMBO J 29(18):3082–3093. https://doi.org/10.1038/emboj.2010.199

Bhagat R, Minaya MA, Renganathan A et al (2023) Long non-coding RNA SNHG8 drives stress granule formation in tauopathies. Mol Psychiatry 28(11):4889–4901. https://doi.org/10.1038/s41380-023-02237-2

Bhatti GK, Khullar N, Sidhu IS et al (2021) Emerging Role of non-coding RNA in health and disease. Metab Brain Dis 36:1119–1134. https://doi.org/10.1007/s11011-021-00739-y

Bink DI, Pauli J, Maegdefessel L et al (2023) Endothelial microRNAs and long noncoding RNAs in cardiovascular ageing. Atherosclerosis 374:99–106. https://doi.org/10.1016/j.atherosclerosis.2023.03.019

Black CM, Braden AA, Nasim S et al (2024) The association between long non-coding RNAs and Alzheimer's disease. Brain Sci 14(8):818. https://doi.org/10.3390/brainsci14080818

Boon RA, Iekushi K, Lechner S et al (2013) MicroRNA-34a regulates cardiac ageing and function. Nature 495(7439):107–110. https://doi.org/10.1038/nature11919

Burzio VA, Villota C, Villegas J et al (2009) Expression of a family of noncoding mitochondrial RNAs distinguishes normal from cancer cells. Proc Natl Acad Sci 106(23):9430–9434. https://doi.org/10.1073/pnas.0903086106

Cai L, Qi B, Wu X et al (2019) Circular RNA Ttc3 regulates cardiac function after myocardial infarction by sponging miR-15b. J Mol Cell Cardiol 130:10–22. https://doi.org/10.1186/s13041-020-00656-8

Cai L-J, Tu L, Huang X-M et al (2020) LncRNA MALAT1 facilitates inflammasome activation via epigenetic suppression of Nrf2 in Parkinson's disease. Mol Brain 13:1–15. https://doi.org/10.1016/j.yjmcc.2019.03.007

Cai J, Wang T, Deng X et al (2023) GM-lncLoc: LncRNAs subcellular localization prediction based on graph neural network with meta-learning. BMC Genomics 24(1):52. https://doi.org/10.1186/s12864-022-09034-1

Cánovas-Cervera I, Nacher-Sendra E, Osca-Verdegal R et al (2023) The intricate Role of non-coding RNAs in sepsis-associated disseminated intravascular coagulation. Int J Mol Sci 24(3):2582. https://doi.org/10.3390/ijms24032582

Cao J (2014) The functional Role of long non-coding RNAs and epigenetics. Biol Proced Online 16(1):42. https://doi.org/10.1186/1480-9222-16-11

Cao B, Wang T, Qu Q et al (2018) Long noncoding RNA SNHG1 promotes neuroinflammation in Parkinson's disease via regulating miR-7/NLRP3 pathway. Neuroscience 388:118–127. https://doi.org/10.1016/j.neuroscience.2018.07.019

Cao M, Li H, Zhao J et al (2019) Identification of age-and gender-associated long noncoding RNAs in the human brain with Alzheimer's disease. Neurobiol Aging 81:116–126. https://doi.org/10.1016/j.neurobiolaging.2019.05.023

Cao Y, Jiang C, Lin H et al (2021) Silencing of long noncoding RNA growth arrest–specific 5 alleviates neuronal cell apoptosis and inflammatory responses through sponging microRNA-93 to repress PTEN expression in spinal cord injury. Front Cell Neurosci 15:646788. https://doi.org/10.3389/fncel.2021.646788

Carvalhal S, Carmona B, Tassin A-M et al (2024) Molecular mechanisms of cilia related diseases. Front Mol Biosci 11:1421419. https://doi.org/10.3389/fmolb.2024.1421419

Cech TR, Steitz JA (2014) The noncoding RNA revolution—trashing old rules to forge new ones. Cell 157(1):77–94. https://doi.org/10.1016/j.cell.2014.03.008

Chen G, Wang Z, Wang D et al (2012) LncRNADisease: a database for long-non-coding RNA-associated diseases. Nucleic Acids Res 41(D1):D983–D986. https://doi.org/10.1093/nar/gks1099

Chen J, Shishkin AA, Zhu X et al (2016a) Evolutionary analysis across mammals reveals distinct classes of long non-coding RNAs. Genome Biol 17:1–17. https://doi.org/10.1186/s13059-016-0880-9

Chen L, Feng P, Zhu X et al (2016b) Long non-coding RNA Malat1 promotes neurite outgrowth through activation of ERK/MAPK signalling pathway in N2a cells. J Cell Mol Med 20(11):2102–2110. https://doi.org/10.1111/jcmm.12904

Chen J, Cui L, Yuan J et al (2017) Circular RNA WDR77 target FGF-2 to regulate vascular smooth muscle cells proliferation and migration by sponging miR-124. Biochem Biophys Res Commun 494(1–2):126–132. https://doi.org/10.1016/j.bbrc.2017.10.068

Chen Y, Lian Y-j, Ma Y-q et al (2018) LncRNA SNHG1 promotes α-synuclein aggregation and toxicity by targeting miR-15b-5p to activate SIAH1 in human neuroblastoma SH-SY5Y cells. Neurotoxicology 68:212–221. https://doi.org/10.1016/j.neuro.2017.12.001

Chen Q, Zhu C, Jin Y (2020) The oncogenic and tumor suppressive functions of the Long non-coding RNA MALAT1: an emerging controversy. Front Genet 11. https://doi.org/10.3389/fgene.2020.00093

Chen Y, Chen X, Li H et al (2021) Serum extracellular vesicles containing MIAT induces atrial fibrosis, inflammation and oxidative stress to promote atrial remodeling and atrial fibrillation via blockade of miR-485-5p-mediated CXCL10 inhibition. Clin Transl Med 11(8):e482. https://doi.org/10.1002/ctm2.482

Chen X, Liu H, Huang R et al (2022) Screening of plasma exosomal lncRNAs to identify potential biomarkers for obstructive sleep apnea. Ann Transl Med 10(17). https://doi.org/10.21037/atm-22-3818

Chen S, Cao Z, Nandi A et al (2024) The global macroeconomic burden of Alzheimer's disease and other dementias: estimates and projections for 152 countries or territories. Lancet Glob Health 12(9):e1534–e1543. https://doi.org/10.1016/S2214-109X(24)00264-X

Chistiakov DA, Sobenin IA, Orekhov AN et al (2015) Human miR-221/222 in physiological and atherosclerotic vascular remodeling. Biomed Res Int 2015(1):354517. https://doi.org/10.1155/2015/354517

Clemson CM, Hutchinson JN, Sara SA et al (2009) An architectural Role for a nuclear noncoding RNA: NEAT1 RNA is essential for the structure of paraspeckles. Mol Cell 33(6):717–726. https://doi.org/10.1016/j.molcel.2009.01.026

Climent M, García-Giménez JL (2024) Special issue "The Role of non-coding RNAS involved in cardiovascular diseases and cellular communication". Int J Mol Sci 25(11):6034. https://doi.org/10.3390/ijms25116034

Congrains A, Kamide K, Oguro R et al (2012) Genetic variants at the 9p21 locus contribute to atherosclerosis through modulation of ANRIL and CDKN2A/B. Atherosclerosis 220(2):449–455. https://doi.org/10.1016/j.atherosclerosis.2011.11.017

Consortium E P (2007) Identification and analysis of functional elements in 1% of the human genome by the ENCODE pilot project. Nature 447(7146):799–816. https://doi.org/10.1038/nature05874

Correia CCM, Rodrigues LF, de Avila Pelozin BR et al (2021) Long non-coding RNAs in cardiovascular diseases: potential function as biomarkers and therapeutic targets of exercise training. Noncoding RNA 7(4):65. https://doi.org/10.3390/ncrna7040065

Crump C, Sundquist K, Winkleby MA et al (2013) Comorbidities and mortality in bipolar disorder: a Swedish national cohort study. JAMA Psychiatry 70(9):931–939. https://doi.org/10.1001/jamapsychiatry.2013.1394

Das S, Shah R, Dimmeler S et al (2020) Noncoding RNAs in cardiovascular disease: current knowledge, tools and Technologies for Investigation, and future directions: a scientific statement from the American Heart Association. Circ Genom Precis Med 13(4):e000062. https://doi.org/10.1161/hcg.0000000000000062

Dhanoa JK, Sethi RS, Verma R et al (2018) Long non-coding RNA: its evolutionary relics and biological implications in mammals: a review. J Anim Sci Technol 60:1–10. https://doi.org/10.1186/s40781-018-0183-7

Di Gesualdo F, Capaccioli S, Lulli M (2014) A pathophysiological view of the long non-coding RNA world. Oncotarget 5(22):10976–10996. https://doi.org/10.18632/oncotarget.2770

Ding Y, Luan W, Wang Z et al (2022) LncRNA BDNF-AS as ceRNA regulates the miR-9-5p/BACE1 pathway affecting neurotoxicity in Alzheimer's disease. Arch Gerontol Geriatr 99:104614. https://doi.org/10.1016/j.archger.2021.104614

Du WW, Xu J, Yang W et al (2021) A neuroligin isoform translated by circNlgn contributes to cardiac remodeling. Circ Res 129(5):568–582. https://doi.org/10.1161/CIRCRESAHA.120.318364

Esteller M (2011) Non-coding RNAs in human disease. Nat Rev Genet 12(12):861–874. https://doi.org/10.1038/nrg3074

Faghihi MA, Modarresi F, Khalil AM et al (2008) Expression of a noncoding RNA is elevated in Alzheimer's disease and drives rapid feed-forward regulation of β-secretase. Nat Med 14(7):723–730. https://doi.org/10.1038/nm1784

Fan Y, Li J, Yang Q et al (2019) Dysregulated Long non-coding RNAs in Parkinson's disease contribute to the apoptosis of human Neuroblastoma cells. Front Neurosci 13. https://doi.org/10.3389/fnins.2019.01320

Fan J, Li H, Xie R et al (2021a) LncRNA ZNF593-AS alleviates contractile dysfunction in dilated cardiomyopathy. Circ Res 128(11):1708–1723. https://doi.org/10.1161/CIRCRESAHA.120.318437

Fan Y, Zhang J, Zhuang X et al (2021b) Epigenetic transcripts of LINC01311 and hsa-miR-146a-5p regulate neural development in a cellular model of Alzheimer's disease. IUBMB Life 73(7):916–926. https://doi.org/10.1002/iub.2472

Fan X, Yang Y, Chen C et al (2022) Pervasive translation of circular RNAs driven by short IRES-like elements. Nat Commun 13(1):3751. https://doi.org/10.1038/s41467-022-31327-y

Fang S, Zhang L, Guo J et al (2018) NONCODEV5: a comprehensive annotation database for long non-coding RNAs. Nucleic Acids Res 46(D1):D308–D314. https://doi.org/10.1093/nar/gkx1107

Farooq Mohi-U-din S, Tariq M, Tariq A (2024) Deep dive into health: harnessing AI and deep learning for brain and heart care. Int J Adv Eng Technol Innov 1(4):248–267. https://doi.org/10.70749/ijbr.v3i2.665

Feng L, Liao Y-T, He J-C et al (2018) Plasma long non-coding RNA BACE1 as a novel biomarker for diagnosis of Alzheimer disease. BMC Neurol 18:1–8. https://doi.org/10.1186/s12883-017-1008-x

Fernandes JCR, Acuña SM, Aoki JI et al (2019) Long non-coding RNAs in the regulation of gene expression: physiology and disease. Noncoding. RNA 5(1). https://doi.org/10.3390/ncrna5010017

Fichtlscherer S, De Rosa S, Fox H et al (2010) Circulating microRNAs in patients with coronary artery disease. Circ Res 107(5):677–684. https://doi.org/10.1161/CIRCRESAHA.109.215566

Fire A, Xu S, Montgomery MK et al (1998) Potent and specific genetic interference by double-stranded RNA in Caenorhabditis elegans. Nature 391(6669):806–811. https://doi.org/10.1038/35888

Fitch MT, Silver J (2008) CNS injury, glial scars, and inflammation: inhibitory extracellular matrices and regeneration failure. Exp Neurol 209(2):294–301. https://doi.org/10.1016/j.expneurol.2007.05.014

Folkersen L, Kyriakou T, Goel A et al (2009) Relationship between CAD risk genotype in the chromosome 9p21 locus and gene expression. Identification of eight new ANRIL splice variants. PLoS One 4(11):e7677. https://doi.org/10.1371/journal.pone.0007677

Fotuhi SN, Khalaj-Kondori M, Hoseinpour Feizi MA et al (2019) Long non-coding RNA BACE1-AS May serve as an Alzheimer's disease blood-based biomarker. J Mol Neurosci 69(3):351–359. https://doi.org/10.1007/s12031-019-01364-2

Francelle L, Galvan L, Gaillard M-C et al (2015) The striatal long noncoding RNA Abhd11os is neuroprotective against an N-terminal fragment of mutant huntingtin in vivo. Neurobiol Aging 36(3):1601.e1607–1601.e1616. https://doi.org/10.1016/j.neurobiolaging.2014.11.014
Fritah S, Niclou SP, Azuaje F (2014) Databases for lncRNAs: a comparative evaluation of emerging tools. RNA 20(11):1655–1665. https://doi.org/10.1261/rna.044040.113
Garikipati VNS, Verma SK, Cheng Z et al (2019) Circular RNA CircFndc3b modulates cardiac repair after myocardial infarction via FUS/VEGF-A axis. Nat Commun 10(1):4317. https://doi.org/10.1038/s41467-019-11777-7
Gluba-Sagr A, Franczyk B, Rysz-Górzyńska A et al (2024) The Role of selected lncRNAs in lipid metabolism and cardiovascular disease risk. Int J Mol Sci 25(17):9244. https://doi.org/10.3390/ijms25179244
Gomes AQ, Nolasco S, Soares H (2013) Non-coding RNAs: multi-tasking molecules in the cell. Int J Mol Sci 14(8):16010–16039. https://doi.org/10.3390/ijms140816010
Gonzalez I, Munita R, Agirre E et al (2015) A lncRNA regulates alternative splicing via establishment of a splicing-specific chromatin signature. Nat Struct Mol Biol 22(5):370–376. https://doi.org/10.1038/nsmb.3005
Good DJ (2023) Non-coding RNAs in human health and diseases. Genes 14(7):1429. https://doi.org/10.3390/genes14071429
Gourraud PA, Henry RG, Cree BA et al (2014) Precision medicine in chronic disease management: the multiple sclerosis BioScreen. Ann Neurol 76(5):633–642. https://doi.org/10.1002/ana.24282
Greco S, Piella S, Made' A et al (2022) LncRNA BACE1-AS: a link between heart failure and Alzheimer's disease. Eur Heart J 43(Supplement_2):ehac544. 2945. https://doi.org/10.1093/eurheartj/ehac544.2945
Grillone K, Caridà G, Luciano F et al (2024) A systematic review of non-coding RNA therapeutics in early clinical trials: a new perspective against cancer. J Transl Med 22(1):731. https://doi.org/10.1186/s12967-024-05554-4
Groszer M, Erickson R, Scripture-Adams DD et al (2001) Negative regulation of neural stem/progenitor cell proliferation by the Pten tumor suppressor gene in vivo. Science 294(5549):2186–2189. https://doi.org/10.1126/science.1065518
Grote P, Wittler L, Hendrix D et al (2013) The tissue-specific lncRNA Fendrr is an essential regulator of heart and body wall development in the mouse. Dev Cell 24(2):206–214. https://doi.org/10.1016/j.devcel.2012.12.012
Gu L, Guo Z (2013) Alzheimer's Aβ42 and Aβ40 peptides form interlaced amyloid fibrils. J Neurochem 126(3):305–311. https://doi.org/10.1111/jnc.12202
Gu C, Chen C, Wu R et al (2018) Long noncoding RNA EBF3-AS promotes neuron apoptosis in Alzheimer's disease. DNA Cell Biol 37(3):220–226. https://doi.org/10.1089/dna.2017.4012
Gu R, Wang L, Tang M et al (2020) LncRNA Rpph1 protects amyloid-β induced neuronal injury in SK-N-SH cells via miR-122/Wnt1 axis. Int J Neurosci 130(5):443–453. https://doi.org/10.1080/00207454.2019.1692834
Gu R, Liu R, Wang L et al (2021) LncRNA RPPH1 attenuates Aβ25-35-induced endoplasmic reticulum stress and apoptosis in SH-SY5Y cells via miR-326/PKM2. Int J Neurosci 131(5):425–432. https://doi.org/10.1080/00207454.2020.1746307
Guo R, Zou B, Liang Y et al (2021) LncRNA RCAT1 promotes tumor progression and metastasis via miR-214-5p/E2F2 axis in renal cell carcinoma. Cell Death Dis 12(7):689. https://doi.org/10.1038/s41419-021-03955-7
Gutschner T, Diederichs S (2012) The hallmarks of cancer: a long non-coding RNA point of view. RNA Biol 9(6):703–719. https://doi.org/10.4161/rna.20481
Guttman M, Rinn JL (2012) Modular regulatory principles of large non-coding RNAs. Nature 482(7385):339–346. https://doi.org/10.1038/nature10887
Haerty W, Ponting CP (2013) Mutations within lncRNAs are effectively selected against in fruitfly but not in human. Genome Biol 14:1–16. https://doi.org/10.1186/gb-2013-14-5-r49

Halasz H, Carpenter S (2022) Challenges and future directions for LncRNAs and inflammation. Adv Exp Med Biol 1363:179–183. https://doi.org/10.1007/978-3-030-92034-0_10
Hall IF, Climent M, Quintavalle M et al (2019) Circ_Lrp6, a circular RNA enriched in vascular smooth muscle cells, acts as a sponge regulating miRNA-145 function. Circ Res 124(4):498–510. https://doi.org/10.1161/CIRCRESAHA.118.314240
Han P, Chang C-P (2015) Long non-coding RNA and chromatin remodeling. RNA Biol 12(10):1094–1098. https://doi.org/10.1080/15476286.2015.1063770
Han P, Li W, Lin C-H et al (2014) A long noncoding RNA protects the heart from pathological hypertrophy. Nature 514(7520):102–106. https://doi.org/10.1038/nature13596
Han Q, Zhang L, Liao R (2024) Diagnostic and prognostic significance of miR-320a-3p in patients with chronic heart failure. BMC Cardiovasc Disord 24(1):308. https://doi.org/10.1186/s12872-024-03966-0
Harismendy O, Notani D, Song X et al (2011) 9p21 DNA variants associated with coronary artery disease impair interferon-γ signalling response. Nature 470(7333):264–268. https://doi.org/10.1038/nature09753
Harries LW (2012) Long non-coding RNAs and human disease. Biochem Soc Trans 40(4):902–906. https://doi.org/10.1042/BST20120020
He C, Jiang B, Ma J et al (2016) Aberrant NEAT1 expression is associated with clinical outcome in high grade glioma patients. APMIS 124(3):169–174. https://doi.org/10.1111/apm.12480
He S-x, Yu C-s, Jia X-d et al (2024) Application and prospects of polygenic risk score (PRS) in genetic disease research: a review of data analysis methods. Prog Biochem Biophys 51(8):1797–1808. https://doi.org/10.16476/j.pibb.2023.0389
Herrera-Rivero M, Hernández-Aguilar ME, Aranda-Abreu GE (2015) A strategy focused on MAPT, APP, NCSTN and BACE1 to build blood classifiers for Alzheimer's disease. J Theor Biol 376:32–38. https://doi.org/10.1016/j.jtbi.2015.03.039
Hoagland MB, Stephenson ML, Scott JF et al (1958) A soluble ribonucleic acid intermediate in protein synthesis. J Biol Chem 231(1):241–257. https://doi.org/10.1016/S0021-9258(19)77302-5
Holdt LM, Beutner F, Scholz M et al (2010) ANRIL expression is associated with atherosclerosis risk at chromosome 9p21. Arterioscler Thromb Vasc Biol 30(3):620–627. https://doi.org/10.1161/ATVBAHA.109.196832
Holley RW, Apgar J, Everett GA et al (1965) Structure of a ribonucleic acid. Science 147(3664):1462–1465. https://doi.org/10.1126/science.147.3664.1462
Hostiuc M, Scafa A, Iancu B et al (2024) Ethical implications of developing RNA-based therapies for cardiovascular disorders. Front Bioeng Biotechnol 12:1370403. https://doi.org/10.3389/fbioe.2024.1370403
Hu G, Niu F, Humburg BA et al (2018) Molecular mechanisms of long noncoding RNAs and their Role in disease pathogenesis. Oncotarget 9(26):18648–18663. https://doi.org/10.18632/oncotarget.24307
Hu Y-W, Guo F-X, Xu Y-J et al (2019) Long noncoding RNA NEXN-AS1 mitigates atherosclerosis by regulating the actin-binding protein NEXN. J Clin Invest 129(3):1115–1128. https://doi.org/10.1172/JCI98230
Hu D-J, Li Z-Y, Zhu Y-T et al (2020) Overexpression of long noncoding RNA ANRIL inhibits phenotypic switching of vascular smooth muscle cells to prevent atherosclerotic plaque development in vivo. Aging (Albany NY) 13(3):4299. https://doi.org/10.18632/aging.202392
Huang Y, Wang L, Mao Y et al (2019) Long noncoding RNA-H19 contributes to atherosclerosis and induces ischemic stroke via the upregulation of acid phosphatase 5. Front Neurol 10:32. https://doi.org/10.3389/fneur.2019.00032
Huang C-K, Kafert-Kasting S, Thum T (2020) Preclinical and clinical development of noncoding RNA therapeutics for cardiovascular disease. Circ Res 126(5):663–678. https://doi.org/10.1161/CIRCRESAHA.119.315856
Huang X-H, Li J-L, Li X-Y et al (2021a) miR-208a in cardiac hypertrophy and remodeling. Front Cardiovasc Med 8. https://doi.org/10.3389/fcvm.2021.773314

Huang Y, Wang J, Zhao Y et al (2021b) cncRNAdb: a manually curated resource of experimentally supported RNAs with both protein-coding and noncoding function. Nucleic Acids Res 49(D1):D65–D70. https://doi.org/10.1093/nar/gkaa791

Huang M, Wu B, Ou X et al (2024) Diagnostic and prognostic values of HCG15 and morrbid in acute myocardial infarction. Front Pharmacol 15:1492746. https://doi.org/10.3389/fphar.2024.1492746

Hui J, Huishan W, Tao L et al (2017) miR-539 as a key negative regulator of the MEK pathway in myocardial infarction. Herz 42(8):781–789. https://doi.org/10.1007/s00059-016-4517-2

Ilieva MS (2024) Non-coding RNAs in neurological and neuropsychiatric disorders: unraveling the hidden players in disease pathogenesis. Cells 13(12):1063. https://doi.org/10.3390/cells13121063

Ishii N, Ozaki K, Sato H et al (2006) Identification of a novel non-coding RNA, MIAT, that confers risk of myocardial infarction. J Hum Genet 51(12):1087–1099. https://doi.org/10.1007/s10038-006-0070-9

Jalali S, Jayaraj GG, Scaria V (2012a) Integrative transcriptome analysis suggest processing of a subset of long non-coding RNAs to small RNAs. Biol Direct 7:1–13. https://doi.org/10.1186/1745-6150-7-25

Jalali S, Ramanathan GK, Parthasarathy PT et al (2012b) Mir-206 regulates pulmonary artery smooth muscle cell proliferation and differentiation. PLoS One 7(10):e4680. https://doi.org/10.1371/journal.pone.0046808

Jathar S, Kumar V, Srivastava J et al (2017) Technological developments in lncRNA biology. Adv Exp Med Biol 1008:283–323. https://doi.org/10.1007/978-981-10-5203-3_10

Jeck WR, Sharpless NE (2014) Detecting and characterizing circular RNAs. Nat Biotechnol 32(5):453–461. https://doi.org/10.1038/nbt.2890

Ji Y, Lu Y, Yang F et al (2010) Acute and gradual increases in BDNF concentration elicit distinct signaling and functions in neurons. Nat Neurosci 13(3):302–309. https://doi.org/10.1038/nn.2505

Jiao L, Li M, Shao Y et al (2019) lncRNA-ZFAS1 induces mitochondria-mediated apoptosis by causing cytosolic Ca2+ overload in myocardial infarction mice model. Cell Death Dis 10(12):942. https://doi.org/10.1038/s41419-019-2136-6

Johnsson P, Lipovich L, Grandér D et al (2014) Evolutionary conservation of long non-coding RNAs; sequence, structure, function. Biochim Biophys Acta 1840(3):1063–1071. https://doi.org/10.1016/j.bbagen.2013.10.035

Jun Y (2019). Meta-analytical strategies for biomarker selection in transcriptomic data, Seoul National University Graduate School. https://hdl.handle.net/10371/152946

Kamps R, Robinson EL (2024) LNC-ing genetics in mitochondrial disease. Noncoding RNA 10(6):57. https://doi.org/10.3390/ncrna10060057

Kang M, Jin S, Lee D et al (2020) MRI visualization of whole brain macro- and microvascular remodeling in a rat model of ischemic stroke: a pilot study. Sci Rep 10(1):4989. https://doi.org/10.1038/s41598-020-61656-1

Kang H, Ga YJ, Kim SH et al (2023) Small interfering RNA (siRNA)-based therapeutic applications against viruses: principles, potential, and challenges. J Biomed Sci 30(1):88. https://doi.org/10.1186/s12929-023-00981-9

Karakikes I, Chaanine AH, Kang S et al (2013) Therapeutic cardiac-targeted delivery of miR-1 reverses pressure overload–induced cardiac hypertrophy and attenuates pathological remodeling. J Am Heart Assoc 2(2):e000078. https://doi.org/10.1161/JAHA.113.000078

Kay MA (2024) Ruvkun and Ambros recognized for miRNAs. Mol Ther Nucleic Acids 35(4). https://doi.org/10.1016/j.omtn.2024.102379

Ke S, Yang Z, Yang F et al (2019) Long noncoding RNA NEAT1 aggravates Aβ-induced neuronal damage by targeting miR-107 in Alzheimer's disease. Yonsei Med J 60(7):640–650. https://doi.org/10.3349/ymj.2019.60.7.640

Khodayi M, Khalaj-Kondori M, Feizi MAH et al (2022) Plasma lncRNA profiling identified BC200 and NEAT1 lncRNAs as potential blood-based biomarkers for late-onset Alzheimer's disease. EXCLI J 21:772. https://doi.org/10.17179/excli2022-4764

Kickstein E, Krauss S, Thornhill P et al (2010) Biguanide metformin acts on tau phosphorylation via mTOR/protein phosphatase 2A (PP2A) signaling. Proc Natl Acad Sci 107(50):21830–21835. https://doi.org/10.1073/pnas.0912793107

Kim Y, Lee M (2023) Deep learning approaches for lncrna-mediated mechanisms: a comprehensive review of recent developments. Int J Mol Sci 24(12):10299. https://doi.org/10.3390/ijms241210299

Kim JO, Song DW, Kwon EJ et al (2015) miR-185 plays an anti-hypertrophic Role in the heart via multiple targets in the calcium-signaling pathways. PLoS One 10(3):e0122509. https://doi.org/10.1371/journal.pone.0122509

Kino T, Hurt DE, Ichijo T et al (2010) Noncoding RNA gas5 is a growth arrest–and starvation-associated repressor of the glucocorticoid receptor. Sci Signal 3(107):ra8. https://doi.org/10.1126/scisignal.2000568

Kornienko AE, Guenzl PM, Barlow DP et al (2013) Gene regulation by the act of long non-coding RNA transcription. BMC Biol 11(1):59. https://doi.org/10.1186/1741-7007-11-59

Kotzin JJ, Spencer SP, McCright SJ et al (2016) The long non-coding RNA Morrbid regulates Bim and short-lived myeloid cell lifespan. Nature 537(7619):239–243. https://doi.org/10.1038/nature19346

Kresge N, Simoni RD, Hill RL (2005) The discovery of tRNA by Paul C. Zamecnik. J Biol Chem 280(40):e37–e39. https://doi.org/10.1016/S0021-9258(20)79029-0

Kumar S, Williams D, Sur S et al (2019) Role of flow-sensitive microRNAs and long noncoding RNAs in vascular dysfunction and atherosclerosis. Vascul Pharmacol 114:76–92. https://doi.org/10.1016/j.vph.2018.10.001

Kumarswamy R, Bauters C, Volkmann I et al (2014) Circulating long noncoding RNA, LIPCAR, predicts survival in patients with heart failure. Circ Res 114(10):1569–1575. https://doi.org/10.1161/CIRCRESAHA.114.303915

Kuthethur R, Shukla V, Mallya S et al (2022) Expression analysis and function of mitochondrial genome-encoded microRNAs. J Cell Sci 135(8):jcs258937. https://doi.org/10.1242/jcs.258937

Lafontaine DL (2015) Noncoding RNAs in eukaryotic ribosome biogenesis and function. Nat Struct Mol Biol 22(1):11–19. https://doi.org/10.1038/nsmb.2939

Ma, Lina, Vladimir B. Bajic, et al (2013) On the classification of long non-coding RNAs. RNA biology 10(6): 924–933. https://doi.org/10.4161/rna.24604

Lan Z, Chen Y, Jin J et al (2022) Long non-coding RNA: insight into mechanisms of Alzheimer's disease. Front Mol Neurosci 14:821002. https://doi.org/10.3389/fnmol.2021.821002

Laura Francés J, Pagiatakis C, Di Mauro V et al (2023) Therapeutic potential of EVs: targeting cardiovascular diseases. Biomedicine 11(7):1907. https://doi.org/10.3390/biomedicines11071907

Lauretti E, Dabrowski K, Praticò D (2021) The neurobiology of non-coding RNAs and Alzheimer's disease pathogenesis: pathways, mechanisms and translational opportunities. Ageing Res Rev 71:101425. https://doi.org/10.1016/j.arr.2021.101425

Le LT, Nhu CX (2023) The Role of long non-coding RNAs in cardiovascular diseases. Int J Mol Sci 24(18):13805. https://doi.org/10.3390/ijms241813805

LeCun Y, Boser B, Denker JS et al (1989) Backpropagation applied to handwritten zip code recognition. Neural Comput 1(4):541–551. https://doi.org/10.1162/neco.1989.1.4.541

Leonenko G, Baker E, Stevenson-Hoare J et al (2021) Identifying individuals with high risk of Alzheimer's disease using polygenic risk scores. Nat Commun 12(1):4506. https://doi.org/10.1038/s41467-021-24082-z

Leptidis S, Papakonstantinou E, Diakou KI et al (2022) Epitranscriptomics of cardiovascular diseases. Int J Mol Med 49(1):1–21. https://doi.org/10.3892/ijmm.2021.5064

Li K, Blum Y, Verma A et al (2010) A noncoding antisense RNA in tie-1 locus regulates tie-1 function in vivo. Blood 115(1):133–139. https://doi.org/10.1182/blood-2009-09-242180

Li W, Notani D, Ma Q et al (2013) Functional Role of enhancer RNAs for oestrogen-dependent transcriptional activation. Nature 498(7455):516–520. https://doi.org/10.1038/nature12210

Li J, Zhu L, Guan F et al (2018) Relationship between schizophrenia and changes in the expression of the long non-coding RNAs Meg3, Miat, Neat1 and Neat2. J Psychiatr Res 106:22–30. https://doi.org/10.1016/j.jpsychires.2018.09.005

Li F, Wang Y, Yang H et al (2019) The effect of BACE1-AS on β-amyloid generation by regulating BACE1 mRNA expression. BMC Mol Biol 20:1–10. https://doi.org/10.1186/s12867-019-0140-0

Li X, Wang S-W, Li X-L et al (2020) Knockdown of long non-coding RNA TUG1 depresses apoptosis of hippocampal neurons in Alzheimer's disease by elevating microRNA-15a and repressing ROCK1 expression. Inflamm Res 69:897–910. https://doi.org/10.1007/s00011-020-01364-8

Li Z, Gao J, Sun D et al (2022) LncRNA MEG3: potential stock for precision treatment of cardiovascular diseases. Front Pharmacol 13:1045501. https://doi.org/10.3389/fphar.2022.1045501

Li DF, Liu QS, Yang MF et al (2023) Nanomaterials for mRNA-based therapeutics: challenges and opportunities. Bioeng Transl Med 8(3):e10492. https://doi.org/10.1002/btm2.10492

Li X, Zhang Y, Ding Z et al (2024) LncRNA H19: a novel biomarker in cardiovascular disease. Acta Cardiol Sin 40(2):172. https://doi.org/10.6515/ACS.202403_40(2).20230925A

Lim JW, Lee J, Pae AN (2020) Mitochondrial dysfunction and Alzheimer's disease: prospects for therapeutic intervention. BMB Rep 53(1):47. https://doi.org/10.5483/BMBRep.2020.53.1.279

Lin B, Chen X, Lu C et al (2021) Loss of exosomal LncRNA HCG15 prevents acute myocardial ischemic injury through the NF-κB/p65 and p38 pathways. Cell Death Dis 12(11):1007. https://doi.org/10.1038/s41419-021-04281-8

Liu J, Lu F (2024) The piRNA size matters. Natl Sci Rev 11(2):nwad297. https://doi.org/10.1093/nsr/nwad297

Liu T, Huang Y, Chen J et al (2014) Attenuated ability of BACE1 to cleave the amyloid precursor protein via silencing long noncoding RNA BACE1-AS expression. Mol Med Rep 10(3):1275–1281. https://doi.org/10.3892/mmr.2014.2351

Liu Y, Zhang M, Liang L et al (2015) Over-expression of lncRNA DANCR is associated with advanced tumor progression and poor prognosis in patients with colorectal cancer. Int J Clin Exp Pathol 8(9):11480

Liu SJ, Nowakowski TJ, Pollen AA et al (2016) Single-cell analysis of long non-coding RNAs in the developing human neocortex. Genome Biol 17(1):67. https://doi.org/10.1186/s13059-016-0932-1

Liu K, Hao Q, Wei J et al (2018a) MicroRNA-19a/b-3p protect the heart from hypertension-induced pathological cardiac hypertrophy through PDE5A. J Hypertens 36(9):1847–1857. https://doi.org/10.1097/HJH.0000000000001769

Liu Y, He X, Li Y et al (2018b) Cerebrospinal fluid CD4+ T lymphocyte-derived miRNA-let-7b can enhances the diagnostic performance of Alzheimer's disease biomarkers. Biochem Biophys Res Commun 495(1):1144–1150. https://doi.org/10.1016/j.bbrc.2017.11.122

Liu R, Li F, Zhao W (2020) Long noncoding RNA NEAT1 knockdown inhibits MPP+-induced apoptosis, inflammation and cytotoxicity in SK-N-SH cells by regulating miR-212-5p/RAB3IP axis. Neurosci Lett 731:135060. https://doi.org/10.1016/j.neulet.2020.135060

Liu Z, Tao C, Li S et al (2021) circFL-seq reveals full-length circular RNAs with rolling circular reverse transcription and nanopore sequencing. Elife 10:e69457. https://doi.org/10.7554/eLife.69457

Long JM, Holtzman DM (2019) Alzheimer disease: an update on pathobiology and treatment strategies. Cell 179(2):312–339. https://doi.org/10.1016/j.cell.2019.09.001

Losko M, Kotlinowski J, Jura J (2016) Long noncoding RNAs in metabolic syndrome related disorders. Mediators Inflamm 2016(1):5365209. https://doi.org/10.1155/2016/5365209

Lu J, Liu L, Chen J et al (2022) LncRNA HOTAIR in exercise-induced neuro-protective function in Alzheimer's disease. Folia Neuropathol 60(4):414–420. https://doi.org/10.5114/fn.2022.118961

Luo Y, Deng X, Chen Q et al (2023) Up-regulation of miR-155 protects against chronic heart failure by inhibiting HIF-1α. Am J Transl Res 15(11):6425

Ma L, Bajic VB, Zhang Z (2013) On the classification of long non-coding RNAs. RNA Biol 10(6):924–933. https://doi.org/10.4161/rna.24604

Ma P, Li Y, Zhang W et al (2019) Long non-coding RNA MALAT1 inhibits neuron apoptosis and neuroinflammation while stimulates neurite outgrowth and its correlation with MiR-125b mediates PTGS2, CDK5 and FOXQ1 in Alzheimer's disease. Curr Alzheimer Res 16(7):596–612. https://doi.org/10.2174/1567205016666190725130134

Magistri M, Velmeshev D, Makhmutova M et al (2015) Transcriptomics profiling of Alzheimer's disease reveal neurovascular defects, altered amyloid-β homeostasis, and deregulated expression of long noncoding RNAs. J Alzheimer's Dis 48(3):647–665. https://doi.org/10.3233/JAD-150398

Mahato RK, Bhattacharya S, Khullar N et al (2024) Targeting long non-coding RNAs in cancer therapy using CRISPR-Cas9 technology: a novel paradigm for precision oncology. J Biotechnol 379:98–119. https://doi.org/10.1016/j.jbiotec.2023.12.003

Marín-Béjar O, Marchese FP, Athie A et al (2013) Pint lincRNA connects the p53 pathway with epigenetic silencing by the Polycomb repressive complex 2. Genome Biol 14:1–17. https://doi.org/10.1186/gb-2013-14-9-r104

Mathy NW, Chen X-M (2017) Long non-coding RNAs (lncRNAs) and their transcriptional control of inflammatory responses. J Biol Chem 292(30):12375–12382. https://doi.org/10.1074/jbc.R116.760884

Mattick JS, Amaral PP, Carninci P et al (2023) Long non-coding RNAs: definitions, functions, challenges and recommendations. Nat Rev Mol Cell Biol 24(6):430–447. https://doi.org/10.1038/s41580-022-00566-8

Maurano MT, Humbert R, Rynes E et al (2012) Systematic localization of common disease-associated variation in regulatory DNA. Science 337(6099):1190–1195. https://doi.org/10.1126/science.1222794

Mazdeh M, Komaki A, Omrani MD et al (2018) Expression analysis of beta-secretase 1 (BACE1) and its naturally occurring antisense (BACE1-AS) in blood of epileptic patients. Neurol Sci 39:1565–1569. https://doi.org/10.1007/s10072-018-3458-3

Meng J, Ding T, Chen Y et al (2021) LncRNA-Meg3 promotes Nlrp3-mediated microglial inflammation by targeting miR-7a-5p. Int Immunopharmacol 90:107141. https://doi.org/10.1016/j.intimp.2020.107141

Michalik KM, You X, Manavski Y et al (2014) Long noncoding RNA MALAT1 regulates endothelial cell function and vessel growth. Circ Res 114(9):1389–1397. https://doi.org/10.1161/CIRCRESAHA.114.303265

Micheletti R, Plaisance I, Abraham BJ et al (2017) The long noncoding RNA Wisper controls cardiac fibrosis and remodeling. Sci Transl Med 9(395):eaai9118. https://doi.org/10.1126/scitranslmed.aai9118

Modarresi F, Faghihi MA, Patel NS et al (2011) Knockdown of BACE1-AS nonprotein-coding transcript modulates beta-amyloid-related hippocampal neurogenesis. Int J Alzheimers Dis 2011(1):929042. https://doi.org/10.4061/2011/929042

Mohammad F, Mondal T, Guseva N et al (2010) Kcnq1ot1 noncoding RNA mediates transcriptional gene silencing by interacting with Dnmt1. Development 137(15):2493–2499. https://doi.org/10.1242/dev.048181

Molla G, Bitew M (2024) Revolutionizing personalized medicine: synergy with multi-omics data generation, Main hurdles, and future perspectives. Biomedicine 12(12):2750. https://doi.org/10.3390/biomedicines12122750

Monteiro JP, Bennett M, Rodor J et al (2019) Endothelial function and dysfunction in the cardiovascular system: the long non-coding road. Cardiovasc Res 115(12):1692–1704. https://doi.org/10.1093/cvr/cvz154

Montgomery RM (2024) Global trends in disease burden, morbidity, and mortality: implications of emerging artificial intelligence (AI) technologies. https://doi.org/10.20944/preprints202409.1130.v1

Mosquera-Heredia MI, Vidal OM, Morales LC et al (2024) Long non-coding RNAs and Alzheimer's disease: towards personalized diagnosis. Int J Mol Sci 25(14):7641. https://doi.org/10.3390/ijms25147641

Mukhopadhyay C, Singh J (2014) MicroRNAs: basics, implications, and modulation in improving disease resistance. DBT sponsored short term training course on molecular tools and bioinformatics approaches for livestock genome analysis, delivered on 13th September

Mus E, Hof PR, Tiedge H (2007) Dendritic BC200 RNA in aging and in Alzheimer's disease. Proc Natl Acad Sci 104(25):10679–10684. https://doi.org/10.1073/pnas.0701532104

Nabi A, Dilekoglu B, Adebali O et al (2023) Discovering misannotated lncRNAs using deep learning training dynamics. Bioinformatics 39(1):btac821. https://doi.org/10.1093/bioinformatics/btac821

Nappi F (2024) Non-coding RNA-targeted therapy: a state-of-the-art review. Int J Mol Sci 25(7):3630. https://doi.org/10.3390/ijms25073630

Ngowi EE, Lu T, Liu Q et al (2024) Biofluid-derived Exosomal LncRNAs: their potential in obesity and related comorbidities. Biology 13(12):976. https://doi.org/10.3390/biology13120976

Niu D-K, Jiang L (2013) Can ENCODE tell us how much junk DNA we carry in our genome? Biochem Biophys Res Commun 430(4):1340–1343. https://doi.org/10.1016/j.bbrc.2012.12.074

Novelli G, Spitalieri P, Murdocca M et al (2022) Organoid factory: the recent Role of the human induced pluripotent stem cells (hiPSCs) in precision medicine. Front Cell Dev Biol 10:1059579. https://doi.org/10.3389/fcell.2022.1059579

O'Brien J, Hayder H, Zayed Y et al (2018) Overview of microRNA biogenesis, mechanisms of actions, and circulation. Front Endocrinol 9:402. https://doi.org/10.3389/fendo.2018.00402

Osbourne A, Calway T, Broman M et al (2014) Downregulation of connexin43 by microRNA-130a in cardiomyocytes results in cardiac arrhythmias. J Mol Cell Cardiol 74:53–63. https://doi.org/10.1016/j.yjmcc.2014.04.024

Palade GE (1955) A small particulate component of the cytoplasm. J Biophys Biochem Cytol 1(1):59. https://doi.org/10.1083/jcb.1.1.59

Pan J, Xu Z, Guo G et al (2021) Circ_nuclear factor IX (circNfix) attenuates pressure overload-induced cardiac hypertrophy via regulating miR-145-5p/ATF3 axis. Bioengineered 12(1):5373–5385. https://doi.org/10.1080/21655979.2021.1960462

Pang KC, Frith MC, Mattick JS (2006) Rapid evolution of noncoding RNAs: lack of conservation does not mean lack of function. Trends Genet 22(1):1–5. https://doi.org/10.1016/j.tig.2005.10.003

Petkovic A, Erceg S, Munjas J et al (2023) LncRNAs as regulators of atherosclerotic plaque stability. Cells 12(14):1832. https://doi.org/10.3390/cells12141832

Plank JL, Dean A (2014) Enhancer function: mechanistic and genome-wide insights come together. Mol Cell 55(1):5–14. https://doi.org/10.1016/j.molcel.2014.06.015

Qu Z, Adelson DL (2012) Evolutionary conservation and functional Role of ncRNA. Front Genet 3:205. https://doi.org/10.3389/fgene.2012.00205

Quan Y, Wang J, Wang S et al (2020) Association of the plasma long non-coding RNA MEG3 with Parkinson's disease. Front Neurol 11:532891. https://doi.org/10.3389/fneur.2020.532891

Quinn JJ, Chang HY (2016) Unique features of long non-coding RNA biogenesis and function. Nat Rev Genet 17(1):47–62. https://doi.org/10.1038/nrg.2015.10

Rakicevic L (2023) DNA and RNA molecules as a foundation of therapy strategies for treatment of cardiovascular diseases. Pharmaceutics 15(8):2141. https://doi.org/10.3390/pharmaceutics15082141

Ran FA, Hsu PD, Wright J et al (2013) Genome engineering using the CRISPR-Cas9 system. Nat Protoc 8(11):2281–2308. https://doi.org/10.1038/nprot.2013.143

Ransohoff JD, Wei Y, Khavari PA (2018) The functions and unique features of long intergenic non-coding RNA. Nat Rev Mol Cell Biol 19(3):143–157. https://doi.org/10.1038/nrm.2017.104

Rao DD, Vorhies JS, Senzer N et al (2009) siRNA vs. shRNA: similarities and differences. Adv Drug Deliv Rev 61(9):746–759. https://doi.org/10.1016/j.addr.2009.04.004

Reddy R, Busch H (1983) Small nuclear RNAs and RNA processing. Prog Nucleic Acid Res Mol Biol 30:127–162. https://doi.org/10.1016/S0079-6603(08)60685-6

Reddy R, Singh R (1991) Synthesis of small nuclear RNAs. Prog Mol Subcell Biol:1–36. https://doi.org/10.1007/978-3-642-76553-7_1

Ribas VT, Costa MR (2017) Gene manipulation strategies to identify molecular regulators of axon regeneration in the central nervous system. Front Cell Neurosci 11:231. https://doi.org/10.3389/fncel.2017.00231

Rinn JL, Kertesz M, Wang JK et al (2007) Functional demarcation of active and silent chromatin domains in human HOX loci by noncoding RNAs. Cell 129(7):1311–1323. https://doi.org/10.1016/j.cell.2007.05.022

Ruvkun G, Ambros V, Coulson A et al (1989) Molecular genetics of the Caenorhabditis elegans heterochronic gene lin-14. Genetics 121(3):501–516. https://doi.org/10.1093/genetics/121.3.501

Salzman J (2016) Circular RNA expression: its potential regulation and function. Trends Genet 32(5):309–316. https://doi.org/10.1016/j.tig.2016.03.002

Sanbonmatsu K (2022) Getting to the bottom of lncRNA mechanism: structure–function relationships. Mamm Genome 33(2):343–353. https://doi.org/10.1007/s00335-021-09924-x

Santos-Pereira JM, Aguilera A (2015) R loops: new modulators of genome dynamics and function. Nat Rev Genet 16(10):583–597. https://doi.org/10.1038/nrg3961

Sasaki YT, Ideue T, Sano M et al (2009) MENε/β noncoding RNAs are essential for structural integrity of nuclear paraspeckles. Proc Natl Acad Sci 106(8):2525–2530. https://doi.org/10.1073/pnas.0807899106

Sasmita BR, Zhu Y, Gan H et al (2022) Leukocyte and its subtypes as predictors of short-term outcome in cardiogenic shock complicating acute myocardial infarction: a cohort study. Shock 57(3):351–359. https://doi.org/10.1097/SHK.0000000000001876

Satpathy AT, Chang HY (2015) Long noncoding RNA in hematopoiesis and immunity. Immunity 42(5):792–804. https://doi.org/10.1016/j.immuni.2015.05.004

Šatrauskienė A, Navickas R, Laucevičius A et al (2021) Mir-1, miR-122, miR-132, and miR-133 are related to subclinical aortic atherosclerosis associated with metabolic syndrome. Int J Environ Res Public Health 18(4):1483. https://doi.org/10.3390/ijerph18041483

Schmidhuber J, Hochreiter S (1997) Long short-term memory. Neural Comput 9(8):1735–1780. https://doi.org/10.1162/neco.1997.9.8.1735

Schmitt AM, Garcia JT, Hung T et al (2016) An inducible long noncoding RNA amplifies DNA damage signaling. Nat Genet 48(11):1370–1376. https://doi.org/10.1038/ng.3673

Schmitz K-M, Mayer C, Postepska A et al (2010) Interaction of noncoding RNA with the rDNA promoter mediates recruitment of DNMT3b and silencing of rRNA genes. Genes Dev 24(20):2264–2269. https://doi.org/10.1101/gad.590910

Sharma AK, Bisht P, Gupta B et al (2023) Investigating miRNA subfamilies: can they assist in the early diagnosis of acute myocardial infarction? Drug Discov Today 103695. https://doi.org/10.1016/j.drudis.2023.103695

Shi X, Sun M, Liu H et al (2013) Long non-coding RNAs: a new frontier in the study of human diseases. Cancer Lett 339(2):159–166. https://doi.org/10.1016/j.canlet.2013.06.013

Shi P, Ji H, Zhang H et al (2020) circANRIL reduces vascular endothelial injury, oxidative stress and inflammation in rats with coronary atherosclerosis. Exp Ther Med 20(3):2245–2251. https://doi.org/10.3892/etm.2020.8956

Shobeiri P, Alilou S, Jaberinezhad M et al (2023) Circulating long non-coding RNAs as novel diagnostic biomarkers for Alzheimer's disease (AD): a systematic review and meta-analysis. PLoS One 18(3):e0281784. https://doi.org/10.1371/journal.pone.0281784

Shore AN, Kabotyanski EB, Roarty K et al (2012) Pregnancy-induced noncoding RNA (PINC) associates with polycomb repressive complex 2 and regulates mammary epithelial differentiation. PLoS Genet 8(7):e1002840. https://doi.org/10.1371/journal.pgen.1002840

Shou F, Li G, Morshedi M (2024) Long non-coding RNA ANRIL and its Role in the development of age-related diseases. Mol Neurobiol 1-11. https://doi.org/10.1007/s12035-024-04074-y

Simone R, Javad F, Emmett W et al (2021) MIR-NATs repress MAPT translation and aid proteostasis in neurodegeneration. Nature 594(7861):117–123. https://doi.org/10.1038/s41586-021-03556-6

Singh DD, Kim Y, Choi SA et al (2023) Clinical significance of MicroRNAs, long non-coding RNAs, and CircRNAs in cardiovascular diseases. Cells 12(12):1629. https://doi.org/10.3390/cells12121629

Skroblin P, Mayr M (2014) "Going long": long non-coding RNAs as biomarkers. Circ Res 115(7):607–609. https://doi.org/10.1161/CIRCRESAHA.114.304839

Sohag MMH, Raqib SM, Akhmad SA (2021) OMICS approaches in cardiovascular diseases: a mini review. Genomics Inform 19(2):e13. https://doi.org/10.5808/gi.21002

Song XW, Li Q, Lin L et al (2010) MicroRNAs are dynamically regulated in hypertrophic hearts, and miR-199a is essential for the maintenance of cell size in cardiomyocytes. J Cell Physiol 225(2):437–443. https://doi.org/10.1002/jcp.22217

Sosińska P, Mikuła-Pietrasik J, Książek K (2015) The double-edged sword of long non-coding RNA: the Role of human brain-specific BC200 RNA in translational control, neurodegenerative diseases, and cancer. Mutat Res Rev Mutat Res 766:58–67. https://doi.org/10.1016/j.mrrev.2015.08.002

Soutschek M, Schratt G (2023) Non-coding RNA in the wiring and remodeling of neural circuits. Neuron 111(14):2140–2154. https://doi.org/10.1016/j.neuron.2023.04.031

Srinivas T, Siqueira E, Guil S (2024) Techniques for investigating lncRNA transcript functions in neurodevelopment. Mol Psychiatry 29(4):874–890. https://doi.org/10.1038/s41380-023-02377-5

Statello L, Guo C-J, Chen L-L et al (2021) Gene regulation by long non-coding RNAs and its biological functions. Nat Rev Mol Cell Biol 22(2):96–118. https://doi.org/10.1038/s41580-020-00315-9

Sun J, Zhang Z, Yang S (2019) Circ_RUSC2 upregulates the expression of miR-661 target gene SYK and regulates the function of vascular smooth muscle cells. Biochem Cell Biol 97(6):709–714. https://doi.org/10.1139/bcb-2019-0031

Sun Y, Huang S, Wan C et al (2021) Knockdown of lncRNA ENST00000609755. 1 confers protection against early oxLDL-induced coronary heart disease. Front Cardiovasc Med 8:650212. https://doi.org/10.3389/fcvm.2021.650212

Sunwoo H, Dinger ME, Wilusz JE et al (2009) MEN ε/β nuclear-retained non-coding RNAs are up-regulated upon muscle differentiation and are essential components of paraspeckles. Genome Res 19(3):347–359. https://doi.org/10.1101/gr.087775.108

Taghvimi S, Soltani Fard E, Khatami SH et al (2024) lncRNA HOTAIR and cardiovascular diseases. Funct Integr Genomics 24(5):1–18. https://doi.org/10.1007/s10142-024-01444-6

Tan J, Li X, Zhang L et al (2022) Recent advances in machine learning methods for predicting LncRNA and disease associations. Front Cell Infect Microbiol 12:1071972. https://doi.org/10.3389/fcimb.2022.1071972

Tanashyan MM, Shabalina AA, Kuznetsova PI et al (2023) miR-33a and its association with lipid profile in patients with carotid atherosclerosis. Int J Mol Sci 24(7):6376. https://doi.org/10.3390/ijms24076376

Tauber CV, Schwarz SC, Rösler TW et al (2023) Different MAPT haplotypes influence expression of total MAPT in postmortem brain tissue. Acta Neuropathol Commun 11(1):40. https://doi.org/10.1186/s40478-023-01534-9

Tian C, Hu G, Gao L et al (2020) Extracellular vesicular MicroRNA-27a* contributes to cardiac hypertrophy in chronic heart failure. J Mol Cell Cardiol 143:120–131. https://doi.org/10.1016/j.yjmcc.2020.04.032

Tian C, Gao L, Rudebush TL et al (2022) Extracellular vesicles regulate sympatho-excitation by Nrf2 in heart failure. Circ Res 131(8):687–700. https://doi.org/10.1161/CIRCRESAHA.122.320916

Torres AG, Reina O, Stephan-Otto Attolini C et al (2019) Differential expression of human tRNA genes drives the abundance of tRNA-derived fragments. Proc Natl Acad Sci 116(17):8451–8456. https://doi.org/10.1073/pnas.1821120116
Uchida S, Dimmeler S (2015) Long noncoding RNAs in cardiovascular diseases. Circ Res 116(4):737–750. https://doi.org/10.1161/CIRCRESAHA.116.30252
Valadi H, Ekström K, Bossios A et al (2007) Exosome-mediated transfer of mRNAs and microRNAs is a novel mechanism of genetic exchange between cells. Nat Cell Biol 9(6):654–659. https://doi.org/10.1038/ncb1596
Vassar R, Kandalepas PC (2011) The β-secretase enzyme BACE1 as a therapeutic target for Alzheimer's disease. Alzheimer's Res Ther 3(3):20. https://doi.org/10.1186/alzrt82
Viereck J, Thum T (2017) Circulating noncoding RNAs as biomarkers of cardiovascular disease and injury. Circ Res 120(2):381–399. https://doi.org/10.1161/circresaha.116.308434
Wang Y, Sun X (2020) The functions of LncRNA in the heart. Diabetes Res Clin Pract 168:108249. https://doi.org/10.1016/j.diabres.2020.108249
Wang K, Liu F, Zhou L-Y et al (2014a) The long noncoding RNA CHRF regulates cardiac hypertrophy by targeting miR-489. Circ Res 114(9):1377–1388. https://doi.org/10.1161/CIRCRESAHA.114.302476
Wang K, Long B, Zhou L-Y et al (2014b) CARL lncRNA inhibits anoxia-induced mitochondrial fission and apoptosis in cardiomyocytes by impairing miR-539-dependent PHB2 downregulation. Nat Commun 5(1):3596. https://doi.org/10.1038/ncomms4596
Wang K, Liu C-Y, Zhou L-Y et al (2015) APF lncRNA regulates autophagy and myocardial infarction by targeting miR-188-3p. Nat Commun 6(1):6779. https://doi.org/10.1038/ncomms7779
Wang J, Ye C, Xiong H et al (2016a) Dysregulation of long non-coding RNA in breast cancer: an overview of mechanism and clinical implication. Oncotarget 8(3):5508. https://doi.org/10.18632/oncotarget.12537
Wang Z, Zhang X-J, Ji Y-X et al (2016b) The long noncoding RNA Chaer defines an epigenetic checkpoint in cardiac hypertrophy. Nat Med 22(10):1131–1139. https://doi.org/10.1038/nm.4179
Wang J, Du S, Wang J et al (2017a) The prognostic value of abnormally expressed lncRNAs in colorectal cancer: a meta-analysis. PLoS One 12(6):e0179670. https://doi.org/10.1371/journal.pone.0179670
Wang W, Zhuang Q, Ji K et al (2017b) Identification of miRNA, lncRNA and mRNA-associated ceRNA networks and potential biomarker for MELAS with mitochondrial DNA A3243G mutation. Sci Rep 7(1):41639. https://doi.org/10.1038/srep41639
Wang L, Shen C, Wang Y et al (2019) Identification of circular RNA Hsa_circ_0001879 and Hsa_circ_0004104 as novel biomarkers for coronary artery disease. Atherosclerosis 286:88–96. https://doi.org/10.1016/j.atherosclerosis.2019.05.006
Wang W, Min L, Qiu X et al (2021) Biological function of Long non-coding RNA (LncRNA) Xist. Front Cell Dev Biol 9. https://doi.org/10.3389/fcell.2021.645647
Wang J, Yue B-L, Huang Y-Z et al (2022a) Exosomal RNAs: novel potential biomarkers for diseases—a review. Int J Mol Sci 23(5):2461. https://doi.org/10.3390/ijms23052461
Wang J, Zhao J, Hu P et al (2022b) Long non-coding RNA HOTAIR in central nervous system disorders: new insights in pathogenesis, diagnosis, and therapeutic potential. Front Mol Neurosci 15:949095. https://doi.org/10.3389/fnmol.2022.949095
Watts R, Johnsen VL, Shearer J et al (2013) Myostatin-induced inhibition of the long noncoding RNA Malat1 is associated with decreased myogenesis. Am J Physiol Cell Physiol 304(10):C995–C1001. https://doi.org/10.1152/ajpcell.00392.2012
Wei L, Yuan M, Zhou R et al (2015) MicroRNA-101 inhibits rat cardiac hypertrophy by targeting Rab1a. J Cardiovasc Pharmacol 65(4):357–363. https://doi.org/10.1097/FJC.0000000000000203
Wei B, Li Z, Wang L et al (2024) miR-200b-3p relieved inflammation in patients with heart failure by regulating ZEB1 expression. J Cardiothorac Surg 19(1):271. https://doi.org/10.1186/s13019-024-02628-8

Weingarten-Gabbay S, Elias-Kirma S, Nir R et al (2016) Systematic discovery of cap-independent translation sequences in human and viral genomes. Science 351(6270):aad4939. https://doi.org/10.1126/science.aad4939
Wilkinson B, Campbell DB (2013) Contribution of long noncoding RNAs to autism spectrum disorder risk. Int Rev Neurobiol 113:35–59. https://doi.org/10.1016/B978-0-12-418700-9.00002-2
Wu C, Arora P (2015) Long noncoding Mhrt RNA: molecular crowbar unravel insights into heart failure treatment. Circ Cardiovasc Genet 8(1):213–215. https://doi.org/10.1161/CIRCGENETICS.115.001019
Wu Y-X, Zhang S-H, Cui J et al (2018) Long noncoding RNA XR007793 regulates proliferation and migration of vascular smooth muscle cell via suppressing miR-23b. Med Sci Monit 24:5895. https://doi.org/10.12659/MSM.908902
Wu W-p, Zhou M-y, Liu D-l et al (2021) circGNAQ, a circular RNA enriched in vascular endothelium, inhibits endothelial cell senescence and atherosclerosis progression. Mol Ther Nucleic Acids 26:374–387. https://doi.org/10.1016/j.omtn.2021.07.020
Wu Y, Li P, Liu L et al (2022) lncRNA <em>Neat1</em> regulates neuronal dysfunction post-sepsis via stabilization of hemoglobin subunit beta. Mol Ther 30(7):2618–2632. https://doi.org/10.1016/j.ymthe.2022.03.011
Xia W, Chen H, Xie C et al (2020) Long-noncoding RNA MALAT1 sponges microRNA-92a-3p to inhibit doxorubicin-induced cardiac senescence by targeting ATG4a. Aging (Albany NY) 12(9):8241. https://doi.org/10.18632/aging.103136
Xuan C, Yang E, Zhao S et al (2023) Regulation of LncRNAs and microRNAs in neuronal development and disease. PeerJ 11:e15197. https://doi.org/10.7717/peerj.15197
Yan Y, Yan H, Teng Y et al (2020) Long non-coding RNA 00507/miRNA-181c-5p/TTBK1/MAPT axis regulates tau hyperphosphorylation in Alzheimer's disease. J Gene Med 22(12):e3268. https://doi.org/10.1002/jgm.3268
Yang C, Chapman AG, Kelsey AD et al (2011) X-chromosome inactivation: molecular mechanisms from the human perspective. Hum Genet 130:175–185. https://doi.org/10.1007/s00439-011-0994-9
Yeh C-F, Chang Y-CE, Lu C-Y et al (2020) Expedition to the missing link: long noncoding RNAs in cardiovascular diseases. J Biomed Sci 27:1–16. https://doi.org/10.1186/s12929-020-00647-w
Yu B, Jiang Y, Wang X et al (2020) An integrated hypothesis for miR-126 in vascular disease. Med Res Arch 8(5). https://doi.org/10.18103/mra.v8i5.2133
Yu Y, Yang H, Li Q et al (2023) Stress-enhanced cardiac lncRNA Morrbid protects hearts from acute myocardial infarction. JCI Insight 8(16). https://doi.org/10.1172/jci.insight.165568
Zeng C, Chen Z, Yang H et al (2022) Advanced high resolution three-dimensional imaging to visualize the cerebral neurovascular network in stroke. Int J Biol Sci 18(2):552–571. https://doi.org/10.7150/ijbs.64373
Zhang J (2024) Non-coding RNAs and angiogenesis in cardiovascular diseases: a comprehensive review. Mol Cell Biochem:1–33. https://doi.org/10.1007/s11010-023-04919-5
Zhang J, Wang R (2021) Deregulated lncRNA MAGI2-AS3 in Alzheimer's disease attenuates amyloid-β induced neurotoxicity and neuroinflammation by sponging miR-374b-5p. Exp Gerontol 144:111180. https://doi.org/10.1016/j.exger.2020.111180
Zhang L-M, Wang M-H, Yang H-C et al (2019a) Dopaminergic neuron injury in Parkinson's disease is mitigated by interfering lncRNA SNHG14 expression to regulate the miR-133b/α--synuclein pathway. Aging (Albany NY) 11(21):9264. https://doi.org/10.18632/aging.102330
Zhang L, Fang Y, Cheng X et al (2019b) Silencing of long noncoding RNA SOX21-AS1 relieves neuronal oxidative stress injury in mice with Alzheimer's disease by upregulating FZD3/5 via the Wnt signaling pathway. Mol Neurobiol 56(5):3522–3537. https://doi.org/10.1007/s12035-018-1299-y
Zhang Y, Zhao Y, Ao X et al (2021) The Role of non-coding RNAs in Alzheimer's disease: from regulated mechanism to therapeutic targets and diagnostic biomarkers. Front Aging Neurosci 13:654978. https://doi.org/10.3389/fnagi.2021.654978

Zhang X, Gao Y, Wu H et al (2022) LncRNA HOX transcript antisense RNA mitigates cardiac function injury in chronic heart failure via regulating microRNA-30a-5p to target KDM3A. J Cell Mol Med 26(5):1473–1485. https://doi.org/10.1111/jcmm.17160

Zhang X, Sun S, Ren G et al (2023) Advances in intercellular communication mediated by Exosomal ncRNAs in cardiovascular disease. Int J Mol Sci 24(22):16197. https://doi.org/10.3390/ijms242216197

Zhao ZH, Hao W, Meng QT et al (2017) Long non-coding RNA MALAT1 functions as a mediator in cardioprotective effects of fentanyl in myocardial ischemia-reperfusion injury. Cell Biol Int 41(1):62–70. https://doi.org/10.1002/cbin.10701

Zhao Y, Zhang Y, Zhang L et al (2018) The potential markers of circulating microRNAs and long non-coding RNAs in Alzheimer's disease. Aging Dis 2019(10):1293–1301. https://doi.org/10.14336/AD.2018.1105

Zhao Q, Li W, Pan W et al (2021) CircRNA 010567 plays a significant Role in myocardial infarction via the regulation of the miRNA-141/DAPK1 axis. J Thorac Dis 13(4):2447. https://doi.org/10.21037/jtd-21-212

Zhou K, Zhang C, Yao H et al (2018) Knockdown of long non-coding RNA NEAT1 inhibits glioma cell migration and invasion via modulation of SOX2 targeted by miR-132. Mol Cancer 17:1–11. https://doi.org/10.1186/s12943-018-0849-2

Zhou B, Li L, Qiu X et al (2020a) Long non-coding RNA ANRIL knockdown suppresses apoptosis and pro-inflammatory cytokines while enhancing neurite outgrowth via binding microRNA-125a in a cellular model of Alzheimer's disease. Mol Med Rep 22(2):1489–1497. https://doi.org/10.3892/mmr.2020.11203

Zhou W-Y, Cai Z-R, Liu J et al (2020b) Circular RNA: metabolism, functions and interactions with proteins. Mol Cancer 19(1):172. https://doi.org/10.1186/s12943-020-01286-3

Zhou B, Ding M, Feng J et al (2023) EVlncRNA-Dpred: improved prediction of experimentally validated lncRNAs by deep learning. Brief Bioinform 24(1). https://doi.org/10.1093/bib/bbac583

Zhu H-C, Wang L-M, Wang M et al (2012) MicroRNA-195 downregulates Alzheimer's disease amyloid-β production by targeting BACE1. Brain Res Bull 88(6):596–601. https://doi.org/10.1016/j.brainresbull.2012.05.018

Zhu L, Lin M, Ma J et al (2019) The Role of LINC 00094/miR-224-5p (miR-497-5p)/Endophilin-1 axis in Memantine mediated protective effects on blood-brain barrier in AD microenvironment. J Cell Mol Med 23(5):3280–3292. https://doi.org/10.1111/jcmm.14214

Zhu L, Li N, Sun L et al (2021) Non-coding RNAs: the key detectors and regulators in cardiovascular disease. Genomics 113(1):1233–1246. https://doi.org/10.1016/j.ygeno.2020.10.024

Zhu W, Luo L, Ye G et al (2023) Potential diagnostic value of N1LR and SNHG1 in acute myocardial infarction. BMC Med Genet 16(1):71. https://doi.org/10.1186/s12920-023-01501-2

Zhu L, Guo M, Li K et al (2024) The association and prognostic implications of long non-coding RNAs in major psychiatric disorders, Alzheimer's diseases and Parkinson's diseases: a systematic review. Int J Mol Sci 25(20):10995. https://doi.org/10.3390/ijms252010995

Zhuang J, Chen Z, Cai P et al (2020) Targeting microRNA-125b promotes neurite outgrowth but represses cell apoptosis and inflammation via blocking PTGS2 and CDK5 in a FOXQ1-dependent way in Alzheimer disease. Front Cell Neurosci 14:587747. https://doi.org/10.3389/fncel.2020.587747

Chapter 3
Role of SLC Transporters and Noncoding RNA in Regulating the Genetic Landscape of the Blood Disorders

Mounika Sarvepalli, Aditya Velidandi, Amit Mishra, and Ravi Kumar Gutti

Contents

M. Sarvepalli
Department of Biochemistry, School of Life Sciences, University of Hyderabad, Hyderabad, Telangana, India

School of Life Sciences, B.S. Abdur Rahman Crescent Institute of Science and Technology, Chennai, India

A. Velidandi
Department of Biotechnology, National Institute of Technology, Warangal, Telangana, India

A. Mishra
Department of Bioscience and Bioengineering, Cellular and Molecular Neurobiology Unit, Indian Institute of Technology Jodhpur, Jodhpur, Rajasthan, India

R. K. Gutti (✉)
Department of Biochemistry, School of Life Sciences, University of Hyderabad, Hyderabad, Telangana, India

M. F. Ullah, A. Ahmad (eds.), *Non-coding RNAs*, Experientia Supplementum 115,
https://doi.org/10.1007/978-3-032-06948-1_3

Abstract The complex interplay between solute carrier (SLC) transporters and noncoding RNAs (ncRNAs) in regulating genetic and metabolic pathways in hematological disorders remains underexplored. This chapter addresses the critical need to understand how these regulators influence the development, progression, and therapeutic resistance of leukemia. By elucidating these mechanisms, this chapter aims to contribute to the development of novel, targeted interventions for improved clinical outcomes. This chapter delves into the multifaceted roles of SLC transporters and ncRNAs in blood cell regulation, highlighting their impact on hematopoiesis and leukemogenesis. Topics include the hierarchical differentiation of hematopoietic stem cells, the molecular subtypes of leukemia, and the regulatory roles of transcription factors and chromatin modifiers in leukemia progression. Additionally, this chapter explores how SLC transporters mediate nutrient uptake, drug resistance, and metabolic adaptations in leukemic cells. It further examines the involvement of ncRNAs—such as microRNAs, long noncoding RNAs, and circular RNAs—in modulating transporter activity, impacting both normal and malignant hematopoietic processes. Through these discussions, this chapter offers critical insights into the therapeutic potential of targeting SLC transporters and ncRNAs in leukemia treatment. It underscores the importance of understanding the genetic and epigenetic landscapes of blood disorders to develop precision medicine approaches. By integrating recent advancements and highlighting gaps in knowledge, this chapter provides a foundation for future research aimed at overcoming the challenges of multidrug resistance and enhancing treatment efficacy in hematological malignancies.

Keywords Epigenetics · Hematopoiesis · Leukemia · Leukemogenesis · Noncoding RNAs · Transporters

Abbreviations

ABC	ATP-binding cassette
ALL	Acute lymphoblastic leukemia
AML	Acute myeloid leukemia
ASIR	Age-standardized incidence rate
ASMR	Age-standardized mortality rate
B-ALL	B-cell precursor acute lymphoblastic leukemia
CircRNAs	Circular RNAs
CLL	Chronic lymphocytic leukemia
CML	Chronic myeloid leukemia
GRNs	Gene regulatory networks
HSCs	Hematopoietic stem cells
LncRNAs	Long noncoding RNAs
LSCs	Leukemia stem cells

MCTs	Monocarboxylate transporters
MDR	Multidrug resistance
MiRNAs	MicroRNAs
NcRNAs	Noncoding RNAs
SLC	Solute carrier
T-ALL	T-cell precursor acute lymphoblastic leukemia
TFs	Transcription factors
TKIs	Tyrosine kinase inhibitors

3.1 Hematopoiesis

Blood is a multifaceted subject that spans various disciplines, which is primarily viewed through its biological and medical functions. The development and differentiation of blood cells, known as hematopoiesis, is a complex process involving the transformation of multipotent hematopoietic stem cells (HSCs) into various specialized blood cell lineages (Fig. 3.1) (Brown et al. 2020). HSCs reside in the bone marrow and are responsible for the production of all blood cell types. These stem cells are multipotent, meaning they have the potential to differentiate into various blood cell lineages, including myeloid and lymphoid cells (Lee and Hong 2020). The differentiation process involves a hierarchical scheme where HSCs first become oligopotent progenitors, which then further differentiate into unipotent progenitors, eventually giving rise to specific blood cell types such as erythrocytes, leukocytes, and platelets (Manso et al. 2024).

In adult humans, it is estimated that between 0.44×10^5 and 2.15×10^5 HSCs are actively involved in producing white blood cells at any given time. By comparison, a single microliter of blood normally contains 4×10^3 to 10×10^3 white blood cells and 3.5×10^6 to 5.2×10^6 red blood cells. Considering the average adult blood volume of 5 to 6 liters, this equates to a total of approximately 2×10^{10} to 6×10^{10} white blood cells and 2×10^{13} to 3×10^{13} red blood cells. On average, about 10^6 new blood cells are generated every second, all of which originate from HSCs. This highlights the rapid turnover of hematopoietic cells, necessitating highly precise biological mechanisms to regulate this dynamic process (Brown et al. 2020).

This process is crucial for maintaining blood homeostasis and involves a series of tightly regulated steps that ensure the proper formation and function of blood cells. During human development, the hierarchy of blood cell lineage differentiation undergoes significant changes. In fetal development, a diverse array of oligopotent progenitors is present, allowing for the intermingling of myeloid, erythroid, and megakaryocytic fates. However, in adults, the hematopoietic hierarchy simplifies into a “two-tier” system dominated by multipotent and unipotent progenitors. This developmental shift reflects changes in the microenvironment and regulatory mechanisms that influence HSCs behavior and lineage commitment over time (Notta et al. 2016).

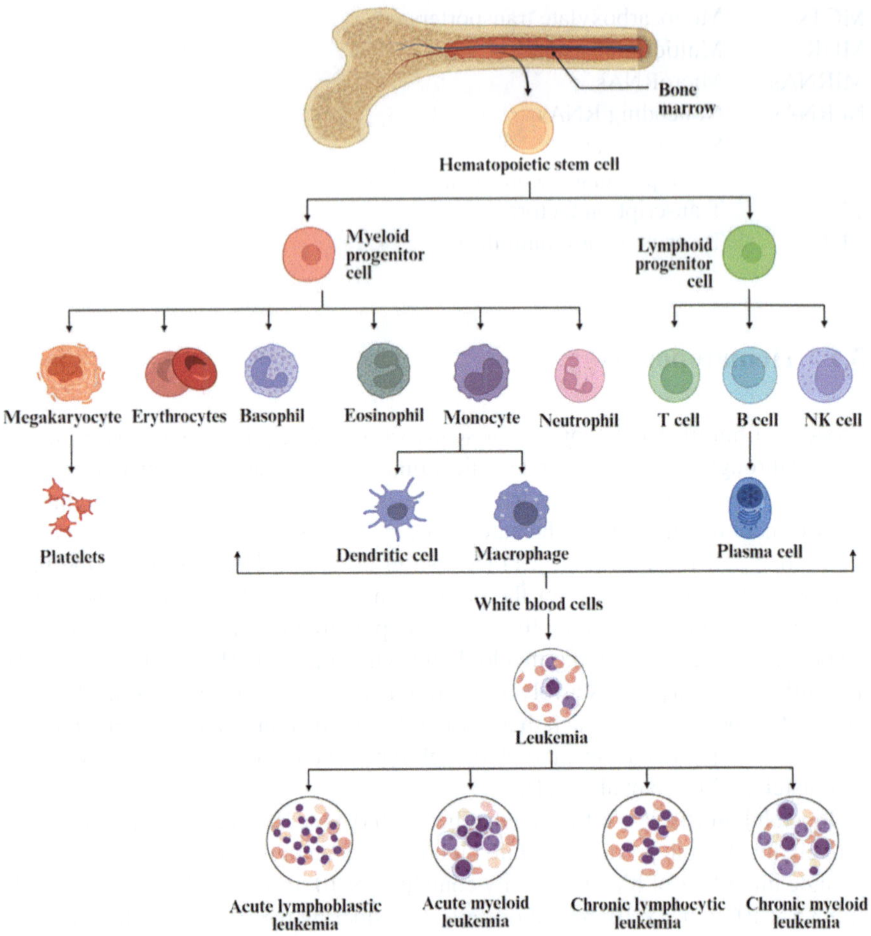

Fig. 3.1 Hematopoiesis depicting the differentiation of HSCs into various blood cell lineages. The figure highlights key intermediate progenitor cell types (e.g., common myeloid progenitors and common lymphoid progenitors) and their roles in normal hematopoiesis. Additionally, it maps the origins of various hematological malignancies. (The image was generated using BioRender)

Lineage commitment is regulated by a combination of intrinsic genetic programs and extrinsic signals from the microenvironment. Key transcription factors, such as stem cell leukemia/T cell acute leukemia 1, play a crucial role in directing HSCs toward specific lineages by activating or repressing target genes (Chagraoui et al. 2018). The process of lineage restriction involves the suppression of alternative cell fates, ensuring that progenitor cells commit to a specific lineage. This is achieved through the action of transcriptional repressors and chromatin-modifying complexes that maintain lineage-specific gene expression patterns (Chagraoui et al. 2018). Some studies suggest that HSCs and their progeny can exhibit a degree of flexibility, allowing them to "change their mind" and adopt alternative developmental pathways under certain conditions. This plasticity is particularly relevant in the context of leukemias, where oncogenic transformations can lock cells into specific lineages, disrupting normal hematopoietic processes (Brown et al. 2020). Understanding these dynamics is crucial for developing therapeutic strategies to address hematological disorders.

3.2 Leukemia

3.2.1 In Numbers

In 2020, the global incidence of leukemia was reported at 474,519 cases, with 311,594 deaths worldwide (Padda et al. 2021), accounting for approximately 2.5% of all new cancer incidence and 3.1% of cancer-related mortality (Huang et al. 2023). The age-standardized incidence rate (ASIR) was 5.4 per 100,000, and the age-standardized mortality rate (ASMR) was 3.3 per 100,000 globally (Huang et al. 2023). Childhood leukemia accounted for 67,008 new cases and 25,080 deaths in 2020, with a global ASIR of 3.4 per 100,000 and an ASMR of 1.3 per 100,000 (Mohammadian-Hafshejani et al. 2024). The burden of leukemia is higher in males than females, with males accounting for a larger proportion of both incidence and mortality rates (Mohammadian-Hafshejani et al. 2024). From 1990 to 2019, there was a decreasing trend in the global ASIR and ASMR for leukemia, with the ASIR showing an estimated annual percentage change of −0.37 and the ASMR decreasing by 15.75% for males and 21.45% for females (Qu et al. 2024). Regions with higher sociodemographic index, such as high-income countries, generally experience a heavier burden of leukemia, particularly in terms of disability-adjusted life years (Li et al. 2023). However, these regions also show a decreasing trend in leukemia-related deaths and disability-adjusted life years, likely due to better healthcare systems (Ou et al. 2020). In contrast, low sociodemographic index regions face a higher proportion of leukemia-related deaths in younger age groups, highlighting the need for targeted healthcare interventions (Lin et al. 2020). Despite the overall decline, some countries, such as Germany, Korea, Japan, Canada, and the United Kingdom, have experienced increasing trends in leukemia incidence (Huang et al. 2023).

3.2.2 Types and Subtypes

Leukemia is a complex group of hematological malignancies characterized by the clonal proliferation of abnormal blood cells. The classification of leukemia into subtypes is crucial for diagnosis, treatment, and prognosis. The four primary subtypes of leukemia are acute lymphoblastic leukemia (ALL), acute myeloid leukemia (AML), chronic lymphocytic leukemia (CLL), and chronic myeloid leukemia

Table 3.1 Types of leukemia (Rafiq et al. 2018; Powsner et al. 2021)

Property	ALL	AML	CLL	CML
Progression	Rapid	Rapid	Slow	Slow
Lineage	Lymphoid	Myeloid	Lymphoid	Myeloid
Cell type affected	Immature B or T cells (Bone marrow) and macrophages	Immature myeloid lineage cells (Bone marrow)	Peripheral B or T cells (Lymph nodes)	Pluripotent hematopoietic stem cells (Bone marrow)
Etiology	Chromosomal aberration resulting in abnormal transcription factors that affect development of B and T cells	Oncogenic mutations impede differentiation, accumulating immature myeloid blasts in bone marrow	Chromosomal deletion or possible somatic hypermutations of postgerminal or naïve B cells	Tyrosine kinase pathway related chromosomal translocation—Philadelphia chromosome
Age group	Children > Adults	Adults > Children	Older adults (55+ age)	Adults (20 to 50), rare in children
Incidence	6000 cases a year	18,000 cases a year	15,000 cases a year	6000 cases a year
Key genetic markers	BCR-ABL1 (Philadelphia chromosome, some cases), TEL-AML1	FLT3-ITD, NPM1 mutations, t(15;17) in APL	TP53 mutations, deletions (13q, 11q), trisomy 12	BCR-ABL1 (Philadelphia chromosome)
Prognosis	Better in children; poorer in adults	Variable; depends on cytogenetics	Variable; poorer with 17p deletion or TP53 mutation	Good with effective tyrosine kinase inhibitors (TKIs) therapy
Prevalence	33%	80%	30%	90%
5-year survival rate	69%	28%	85%	69%

(CML) (Table 3.1). Each subtype has distinct clinical features, epidemiology, and treatment options. Furthermore, each subtype has further subdivisions based on genetic and molecular characteristics.

3.2.2.1 Acute Lymphoblastic Leukemia

ALL is more prevalent in children, making it the most common childhood leukemia (Karrar et al. 2022). In ALL, chromosomal translocations such as TEL/AML1 and PML/RARα are common. These translocations result in fusion proteins that disrupt normal transcriptional regulation, leading to leukemogenesis. ALL is primarily divided into B-cell precursor ALL (B-ALL) and T-cell ALL (T-ALL), each with unique molecular characteristics. Gene expression markers have been identified that differentiate these subtypes and have implications for prognosis and treatment strategies (Nourbakhsh et al. 2024).

3.2.2.2 Acute Myeloid Leukemia

AML prognosis is generally poor, with a 5-year overall survival rate of approximately 22.6% (Dawod and Hanbali 2012). In AML, mutations such as those in FLT3, NPM1, CEBPA, and DNMT3A are common and significantly influence disease progression and prognosis. FLT3 mutations, for instance, lead to constitutive activation of tyrosine kinase signaling, promoting cell proliferation and survival (Singh et al. 2023). Mutations in genes like TET2, IDH1/2, and DNMT3A affect epigenetic regulation, contributing to leukemogenesis by altering DNA methylation and histone modification patterns (Blau 2016). AML is traditionally classified into subtypes M0 to M7 based on the morphology and cytochemistry of leukemic cells. Subtypes M1, M2, and M3 are characterized by the presence of myeloblasts, with specific features distinguishing each subtype. For instance, M3 is associated with promyelocytic leukemia, which has a distinct treatment approach due to its sensitivity to all-*trans* retinoic acid (Ladines-Castro et al. 2016; Azzahra et al. 2023).

3.2.2.3 Chronic Lymphocytic Leukemia

CLL is the most common leukemia in adults, particularly affecting the elderly, with a generally indolent course (Karrar et al. 2022). CLL has a more favorable prognosis, with a median survival of about 10 years (Rodriguez et al. 2015). CLL is characterized by a high frequency of chromosomal aberrations and point mutations, particularly in genes involved in cell cycle regulation and apoptosis, such as TP53, NOTCH1, and SF3B1.

3.2.2.4 Chronic Myeloid Leukemia

CML is more common in adults and progresses slowly, is characterized by the overproduction of mature myeloid cells and by the presence of the Philadelphia chromosome (Karrar et al. 2022). The hallmark of CML is the BCR/ABL fusion gene, resulting from a translocation between chromosomes 9 and 22. This fusion leads to the production of a constitutively active tyrosine kinase that drives uncontrolled cell proliferation.

The incidence of leukemia in India varies significantly with age and type. The incidence of leukemia is notably higher among children aged 0–14 and adults aged 60–85, indicating age as a significant factor in leukemia risk (Rifat et al. 2023). CML and ALL have seen a decrease in mortality rates, whereas AML and CLL have experienced an increase in deaths over the past decades (Nalage et al. 2024). This suggests a shift in the burden of leukemia types, with AML and CLL becoming more prominent concerns.

3.3 Regulators in Leukemia

Leukemia cells exhibit a sparsely interconnected transcriptional hierarchy stabilized by incoherent feed-forward loops. This network is composed of core transcriptional regulators with narrow, nonoverlapping direct transcriptional programs. Disruption of these programs by BET bromodomain and CDK7 inhibitors can alter gene expression dynamics (Harada et al. 2023). Regulators in leukemia play a crucial role in the pathogenesis, progression, and treatment of the disease. These regulators include gene regulatory networks (GRNs), transcription factors (TFs), chromatin regulators, immune regulators, and RNA modifications, each contributing to the complex biology of leukemia.

Leukemia stem cells (LSCs) and HSCs are governed by distinct GRNs. In AML, LSC-specific GRNs include transcription factors like FOXK1 and MEIS1, which differ from the HSC network comprising ETS1, EGR1, RUNX2, and ZNF683. These networks are crucial for maintaining stemness and can be targeted to reduce LSC numbers, as demonstrated by the inhibition of CDK6 in AML samples (Zhang et al. 2024a). Myeloid master regulators such as PU.1 and CEBPA play a role in suppressing leukemogenesis by downregulating cell cycle-related genes in NPM1-mutated leukemia. Activation of these transcription factors induces myeloid differentiation and cell cycle arrest, reducing leukemic cell proliferation (Hamamoto et al. 2023).

Chromatin regulators, including those involved in the nucleosome remodeling and deacetylase complex, are frequently mutated in leukemia. Combinatorial CRISPR knockout screens have identified dependencies within chromatin complexes, highlighting potential therapeutic targets such as the ING5 PHD domain in the KAT7 complex (Najm et al. 2023). N6-methyladenosine (m6A) RNA methylation is another epigenetic modification implicated in AML. Specific m6A regulators, such as METTL14 and YTHDC2, are associated with patient prognosis and can serve as biomarkers for survival outcomes (Liao et al. 2022; Wang et al. 2022). The bone marrow microenvironment in AML is characterized by immune dysregulation, particularly affecting CD8+ T cells. Immune-suppressive factors and downregulation of the NF-κB pathway contribute to T-cell dysfunction. Restoring NF-κB signaling can enhance T-cell effector functions, offering a potential strategy to improve immune responses against AML (Zhang et al. 2023). Noncoding RNAs (ncRNAs) (such as long noncoding RNAs (lncRNAs), microRNAs (miRNAs), and circular RNAs (circRNAs)) are involved in leukemia by modulating gene expression and contributing to therapeutic resistance. Targeting these RNA elements could improve treatment outcomes and provide new avenues for leukemia therapy (Elcheva and Spiegelman 2020; Shi et al. 2023). Posttranscriptional regulation, mediated by RNA-binding proteins, is crucial in leukemia. These proteins control mRNA splicing, stability, and translation, contributing to the disease's development and progression (Prieto and Kharas 2020; Podszywalow-Bartnicka et al. 2021). Epigenetic regulators also play a significant role in leukemia. For instance, the Tudor-domain containing protein SGF29 is involved in the transcription of key AML oncogenes. Its deletion impairs leukemogenesis (Barbosa et al. 2023).

3.4 Role of Transporters in Leukemia

To ensure survival and proper cellular function, cells must regulate their internal environment by controlling the movement of specific molecules, such as metabolites and ions, across the lipid bilayer. This selective transport is crucial for maintaining nutrient levels, waste discharge, and regulating cell volume. Given the diversity of compounds that need to be transported, various transporter proteins have evolved, which can be categorized into four primary superfamilies: (a) ATP-binding cassette (ABC) transporters, (b) ion pumps (ATPases), a(c) ion channels, and (d) solute carrier (SLC) transporters. These transporter systems are essential for maintaining cellular homeostasis and metabolic regulation across different physiological conditions (Hediger et al. 2013; Pizzagalli et al. 2021).

Transporters play a significant role in leukemia, particularly in influencing drug resistance and treatment outcomes. In leukemia, the ABC transporters and SLC transporters are crucial in mediating drug efflux and influx, respectively, which can affect the efficacy of chemotherapy. These transporters are implicated in various types of leukemia, including AML, CML, and ALL. ABC transporters, such as ABCC1, ABCG2, and ABCB1, are known to contribute to multidrug resistance (MDR) in AML by expelling chemotherapeutic drugs from leukemic cells, thereby reducing drug efficacy. High expression levels of these transporters are associated with treatment failure and poor prognosis in AML patients (Vasconcelos et al. 2021; Cerovska et al. 2024). In CML, ABC transporters can affect the distribution and efficacy of TKIs, which are crucial for treatment. These transporters can sequester TKIs in lysosomes, reducing their cytosolic concentrations and contributing to drug resistance (Verhagen et al. 2023). SLC transporters, such as SLC29A1, are involved in the uptake of chemotherapeutic agents like cytarabine in AML. Variants in these transporters can influence drug absorption and efficacy, impacting treatment outcomes (Megías-Vericat et al. 2022). In B-ALL, genetic variants in SLC transporters have been associated with hematologic toxicities due to chemotherapy. These variants can serve as risk markers for adverse events, aiding in risk stratification and therapeutic decision-making (Escalante-Bautista et al. 2023). Figure 3.2 presents five main alterations that ultimately lead to the malignant transformation.

While the role of transporters in leukemia is primarily associated with drug resistance, their involvement in leukemic cell biology suggests potential therapeutic targets. Inhibiting transporter activity could enhance chemotherapy efficacy and disrupt leukemic cell survival mechanisms, offering a dual approach to treatment. Targeting transporters presents a promising therapeutic strategy for the treatment of leukemia, particularly AML. This approach leverages the role of transporters in cellular processes, including drug resistance and cellular homeostasis, to improve treatment outcomes. However, the precise mechanisms and broader implications of transporter activity in leukemia require further investigation.

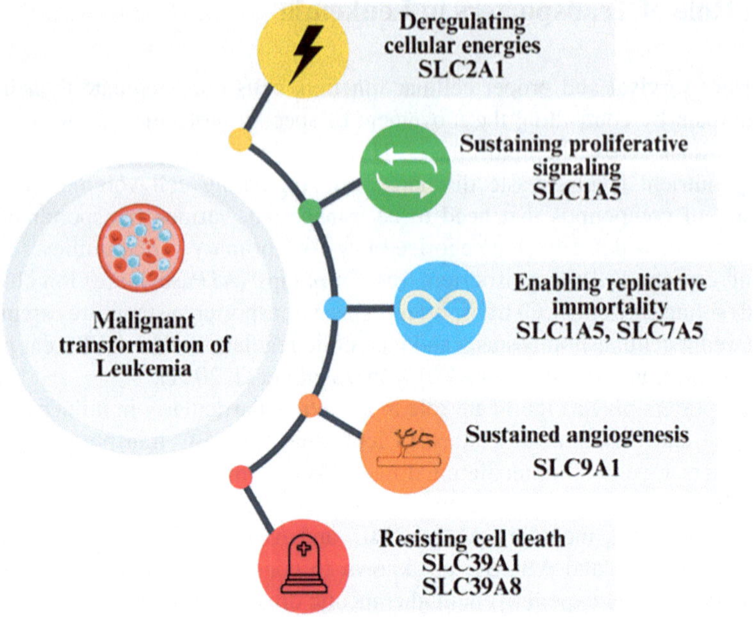

Fig. 3.2 Hallmarks of leukemia and role of SLC transporters. (Adopted with permission El-Gebali et al. 2013)

3.5 Solute Carrier Transporters

As the second largest (more than 450 members organized into 65 families) superfamily of membrane transporters, SLCs facilitate the movement of a wide range of substrates, including ions, neurotransmitters, amino acids, and drugs, across cellular membranes (Table 3.2) (Bai 2021; Alam et al. 2023). SLCs govern chemical exchange across cellular membranes, influencing metabolic regulation and cellular identity (Wiedmer et al. 2024). They are involved in the uptake and efflux of nutrients, affecting the cell's metabolic state (Alam et al. 2023). SLCs can modulate cellular programs, impacting osmolyte balancing and glycosylation (Wiedmer et al. 2024). SLCs at the plasma membrane act as gatekeepers, responding dynamically to metabolic states and influencing cancer progression by regulating metabolism (Pizzagalli et al. 2021). In cancer, SLCs contribute to biomass accumulation, redox homeostasis, and metabolic crosstalk within the tumor microenvironment (Alam et al. 2023). These functions are vital for various physiological processes and is implicated in numerous diseases when dysregulated, highlighting the importance of understanding their structure and function (Bai 2021).

The molecular mechanisms by which SLCs facilitate transport involve specific structural features and transport modes (facilitated diffusion or active transport) that are adapted to their substrate specificity and cellular function. SLCs function as symporters, antiporters, or channels, utilizing ion gradients to facilitate substrate

Table 3.2 Key SLC transporters: substrates, functions, and implications

SLC	Substrate	Function	Remarks	Ref.
SLC1A5	Glutamine	Amino acid transport to support rapid cell proliferation	Overexpression linked to leukemic cell metabolism	Kanai et al. (2013)
SLC2A1	Glucose	Facilitates glucose uptake for glycolysis	High expression in glycolytic phenotypes of leukemia	Mueckler and Thorens (2013)
SLC5A3	Myo-inositol	Regulates inositol transport for cell signaling and survival	Upregulated in certain leukemias, contributing to cell proliferation	Wright (2013)
SLC6A6	Taurine	Maintains cell volume and antioxidant activity	Overexpression supports leukemic cell survival under oxidative stress	Pramod et al. (2013)
SLC7A5	Large neutral amino acids	Supports protein biosynthesis and metabolic signaling	Critical for leukemic cell survival	Fotiadis et al. (2013)
SLC19A1	Folate	Mediates folate uptake for nucleotide biosynthesis	Overexpression linked to antifolate resistance in leukemia	Zhao and Goldman (2013)
SLC22A5	Carnitine	Transports carnitine for fatty acid oxidation	Dysregulation affects energy metabolism in leukemia	Koepsell (2013)
SLC25A1	Citrate	Mitochondrial citrate export for lipid biosynthesis	Associated with metabolic rewiring in leukemia	Palmieri (2013)
SLC25A3	Phosphate	Transports phosphate into mitochondria for ATP production	Alters mitochondrial metabolism in leukemic cells	Palmieri (2013)

transport (He et al. 2009; Bai 2021). Many SLCs function as symporters (or active or cotransporters), utilizing ion gradients (such as H^+ or Na^+) to drive the transport of substrates against their concentration gradient. For example, sodium-glucose cotransporters move both Na^+ ions and glucose into cells. Others act as antiporters or exchangers, exchanging one substrate for another across the membrane (Bai 2021). For example, the Na^+/K^+ ATPase pump actively exchanges sodium ions out of the cell for potassium ions into the cell, maintaining proper ion balance. Some SLCs, such as the SLC23 family, function as uniporters (or channel-like or facilitative or passive transporters), facilitating the passive transport of substrates without coupling to ion gradients (Holzhüter 2022). For instance, glucose transporters facilitate the movement of glucose across the cell membrane without the need for coupling with other ions or molecules.

SLCs have specific substrate-binding sites that determine their substrate specificity. A conserved negatively charged residue in the binding site is crucial for protein stability rather than substrate binding, as seen in SLC23, SLC4, and SLC26 families (Holzhüter 2022). The oligomerization of SLCs, such as the formation of dimers in the SLC26 family, is essential for their activity (Holzhüter 2022). Additionally,

SLCs exhibit various protein folds, such as the Major Facilitator Superfamily and Leucine Transporter folds, which are critical for their transport function. These structures allow for the alternating access model, where the binding site is alternately exposed to either side of the membrane (Bai 2021).

The SLC superfamily is organized into multiple families based on sequence homology and transport function, with each family having distinct roles in cellular transport. For example, SLC16A6 is involved in the transport of monocarboxylates, which are crucial for cellular metabolism (Frommelt et al. 2024). The SLC4 family is responsible for the exchange of ions like HCO^{3-} and Cl^-, playing a role in processes such as CO_2 transport in erythrocytes and pH regulation (Dong et al. 2023). SLC18B1 and SLC22A4 are involved in the transport of polyamines, which are essential for cellular growth and function. These transporters help distribute polyamines to various tissues and intracellular organelles, contributing to cellular homeostasis by regulating polyamine levels (Pochini 2024). SLC22 family, including organic cation and anion transporters, plays a significant role in maintaining central nervous system homeostasis by regulating the transport of neurotransmitters and other compounds across the blood–brain barrier (Farthing and Sweet 2014). SLC26 transporters interact with cystic fibrosis transmembrane conductance regulator to regulate anions (Cl^-, HCO^{3-}, SCN^-, and I^-) transport in epithelial tissues, which is essential for maintaining the ionic balance and proper function of these tissues (Fong 2012). SLC30 and SLC39 families are involved in zinc homeostasis, with SLC30 mediating zinc efflux and SLC39 facilitating zinc uptake. These transporters ensure the proper distribution of zinc within cells (Palmiter and Huang 2004; Schweigel-Röntgen 2014). SLC31 family regulates copper homeostasis, with SLC31A1 acting as a high-affinity copper importer (Schweigel-Röntgen 2014). Furthermore, SLC transporters can adapt to environmental nutrient restrictions by upregulating specific transporters to sustain cellular fitness. For instance, SLC7A3 and SLC7A5 help cells overcome amino acid limitations by increasing uptake, demonstrating the transporters' role in adapting to metabolic bottlenecks (Rebsamen et al. 2022).

3.6 Role of Solute Carrier Transporters in Leukemia

SLC transporters play a key role in regulating what enters and exits cells, making them essential to numerous cellular and physiological processes. These processes include the absorption of glucose, the uptake of water-soluble vitamins essential for critical functions, the transport of metal ions, the intestinal intake of amino acids, and the reabsorption of neurotransmitters into presynaptic neurons (Bharadwaj et al. 2024). SLC transporters are crucial for the uptake of nutrients, such as amino acids and glucose, which are essential for the metabolic demands of leukemic cells (Fig. 3.3) (Hushmandi et al. 2024). Therefore, SLCs transporters play significant roles in the development and progression of leukemia, particularly through their influence on drug resistance, nutrient transport, and cellular metabolism.

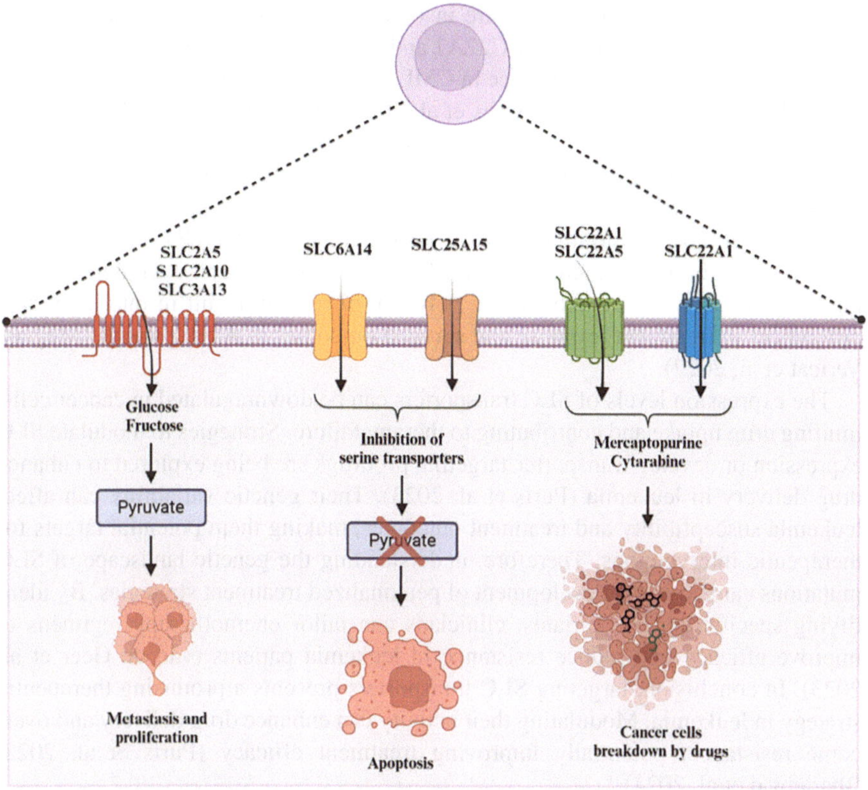

Fig. 3.3 Schematic representation of SLC transporter mechanisms in leukemia metabolism. (The image was generated using BioRender)

These transporters are involved in the uptake and efflux of various substances, including nutrients and chemotherapeutic agents, which are crucial for the survival and proliferation of leukemic cells. Dysregulation of these transporters can promote tumorigenesis by supporting the metabolic needs of cancer cells (Hushmandi et al. 2024). Specific SLC family members, such as SLC2A5, S LC2A10, and SLC3A13, are significantly associated with prognosis of AML patients. SLC2 are involved in glucose uptake and have been identified as potential prognostic biomarkers in AML, highlighting their role in cancer cell survival and proliferation (Bai 2021). Inhibition of serine transporters like SLC6A14 and SLC25A15 can reduce serine uptake and impair cancer cell proliferation, suggesting a therapeutic avenue for targeting metabolic pathways in leukemia (Papalazarou et al. 2023). Monocarboxylate transporters (MCTs) like MCT1 and MCT4 are involved in lactate metabolism, which is essential for the high proliferation rate of leukemic cells. Inhibiting these transporters can impair cell proliferation and increase sensitivity to chemotherapy (Saulle et al. 2021). MCT1 has been identified as a potential target for stratifying AML patients and delivering cytotoxic drugs, highlighting the role of SLC transporters in precision oncology (Saulle et al. 2021).

Additionally, SLC transporters are involved in the uptake of anticancer drugs. Variants in SLC genes, such as SLC22A1 and SLC22A5, have been associated with altered drug response and resistance in CML, affecting the efficacy of TKIs and the risk of BCR-ABL1 mutations (Alves et al. 2022; Marin et al. 2024). Mutations in SLC genes can lead to decreased uptake and retention of chemotherapeutic drugs. For example, downregulation of SLC29A1 has been linked to resistance to azacitidine in AML cell lines, as it reduces the intracellular concentration of the drug (Kutyna et al. 2023). Variations in SLC genes, such as SLCO1B1 and SLC29A1, influence the pharmacokinetics of drugs like anthracyclines and cytarabine, affecting their efficacy and toxicity. These polymorphisms can result in interindividual differences in treatment response and survival outcomes in AML patients (Megías-Vericat et al. 2022).

The expression levels of SLC transporters can be downregulated in cancer cells, limiting drug uptake and contributing to therapy failure. Strategies to modulate SLC expression or develop transporter-targeting prodrugs are being explored to enhance drug delivery in leukemia (Puris et al. 2023). Their genetic variability can affect leukemia susceptibility and treatment outcomes, making them potential targets for therapeutic interventions. Therefore, understanding the genetic landscape of SLC mutations can aid in the development of personalized treatment strategies. By identifying specific genetic variants, clinicians can tailor chemotherapy regimens to improve efficacy and reduce resistance in leukemia patients (van de Geer et al. 2023). In conclusion, targeting SLC transporters presents a promising therapeutic strategy in leukemia. Modulating their activity can enhance drug delivery and overcome resistance, potentially improving treatment efficacy (Puris et al. 2023; Bharadwaj et al. 2024).

3.7 Noncoding RNAs

3.7.1 Types and Importance

NcRNAs are crucial regulators of gene expression, playing significant roles in various biological processes (regulating gene expression, cellular homeostasis, etc.) and diseases (cell signaling in multiple myeloma, with aberrant expression implicated in carcinogenesis, tumor suppression, and disease progression). They function through diverse mechanisms, including chromatin remodeling, transcriptional regulation, and posttranscriptional modulation. The complexity and diversity of ncRNAs, such as miRNAs, lncRNAs, and circRNAs, allow them to interact with DNA, RNA, and proteins, influencing cellular pathways and phenotypic outcomes. MiRNAs are pivotal in gene silencing by inhibiting translation or destabilizing target mRNAs. They are extensively studied in cancer, where they can act as oncogenes or tumor suppressors (Kimura and Tartaglia 2023; Lei et al. 2024). LncRNAs are involved in complex gene regulatory functions, including chromatin remodeling and transcriptional regulation. They can form 3D structures to interact with proteins and nucleic acids, influencing gene expression networks (Kimura and Tartaglia 2023; Zhang et al. 2024b). CircRNAs modulate gene expression by acting as miRNA sponges, thereby influencing disease progression, particularly in tumors (Zhang et al. 2024c). Furthermore, ncRNAs, such as enhancer RNAs and PROMoter uPstream Transcripts (PROMPTs) are produced at transcriptional regulatory elements and can regulate transcription and chromatin structure, impacting cellular fate and genomic stability (Fosseprez and Cuvier 2024).

While ncRNAs are recognized for their regulatory roles, challenges remain in fully understanding their mechanisms and functions. The complexity of ncRNAs interactions and their diverse roles in cellular processes necessitate further research to elucidate their potential in therapeutic applications. Additionally, the development of advanced sequencing technologies continues to uncover new ncRNAs, expanding our understanding of their regulatory networks and biological significance.

3.7.2 Role in Transporters Regulation

NcRNAs play a significant role in the regulation of transporters, particularly in the context of MDR in leukemia. MiRNAs are small ncRNAs that regulate gene expression posttranscriptionally by binding to complementary sequences on target mRNAs, leading to their degradation or translational repression (Carlberg et al. 2023). They have been shown to directly or indirectly regulate the expression of ABC transporters, such as ABCG2 and ABCB1, which are key players in MDR in leukemic cells (Chae et al. 2022; To et al. 2024). MiRNAs can alter the pharmacokinetics of drugs by modulating transporter expression, thereby affecting drug efficacy and toxicity (Chae et al. 2022). LncRNAs can influence the expression of ABC transporters by acting as molecular sponges for miRNAs, thereby preventing miRNAs from binding to their target mRNAs (To et al. 2024). LncRNAs have been implicated in the upregulation of ABC transporters in MDR leukemic cells, contributing to the reduced efficacy of chemotherapy (To et al. 2024).

Furthermore, ncRNAs also play a significant role in the regulation of SLC transporters in leukemia cells, influencing various cellular processes such as proliferation, differentiation, drug transport, drug resistance, and cellular detoxification. These ncRNAs are involved in the modulation of gene expression, which is crucial for the pathogenesis and progression of leukemia. Their regulatory functions extend to the control of SLC transporters, which are essential for maintaining cellular homeostasis and mediating drug transport, thereby impacting the efficacy of leukemia treatments. The regulation of SLC transporters by ncRNAs is complex and involves interactions at multiple levels, including epigenetic, transcriptional, and posttranscriptional modifications.

DNA methylation is a key epigenetic mechanism that affects the expression of SLC transporters. For instance, hypermethylation of SLC22A4 and SLC22A3 genes leads to their downregulation in AML, contributing to drug resistance and poor prognosis. Treatment with hypomethylating agents can restore the expression of these transporters, enhancing drug uptake and sensitivity (Buelow et al. 2021; Gu et al. 2022). The methylation status of SLC transporters such as SLC22A3 is a predictor of prognosis in AML. Demethylation treatments can activate these transporters, enhancing drug uptake and potentially improving patient outcomes (Gu et al. 2022). In CLL, the expression of SLC transporters like SLC22A1 (OCT1) is higher in cancer cells compared to healthy lymphocytes. This differential expression can be targeted to overcome chemoresistance by using transporter-specific cytostatics, thereby improving drug efficacy (Gupta 2009).

MicroRNAs are small ncRNAs that regulate gene expression posttranscriptionally by binding to the 3′ untranslated regions of target mRNAs, leading to their degradation or translational repression. Table 3.3 presents few examples. In leukemia, miRNAs have been shown to modulate the expression of genes involved in drug resistance, including those encoding SLC transporters. The dysregulation of miRNAs can lead to altered expression of SLC transporters, contributing to MDR in leukemia cells. For instance, the loss of miR-15a/16–1 in CLL results in the accumulation of antiapoptotic proteins, which can indirectly affect the function of SLC transporters involved in drug efflux (Balatti and Croce 2022).

LncRNAs can influence the transcription of SLC transporter genes by acting as molecular scaffolds or decoys for transcription factors and chromatin-modifying complexes. This regulation is crucial for maintaining the balance of solutes and drugs within leukemia cells (Schwarzer et al. 2017; Ng et al. 2019). LncRNAs are also involved in the pathogenesis of AML by modulating the expression of genes that control cell proliferation and differentiation, including those encoding SLC transporters. This modulation can affect the sensitivity of leukemia cells to chemotherapeutic agents (Ng et al. 2019; Wang et al. 2024).

Table 3.3 SLC and their associated miRNA in leukemia and other diseases reported in literature

SLC	miRNA	Disease	Remarks	Ref.
SLC2A1	miR-144	Leukemia and breast cancer	miR-144 inhibits SLC2A1 expression, impacting glucose uptake and glycolysis	Xu et al. (2022)
SLC25A1	miR-33a/b	Leukemia and metabolic disorders	miR-33a/b downregulates SLC25A1, altering citrate metabolism	Koppula et al. (2021)
SLC19A1	miR-21	ALL	miR-21 upregulation linked to antifolate resistance by targeting SLC19A1	Wang et al. (2018)
SLC43A2	miR-193a-3p	AML	miR-193a-3p suppresses SLC43A2, impacting amino acid transport and growth	Grossi et al. (2017)
SLC16A3	miR-124	Glioblastoma and leukemia	miR-124 suppresses SLC16A3, altering lactate export and metabolic adaptation	Wu et al. (2018), Hou et al. (2019)

CircRNAs can act as miRNA sponges, sequestering miRNAs and preventing them from binding to their target mRNAs. This mechanism can indirectly regulate the expression of SLC transporters by modulating the availability of miRNAs that target these transporters (Bhattacharya and Gutti 2022). By affecting the expression of miRNAs and their target genes, circRNAs can influence the homeostasis of leukemia cells, impacting processes such as drug uptake and resistance (Bhattacharya and Gutti 2022).

While the role of ncRNAs in regulating SLC transporters in leukemia is not yet well-documented, it is important to consider the broader implications of these findings. The dysregulation of ncRNAs can lead to altered cellular processes beyond drug resistance, affecting the overall pathogenesis and progression of leukemia. While the therapeutic potential of targeting ncRNA-SLC transporter interactions is promising, challenges remain. The complexity of ncRNA regulation and the diverse roles of SLC transporters in different tissues necessitate a nuanced approach to therapy development. Additionally, the poor genetic conservation of lncRNAs between species complicates the translation of findings from model organisms to humans (Ye et al. 2021). Nonetheless, continued research into the molecular mechanisms governing ncRNA-SLC interactions could unlock new therapeutic possibilities for diseases characterized by dysregulated nutrient and ion homeostasis.

3.8 Conclusion

The intricate role of SLC transporters and ncRNAs in blood disorders presents a promising yet challenging research frontier. Despite significant advances, the current understanding of these molecular regulators remains incomplete, particularly regarding their specific interactions and downstream effects in different leukemia subtypes. Moreover, variability in experimental conditions, models, and genetic backgrounds has contributed to inconsistent findings across studies. Limitations in the availability of reliable, high-throughput functional assays and challenges in translating findings from preclinical models to clinical settings further hinder progress. Addressing these gaps is crucial for realizing the full therapeutic potential of SLC transporters and ncRNA-targeted strategies.

Looking forward, a multifaceted approach integrating advanced technologies such as CRISPR-based gene editing, single-cell sequencing, and multi-omics profiling is essential to unravel the complexity of transporter regulation and ncRNA functionality. Investigations into the epigenetic and posttranscriptional mechanisms that modulate transporter activity in leukemic cells could uncover novel biomarkers and therapeutic targets. Additionally, collaborative efforts to standardize experimental protocols and validate findings in clinically relevant models will be pivotal. Expanding the scope to include underexplored transporters and ncRNAs, coupled with a focus on patient-specific variations, holds the promise of paving the way for precision medicine in leukemia treatment.

Acknowledgments This work was supported by University of Hyderabad-Institution of Eminence Grant (No: UOH-IOE-RC3-21-006); Science and Engineering Research Board (SERB, EEQ/2022/00211); Indian Council of Medical Research (ICMR, No: EMDR/SG/12/2023/-1701); DBT-BUILDER, DST-FIST, ANRF-PAIR and Council of Scientific and Industrial Research (CSIR, No: 27(0343)/19/EMR-II) grants of Government of India.

Conflict of Interest All authors declare that they have no conflict of interest.

Ethical Approval This chapter does not contain any studies with human participants or animals performed by any of the authors.

Author Contributions Conceptualization, M.S., A.M. and R.K.G.; writing—original draft preparation, M.S. and A.V.; writing—review and editing, M.S., A.V., A. M., and R.K.G.; visualization, M.S.; supervision, R.K.G.; project administration, R.K.G. All authors have read and agreed to the submitted version of the manuscript.

References

Alam S et al (2023) Membrane transporters in cell physiology, cancer metabolism and drug response. Dis Model Mech 16(11):dmm050404. https://doi.org/10.1242/dmm.050404

Alves R et al (2022) Genetic variants of ABC and SLC transporter genes and chronic myeloid leukaemia: impact on susceptibility and prognosis. Int J Mol Sci 23(17):9815–9815. https://doi.org/10.3390/ijms23179815

Azzahra SJ et al (2023) Identification of acute myeloid leukemia (AML) subtypes: M1, M2, M3 on white blood cells using microscopic images. In: IES 2023 - international electronics symposium: unlocking the potential of immersive technology to live a better life, proceeding. Institute of Electrical and Electronics Engineers Inc., pp 454–459. https://doi.org/10.1109/IES59143.2023.10242476

Bai X (2021) Progress in structural biology of solute carriers. Curr Mol Biol Rep 7(2):9–19. https://doi.org/10.1007/S40610-021-00144-5

Balatti V, Croce CM (2022) Small non-coding RNAs in leukemia. Cancers, (Vol 14)14:509–509. https://doi.org/10.3390/cancers14030509

Barbosa K et al (2023) High density domain-focused CRISPR screens reveal novel epigenetic regulators of HOX/MEIS activation in acute myeloid leukemia. bioRxiv:2022.12.12.519332. https://doi.org/10.1101/2022.12.12.519332

Bharadwaj R et al (2024) Targeting solute carrier transporters (SLCs) as a therapeutic target in different cancers. Diseases 12:63. https://doi.org/10.3390/diseases12030063

Bhattacharya M, Gutti RK (2022) Non-coding RNAs: are they the protagonist or antagonist in the regulation of leukemia? Am J Transl Res 14(3):1406–1432. Available at: https://pmc.ncbi.nlm.nih.gov/articles/PMC8991171/. Accessed 15 January 2025

Blau OV (2016) Genetic mutations in acute myeloid leukemia. Clin Oncohematol 9(3):245–256. https://doi.org/10.21320/2500-2139-2016-9-3-245-256

Brown G, Sánchez L, Sánchez-García I (2020) Lineage decision-making within normal haematopoietic and leukemic stem cells. Int J Mol Sci 21:2247. https://doi.org/10.3390/ijms21062247

Buelow DR et al (2021) DNA methylation-based epigenetic repression of SLC22A4 promotes resistance to cytarabine in acute myeloid leukemia. Clin Transl Sci 14(1):137–142. https://doi.org/10.1111/cts.12861

Carlberg C, Velleuer E, Molnár F (2023) Regulatory impact of non-coding RNA. Mol Med:131–144. https://doi.org/10.1007/978-3-031-27133-5_9

Cerovska E et al (2024) ABC transporters are predictors of treatment failure in acute myeloid leukaemia. Biomed Pharmacother 170:115930–115930. https://doi.org/10.1016/j.biopha.2023.115930
Chae YJ et al (2022) Regulation of drug transporters by microRNA and implications in disease treatment. J Pharm Investig 52:23–47. https://doi.org/10.1007/s40005-021-00538-8
Chagraoui H et al (2018) SCL/TAL1 cooperates with Polycomb RYBP-PRC1 to suppress alternative lineages in blood-fated cells. Nat Commun 9(1):5375–5375. https://doi.org/10.1038/s41467-018-07787-6
Dawod M, Hanbali A (2012) Prognosis and survival in acute myelogenous leukemia. In: Myeloid leukemia - clinical diagnosis and treatment. Intech. https://doi.org/10.5772/27092
Dong J et al (2023) Research progress on structure, function and disease correlation of solute carrier family 4. Sheng Li Xue Bao:137–150. https://doi.org/10.13294/j.aps.2023.0008
Elcheva IA, Spiegelman VS (2020) The role of cis-and trans-acting rna regulatory elements in leukemia. Cancers:1–30. https://doi.org/10.3390/cancers12123854
El-Gebali S et al (2013) Solute carriers (SLCs) in cancer. Mol Aspects Med:719–734. https://doi.org/10.1016/j.mam.2012.12.007
Escalante-Bautista D et al (2023) Association between genetic variants of membrane transporters and the risk of high-grade hematologic adverse events in a cohort of Mexican children with B-cell acute lymphoblastic leukemia. Front Oncol 13. https://doi.org/10.3389/fonc.2023.1276352
Farthing C, Sweet D (2014) Expression and function of organic cation and anion transporters (SLC22 family) in the CNS. Curr Pharm Des 20(10):1472–1486. https://doi.org/10.2174/13816128113199990456
Fong P (2012) CFTR-SLC26 transporter interactions in epithelia. Biophys Rev:107–116. https://doi.org/10.1007/s12551-012-0068-9
Fosseprez O, Cuvier O (2024) Uncovering the functions and mechanisms of regulatory elements-associated non-coding RNAs. Biochim Biophys Acta:195059–195059. https://doi.org/10.1016/j.bbagrm.2024.195059
Fotiadis D, Kanai Y, Palacín M (2013) The SLC3 and SLC7 families of amino acid transporters. Mol Aspects Med:139–158. https://doi.org/10.1016/j.mam.2012.10.007
Frommelt F et al (2024) The solute carrier superfamily interactome. Cold Spring Harb Lab. https://doi.org/10.1101/2024.09.30.615192
Grossi I et al (2017) Biological function of MicroRNA193a-3p in health and disease. Int J Genom 2017. https://doi.org/10.1155/2017/5913195
Gu Y et al (2022) SLC22A3 methylation-mediated gene silencing predicts adverse prognosis in acute myeloid leukemia. Clin Epigenetics 14(1). https://doi.org/10.1186/s13148-022-01373-w
Gupta S (2009) Expression of SLC transporters in chronic lymphocytic leukaemia cells and their interaction with cytostatics. Georg-August-Universität Göttingen. https://doi.org/10.53846/goediss-3480
Hamamoto K et al (2023) Myeloid master regulators PU.1 and CEBPA suppress leukemogenesis through downregulating cell cycle related genes in NPM1 mutated leukemia. Blood 142(Supplement 1):5690–5690. https://doi.org/10.1182/blood-2023-189988
Harada T et al (2023) Leukemia core transcriptional circuitry is a sparsely interconnected hierarchy stabilized by incoherent feed-forward loops. bioRxiv [Preprint]. https://doi.org/10.1101/2023.03.13.532438
He L, Vasiliou K, Nebert DW (2009) Analysis and update of the human solute carrier (SLC) gene superfamily. Hum Genomics 3(2):195–206. https://doi.org/10.1186/1479-7364-3-2-195
Hediger MA et al (2013) The ABCs of membrane transporters in health and disease (SLC series): introduction. Mol Aspects Med:95–107. https://doi.org/10.1016/j.mam.2012.12.009
Holzhüter KEA (2022) Substrate translocation in SLC23 and SLC26 transporters. Universitätsbibliothek Johann Christian Senckenberg. https://doi.org/10.21248/gups.65827

Hou L et al (2019) Interfering cellular lactate homeostasis overcomes Taxol resistance of breast cancer cells through the microRNA-124-mediated lactate transporter (MCT1) inhibition. Cancer Cell Int 19(1):1–12. https://doi.org/10.1186/s12935-019-0904-0
Huang J et al (2023) Disease burden, risk factors, and trends of lip, oral cavity, pharyngeal cancers: a global analysis. Cancer Med 12(17):18153–18164. https://doi.org/10.1002/cam4.6391
Hushmandi K et al (2024) Amino acid transporters within the solute carrier superfamily: under-appreciated proteins and novel opportunities for cancer therapy. Mol Metab:101952–101952. https://doi.org/10.1016/j.molmet.2024.101952
Kanai Y et al (2013) The SLC1 high-affinity glutamate and neutral amino acid transporter family. Mol Aspects Med 34(2–3):108–120. https://doi.org/10.1016/J.MAM.2013.01.001
Karrar HR et al (2022) Leukemia perspective in current practice. World Fam Med J 20(13). https://doi.org/10.5742/mewfm.2023.95251581
Kimura T, Tartaglia GG (2023) Editorial: mechanisms of action of natural antisense transcripts on the post-transcriptional regulation of sense protein coding gene expression during development and in cancer. Front Mol Biosci. https://doi.org/10.3389/fmolb.2023.1268124
Koepsell H (2013) The SLC22 family with transporters of organic cations, anions and zwitterions. Mol Aspects Med:413–435. https://doi.org/10.1016/j.mam.2012.10.010
Koppula P, Zhuang L, Gan B (2021) Cystine transporter SLC7A11/xCT in cancer: ferroptosis, nutrient dependency, and cancer therapy. Protein Cell 12(8):599–620. https://doi.org/10.1007/s13238-020-00789-5
Kutyna MM et al (2023) Solute carrier family 29A1 mediates in vitro resistance to Azacitidine in acute myeloid leukemia cell lines. Int J Mol Sci 24(4):3553–3553. https://doi.org/10.3390/ijms24043553
Ladines-Castro W et al (2016) Morphology of leukaemias. Revista Médica del Hospital General de México 79(2):107–113. https://doi.org/10.1016/j.hgmx.2015.06.007
Lee JY, Hong SH (2020) Hematopoieticstem cells and their roles in tissueregeneration. Int J Stem Cells 13(1):1. https://doi.org/10.15283/ijsc19127
Lei M et al (2024) Roles of noncoding RNAs in multiple myeloma. Leuk Res 146:107593. https://doi.org/10.1016/j.leukres.2024.107593
Li Y et al (2023) Global burden and disparities in hematologic malignancies: a comparative analysis from 1990 to 2019. Blood 142(Supplement 1):7280–7280. https://doi.org/10.1182/blood-2023-173597
Liao X et al (2022) Comprehensive analysis of the landscape characteristics of m6A regulators in acute myeloid leukemia. Res Square. https://doi.org/10.21203/rs.3.rs-1511744/v1
Lin Q et al (2020) Global, regional, and National Burden of chronic myeloid leukemia, 1990–2017: a systematic analysis for the global burden of disease study 2017. Front Oncol 10:580759. https://doi.org/10.3389/fonc.2020.580759
Manso BA, Rodriguez y Baena A, Forsberg EC (2024) From hematopoietic stem cells to platelets: unifying differentiation pathways identified by lineage tracing mouse models. Cells:704. https://doi.org/10.3390/cells13080704
Marin JJG et al (2024) Impact of genetic variants in the solute carrier (SLC) genes encoding drug uptake transporters on the response to anticancer chemotherapy. Cancer Drug Resistance [Preprint]. https://doi.org/10.20517/cdr.2024.42
Megías-Vericat JE et al (2022) Systematic review of pharmacogenetics of ABC and SLC transporter genes in acute myeloid leukemia. Pharmaceutics:878–878. https://doi.org/10.3390/pharmaceutics14040878
Mohammadian-Hafshejani A, Farber IM, Kheiri S (2024) Global incidence and mortality of childhood leukemia and its relationship with the Human Development Index. PLoS One 19:e0304354. https://doi.org/10.1371/journal.pone.0304354
Mueckler M, Thorens B (2013) The SLC2 (GLUT) family of membrane transporters. Mol Aspects Med:121–138. https://doi.org/10.1016/j.mam.2012.07.001
Najm FJ et al (2023) Chromatin complex dependencies reveal targeting opportunities in leukemia. Nat Commun 14(1). https://doi.org/10.1038/s41467-023-36150-7

Nalage DN et al (2024) Leukemia in India: insights into incidence, prevalence, mortality, and disability-adjusted life years. Cureus [Preprint]. https://doi.org/10.7759/cureus.62557

Ng M, Heckl D, Klusmann JH (2019) The regulatory roles of long noncoding RNAs in acute myeloid leukemia. Front Oncol 9:570–570. https://doi.org/10.3389/FONC.2019.00570/FULL

Notta F et al (2016) Distinct routes of lineage development reshape the human blood hierarchy across ontogeny. Science 351(6269). https://doi.org/10.1126/science.aab2116

Nourbakhsh M et al (2024) Data-driven discovery of gene expression markers distinguishing pediatric acute lymphoblastic leukemia subtypes. bioRxiv:2024.02.26.582026. https://doi.org/10.1101/2024.02.26.582026

Ou Z et al (2020) Analysis of the Global Burden of Disease study highlights the trends in death and disability-adjusted life years of leukemia from 1990 to 2017. Cancer Commun 40(11):598–610. https://doi.org/10.1002/cac2.12094

Padda J et al (2021) Metabolic acidosis in leukemia. Cureus 13(9). https://doi.org/10.7759/cureus.17732

Palmieri F (2013) The mitochondrial transporter family SLC25: identification, properties and physiopathology. Mol Aspects Med:465–484. https://doi.org/10.1016/j.mam.2012.05.005

Palmiter RD, Huang L (2004) Efflux and compartmentalization of zinc by members of the SLC30 family of solute carriers. Pflugers Arch:744–751. https://doi.org/10.1007/s00424-003-1070-7

Papalazarou V et al (2023) Phenotypic profiling of solute carriers characterizes serine transport in cancer. Nat Metab 5(12):2148–2168. https://doi.org/10.1038/s42255-023-00936-2

Pizzagalli MD, Bensimon A, Superti-Furga G (2021) A guide to plasma membrane solute carrier proteins. FEBS J:2784–2835. https://doi.org/10.1111/febs.15531

Pochini L (2024) Involvement of mammalian SoLute Carriers (SLC) in the traffic of polyamines. Front Mol Biosci. https://doi.org/10.3389/fmolb.2024.1452184

Podszywalow-Bartnicka P, Wolczyk M, Piwocka K (2021) Targeting of post-transcriptional regulation as treatment strategy in acute leukemia. Acute Leuk. https://doi.org/10.5772/intechopen.94421

Powsner EH, Harris JC, Day ES (2021) Biomimetic nanoparticles for the treatment of hematologic malignancies. Adv NanoBiomed Res:2000047. https://doi.org/10.1002/anbr.202000047

Pramod AB et al (2013) SLC6 transporters: structure, function, regulation, disease association and therapeutics. Mol Aspects Med:197–219. https://doi.org/10.1016/j.mam.2012.07.002

Prieto C, Kharas MG (2020) RNA regulators in leukemia and lymphoma. Cold Spring Harb Perspect Med 10(5). https://doi.org/10.1101/cshperspect.a034967

Puris E, Fricker G, Gynther M (2023) The role of solute carrier transporters in efficient anti-cancer drug delivery and therapy. Pharmaceutics:364–364. https://doi.org/10.3390/pharmaceutics15020364

Qu X et al (2024) Global, regional, and national burdens of leukemia from 1990 to 2019: a systematic analysis of the global burden of disease in 2019 based on the APC model. Cancer Med 13(17). https://doi.org/10.1002/cam4.7150

Rafiq S et al (2018) Molecular targets of curcumin and future therapeutic role in leukemia. J Biosci Med 06(04):33–50. https://doi.org/10.4236/jbm.2018.64003

Rebsamen M et al (2022) Gain-of-function genetic screens in human cells identify SLC transporters overcoming environmental nutrient restrictions. Life Sci Alliance 5(11):e202201404–e202201404. https://doi.org/10.26508/LSA.202201404

Rifat RH et al (2023) Incidence, mortality, and epidemiology of leukemia in South Asia: an ecological study. Open J Epidemiol 13(01):73–82. https://doi.org/10.4236/ojepi.2023.131006

Rodriguez MA, Ballas L, Simar K (2015) Hematologic cancer survivorship management: lymphoma. In: Advances in cancer survivorship management. Springer, New York, pp 201–218. https://doi.org/10.1007/978-1-4939-0986-5_12

Saulle E et al (2021) Targeting lactate metabolism by inhibiting MCT1 or MCT4 impairs leukemic cell proliferation, induces two different related death-pathways and increases chemotherapeutic sensitivity of acute myeloid leukemia cells. Front Oncol 10:621458–621458. https://doi.org/10.3389/fonc.2020.621458

Schwarzer A et al (2017) The non-coding RNA landscape of human hematopoiesis and leukemia. Nat Commun 8(1):218–218. https://doi.org/10.1038/s41467-017-00212-4

Schweigel-Röntgen M (2014) The families of zinc (SLC30 and SLC39) and copper (SLC31) transporters. Curr Top Membr:321–355. https://doi.org/10.1016/B978-0-12-800223-0.00009-8

Shi H et al (2023) Long non-coding RNAs regulate treatment outcome in leukemia: what have we learnt recently? Cancer Med:13966–13977. https://doi.org/10.1002/cam4.6027

Singh H, Kumar M, Kanungo H (2023) Role of gene mutations in acute myeloid leukemia: a review article. Global Med Genet 10(02):123–128. https://doi.org/10.1055/s-0043-1770768

To KKW et al (2024) Utilizing non-coding RNA-mediated regulation of ATP binding cassette (ABC) transporters to overcome multidrug resistance to cancer chemotherapy. Drug Resist Updat 73. https://doi.org/10.1016/j.drup.2024.101058

van de Geer WS et al (2023) Identifying somatic changes in drug transporters using whole genome and transcriptome sequencing data of advanced tumors. Biomed Pharmacother 159:114210. https://doi.org/10.1016/j.biopha.2022.114210

Vasconcelos FC et al (2021) Update on drug transporter proteins in acute myeloid leukemia: pathological implication and clinical setting. Crit Rev Oncol Hematol:103281–103281. https://doi.org/10.1016/j.critrevonc.2021.103281

Verhagen NE et al (2023) Transporter-mediated cellular distribution of tyrosine kinase inhibitors as a potential resistance mechanism in chronic myeloid leukemia. Pharmaceutics. https://doi.org/10.3390/pharmaceutics15112535

Wang SM et al (2018) MiR-595 suppresses the cellular uptake and cytotoxic effects of methotrexate by targeting SLC19A1 in CEM/C1 cells. Basic Clin Pharmacol Toxicol 123(1):8–13. https://doi.org/10.1111/bcpt.12966

Wang L et al (2022) A novel prognostic model based on N6-methyladenosine regulators predicts the prognosis of acute myeloid leukemia. Res Square. https://doi.org/10.21203/rs.3.rs-1518379/v1

Wang X et al (2024) Functions, mechanisms, and therapeutic implications of noncoding RNA in acute myeloid leukemia. Fundam Res [Preprint]. https://doi.org/10.1016/j.fmre.2023.04.012

Wiedmer T et al (2024) Metabolic mapping of the human solute carrier superfamily. Cold Spring Harbor Laboratory. https://doi.org/10.1101/2024.09.23.614124

Wright EM (2013) Glucose transport families SLC5 and SLC50. Mol Aspects Med:183–196. https://doi.org/10.1016/j.mam.2012.11.002

Wu DH et al (2018) MiR-124 suppresses pancreatic ductal adenocarcinoma growth by regulating monocarboxylate transporter 1-mediated cancer lactate metabolism. Cell Physiol Biochem 50(3):936–951. https://doi.org/10.1159/000494477

Xu Y et al (2022) miRNA-199a-5p/SLC2A1 axis regulates glucose metabolism in non-small cell lung cancer. J Cancer 13(7):2352–2361. https://doi.org/10.7150/jca.67990

Ye WC et al (2021) Potential therapeutic targeting of lncRNAs in cholesterol homeostasis. Front Cardiovasc Med:688546–688546. https://doi.org/10.3389/fcvm.2021.688546

Zhang Y et al (2023) Dysregulation of immune regulators in AML bone marrow promotes dysfunction of CD8 + T cells. Blood 142(Supplement 1):5616–5616. https://doi.org/10.1182/blood-2023-187900

Zhang B et al (2024a) Distinct gene regulatory networks govern hematopoietic and leukemia stem cells. Cold Spring Harb Lab. https://doi.org/10.1101/2024.09.28.614980

Zhang J et al (2024b) New mechanism of LncRNA: in addition to act as a ceRNA. Non-coding RNA Res:1050–1060. https://doi.org/10.1016/j.ncrna.2024.06.002

Zhang YJ et al (2024c) Comprehensive review for non-coding RNAs: from mechanisms to therapeutic applications. Biochem Pharmacol:116218. https://doi.org/10.1016/j.bcp.2024.116218

Zhao R, Goldman ID (2013) Folate and thiamine transporters mediated by facilitative carriers (SLC19A1-3 and SLC46A1) and folate receptors. Mol Aspects Med:373–385. https://doi.org/10.1016/j.mam.2012.07.006

Chapter 4
Reconnaissance of the Good, the Bad, and the Ugly Role of tRNAs in Carcinogenesis and Metastasis

Malik Ali Asghar, Mehraj Abbasov, Rukset Attar, and Ammad Ahmad Farooqi

Contents

Abstract Our current knowledge of the translational mechanisms and the detailed structural insights of its components have highlighted the characteristically exclusive role of tRNAs and aminoacyl-tRNA synthetase diversity in the evolution of the genetic codes. Phenomenal advancements in mass spectrometry and high-throughput sequencing have enabled the researchers in developing a better understanding of the complex landscape of tRNA modifications. Emerging evidence has started to shed light on linchpin role of tRNAs in carcinogenesis and metastatic dissemination. In this chapter, we have provided an overview of the role of tRNAs, aminoacyl-tRNA synthetases, and tRNA-derived small RNAs in the onset and progression of cancer.

Keywords Carcinogenesis · tRNAs · Metastasis · Noncoding RNA

M. A. Asghar · A. A. Farooqi (✉)
Institute of Biomedical and Genetic Engineering (IBGE), Islamabad, Pakistan

M. Abbasov
Department of Molecular Genetics and Genomics, Genetic Resources Institute of the Ministry of Science and Education of the Republic of Azerbaijan, Baku, Azerbaijan

R. Attar (✉)
Department of Obstetrics and Gynecology, Faculty of Medicine, Yeditepe University Hospital, Yeditepe University, Istanbul, Turkey

M. F. Ullah, A. Ahmad (eds.), *Non-coding RNAs*, Experientia Supplementum 115, https://doi.org/10.1007/978-3-032-06948-1_4

Abbreviations

4-NQO-4	Nitroquinoline-1-oxide
KEOPS	Kinase, endopeptidase, and other proteins of small size
MLASA	Myopathy, lactic acidosis, and sideroblastic anemia
NAT10	Acetyltransferase 10
SerRS	Seryl-tRNA synthetase
tRNA	Transfer RNA

4.1 Introduction

With a growing prevalence of tRNA defects in different diseases, tRNAs have gradually occupied the central stage because of their fundamental role in the translational machinery. tRNA is a group of noncoding RNA first discovered in the 1950s by Paul C. Zamecnick (Kresge et al. 2005). They act as an adaptor molecule bridging the transcriptome to the proteome by transporting amino acids to the messenger RNA relevant to the three base codons on ribosomes (Schimmel 2018). tRNA molecules are highly conserved across species owing to their functional significance (Guy and Phizicky 2014).

tRNA molecules form a clover leaf-shaped structure consisting of anticodon loop, acceptor stem, D arm, dihydrouridine (D) loop, anticodon arm, variable loop, TΨC (T), and T arm (Chujo and Tomizawa 2021). The D loop and the T loop in the clover leaf-like structure fold together to form an L-shaped-like structure (Barends et al. 2002). The acceptor arm accepts amino acid at the 3′ end is responsible for carrying it to the ribosome (Carter and Wolfenden 2015) where the anticodon loop is responsible for the recognizing and binding to the specific codon on mRNA at the A site on ribosome (Murakami et al. 2009).

The main function of tRNA is to facilitate translations via deciphering the three-letter codon on mRNA and to transport specific amino acid against it. tRNA achieves this feat via aminoacyl transferases, which recognize the codon triplet on mRNA and match the correct amino acid on tRNA. tRNA bound with the amino acid is captured by the elongation complex, where it facilitates the transfer of amino acid. The first and second position base of codon of mRNA strictly obey Watson crick base pairing with second and third base of anticodon while the third base of codon and the first base of tRNA may be modified to obey wobble base pairing, which allows for 61 combinations of codons to be read by limited number of tRNAs (Crick 1966).

Functionality of tRNA depends on several modifications catalyzed via tRNA modifying enzymes. The structure of tRNA is such that it can accommodate over 170 modifications (Barbieri and Kouzarides 2020). Most important of these modifications include pseudouridylation, 2′-O-methylation, cytosine-5-methylation, cytosine-3-modification, queuosinylation, N6-isopentenyladenosine and

N6-methyl-threonylcarbamoyladenosine modifications, N1-methyladenosine, N7-methylguanosine, and 5-methoxycarbonylmethyl-2-thiouridine (Wu et al. 2024b).

tRNA modifications are functionally significant for normal functions of tRNAs. However, dysregulation of these modifications via defects in tRNA structure or tRNA modifying enzymes termed as tRNA modopathies have clinically significant consequences in the onset and progression of many cancers, malignancies, and neurological disorders (Chujo and Tomizawa 2021). Functional studies using animal models have identified various defects in tRNA modification enzymes. The aberrations could be due to defects in methylation, aminoacylation, or the genes encoding tRNAs (Suzuki 2021). Landmark discoveries continue to push the boundaries and chemical space of tRNA modifications.

Mutations in elongator complex proteins (ELP1 ELP2, ELP3, and ELP4) that are necessary for biosynthesis and biogenesis of mcm^5U, which is important for protein synthesis via codon decoding, has been observed in patients with familial dysautonomia (Karlsborn et al. 2014), intellectual disability (Cohen et al. 2015), amyotrophic lateral sclerosis (Simpson et al. 2008), and autism spectrum disorder (Addis et al. 2015), respectively. N6-threonyl-carbamoylation (t6A) is a universal tRNA modification to ensure the translational accuracy. The pivotal role of Kae1 enzymes is to catalyze the final step of the universal tRNA modification, t6A. Kae1 resides within KEOPS (kinase, endopeptidase, and other proteins of small size) complex. Similarly, defects in OSGEP, YRDC, and genes responsible for coding components of KEOPS complex, which are responsible for t^6A modification, are associated with global developmental delay with renal defects and Galloway-Mowat syndrome, respectively (Arrondel et al. 2019; Edvardson et al. 2017). Defects in methyltransferase NSUN2 (Abbasi-Moheb et al. 2012), TRMT1 (Davarniya et al. 2015), FTSJ1 (Freude et al. 2004), SDR5C1, DALRD3 (Lentini et al. 2020), TRMT10A (Igoillo-Esteve et al. 2013), and HSD17B10 (Oerum et al. 2017) can lead to aberrant methylations, which can disrupt translational initiation and progression. Mutations in THG1L (Edvardson et al. 2016), CTU2 (Shaheen et al. 2016), ADAT3 (Alazami et al. 2013), PUS3 (de Paiva et al. 2019), and PUS 7 (Darvish et al. 2019) have also been found in patients with intellectual disability.

In addition to intellectual disability, several of the tRNA mutations have also been reported in mitochondrial deficiencies. Mutations in the mitochondrial tRNA genes that are responsible for transport of leucine (mt-$tRNA^{Leu}$) and lysine (mt-$tRNA^{Lys}$) have been reported in majority of patients with MELAS and MERRF, respectively (Goto et al. 1990, 1991; Shoffner et al. 1990). In addition, mutations in MTO1 and GTPBP2 gene, which code for subunits of enzymes catalyzing uridine modifications (τm^5U), lead to mitochondrial dysfunction and have been associated with hypertrophic cardiomyopathy and lactic acidosis (Fakruddin et al. 2018; Kopajtich et al. 2014). MTU1 gene, coding for mitochondrial tRNA-specific 2-thiouridylase 1, has been reported in severe liver failure. Mutations in this gene impair enzyme production leading to lower τm^5s^2U modifications (Zeharia et al. 2009). Mutations in NSUN3 and ALKBH5 have been reported to decrease mitochondria translational efficiency. The two genes code for enzymes that form a chain reaction to modify C34 of mt-tRNA m^5C34 and then to f^5C34, which is necessary to start normal translation. Defects in these genes impair the ability of tRNA to start translation, leading to several encephalopathies and seizures (Takemoto et al. 2009; Van Haute et al. 2016; Haag et al. 2016). Mutations in PUS1 leads to lower pseudouridine modification (Ψ) and has been reported to cause myopathy, lactic acidosis, and sideroblastic anemia (MLASA) (Zhao et al. 2004).

NSUN2 and METTL1, which mediate m^5C in tRNA, have been found to be overexpressed in various cancers (Delaunay and Frye 2019; Lin et al. 2018). The Elongator complex is involved in tRNA modifications in the cytoplasm. ELP1 overexpression has been reported in drug-resistant melanomas. Elp1 or Elp3-depleted B16-F10 cells lost their colony-forming and metastasizing properties (Rapino et al. 2018). N-acetyltransferase 10 (NAT10) has been reported in several malignant tumors. NAT10 mediates N4-acetylcytidine (ac4C) modification of tRNAs. RNPS1 (RNA-binding protein with serine-rich domain 1) inhibits the ubiquitination and degradation of NAT10 by E3 ubiquitin ligase, ZSWIM6 (Zinc finger SWIM domain-containing protein 6). NAT10 stability triggers an increase in specific tRNA ac4C modifications, thus boosting the translational process of oncogenes. There was a marked reduction in the tumorigenesis and metastatic dissemination to the cervical lymph nodes in mice orthotopically implanted with RNPS1 knockdown SCC-15 cells in tongue (Wang et al. 2024a). Conditional knock-in of NAT10 significantly promoted esophageal cancer progression, as evidenced by enlarged esophagus and significant increase in the number of lesion areas. Esophageal cancer progression was induced by the addition of 4-nitroquinoline-1-oxide (4-NQO) to the drinking water of 2-month-old mice for 16 weeks. Moreover, conditional knock-out of NAT10 significantly impaired the progression of esophageal cancer. Remodelin (Small-molecule inhibitor of NAT10) efficiently induced regression of the tumors in mice subcutaneously xenografted with TE9 or OE19 cells. NAT10-mediated ac4C tRNA modifications are essential for stability of oncogenic EGFR (Epidermal Growth Factor Receptor). Gefitinib, an EGFR inhibitor worked synergistically with remodelin and induced shrinkage of the tumors in xenografted models (Wei et al. 2023).

CDKAL1 (CDK5 regulatory subunit-associated protein 1-like 1) is a tRNA-modifying enzyme. CDKAL1 gene is responsible for 2-methylthio-N6-threonylcarbamoyladenosine (ms^2t^6A), in cytoplasmic $tRNA^{Lys}$, which are responsible for accurate translation of lysine codons in proinsulin (Wei et al. 2011). Mutation in CDKAL1 or in genes such as Irp2 (Iron regulatory protein 2), which supplies iron required for normal functioning of CDKAL1 has been reported in diabetes owing to less translation of proinsulin mediated via hypomethylation of $tRNA^{Lys}$. Reduction in the functions of CDKAL1 in $Irp2^{-/-}$ β cells leads to misreading of lysine codons in proinsulin and consequent impairment of proinsulin processing (Santos et al. 2020).

4.2 Mechanism-Based Role of tRNA-Associated Machinery in the Regulation of Carcinogenesis and Metastasis

4.2.1 Aminoacyl-tRNA Synthetase

Aminoacyl-tRNA synthetases catalyze the ligation of amino acids to their corresponding tRNAs in protein synthesis. Isoleucyl-tRNA synthetase 2 (IARS2) is an enzyme involved in the ligation of isoleucine to the corresponding tRNA. Phosphorylation of β-Catenin leads to its degradation via the ubiquitin/proteasome pathway. IARS2 inhibits β-Catenin phosphorylation and proteasomal degradation in pancreatic cancer cells (Fig. 4.1a–c). Tumor weight was significantly higher in mice inoculated with IARS2-overexpressing BxPC-3 cancer cells. Conversely, the average tumor weight was reduced in mice xenografted with IARS2-silenced PANC-1 cancer cells. Moreover, there was a remarkable increase in the

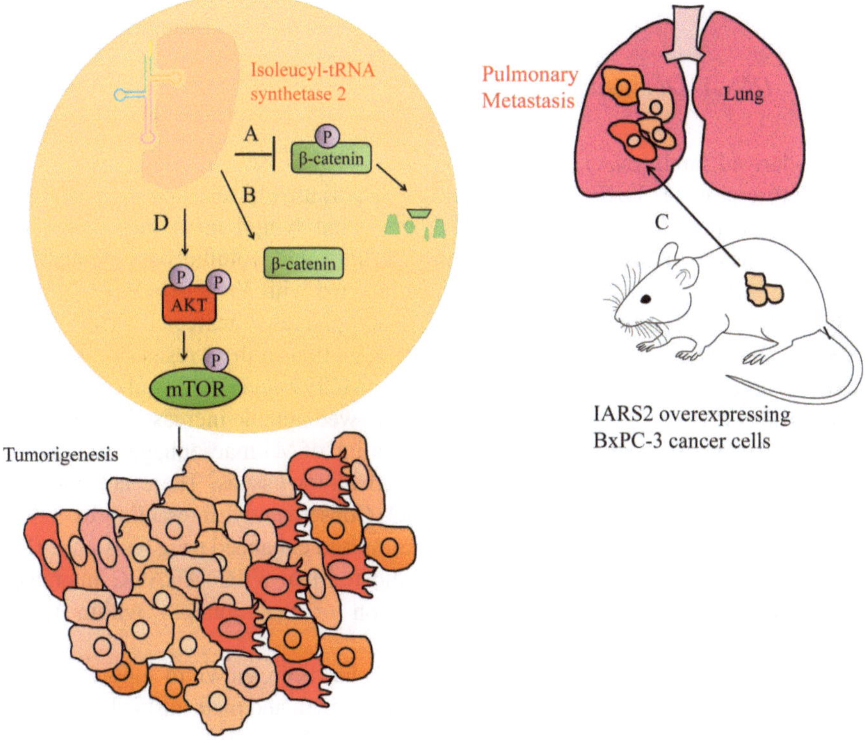

Fig. 4.1 (**a–c**) Isoleucyl-tRNA synthetase 2 (IARS2) is involved in enhancing the stability of β-catenin by inhibition of phosphorylation and proteasomal degradation of β-catenin. IARS2 overexpression potently enhanced the pulmonary metastasis in mice. (**d**) AKT/mTOR signaling is active in IARS2-overexpressing cancer cells and promotes tumor progression

pulmonary metastatic nodules in mice injected with IARS2-overexpressing BxPC-3 cancer cells (Jin et al. 2024).

IARS2 also induced activation of pro-survival signaling in cancer cells. Levels of p-AKT (Ser-473 and Thr-308) and p-mTOR were found to be reduced in IARS2-silenced A549 and H1299 cells (Fig. 4.1d). There was a significant shrinkage of the tumor mass in mice subcutaneously injected with IARS2-silenced-A549 cells (Di et al. 2019).

Seryl-tRNA synthetase (SerRS) has been shown to inhibit breast cancer progression. SerRS overexpression inhibited the growth of primary tumors and metastasis in a spontaneous metastasis model. SerRS significantly reduced the pulmonary metastatic foci in immunodeficient NSG mice intravenously injected with SerRS WT-overexpressing MDA-MB-231 cancer cells (Jiang et al. 2025).

The potential of ARSs as therapeutic targets has started to attract the attention of medicinal chemists. Therefore, previously unprecedented discoveries regarding their roles in multifaceted regulatory pathways are rapidly opening new possibilities for ARSs as druggable targets.

4.2.2 tRNA-Derived Fragments

tRNA-derived small RNAs have been shown to play an essential role in carcinogenesis. Expression of 3′tRF-AlaAGC is high in adriamycin-resistant breast cancer cells. Overexpression of 3′tRF-AlaAGC promoted lymph node metastases and deeper invasion in breast cancer patients. Weight and volume of subcutaneous tumors were enhanced significantly in mice injected with 3′tRF-AlaAGC agomirs. MCF-7/ADR cells cocultured with M2 macrophages transfected with 3′tRF-AlaAGC inhibitors demonstrated significant reduction in drug resistance and evident increase in apoptosis. However, MCF-7 cells cocultured with 3′tRF-AlaAGC-expressing M2 macrophages showed notable increase in drug resistance. 3′tRF-AlaAGC agomirs induced polarization of M1 macrophages to M2 phenotype within the subcutaneously transplanted tumor tissues. There are potential binding sites for TRADD on 3′tRF-AlaAGC. Consequently, interaction of 3′tRF-AlaAGC and TRADD promotes M2 polarization and enhanced drug resistance (Mo et al. 2024). Collectively, these findings indicated that 3′tRF-AlaAGC interacted with TRADD and promoted the polarization of macrophages within tumor microenvironment.

Transcription elongation factor A3 (TCEA3) has been shown to suppress tumor progression. tRF-Leu-CAG promotes the tumor growth and metastasis of non-small cell lung cancer cells by targeting TCEA3 (Wu et al. 2024a, b).

tRF-34-P4R8YP9LON4VHM has an oncogenic role. It directly binds to 3′-UTR of DAB2IP and silences its expression, thus activating the MEK/ERK signaling cascade and promoting the secretion of VEGFA. Tumor volume was significantly smaller in mice subcutaneously injected with tRF-34-antagomir-transfected PLC/PRF/5 cells. tRF-34-antagomirs suppressed the expression of N-cadherin and increased the expression of DAB2IP and E-cadherin in the tumor xenografts. Thus, these findings suggested that the inhibition of tRF-34-P4R8YP9LON4VHM caused regression of the tumors in xenografted mice (Xu et al. 2025).

tRF-Val-CAC-024 promoted the oligomerization of Aldolase A (ALDOA) and enhanced its enzyme activities. Moreover, tRF-Val-CAC-024 efficiently enhanced aerobic glycolysis in cancer cells. Inhibition of ALDOA abolished the promotional effects of tRF-Val-CAC-024 on tumor metastasis and growth. tRF-Val-CAC-024 antagomir-loaded lipid nanoparticles efficiently induced regression of the tumors in the mice orthotopically implanted with PC9 cancer cells within the lungs. In lung metastasis model, tRF-Val-CAC-024 antagomir-loaded nanoparticles markedly reduced the pulmonary metastatic foci (Wang et al. 2024a, b).

Pancreatic ductal adenocarcinoma (PDAC)-derived exosomes have been shown to activate hepatic stellate cells (HSCs) during liver metastasis of pancreatic cancer. tRF-GluCTC-0005 (tRF-GluCTC) is a cleavage fragment of tRNA-GluCTC. 3′UTR of WDR1 mRNA contains a sequence complementary to tRF-GluCTC. Importantly, the percentage of PDAC liver metastasis lesions was found to be considerably smaller, and the size of the hepatic metastases was also reduced in the WDR1-silenced group. Residues Ser-135 and Ser-138 in WDR1 were found to interact with the YAP and enhanced its stability. WDR1 binds to YAP and promotes the Hippo signaling pathway in HSCs (shown in Fig. 4.2). Exosomes derived from PDAC promoted the ingression of MDSCs in the liver, leading to the formation of pre-metastatic niches. Targeted inhibition of WDR1 significantly reduced the

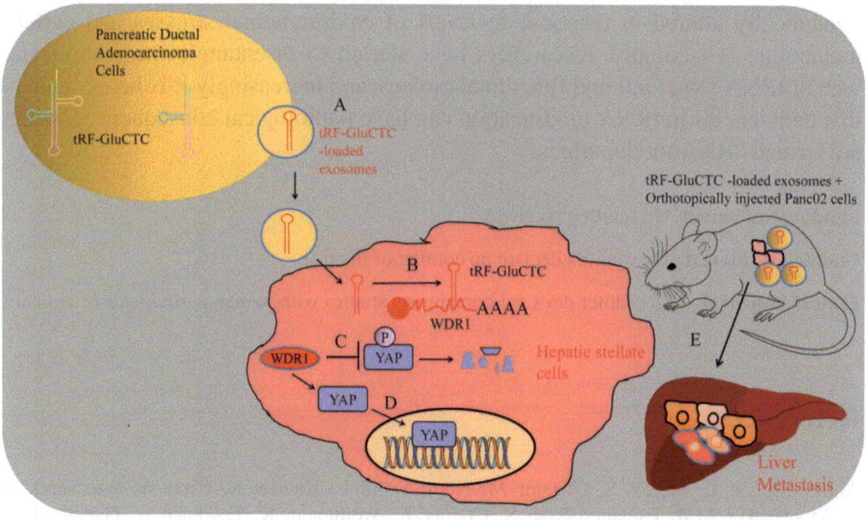

Fig. 4.2 (**a, b**) tRF-GluCTC-0005 (tRF-GluCTC) is a cleavage fragment of tRNA-GluCTC. tRNA-GluCTC interacts with 3′UTR of WDR1 mRNA and increase its stability. (**c, d**) WDR1 inhibits phosphorylation and degradation of YAP. WDR1 promotes nuclear accumulation of YAP proteins for transcriptional upregulation of interleukins. (**e**) tRF-GluCTC-loaded exosomes enhance the infiltration of MDSCs in tumor tissues in mice.

infiltration of MDSCs and macrophages in cancer tissues. tRF-GluCTC-loaded exosomes markedly increased the infiltration of MDSCs in tumor tissues in mice orthotopically injected with Panc02 cells within the spleen (Chen et al. 2024).

tRNA genes sharing similar anticodon having different body sequences are termed as "isodecoders." Selenocysteine-containing proteins are unique as they are translationally modulated by two discrete isodecoders of the Sec tRNA. Two species of Sec tRNA differ by a single 2′-O-hydroxymethyl group on the ribosyl moiety at position 34 (Um34). Modifications in the anticodon loop of selenocysteine tRNA, Um34, mediate translation of selenoproteins, thus promoting metastasis. FTSJ1 is an Um34 methyltransferase as it directly interacts with Sec tRNA as well as translational machinery of selenocysteine. Loss of FTSJ1 significantly reduced metastatic dissemination without restricting the growth of the primary tumors. FTSJ1 overexpression although did not influence the growth of the primary tumors but significantly enhanced the metastatic burden in the organs of the tumor-bearing mice (Nease et al. 2024).

4.3 Concluding Remarks

The field of tRNA is evolving exponentially and we have started to witness pleiotropic roles of tRNAs in carcinogenesis and metastasis. Transdisciplinary researchers have started to scratch the surface of intricate nature and functionalities of tRNAs. Recent technological developments underscored that tRNA modifications can be dynamically altered in response to levels of environmental stresses and cellular metabolites. Essentially, researchers have started to disentangle the complicated web of tRNA structural and functional biology and increasingly it is being realized that deficiencies in tRNA modification can have pathological consequences, which are termed "RNA modopathies."

Funding Statement No funding received.

Conflict of Interest The authors declare no conflict of interest.

Ethical Approval This chapter does not contain any studies with human participants or animals.

References

Abbasi-Moheb L, Mertel S, Gonsior M, Nouri-Vahid L, Kahrizi K, Cirak S, Wieczorek D, Motazacker MM, Esmaeeli-Nieh S, Cremer K, Weißmann R, Tzschach A, Garshasbi M, Abedini SS, Najmabadi H, Ropers HH, Sigrist SJ, Kuss AW (2012) Mutations in NSUN2 cause autosomal-recessive intellectual disability. Am J Hum Genet 90(5):847–855. https://doi.org/10.1016/j.ajhg.2012.03.021

Addis L, Ahn JW, Dobson R, Dixit A, Ogilvie CM, Pinto D, Vaags AK, Coon H, Chaste P, Wilson S, Parr JR, Andrieux J, Lenne B, Tumer Z, Leuzzi V, Aubell K, Koillinen H, Curran S, Marshall

CR, Scherer SW (2015) Microdeletions of ELP4 are associated with language impairment, autism spectrum disorder, and mental retardation. Hum Mutat 36(9):842–850. https://doi.org/10.1002/humu.22816

Alazami AM, Hijazi H, Al-Dosari MS, Shaheen R, Hashem A, Aldahmesh MA, Mohamed JY, Kentab A, Salih MA, Awaji A, Masoodi TA, Alkuraya FS (2013) Mutation in ADAT3, encoding adenosine deaminase acting on transfer RNA, causes intellectual disability and strabismus. J Med Genet 50(7):425–430. https://doi.org/10.1136/jmedgenet-2012-101378

Arrondel C, Missoury S, Snoek R, Patat J, Menara G, Collinet B, Liger D, Durand D, Gribouval O, Boyer O, Buscara L, Martin G, Machuca E, Nevo F, Lescop E, Braun DA, Boschat A-C, Sanquer S, Guerrera IC, Revy P (2019) Defects in t6A tRNA modification due to GON7 and YRDC mutations lead to Galloway-Mowat syndrome. Nat Commun 10(1):3967. https://doi.org/10.1038/s41467-019-11951-x

Barbieri I, Kouzarides T (2020) Role of RNA modifications in cancer. Nat Rev Cancer 20(6):303–322. https://doi.org/10.1038/s41568-020-0253-2

Barends S, Björk K, Gultyaev AP, de Smit MH, Pleij CWA, Kraal B (2002) Functional evidence for D- and T-loop interactions in tmRNA. FEBS Lett 514(1):78–83. https://doi.org/10.1016/S0014-5793(02)02306-2

Carter CW, Wolfenden R (2015) tRNA acceptor-stem and anticodon bases embed separate features of amino acid chemistry. RNA Biol 13(2):145–151. https://doi.org/10.1080/15476286.2015.1112488

Chen W, Peng W, Wang R, Bai S, Cao M, Xiong S, Li Y, Yang Y, Liang J, Liu L, Yazdani HO, Zhao Y, Cheng B (2024) Exosome-derived tRNA fragments tRF-GluCTC-0005 promotes pancreatic cancer liver metastasis by activating hepatic stellate cells. Cell Death Dis 15(1):102. https://doi.org/10.1038/s41419-024-06482-3

Chujo T, Tomizawa K (2021) Human transfer RNA modopathies: diseases caused by aberrations in transfer RNA modifications. FEBS J 288(24):7096–7122. https://doi.org/10.1111/febs.15736

Cohen JS, Srivastava S, Farwell KD, Lu H-M, Zeng W, Lu H, Chao EC, Fatemi A (2015) ELP2 is a novel gene implicated in neurodevelopmental disabilities. Am J Med Genet A 167(6):1391–1395. https://doi.org/10.1002/ajmg.a.36935

Crick FH (1966) Codon-anti codon pairing: the wobble hypothesis. J Mol Biol 19:548–555

Darvish H, Azcona LJ, Alehabib E, Jamali F, Tafakhori A, Ranji-Burachaloo S, Jen JC, Paisán-Ruiz C (2019) A novel PUS7 mutation causes intellectual disability with autistic and aggressive behaviors. Neurol Genet 5(5):e356. https://doi.org/10.1212/nxg.0000000000000356

Davarniya B, Hu H, Kahrizi K, Musante L, Fattahi Z, Hosseini M, Maqsoud F, Farajollahi R, Wienker TF, Ropers HH, Najmabadi H (2015) The role of a novel TRMT1 gene mutation and rare GRM1 gene defect in intellectual disability in two Azeri families. PLoS One 10(8):e0129631. https://doi.org/10.1371/journal.pone.0129631

de Paiva ARB, Lynch DS, Melo US, Lucato LT, Freua F, de Assis BDR, Barcelos I, Listik C, de Castro Dos Santos D, Macedo-Souza LI, Houlden H, Kok F (2019) PUS3 mutations are associated with intellectual disability, leukoencephalopathy, and nephropathy. Neurol Genet 5(1):e306. https://doi.org/10.1212/NXG.0000000000000306

Delaunay S, Frye M (2019) RNA modifications regulating cell fate in cancer. Nat Cell Biol 21(5):552–559. https://doi.org/10.1038/s41556-019-0319-0

Di X, Jin X, Ma H, Wang R, Cong S, Tian C, Liu J, Zhao M, Li R, Wang K (2019) The oncogene IARS2 promotes non-small cell lung cancer tumorigenesis by activating the AKT/MTOR pathway. Front Oncol 9:393. https://doi.org/10.3389/fonc.2019.00393

Edvardson S, Elbaz-Alon Y, Jalas C, Matlock A, Patel K, Labbé K, Shaag A, Jackman JE, Elpeleg O (2016) A mutation in the THG1L gene in a family with cerebellar ataxia and developmental delay. Neurogenetics 17(4):219–225. https://doi.org/10.1007/s10048-016-0487-z

Edvardson S, Prunetti L, Arraf A, Haas D, Bacusmo JM, Hu JF, Ta-Shma A, Dedon PC, de Crécy-Lagard V, Elpeleg O (2017) tRNA N6-adenosine threonylcarbamoyltransferase defect due to KAE1/TCS3 (OSGEP) mutation manifest by neurodegeneration and renal tubulopathy. Eur J Hum Genet 25(5):545–551. https://doi.org/10.1038/ejhg.2017.30

Fakruddin M, Wei FY, Suzuki T, Asano K, Kaieda T, Omori A, Izumi R, Fujimura A, Kaitsuka T, Miyata K, Araki K, Oike Y, Scorrano L, Tomizawa K (2018) Defective mitochondrial tRNA taurine modification activates global proteostress and leads to mitochondrial disease. Cell Rep 22(2):482–496. https://doi.org/10.1016/j.celrep.2017.12.051

Freude KK, Hoffmann K, Jensen LJ, Delatycki MB, des Portes V, Moser BA, Hamel BCJ, van Bokhoven H, Moraine C, Fryns J-P, Chelly J, Gecz J, Lenzner S, Kalscheuer VM, Ropers H-H (2004) Mutations in the FTSJ1 gene coding for a novel S-adenosylmethionine–binding protein cause nonsyndromic X-linked mental retardation. Am J Hum Genet 75(2):305–309. https://doi.org/10.1086/422507

Goto Y, Nonaka I, Horai S (1990) A mutation in the tRNA Leu(UUR) gene associated with the MELAS subgroup of mitochondrial encephalomyopathies. Nature 348(6302):651–653. https://doi.org/10.1038/348651a0

Goto Y, Nonaka I, Horai S (1991) A new mtDNA mutation associated with mitochondrial myopathy, encephalopathy, lactic acidosis and stroke-like episodes (MELAS). Biochim Biophys Acta 1097(3):238–240. https://doi.org/10.1016/0925-4439(91)90042-8

Guy MP, Phizicky EM (2014) Conservation of an intricate circuit for crucial modifications of the tRNAPhe anticodon loop in eukaryotes. RNA 21(1):61–74. https://doi.org/10.1261/rna.047639.114

Haag S, Sloan KE, Ranjan N, Warda AS, Kretschmer J, Blessing C, Hübner B, Seikowski J, Dennerlein S, Rehling P, Rodnina MV, Höbartner C, Bohnsack MT (2016) NSUN 3 and ABH 1 modify the wobble position of mt-t RNA Met to expand codon recognition in mitochondrial translation. EMBO J 35(19):2104–2119. https://doi.org/10.15252/embj.201694885

Igoillo-Esteve M, Genin A, Lambert N, Désir J, Pirson I, Abdulkarim B, Simonis N, Drielsma A, Marselli L, Marchetti P, Vanderhaeghen P, Eizirik DL, Wuyts W, Julier C, Chakera AJ, Ellard S, Hattersley AT, Abramowicz M, Cnop M (2013) tRNA Methyltransferase homolog gene TRMT10A mutation in young onset diabetes and primary microcephaly in humans. PLoS Genet 9(10):e1003888. https://doi.org/10.1371/journal.pgen.1003888

Jiang L, Wang J, Liu Z, Zhang Q, Yang XL (2025) Seryl-tRNA synthetase inhibits Wnt signaling and breast cancer progression and metastasis. FASEB J 39(1):e70294. https://doi.org/10.1096/fj.202401720R

Jin Y, Huang X, Wang Z, Kouken B, Wang Q, Wang L (2024) Isoleucyl-tRNA synthetase 2 promotes pancreatic ductal adenocarcinoma proliferation and metastasis by stabilizing β-catenin. Genes Dis 12(3):101382. https://doi.org/10.1016/j.gendis.2024.101382

Karlsborn T, Tükenmez H, Chen C, Byström AS (2014) Familial dysautonomia (FD) patients have reduced levels of the modified wobble nucleoside mcm5s2U in tRNA. Biochem Biophys Res Commun 454(3):441–445. https://doi.org/10.1016/j.bbrc.2014.10.116

Kopajtich R, Nicholls TJ, Rorbach J, Metodiev MV, Freisinger P, Mandel H, Vanlander A, Ghezzi D, Carrozzo R, Taylor RW, Marquard K, Murayama K, Wieland T, Schwarzmayr T, Mayr JA, Pearce SC, Powell CF, Saada A, Ohtake A, Invernizzi F (2014) Mutations in GTPBP3 cause a mitochondrial translation defect associated with hypertrophic cardiomyopathy, lactic acidosis, and encephalopathy. Am J Hum Genet 95(6):708–720. https://doi.org/10.1016/j.ajhg.2014.10.017

Kresge N, Simoni RD, Hill RL (2005) The discovery of tRNA by Paul C. Zamecnik J Biol Chem 280(40):e37–e39. https://doi.org/10.1016/S0021-9258(20)79029-0

Lentini JM, Alsaif HS, Faqeih E, Alkuraya FS, Fu D (2020) DALRD3 encodes a protein mutated in epileptic encephalopathy that targets arginine tRNAs for 3-methylcytosine modification. Nat Commun 11(1):2510. https://doi.org/10.1038/s41467-020-16321-6

Lin S, Liu Q, Lelyveld VS, Choe J, Szostak JW, Gregory RI (2018) Mettl1/Wdr4-mediated m7G tRNA methylome is required for normal mRNA translation and embryonic stem cell self-renewal and differentiation. Mol Cell 71(2):244–255.e5. https://doi.org/10.1016/j.molcel.2018.06.001

Mo D, Tang X, Ma Y, Chen D, Xu W, Jiang N, Zheng J, Yan F (2024) tRNA-derived fragment 3′tRF-AlaAGC modulates cell chemoresistance and M2 macrophage polarization via binding to TRADD in breast cancer. J Transl Med 22(1):706. https://doi.org/10.1186/s12967-024-05513-z

Murakami H, Ohta A, Suga H (2009) Bases in the anticodon loop of tRNAAlaGGC prevent misreading. Nat Struct Mol Biol 16(4):353–358. https://doi.org/10.1038/nsmb.1580
Nease LA, Church KP, Delclaux I, Murakami S, Astorkia M, Zerhouni M, Cascio G, Hughes RO, Aguirre KN, Zumbo P, Dow LE, Jaffrey S, Betel D, Piskounova E (2024) Selenocysteine tRNA methylation promotes oxidative stress resistance in melanoma metastasis. Nat Cancer 5(12):1868–1884. https://doi.org/10.1038/s43018-024-00844-8
Oerum S, Roovers M, Leichsenring M, Acquaviva-Bourdain C, Beermann F, Gemperle-Britschgi C, Fouilhoux A, Korwitz-Reichelt A, Bailey HJ, Droogmans L, Oppermann U, Sass JO, Yue WW (2017) Novel patient missense mutations in the HSD17B10 gene affect dehydrogenase and mitochondrial tRNA modification functions of the encoded protein. Biochim Biophys Acta (BBA) – Mol Basis Dis 1863(12):3294–3302. https://doi.org/10.1016/j.bbadis.2017.09.002
Rapino F, Delaunay S, Rambow F, Zhou Z, Tharun L, De Tullio P, Sin O, Shostak K, Schmitz S, Piepers J, Ghesquière B, Karim L, Charloteaux B, Jamart D, Florin A, Lambert C, Rorive A, Jerusalem G, Leucci E, Dewaele M (2018) Codon-specific translation reprogramming promotes resistance to targeted therapy. Nature 558(7711):605–609. https://doi.org/10.1038/s41586-018-0243-7
Santos MCFD, Anderson CP, Neschen S, Zumbrennen-Bullough KB, Romney SJ, Kahle-Stephan M, Rathkolb B, Gailus-Durner V, Fuchs H, Wolf E, Rozman J, de Angelis MH, Cai WM, Rajan M, Hu J, Dedon PC, Leibold EA (2020) Irp2 regulates insulin production through iron-mediated Cdkal1-catalyzed tRNA modification. Nat Commun 11(1):296. https://doi.org/10.1038/s41467-019-14004-5
Schimmel P (2018) The emerging complexity of the tRNA world: mammalian tRNAs beyond protein synthesis. Nat Rev Mol Cell Biol 19(1):45–58. https://doi.org/10.1038/nrm.2017.77
Shaheen R, Al-Salam Z, El-Hattab AW, Alkuraya FS (2016) The syndrome dysmorphic facies, renal agenesis, ambiguous genitalia, microcephaly, polydactyly and lissencephaly (DREAM-PL): report of two additional patients. Am J Med Genet A 170(12):3222–3226. https://doi.org/10.1002/ajmg.a.37877
Shoffner JM, Lott MT, Lezza AMS, Seibel P, Ballinger SW, Wallace DC (1990) Myoclonic epilepsy and ragged-red fiber disease (MERRF) is associated with a mitochondrial DNA tRNALys mutation. Cell 61(6):931–937. https://doi.org/10.1016/0092-8674(90)90059-n
Simpson CL, Lemmens R, Miskiewicz K, Broom W, Hansen V, van Vught WJP, Landers JE, Sapp PC, Den V, Knight J, Neale BM, Turner MR, Veldink JH, Ophoff RA, Tripathi V, Beleza A, Shah MN, Proitsi P, Van Hoecke A, Carmeliet P (2008) Variants of the elongator protein 3 (ELP3) gene are associated with motor neuron degeneration. Hum Mol Genet 18(3):472–481. https://doi.org/10.1093/hmg/ddn375
Suzuki T (2021) The expanding world of tRNA modifications and their disease relevance. Nat Rev Mol Cell Biol 22(6):375–392. https://doi.org/10.1038/s41580-021-00342-0
Takemoto C, Spremulli LL, Benkowski LA, Ueda T, Yokogawa T, Watanabe K (2009) Unconventional decoding of the AUA codon as methionine by mitochondrial tRNAMet with the anticodon f5CAU as revealed with a mitochondrial in vitro translation system. Nucleic Acids Res 37(5):1616–1627. https://doi.org/10.1093/nar/gkp001
Van Haute L, Dietmann S, Kremer L, Hussain S, Pearce SF, Powell CA, Rorbach J, Lantaff R, Blanco S, Sauer S, Kotzaeridou U, Hoffmann GF, Memari Y, Kolb-Kokocinski A, Durbin R, Mayr JA, Frye M, Prokisch H, Minczuk M (2016) Deficient methylation and formylation of mt-tRNAMet wobble cytosine in a patient carrying mutations in NSUN3. Nat Commun 7(1):12039. https://doi.org/10.1038/ncomms12039
Wang X, Ling R, Peng Y, Qiu W, Chen D (2024a) RNPS1 stabilizes NAT10 protein to facilitate translation in cancer via tRNA ac4C modification. Int J Oral Sci 16(1):6. https://doi.org/10.1038/s41368-023-00276-7
Wang Q, Song X, Zhang Y, Liang S, Zhang M, Wang H, Feng Y, Li R, Ding H, Chen Y, Xia W, Dong G, Xu L, Mao Q, Jiang F (2024b) A pro-metastatic tRNA fragment drives aldolase A oligomerization to enhance aerobic glycolysis in lung adenocarcinoma. Cell Rep 43(8):114550. https://doi.org/10.1016/j.celrep.2024.114550

Wei F-Y, Suzuki T, Watanabe S, Kimura S, Kaitsuka T, Fujimura A, Matsui H, Atta M, Michiue H, Fontecave M, Yamagata K, Suzuki T, Tomizawa K (2011) Deficit of tRNALys modification by Cdkal1 causes the development of type 2 diabetes in mice. J Clin Invest 121(9):3598–3608. https://doi.org/10.1172/jci58056

Wei W, Zhang S, Han H, Wang X, Zheng S, Wang Z, Yang C, Wang L, Ma J, Guo S, Wang J, Liu L, Choe J, Lin S (2023) NAT10-mediated ac4C tRNA modification promotes EGFR mRNA translation and gefitinib resistance in cancer. Cell Rep 42(7):112810. https://doi.org/10.1016/j.celrep.2023.112810

Wu F, Chai B, Qi P, Han Y, Gu Z, Pan W, Zhang H, Wang X, Liu X, Zou H, Liang C, Li Y, Fang W, Ma Z (2024a) Oncogenic tRNA-derived fragment tRF-Leu-CAG promotes tumorigenesis of lung cancer via targeting TCEA3 and increasing autophagy. J Gene Med 26(9):e3737. https://doi.org/10.1002/jgm.3737

Wu D, Li X, Khan FA, Yuan C, Pandupuspitasari NS, Huang C, Sun F, Guan K (2024b) tRNA modifications and tRNA-derived small RNAs: new insights of tRNA in human disease. Cell Biol Toxicol 40(1):76. https://doi.org/10.1007/s10565-024-09919-9

Xu T, Hua H, Song F, Zhang N, Gao C, Chen Z (2025 Apr) tRF-34-P4R8YP9LON4VHM promotes hepatocellular carcinoma progression and tumour cell-induced angiogenesis via the MEK/ERK pathway. J Cell Mol Med 29(8):e70560. https://doi.org/10.1111/jcmm.70560

Zeharia A, Shaag A, Pappo O, Mager-Heckel A-M, Saada A, Beinat M, Karicheva O, Mandel H, Ofek N, Segel R, Marom D, Rötig A, Tarassov I, Elpeleg O (2009) Acute infantile liver failure due to mutations in the TRMU gene. Am J Hum Genet 85(3):401–407. https://doi.org/10.1016/j.ajhg.2009.08.004

Zhao X, Patton JR, Davis SN, Florence B, Ames SJ, Spanjaard RA (2004) Regulation of nuclear receptor activity by a pseudouridine synthase through posttranscriptional modification of steroid receptor RNA activator. Mol Cell 15(4):549–558. https://doi.org/10.1016/j.molcel.2004.06.044

Chapter 5
The Potential of Noncoding RNAs-siRNAs in Cancer Research and Therapy: Challenges and Solutions

Himanshu, Mudasir Bashir, Moien Lone, and Mohsin Maqbool

Contents

Mudasir Bashir and Moien Lone contributed equally to this work as co-second authors.

Himanshu
Department of Pathology, All India Institute of Medical Sciences (AIIMS), New Delhi, India

M. Bashir
Department of Physiology, Graphic Era Institute of Medical Sciences (GEIMS), Dehradun, Uttarakhand, India

M. Lone
Department of Pathology and Laboratory Medicine, University of California, Los Angeles, CA, USA

M. Maqbool (✉)
Department of Cancer Biology, Jefferson University, Philadelphia, PA, USA

M. F. Ullah, A. Ahmad (eds.), *Non-coding RNAs*, Experientia Supplementum 115,
https://doi.org/10.1007/978-3-032-06948-1_5

Abstract Noncoding RNAs (ncRNAs) are RNA molecules, which play critical roles in regulating gene expression and cellular activities, influencing processes like differentiation, proliferation, and cell survival, unlike messenger RNAs (mRNAs) that act as templates for protein production. Examples of ncRNAs include small interfering RNAs (siRNAs), microRNAs (miRNAs), ribosomal RNAs, transfer RNAs, small nucleolar RNAs (sno RNAs), and small nuclear RNAs (snRNAs). Small interfering RNAs (siRNAs) are a type of noncoding RNA that function primarily by silencing specific genes through a process called RNA interference (RNAi), in which they bind to complementary messenger RNA (mRNA) molecules, causing their degradation and preventing the translation of that mRNA into protein; essentially, they act as a "gene silencing" mechanism by targeting and destroying specific transcripts. In this chapter, we review the biogenesis, functions, and role of RNAs (siRNAs), in cancer research and role in therapy. siRNA (small interfering RNA) denotes small interfering RNA, consisting of 21–25 nucleotides. Discovery of siRNA (small interfering RNA) has been a significant breakthrough in biology. Small interfering RNAs (siRNAs) are single-stranded RNAs that are formed by the cleavage of longer double-stranded RNAs by the enzyme DICER1 within the RISC loading complex, which includes DICER1, an Argonaute protein, and either TARBP2 or PRKRA (PACT). Small interfering RNA (siRNA) is essential for health as it serves as a natural gene silencing tool, controlling gene expression posttranscriptionally, and is involved in several cellular processes such as development, immune response, and stress response; however, when not regulated properly, siRNAs may lead to diseases like cancer and viral infections, positioning them as a promising therapeutic target for targeted gene silencing therapies. siRNAs play a pivotal role in RNA interference (RNAi), a natural cellular process where they degrade complementary mRNA targets, preventing protein synthesis. This gene silencing mechanism has proven to be a valuable tool for controlling gene expression in research and therapeutic contexts. In cancer, siRNAs offer a promising approach to selectively silence oncogenes or other genes involved in tumor progression, thus hindering the development and spread of malignancies. However, the therapeutic potential of siRNAs faces several challenges, including efficient delivery to target cells, off-target effects, and stability issues. This chapter focuses on the biogenesis and functional significance of siRNAs, exploring their roles in cancer research and their promising therapeutic potential. With the continued advancement of RNA-based technologies, siRNAs hold considerable promise as a powerful tool for cancer treatment, offering new avenues for targeted therapies and personalized medicine.

Keywords Noncoding RNAs · Small interfering RNAs · siRNAs · RNA interference · Gene silencing · Cancer research · Gene expression · RNAi · Oncogenes · Molecular therapy

Abbreviations

circRNAs	Circular RNAs
HIFs	Hypoxia-inducible factors
PATs	Promoter-associated RNA
RISC	RNA-induced silencing complex
siRNA	Short interfering RNA
TGFβ	Growth factor-β
TRNA	Telomere RNA

5.1 Introduction

Noncoding RNAs (ncRNAs) are RNA molecules that have crucial roles in cellular processes despite not encoding proteins. Noncoding RNAs (ncRNAs) are broadly categorized into two main types based on their size: small noncoding RNAs and long noncoding RNAs (lncRNAs) (ENCODE Project Consortium 2012). Small ncRNAs are typically less than 200 nucleotides (nt) in length, while lncRNAs exceed this threshold and can span several kilobases (ENCODE Project Consortium 2012; Ha and Kim 2014). Small interfering RNAs (siRNAs) are small 21–25 nucleotide noncoding RNAs, which include those originating from introns, play crucial roles in the regulation of gene expression and cellular functions. They affect

Table 5.1 Overview of the classification of noncoding RNAs along with their associated functions

Types of RNA	Full form	Functions
rRNA	Ribosomal RNA	Translational machinery
tRNA	Transfer RNA	Amino acid carriers
snRNA	Small nuclear RNA	RNA processing
snoRNA	Small nucleolar RNA	RNA modifications
TRNA	Telomere RNA	Chromosome end synthesis
miRNA	MicroRNA	RNA stability and translation control
lncRNA	Long noncoding RNA	Imprinting, epigenetics, nuclear structure
circRNA	Circular RNA	Inhibiting miRNA activity
Endo-siRNA	Endogenous siRNA	RNA degradation
rasiRNA	Repeat associated-derived RNA	Transcriptional control
eRNA	Enhancer-derived RNA	Regulation of gene expression
piRNA	PIWI-interacting RNA	Silencing transposon and mRNA decay
PATs	Promoter-associated RNA	Transcription initiation and pause release

The classification and functional descriptions have been adapted from Fu (2014)

transcription, RNA splicing, editing, translation, chromatin structure, epigenetic memory, and RNA degradation. Grasping these networks can reveal new mechanisms that drive pathological, developmental, and physiological processes, providing exciting opportunities for groundbreaking research and treatment approaches (Castel and Martienssen 2013; Fu 2014).

Noncoding RNAs (ncRNAs) including siRNA can be produced through various mechanisms, like a mechanism, known as back splicing, generates circular RNAs (circRNAs). MicroRNAs (miRNAs) are already recognized as reliable biomarkers and therapeutic targets for various disorders. Emerging evidence suggests that circRNAs, long noncoding RNAs (lncRNAs), as well as Small interfering RNAs (siRNAs) exhibit similar potential, paving the way for innovative approaches to disease management and treatment (Kaikkonen et al. 2011). A brief overview of noncoding RNAs along with their associated functions is found in Table 5.1.

5.2 Noncoding-RNAs in Health and Diseases

Noncoding RNAs (ncRNAs) are increasingly recognized as key regulators of physiological and pathological processes, including genetics, epigenetics, cancer, and developmental disorders. Growing evidence links abnormal ncRNA expression to various human diseases. Advances in RNA sequencing have revealed numerous noncoding transcripts with previously unknown functions. Major types of noncoding linear RNAs include circular RNAs (circRNAs), small interfering RNAs (siRNAs), long noncoding RNAs (lncRNAs), and microRNAs (miRNAs) (Li et al. 2021a) (Table 5.2).

Table 5.2 RNA-based therapeutic approaches

Aspect	Solid tumors	Hematological malignancies
Target accessibility	Requires advanced delivery systems for localization	Relatively easier due to blood circulation
Microenvironment	Complex and hypoxic	Homogeneous, with easier drug diffusion
Key targets	VEGF, KRAS, PD-L1	BCL2, MYC, STAT3
Delivery systems	LNPs, SNALPs, polymer nanoparticles	LNPs, antibody conjugates, polymeric carriers
Major challenges	Endosomal escape, heterogeneity	Off-target effects, immune activation

5.2.1 Noncoding RNA-Based Therapeutics

Noncoding RNAs (ncRNAs), such as long noncoding RNAs (lncRNAs) and microRNAs (miRNAs), have surfaced as potential therapeutic targets in cancer and various other diseases. In the last 10 years, RNA therapies have progressed notably, with multiple antisense oligonucleotides (ASOs) and small interfering RNAs (siRNAs) gaining FDA approval. This chapter also provides an overview of existing RNA-based therapeutic agents and highlight clinical trial drugs, summarized in Table 5.3 as given elsewhere (Loganathan and Doss 2023). siRNA treatments present a hopeful method for addressing different diseases through the silencing of certain genes. The introduction of synthetic siRNA into cells permits the RNA-induced silencing complex (RISC) to break down target mRNA, thereby stopping the synthesis of proteins associated with diseases (Zhang et al. 2021a; Loganathan and Doss 2023).

5.2.2 Using CRISPR/Cas9 to Edit Noncoding RNA for Targeted Treatment of Human Diseases

Recent advancements in CRISPR/Cas9 technology have introduced innovations aimed at improving precision and reducing unintended effects. These include the development of inducible Cas9 systems, anti-CRISPR proteins, highly specific Cas9 variants, and strategies to minimize the duration of CRISPR/Cas9 activity (Zhang et al. 2021a, b). The CRISPR/Cas9 system can be utilized to modify genes that affect siRNA (small interfering RNA) pathways, providing a precise method for addressing human diseases (Li et al. 2021b).

Table 5.3 Current therapeutic drugs and clinical trial candidates targeting noncoding RNAs

Noncoding RNAs as biomarkers and therapeutic targets in human health and disease						
Drug_name	Mode of drug admin	Target	Mechanism of action	Disease	Company	Status of the drug
Fomivirsen	IVT (intravitreal administration)	CMV mRNA	Downregulates IE2	Cytomegalovirus (CMV) retinitis with acquired AIDS	Ionis Pharmaceutical, Novartis	FDA-approved (1998)
Mipomersen	SC (subcutaneous injection)	Apo-B-100 mRNA	Downregulates ApoB	Homozygous familial hypercholesterolemia	Kastle Therapeutics, Ionis Pharmaceuticals, Genzyme	FDA approved (2013)
Nusinersen	ITH	SMN2 pre-mRNA	Splicing modulation	Spinal muscular atrophy	Ionis Pharmaceuticals, Biogen	FDA approved (2016)
Eteplirsen	IV (intravenous)	Exon 51 of DMD	Splicing modulation	Duchenne muscular dystrophy	Sarepta Therapeutics	FDA approved
Inotersen	SC (subcutaneous injection)	TTR mRNA	Downregulates transthyretin mRNA	Familial amyloid polyneuropathy	Ionis Pharmaceuticals	FDA approved (2018)
Golodirsen	IV (intravenous)	Exon 53 of DMD	Splicing modulation	Duchenne muscular dystrophy	Sarepta Therapeutics	FDA approved (2019)
Milasen	Intrathecal	CLN7	Splicing modulation	Batten disease	Boston Children's Hospital	FDA approved (2018)
Casimersen	IV (intravenous)	Exon 45 of DMD	Splicing modulation	Duchenne muscular dystrophy	Sarepta Therapeutics	FDA approved (2021)
1018 ISS	IV (intravenous)	TLR9	Enhancement of cytotoxic effector mechanisms	Non-Hodgkin's lymphoma	Dana-Farber Cancer Institute, Brigham and Women's Hospital, Massachusetts General Hospital, University of Rochester	NCT00251394 (phase II)

Apatorsen (OGX-427)	IV (intravenous)	HSP27	Inhibits expression of heat shock protein (Hsp27)	Urologic cancer, bladder cancer, prostate cancer, urothelial cancer, non-small cell lung cancer, and other cancers	Achieve Life Sciences PRA Health Sciences	NCT00487786 (phase-I), NCT01454089 (phase-II)
Cenersen (EL625)	IV (intravenous)	TP53	Blocks the effects of p53	Acute myelogenous leukemia, lymphoma	Eleos, Inc	NCT00074737 (phase II)
ARRx (AZD5312)	IV (intravenous)	AR	Suppression of human AR expression	Prostate Cancer	AstraZeneca	NCT02144051 (phase I/II)
Custirsen (OGX-011)	IV (intravenous)	ApoJ	Inhibition of clusterin expression	Prostate cancer, breast cancer, non-small cell lung cancer	NCIC Clinical Trials Group, Achieve Life Sciences	NCT00054106 (phase I), NCT00138658 (phase-II)
Patisiran	IV (intravenous)	TTR mRNA	Transthyretin activity is decreased	Polyneuropathy caused by hATTR amyloidosis	Alnylam	FDA approved (2018)
Givosiran	SC (subcutaneous injection)	ALS1 mRNA	ALAS1 is decreased in expression	Acute hepatic porphyria	Alnylam	FDA approved (2020)
Lumasiran	SC (subcutaneous injection)	HAO1 mRNA	Reduced glycolate oxidase activity	Primary hyperoxaluria type 1	Alnylam	FDA approved (2020)
Inclisiran	SC (subcutaneous injection)	PCSK9	Proprotein convertase subtilisin/kexin type 9 is downregulated	Atherosclerotic cardiovascular disease	Novartis	FDA approved (2021)
TKM-080301	Intra-arterial/IV	PLK1	Activation of PLK1 is reduced	Hepatocellular cancer, cancer with hepatic metastases,	National Cancer Institute, Arbutus Biopharma Corporation	NCT01437007 (phase-I), NCT02191878 (phase-II)

(continued)

Table 5.3 (continued)

Noncoding RNAs as biomarkers and therapeutic targets in human health and disease						
Drug_name	Mode of drug admin	Target	Mechanism of action	Disease	Company	Status of the drug
Atu027	IV (intravenous)	PNK3	The expression of PNK3 is silenced	Solid tumors, pancreatic cancer	Silence Therapeutics GmbH, Granzer Regulatory Consulting & Services	NCT00938574 (phase-I), NCT01808638 (phase-II)
siG12D LODER	Locally implanted through EUS biopsy procedure	KRASG12D	Blocks the expression of KRAS	Pancreatic cancer	Silenseed Ltd	NCT01676259 (phase-I), NCT01188785 (phase-II)
ARO-HIF2	IV (intravenous)	HIF2A	HIF2A deregulation	Clear cell renal cell carcinoma	Arrowhead Pharmaceuticals	NCT04169711 (phase-I)
APN401	IV (intravenous)	CBLB	Inhibition of Cbl-b enhances natural killer cell and T cell-mediated antitumor activity	Brain cancer, melanoma, pancreatic cancer, renal cell cancer	Wake Forest University Health Sciences, National Cancer Institute	NCT03087591, NCT02166255 (phase-I)
Vutrisiran	SQ (subcutaneous injection)	TTR	It lowers the expression of the TTR protein	Transthyretin-mediated amyloidosis with or without cardiomyopathy	Alnylam	NCT03759379, NCT04153149 (phase-III)
Pegaptanib	Intravitreal	The heparin-binding domain of VEGF-165	Inhibiting VEGF-165	Neovascular age-related macular degeneration	OSI Pharmaceuticals	FDA-approved (2004)
Defibrotide	IV (intravenous)	Adenosine A1/ A2receptor	Adenosine A1/A2 receptor activation	Veno-occlusive disease in liver	Jazz Pharmaceuticals	FDA-approved (2020)

NOX-A12	IV (intravenous)	CXCL12	Interferes with CXCR4-CXCL12 interactions	Pancreatic cancer, colorectal cancer, multiple myeloma	NOXXON Pharma AG, Merck Sharp & Dohme Corp	NCT01521533 (phase-I), NCT01521533 (phase-II), NCT03168139 (phase-I/II)
NOX-E36	IV (intravenous)/ SQ (subcutaneous)	CCL2	Suppresses the pro-inflammatory chemokine CCL2 by specifically binding to it	Diabetic nephropathy	NOXXON Pharma AG	Phase-II
BNT162b2	IM (intramuscular injection)	Immunogenicity and antibody response to SARS-CoV-2S antigens	Expression of SARS-CoV-2S antigens	COVID-19	BioNTech and Pfizer	FDA-approved (2020)
mRNA-1273	IM (intramuscular injection)	Immunogenicity and antibody response to SARS-CoV-2 S antigens	Expression of SARS-CoV-2S (spike protein) antigens	COVID-19	Moderna	FDA-approved (2020)
CVnCoV	IM (intramuscular injection)	Immunogenicity and antibody response to SARS-CoV-2S antigens	Expression of SARS-CoV-2S antigens	COVID-19	CureVac AG	NCT04652102 (phase-III)
AZD8601	Epicardial	VEGF-A	VEGF-A expression is restored	Ischemic heart disease	AstraZeneca	NCT03370887 (phase-II)
MRT5005	Inhalation	CFTR	CFTR expression is restored	Cystic fibrosis	Translate Bio	NCT03375047 (phase-I/II)

(continued)

Table 5.3 (continued)

Noncoding RNAs as biomarkers and therapeutic targets in human health and disease

Drug_name	Mode of drug admin	Target	Mechanism of action	Disease	Company	Status of the drug
mRNA-3704	IV (intravenous)/ SQ (subcutaneous)	MUT	MUT expression is restored	Methylmalonic aciduria	Moderna	NCT03810690 (phase-I/II)
BNT111	IV (intravenous)/ SQ (subcutaneous)	Targets four non-mutated TAAs (NY-ESO-1, MAGEA3, tyrosinase, and TPTE)	Immune response induction against the four malignant melanoma-associated antigens (tyrosinase, New York-ESO-1 (NY-ESO-1), melanoma-associated antigen A3 (MAGE-A3), and transmembrane phosphatase with tensin homology (TPTE))	Advanced melanoma	BioNTech SE	NCT02410733 (phase-I)
Miravirsen	SC (subcutaneous injection)	miR-122	miRNA-inhibitor	HCV	Roche/Santaris	NCT01200420 (phase-II)
RG-012 (lademirsen)	SC (subcutaneous injection)	miR-21	miRNA-inhibitor	Alport syndrome	Sanofi	NCT03373786 (phase-II)
Cobomarsen	IV/SQ	miR-155	miRNA-inhibitor	Cutaneous T cell lymphoma/mycosis Fungoides	miRagen	NCT03713320, NCT02580552 (phase-II)
MRG-110	Intradermal	miR-92a	miRNA-inhibitor	Wound healing	miRagen	NCT03603431 (phase-I)
AZD4076	SC (subcutaneous injection)	miR-103/107	miRNA-inhibitor	T2D with NAFLD	AstraZeneca	NCT02826525 (phase-I/IIa)

RGLS4326	SC (subcutaneous injection)	miR-17	miRNA-inhibitor	Autosomal dominant polycystic kidney disease	Regulus Therapeutics Inc	NCT04536688 (phase-I)
CDR132L	IV (intravenous)/ SQ (subcutaneous)	miR-132	miRNA-inhibitor	Heart failure	Cardio Pharmaceuticals GmbH	NCT04045405 (phase-I)
TargomiRs	IV (intravenous)/ SQ (subcutaneous)	miR-16	miRNA-mimic	Malignant pleural mesothelioma	EnGeneIC Limited	NCT02369198 (phase-I)
Remlarsen	Intradermal	miR-29	miRNA-mimic	Keloids, scleroderma	miRagen	NCT03601052 (phase-II)
MRX34	IV (intravenous)/ SQ (subcutaneous)	miR-34a	miRNA-mimic	Melanoma, liver cancer	miRNA Therapeutics, Inc	NCT01829971 (phase-I)

5.2.3 Challenges in Using ncRNA for Diagnosis and Treatment

Noncoding RNAs (ncRNAs), such as long noncoding RNAs (lncRNAs) and microRNAs (miRNAs), are crucial in gene regulation and are becoming potential biomarkers and therapeutic targets for numerous diseases. Even with progress in detection methods such as RT-PCR and next-generation sequencing, issues like specificity, delivery, and tolerability obstruct their use in clinical settings. RNA-centric therapies, which include antisense oligonucleotides (ASOs), small interfering RNAs (siRNAs), RNA aptamers, and guide RNAs (gRNAs), provide innovative approaches to regulate gene expression, showcasing possible treatments for various diseases, such as monogenic conditions and cancers. Ongoing research and technological advancements are vital for addressing current obstacles and completely harnessing the therapeutic capabilities of ncRNAs (Sánchez and Huarte 2013).

5.2.4 Small Interfering RNAs (siRNAs) in Stress Responses and Cancer

The double-stranded RNAs (dsRNAs), derived from various sources, are processed into small interfering RNAs (siRNAs). These sources include RNA-dependent RNA polymerases (RDRs) that synthesize dsRNA from single-stranded RNA, replication of RNA viruses, RNAs generated from inverted repeats, naturally occurring *cis*-antisense transcript pairs, and genomic regions rich in retroelements (Khraiwesh et al. 2012).

Dicer-like (DCL) proteins then cleave these dsRNAs into siRNAs, typically 21–24 nucleotides in length. The size of siRNAs varies depending on the specific DCL protein responsible for cleaving double-stranded RNA (dsRNA) into siRNA fragments of different lengths. Often, multiple DCL proteins target the same dsRNA, resulting in the production of siRNAs of varying sizes. Similar to microRNAs (miRNAs), siRNAs are incorporated into the RNA-induced silencing complex (RISC), which contains Argonaute (AGO) proteins. This complex regulates target genes at either the transcriptional or posttranscriptional level through a mechanism known as RNA-directed DNA methylation (RdDM) (Khraiwesh et al. 2012).

5.2.5 siRNAs and Cancer

Precision medicine introduced in 2015 is driven by developments in genome sequencing and bioinformatics and has revolutionized healthcare by facilitating personalized cancer therapies tailored to unique genetic profiles. This method enhances treatment effectiveness and lowers side effects by focusing on mutations specific to the disease (Jameson and Longo 2015; Koch 2016; Mateo et al. 2022).

Small interfering RNA (siRNA) is a nucleic acid therapy that inhibits genes associated with diseases by binding in a sequence-specific manner. Improvements in synthesis technology have rendered siRNA more accurate, quicker, and economical compared to conventional therapies, and its design enables the targeting of genes that were previously deemed "undruggable" (Yoon et al. 2022; Beaucage 2008; Bumcrot et al. 2006; Dong et al. 2019).

The adaptability of siRNA has promoted its swift advancement in multiple diseases. In 2018, the FDA authorized ONPATTRO™ (patisiran), the inaugural siRNA medication, for transthyretin-mediated amyloidosis (Garber 2018; Adams et al. 2018). Since that time, siRNA therapies have grown to address viral infections, genetic conditions, heart diseases, and cancer (Idris et al. 2021; Wong et al. 2012). As of now, four siRNA medications have received FDA approval, with more than 20 undergoing clinical trials, highlighting their increasing importance in precision medicine. Cancer cells exhibit several distinct characteristics compared to normal cells, making them ideal targets for therapy. siRNAs are powerful because they can specifically target and silence any gene related to disease, including genes involved in cancer. This makes siRNAs a promising tool for cancer treatment (Fitzgerald et al. 2017; Subhan and Torchilin 2019; Zogg et al. 2022; Sajid et al. 2020a).

5.2.6 Interaction of siRNAs with Other Noncoding RNAs

siRNAs silence genes via RNA interference and also interact with noncoding RNAs like circRNAs, lncRNAs, and miRNAs, influencing gene regulation and disease progression, particularly in cancer. They can compete with miRNAs for AGO binding, altering oncogene or tumor suppressor regulation. Targeting lncRNAs like MALAT1 or circRNAs with siRNAs shows promise in disrupting metastasis and oncogenic pathways. However, these interactions raise concerns about specificity and off-target effects, highlighting the need for deeper insight into siRNA-ncRNA dynamics to improve therapeutic outcomes (Slack and Chinnaiyan 2019; Setten et al. 2019a).

5.2.7 Small Interfering RNAs (siRNAs) in Cancer Therapy

In eukaryotic cells, RNA interference (RNAi) serves as a protective response against invasive genetic elements such as viral RNA. siRNA-mediated RNAi initiates when double-stranded RNA (dsRNA), derived from external sources or synthetic creation—is trimmed by the enzyme Dicer into approximately 20-nucleotide siRNA segments. These are incorporated into the RNA-induced silencing complex (RISC), where the guide strand leads the complex to matching mRNA. Argonaute 2 (Ago2) subsequently cuts the target mRNA, thereby silencing the gene after transcription. This specific mechanism serves as an effective research tool and possesses substantial therapeutic potential for addressing disease-associated genes (Setten et al. 2019a; Kanasty et al. 2013; Singh et al. 2018; Hanahan 2022; Lim et al. 2022).

5.2.8 Advantages of siRNAs for Cancer Therapy

Small interfering RNA (siRNA) demonstrates high specificity in gene silencing due to their 20-nucleotide sequence, which enables precise base-pair complementarity, thereby minimizing off-target effects. Second, siRNAs exhibit a superior safety profile as their activity is confined to the cytoplasm, avoiding nuclear entry and the risk of genomic integration. This feature significantly reduces the likelihood of mutagenesis. Lastly, siRNAs are highly efficient; even at low concentrations, they can induce potent gene silencing effects. Furthermore, the range of potential targets for siRNA therapy is immense (Singh et al. 2018; Hu et al. 2020).

5.3 Potential Targets for siRNA Drugs for Cancer Therapy

Cancer cells exhibit distinct features that differentiate them from normal tissue cells, playing a pivotal role in their formation and progression. As outlined by Hanahan, cancer is characterized by 14 hallmark traits that present potential targets for molecular therapies. These include sustained proliferative signaling, evasion of growth suppressors, resistance to cell death (apoptosis), unlimited replicative potential, induction of angiogenesis, activation of invasion and metastasis, genomic instability, and immune system evasion. Additional features involve cancer-associated inflammation, reprogramming of cellular energy metabolism, phenotypic plasticity, non-mutational epigenetic alterations, senescence of surrounding stromal cells, and interactions with polymorphic microbiomes. Collectively, these characteristics not only drive tumorigenesis but also provide diverse avenues for therapeutic interventions (Hanahan and Weinberg 2000; Shakil et al. 2019; Mukai 2003).

Targeted siRNA therapy holds immense promise for silencing genes involved in cancer pathogenesis, offering various therapeutic opportunities. Key targets for siRNA-based drugs include: (1) cell cycle regulators like cyclins, cyclin-dependent kinases (CDKs), polo-like kinases (Plks), and mitosis-related molecules that drive uncontrolled proliferation signaling pathways promoting cell proliferation, such as those involving protein kinase N3 (PKN3), ephrin type-A receptor 2 (EphA2), and AKT isoforms molecules implicated in metastasis, like hepatocyte growth factor receptor (HGFR), transforming growth factor-β (TGFβ), and chemokine receptors (CCR2, CXCR2), angiogenesis-related targets, including vascular endothelial growth factors (VEGFs), their receptors (VEGFRs), matrix metalloproteinase 9 (MMP9), and focal adhesion kinase (FAK) antiapoptotic proteins like Survivin,

Bcl2, and Bcl-xL that enable cancer cells to evade programmed cell death (Shakil et al. 2019; Mukai 2003; Winkle et al. 2021). DNA damage repair enzymes such as MGMT and PARP help maintain cancer cell survival immune checkpoint molecules like PD1, PDL1, and CTLA4 that facilitate immune escape gene regulating cancer metabolism, including glucose transporters (GLUTs), lactate dehydrogenase A (LDH-A), hypoxia-inducible factors (HIFs), and glutaminases oncogenes such as c-myb, c-Myc, and RAS that are central to tumor progression cancer-related tyrosine kinases like EGFR, IGF-1R, Ras proteins, and BCR-ABL, as well as multidrug resistance genes (MDRs) that contribute to therapeutic resistance (Loganathan and Doss 2023; Dietlein et al. 2014; Sharma and Allison 2015; Bollag et al. 2010; Zuo et al. 2023). These pathways and molecules provide a comprehensive framework for developing effective siRNA-based cancer treatments (Winkle et al. 2021).

5.4 The Growing Interest in RNA-Based Therapies in Oncology, Its Modifications and Challenges

RNA molecules have shown significant potential as clinical therapeutics due to their ability to specifically target proteins or compounds previously considered "undruggable," often with minimal side effects. They are integral to numerous cellular functions, including facilitating differentiation, regulating cellular mechanisms, and mediating the translation of genetic information. While coding RNAs, which constitute less than 20% of the total RNA population, have been extensively studied, the expansive domain of noncoding RNAs (ncRNAs) remains largely unexplored. Among these, long noncoding RNAs (lncRNAs) demonstrate diverse functions depending on their subcellular localization (Wei et al. 2023; Hahne et al. 2021). Ribosomal RNAs (rRNAs), transfer RNAs (tRNAs), and microRNAs (miRNAs) play vital roles in the translation process, facilitating protein synthesis. Similarly, small nuclear RNAs (snRNAs) and small nucleolar RNAs (snoRNAs) are key components in RNA processing. Additionally, noncoding RNAs like 7SK RNA contribute to the regulation of transcription. Other types of noncoding RNAs, including circular RNAs (circRNAs), RNase P RNAs, vault RNAs, Y RNAs, Piwi-interacting RNAs (piRNAs), and glycan RNAs, have been identified, but their functions remain less well characterized. Some of these RNAs' roles are yet to be fully elucidated (Gulìa et al. 2020; Iwasaki et al. 2015; Li et al. 2022).

Approximately 15% of the estimated 20,000 human proteins are currently considered druggable. To date, around 700 FDA-approved small-molecule drugs target these proteins, highlighting significant opportunities to develop novel therapies that extend beyond traditional druggable targets. Since 2018, nearly 30% of FDA-approved therapies have been biologics including nucleic acid-based treatments such as antisense oligonucleotides (ASOs), aptamers, messenger RNAs (mRNAs), siRNAs, and miRNAs (de la Torre and Albericio 2019; Bennett 2019; Kowalski et al. 2019; Hanna et al. 2019; Saw and Song 2020). This progress signifies a transformative era in RNA-based therapeutics. The clinical viability of siRNA-based treatments has been exemplified by the FDA's approval of drugs like Patisiran and Givosiran (Saw and Song 2020). These advancements reflect a paradigm shift, moving RNA therapeutics from conceptual exploration to tangible clinical applications. Currently, a growing pipeline of siRNA- and mRNA-based therapies is under active investigation, converting the initial excitement surrounding RNA-based approaches into a grounded and realistic optimism for their therapeutic potential (Burnett and Rossi 2012).

Clinical testing of RNA-based therapies faces numerous challenges; however, prior research has provided valuable insights into their potential for future development. To enhance the efficacy of siRNAs in reducing off-target inflammation, chemical modifications are often necessary. Additionally, achieving effective and tissue-specific delivery of these molecules requires the incorporation of natural or synthetic carriers. Several siRNA-based therapeutics, such as CALAA-01, TD101, ALN-VSP02, and ALN-RSV01, have shown promising clinical outcomes, largely due to these strategies. These cases underscore the potential of siRNA therapies (Fig. 5.1) but also highlight the critical role of precisely engineered carriers that can selectively target the intended cells (Burnett and Rossi 2012; Setten et al. 2019b; Zhu et al. 2022).

5.4.1 Challenges in RNA Therapeutics

Precise RNA screening in clinical laboratories relies on analytical validity, which includes comparison to a gold standard, reproducibility, and resilience to slight changes in testing conditions. Guaranteeing these factors is essential for analytical dependability. Nevertheless, incorporating RNA-seq into clinical processes is made challenging by bioinformatics obstacles, frequently referred to as "analysis paralysis." Key problems consist of the absence of uniform best practices, an excessive variety of software tools, and the intricate nature of RNA-seq pipelines, leading to difficulties with integration and compatibility. In therapeutic settings, RNA stability continues to be a significant obstacle. Successful delivery depends on two methods: formulation-driven strategies (such as nanoparticles, liposomes) and chemical alterations that improve stability and cellular absorption while reducing degradation. In siRNA therapies, alterations to the phosphate backbone or nucleotides minimize immunogenicity, enhance bioavailability, and boost therapeutic precision, tackling essential challenges in clinical application (Dong et al. 2019; Saw and Song 2024).

This chapter explores the therapeutic possibilities of small interfering RNAs (siRNAs) in cancer, emphasizing their function in gene silencing, effects on cancer-related pathways, and relationships with noncoding RNAs. It examines progress in siRNA-focused treatments, methods to improve delivery, stability, and precision, and discusses major issues like immunogenicity and unintended targeting effects. This chapter further emphasizes ongoing advancements and upcoming paths in RNA-cantered cancer therapies.

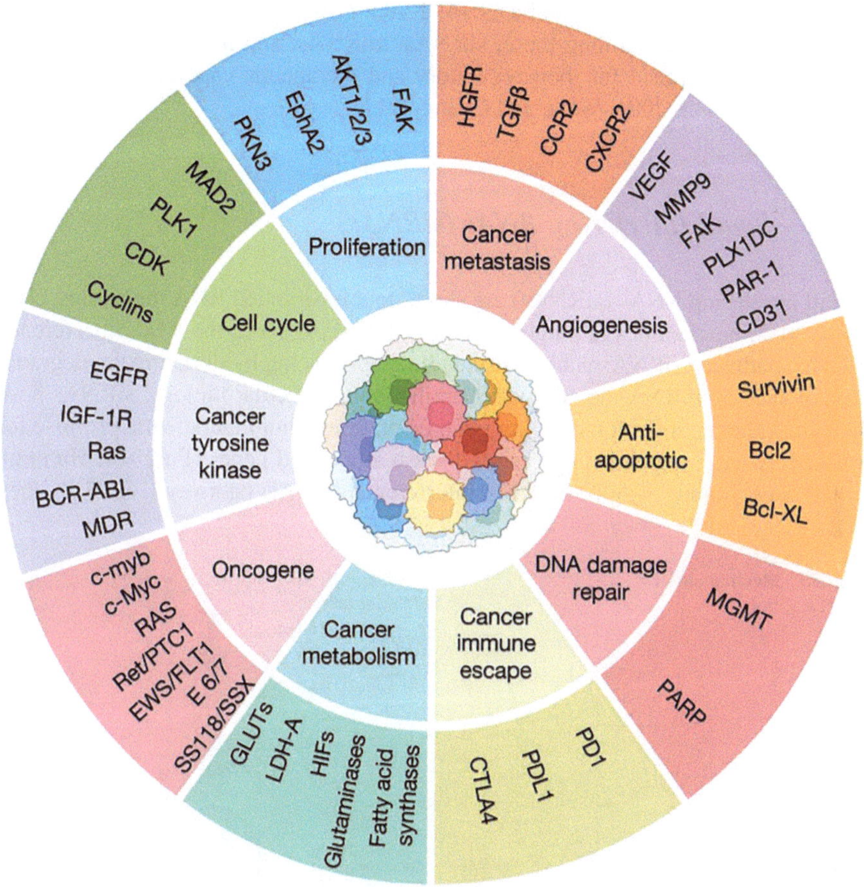

Fig. 5.1 Illustration of the pathways targeted by siRNA in cancer treatment. The inner circle highlights various key pathways implicated in cancer therapy, including cell cycle regulation, proliferation, metastasis, angiogenesis, antiapoptotic mechanisms, DNA damage repair, immune escape, cancer metabolism, oncogenes, and tyrosine kinase signaling. The outer circle identifies specific siRNA interference targets associated with these pathways, which have been investigated in preclinical studies. (This figure is adapted from Zhang et al.)

5.4.2 Understanding Noncoding RNAs and siRNAs

Noncoding RNAs (ncRNAs) are RNA molecules that are transcribed from DNA but do not produce proteins. This various collection encompasses rRNA, tRNA, snRNA, snoRNA, miRNA, lncRNA, and circRNA (Anastasiadou et al. 2018; Li et al. 2020). Although ncRNAs do not possess coding capabilities, they are crucial in regulating gene expression across various levels—transcription, RNA processing, chromatin remodeling, epigenetics, translation, and signaling. They are crucial for numerous biological functions and are associated with illnesses like cancer and neurodegenerative conditions (Fu 2014; Kaikkonen et al. 2011; Li et al. 2021a; Zhang et al. 2021a). Among these, siRNAs, miRNAs, and lncRNAs are the most extensively researched for their regulatory and therapeutic capabilities (Bai et al. 2024; Tuschl et al. 2001; Wang and Chang 2011).

5.4.3 *Small Interfering RNAs (siRNAs)*

Small interfering RNAs (siRNAs) are short, double-stranded RNA molecules, typically 20–25 nucleotides in length, that play a critical role in the RNA interference (RNAi) pathway. siRNAs achieve efficient gene silencing by directing the degradation of specific mRNA targets. Widely utilized in molecular biology, siRNAs have become valuable tools for studying gene function and hold therapeutic promise for selectively targeting and suppressing disease-associated genes (Fu 2014; Hu et al. 2004). Functioning within the RNA interference (RNAi) pathway (Fig. 5.2), they

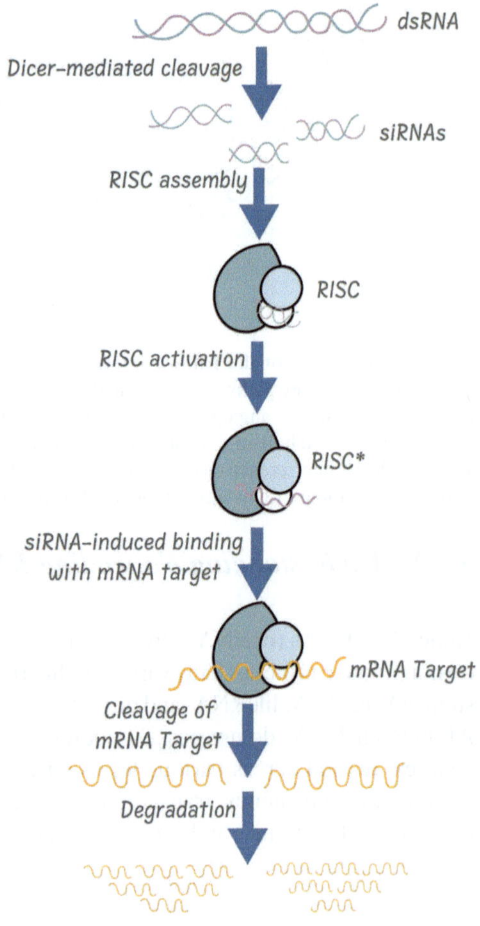

Fig. 5.2 Mechanism of siRNA action

share similarities with microRNAs (miRNAs). siRNAs regulate gene expression by binding to messenger RNA (mRNA) sequences with complementary bases, leading to mRNA degradation and inhibition of translation (Monga et al. 2017). RNA interference (RNAi) is a biological mechanism where double-stranded RNA (dsRNA) induces gene silencing by promoting the degradation of matching mRNA. Initially discovered in plants, this process was later demonstrated in the roundworm *Caenorhabditis elegans*, where long dsRNA effectively downregulated gene expression. Today, RNAi serves as a powerful tool for studying gene functions and has significantly advanced therapeutic research (Monga et al. 2017; Fire et al. 1998; Endoh and Ohtsuki 2009; Laganà et al. 2015).

5.4.4 Structure and Biogenesis of siRNA

Naturally occurring siRNAs have a well-defined structure comprising a sense (passenger) strand and an antisense (guide) strand, typically spanning 20–24 nucleotides. These molecules are characterized by a unique two-nucleotide overhang at the 3′ end of each strand, a result of double-stranded RNA (dsRNA) processing. The siRNA duplex features hydroxyl groups at the 3′ ends and phosphate groups at the 5′ ends. The enzyme Dicer plays a key role in generating siRNAs from long dsRNA or short hairpin RNAs (Bernstein et al. 2001). Synthetic siRNAs with complementary sequences can be introduced into cells to silence almost any gene. In the post-genomic era, siRNAs have become indispensable for validating gene functions and uncovering potential therapeutic targets (Dana et al. 2017; Zhang et al. 2021b).

5.4.5 Biogenesis

The production of siRNAs starts with the enzymatic cleavage of long double-stranded RNA (dsRNA) precursors by the RNase III enzyme Dicer. This process generates siRNA duplexes approximately 20–24 base pairs in length. These siRNA duplexes are then loaded into the RNA-induced silencing complex (RISC). Within RISC, the passenger strand is usually degraded, leaving the guide strand bound to the Argonaute protein, a key component of the complex (Pratt and MacRae 2009).

5.4.6 siRNA-Mediated Gene Silencing

Posttranscriptional Gene Silencing (PTGS). The RNA-induced silencing complex (RISC) plays a pivotal role in the siRNA-mediated posttranscriptional gene silencing (PTGS) process. This mechanism silences gene expression by targeting and cleaving specific mRNA molecules, thereby preventing their translation into proteins (Bhattacharjee et al. 2019). The RNA interference (RNAi) pathway begins with the enzyme Dicer, which processes long double-stranded RNA (dsRNA) into small interfering RNAs (siRNAs) that are 21–23 nucleotides in length, featuring 2-nucleotide overhangs at their 3′ ends (Agrawal et al. 2003). These siRNAs are subsequently loaded into RISC, where the siRNA duplex is unwound to generate a guide (antisense) strand and a passenger (sense) strand. The passenger strand is degraded, while the guide strand, characterized by greater 5′ end stability, is selectively incorporated into the active RISC, enabling precise gene silencing (Lee et al. 2004). Once the guide strand is loaded into RISC, it identifies a fully complementary sequence on the target mRNA and directs the complex to it. The catalytic component of RISC, Argonaute (primarily Ago2), facilitates the endonucleolytic cleavage of the target mRNA (Carthew and Sontheimer 2009). This cleavage occurs precisely at the phosphodiester bond between the nucleotides corresponding to positions 10 and 11 of the siRNA, counting from its 5′ end (Tomari and Zamore 2005). The resulting mRNA is divided into two fragments: a 5′ fragment and a 3′ fragment. These fragments are subsequently degraded by cellular exonucleases—5′-3′ exoribonuclease 1 (XRN1) degrades the 3′ fragment, while the 3′-5′ exosome complex breaks down the 5′ fragment (Hu et al. 2020). The dissociation of cleaved mRNA fragments from the RISC allows the complex to target additional mRNA molecules, amplifying the silencing effect. This separation is thought to be driven by ATP hydrolysis (Gredell et al. 2008). However, mismatches between the siRNA and the target mRNA near the cleavage site can sometimes inhibit the endonucleolytic activity of Argonaute proteins. Additionally, even with complete complementarity, Argonaute proteins within RISC may lack endonuclease activity. In such cases, gene silencing occurs through a mechanism similar to microRNA (miRNA)-mediated suppression. Instead of cleaving the mRNA, RISC either blocks its translation or promotes its degradation (Tomari and Zamore 2005). The efficiency and specificity of siRNA-mediated gene silencing are influenced by factors such as the sequence similarity between the siRNA and target mRNA, the presence of secondary structures in the mRNA, and the accessibility of the target site. As a result, meticulous design of siRNA sequences is essential for achieving effective and precise gene knockdown. With its highly regulated mechanism, siRNA-mediated RNA interference (RNAi) has become a valuable tool in both research and therapeutic contexts. It allows for targeted mRNA degradation, leading to reduced protein levels within cells and playing a critical role in posttranscriptional regulation (Gredell et al. 2008; Yeung et al. 2005).

5.4.7 siRNA as Therapeutic Tools

A decade after the discovery of RNA interference (RNAi) in 1993, the pharmaceutical industry began making significant investments in the development of siRNA-based therapies. SiRNA offers several advantages over monoclonal antibodies and traditional small molecules, making it an appealing therapeutic option (Zhang et al.

2021b; Hu et al. 2020). One of the key benefits of siRNA therapy is its dosing flexibility; it can be administered as infrequently as once every few months, offering a convenient solution for the treatment of chronic diseases. Unlike small molecules and monoclonal antibodies, which require specific protein conformation recognition, siRNA operates through Watson–Crick base pairing with target mRNA. This mechanism enables siRNA to silence virtually any gene with high specificity, provided the corresponding nucleotide sequence is available (Hu et al. 2020), thus expanding the range of therapeutic targets to include previously undruggable proteins. Despite its potential, siRNA therapy encountered numerous challenges during its early research stages, particularly in developing efficient methods for delivering siRNA to target cells (Friedrich and Aigner 2022). Additionally, RNAi therapies were often identified by the immune system as remnants of infectious agents, triggering immune responses that compromised their efficacy and safety. This led to disappointing therapeutic results, as animal models often failed to predict the human immune response, causing many investors to reconsider further investment in RNAi research (Eisenstein 2019; Sajid et al. 2020b). Despite these challenges, companies like Sirna Therapeutics, Dicerna Pharmaceuticals, and Alnylam Pharmaceuticals continued to advance RNAi therapies for human use. As they discovered that most systemically delivered siRNA therapies tend to accumulate in the liver, drug development efforts were redirected toward liver-related diseases, where siRNA treatment showed the most promise. Improvements in the chemical modification of RNA molecules have enhanced siRNA stability, pharmacokinetics, and safety profiles, while also reducing immune activation. These innovations have broadened the therapeutic potential of siRNA-based drugs (Ali Zaidi et al. 2023; Tiemann and Rossi 2009). As a result, numerous siRNA therapies have gained approval, with many more under clinical development. These advancements underscore the significant potential of siRNA as a versatile therapeutic option for treating various diseases, especially those related to the liver. With continued research, siRNA therapies (Table 5.4) are poised to transform the treatment of genetic, infectious, and metabolic disorders, representing a major step toward personalized medicine (Ebenezer et al. 2025).

5.5 The Role of siRNA in Cancer Therapy

Traditional cancer therapies, including surgery, radiation therapy, and chemotherapy, continue to be standard choices. Surgery efficiently eliminates localized tumors but has limitations when it comes to metastasis (Ganesh and Massague 2021; Ji et al. 2022). Radiotherapy employs high-energy radiation to eliminate cancer cells, whereas chemotherapy restricts cell growth but frequently leads to resistance and adverse effects in healthy tissues (Grimes 2022; Nussinov et al. 2021). To address these limitations, immunotherapy has become more prominent. Approaches such as monoclonal antibodies, pericyte-centered treatments, and immune checkpoint blockade (ICB) have demonstrated effectiveness, particularly in melanoma and lung

Table 5.4 RNA-based therapeutic delivery methods

Delivery method	Composition	Payload	Example	Advantage	Limitation
Lipid nanoparticles	DLin-MC3-DMA (MC3)	siRNA	siRNA drug (Patisiran)	Increased half-life; protection from nucleases acids	Elevated risk of immunotoxicity and immunogenicity
	ALC-0315	mRNA, gRNA	SARS-CoV-2 mRNA vaccine (BNT162b2)	Facilitate the endosomal escape of RNA molecules; increase in protein expression or immune responses in mice compared to MC3	
	SM-102	Donor RNA	SARS-CoV-2 mRNA vaccine (mRNA-1273)		
Cationic polymer-based polyplexes	Polyethyleneimine (PEI)	mRNA, siRNA, miRNA, gRNA	SiG12D LODER; CRM197-PEG-PEI-based complexes	High charge density and pH buffer capacity	Potential toxicity and plasma instability
	Chitosan	mRNA, siRNA, (DNA)	siRNA-loaded chitosan lactate nanoparticles (CL-TAT-HA)	Low toxicity, biodegradability, biocompatibility, and permeability enhancing	Low transfection efficiency
	Cyclodextrin polymer (CDP)	siRNA	CALAA-01		
	Poly(β-amino esters) (PBAEs)	mRNA, gRNA, donor RNA	HPV16 E7-targeting CRISPR/short hairpin RNA (shRNA)	Low toxicity	Low charge density
Exosomes	Derived sources: Mesenchymal stem cells	miRNA, siRNA, mRNA, other ncRNAs	miR-124; *KRAS G12D* siRNA	With minimal immune clearance and adverse effects	

Spherical nucleic acids	AuNP, quantum dots (QDs), SiO2, Ag, Fe3O4,	mRNA, siRNA, gRNA, donor RNA	NU-0129 (Bcl2Like12 (Bcl2L12) siRNA)	Show rapid cellular uptake kinetics and intracellular transport; induce a negligible immune response	
			AST-005 (inhibiting TNF-α mRNA via ASOs); XCUR17 (interleukin-17 receptor-α through ASOs)		
DNA nanostructures		ASOs, siRNAs, aptamers, CRISPR-Cas9	AS1411 aptamers	The object's size, shape, and plasticity can be fine-tuned	Enzymatic hydrolysis, low cellular uptake, immune cell recognition and degradation, and unclear biodistribution profiles

Zhu et al. (2022)

cancer. CAR-T cell treatment is effective for drug-resistant leukemias, but it poses risks such as cytokine storms (Khalil et al. 2016; Zhang et al. 2022). Precision medicine, facilitated by advancements in genomics and bioinformatics, customizes treatment by focusing on individual patient genetic changes, enhancing results and reducing toxicity (Ma et al. 2019) One promising avenue within precision medicine is the use of short interfering RNA (siRNA), a nucleic acid-based therapy capable of selectively silencing disease-related genes through sequence-specific targeting (Mateo et al. 2022; Yoon et al. 2022; Hu et al. 2020). Furthermore, advancements in nucleic acid synthesis technology have made siRNA production more precise, faster, and cost-effective compared to small molecule and antibody-based therapies. The development of siRNA sequences has overcome a major challenge in personalized medicine by allowing the targeting of genetic loci previously considered undruggable (Bumcrot et al. 2006; Dong et al. 2019). siRNA therapies have gained attention for their versatility and rapid development. In 2018, the FDA approved the first siRNA drug, ONPATTRO™, for treating hereditary amyloidosis. Since then, siRNAs have been explored for viral infections, genetic disorders, cardiovascular disease, and cancer, with over 20 candidates in clinical trials and 4 FDA approved (Fitzgerald et al. 2017; Subhan and Torchilin 2019; Zogg et al. 2022; Kosmas et al. 2018). Despite their promise, siRNAs face major delivery challenges. Extracellular barriers include degradation by nucleases, rapid renal clearance, immune activation, and poor membrane permeability (Fedorov et al. 2006). Intracellularly, issues include off-target effects and endosomal trapping. To address these, chemical modifications (e.g., backbone, ribose, base, terminal) improve stability and specificity, while delivery systems like viral vectors, polymers, lipids, and conjugates enhance uptake and bioavailability (Corey 2007; Khvorova and Watts 2017; Bramsen and Kjems 2013).

Additionally, for siRNAs to effectively perform their gene-silencing function, they must detach from the delivery vector and remain free within the cytoplasm. To achieve this, various bio-responsive materials have been utilized under different conditions to regulate the release of siRNAs (Zhou et al. 2022; Neuberg and Kichler 2014).

5.6 Mechanisms of siRNA Action in Cancer Cells

Small interfering RNAs (siRNAs) are double-stranded RNA molecules consisting of 21–23 nucleotides that silence genes by specifically targeting certain mRNAs for degradation or preventing translation. Their capability to inhibit oncogenes makes them valuable assets in cancer treatment. Upon entering cells, siRNAs are incorporated into the RNA-induced silencing complex (RISC), where the guiding strand attaches to the target mRNA. Argonaute proteins subsequently cut the mRNA, thereby silencing the gene. This method has proven successful in decreasing KRAS-driven tumor growth in pancreatic cancer while also inhibiting STAT3, a crucial factor in tumor advancement and immune evasion (Takahashi et al. 2020; Gao Lf et al. 2005). In certain instances, siRNAs inhibit translation rather than breaking down mRNA. For instance, focusing on cyclin D1 stops the cell cycle and inhibits the growth of breast cancer cells, whereas siRNAs targeting EGFR interfere with essential signaling pathways for proliferation. Cancer cells often evade apoptosis, promoting unchecked growth. siRNAs help restore apoptosis by silencing antiapoptotic genes. For instance, targeting BCL-2 in leukemia cells enhances caspase activation and sensitizes cells to chemotherapy. Similarly, silencing survivin in cervical cancer increases apoptosis and improves treatment response. siRNAs also inhibit angiogenesis by targeting factors like VEGF. In glioblastoma models, VEGF silencing reduced blood vessel formation, slowed tumor growth, and improved survival. Combining siRNAs with antiangiogenic therapies further boosts efficacy (Miyashita and Reed 1993; Carpenter 2000; Niola et al. 2006).

Cancer cells evade immune detection by exploiting checkpoints like PD-L1. siRNAs targeting PD-L1 can restore immune surveillance and enhance T-cell cytotoxicity, especially when combined with checkpoint inhibitors. siRNAs also help reprogram the tumor microenvironment by reducing pro-inflammatory cytokines. Despite their promise, siRNA therapies face delivery and specificity challenges. Advances like lipid nanoparticles, peptide conjugates, and chemical modifications (e.g., 2′-O-methyl, phosphorothioate) improve stability, reduce immune activation, and enhance therapeutic precision (Won et al. 2022).

5.7 Targeting Oncogenes and Tumor Suppressor Genes

Short double-stranded RNAs known as small interfering RNAs (siRNAs) silence genes via the RNA interference (RNAi) pathway. After being processed by Dicer and incorporated into the RNA-induced silencing complex (RISC), the guide strand helps RISC locate complementary mRNA, at which point Argonaute-2 facilitates mRNA cleavage and silencing (Hannon 2002).

In therapeutic applications, siRNAs can aim at oncogenes such as EGFR, KRAS, MYC, and BCL2, inhibiting tumor growth and improving responses to chemotherapy. Additionally, siRNAs indirectly reinstate tumor suppressor function by silencing negative regulators like MDM2 (an inhibitor of p53) or modulators of the BRCA1/2 and PTEN pathways (Vlassov and An 2012; De La and McDonald 2011; Chappell and Lasky 2014). To improve specificity and effectiveness, siRNAs undergo chemical modifications (such as 2′-O-methyl and LNAs) and are transported using lipid or polymeric nanoparticles. Targeted delivery systems that utilize antibody or aptamer conjugates take advantage of tumor-specific markers such as PD-L1 or CD71. In spite of the challenges related to delivery and immune activation, progress in bioresponsive systems and combination therapies is enhancing the clinical possibilities of siRNA-based cancer therapies (Ziemer and Kurrle 2017; Whitehead and Langer 2009; Manoharan 2017).

5.8 siRNAs in Tumor Growth Inhibition and Apoptosis Induction

siRNAs inhibit tumor growth by silencing genes critical to cancer progression. For instance, targeting STAT3, a transcription factor involved in cell survival and immune evasion, significantly reduced tumor growth and induced apoptosis in Hep2 cells (Gao Lf et al. 2005). Similarly, silencing RPL21, an overexpressed ribosomal protein in pancreatic cancer, disrupted DNA replication and inhibited proliferation (Corey 2007). siRNA-mediated knockdown of VEGF in breast cancer reduced angiogenesis and tumor growth, demonstrating the utility of targeting tumor vasculature (Ambrosini et al. 1997).

siRNAs also restore apoptotic signaling in cancer cells. Silencing antiapoptotic genes such as RHBDF1 in breast cancer, survivin in cervical cancer, and BCL-2 in hepatocellular carcinoma increased apoptosis and improved chemotherapy response. These findings highlight the therapeutic potential of siRNAs in reactivating cell death pathways and suppressing tumor growth (Sharma and Allison 2015; Yan et al. 2008).

5.9 siRNAs for Specific Cancer Types

Small interfering RNAs (siRNAs) have emerged as a powerful tool in cancer therapy due to their ability to silence genes involved in the development of various cancers (Fig. 5.3). By specifically targeting individual mRNAs, siRNAs allow researchers to focus on the unique genetic and molecular weaknesses of different cancer types.

5.10 Application of siRNAs in Solid Tumors Versus Hematological Cancers

Small interfering RNAs (siRNAs) offer targeted gene silencing and have emerged as powerful tools in cancer therapy. While their application spans both solid tumors and hematological malignancies, each presents unique challenges and opportunities (Mirzaei et al. 2021).

In solid tumors such as breast, lung, pancreatic, and liver cancers, siRNAs target genes involved in angiogenesis, drug resistance, and proliferation. Notable targets include EGFR, KRAS, VEGF, and PD-L1, with siRNAs showing efficacy in reducing tumor growth and enhancing immune responses. However, effective delivery remains a major hurdle due to tumor heterogeneity, poor vascularization, and immune clearance. Advanced delivery platforms—like lipid nanoparticles (LNPs), stable nucleic acid-lipid particles (SNALPs), and polymeric carriers—have shown

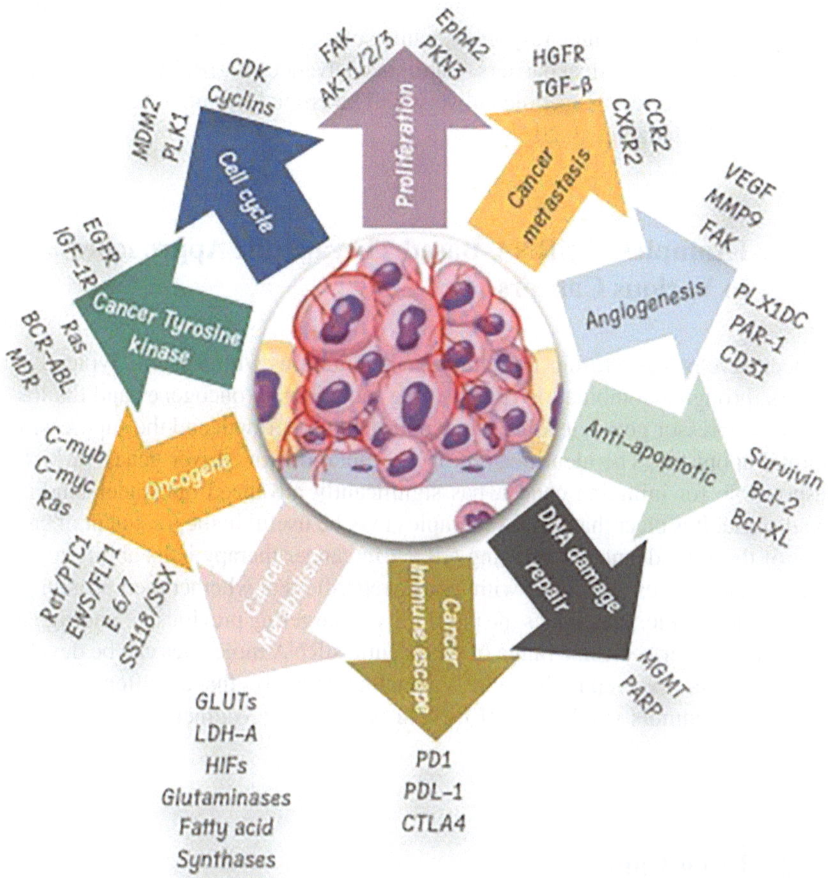

Fig. 5.3 The targeting pathways of siRNA in cancer treatment. The arrows illustrate the diverse pathways involved in cancer treatment, encompassing: cell cycle regulation, proliferation, cancer metastasis, angiogenesis, antiapoptotic processes, DNA damage repair, cancer immune escape, cancer metabolism, oncogene, and cancer tyrosine kinase. The outer represents the specific siRNA interference targets corresponding to these pathways, which have been explored in preclinical research

promise, as seen with ALN-VSP02 in clinical trials (Sharma and Allison 2015; Zuo et al. 2023; Mirzaei et al. 2021). In contrast, hematological malignancies such as leukemia, lymphoma, and myeloma offer more accessible targets due to the circulatory nature of blood cancers. siRNAs targeting BCL2, MYC, and STAT3 have demonstrated effectiveness in promoting apoptosis and overcoming resistance. Delivery methods benefit from easier systemic access, with lipid nanoparticles and antibody-siRNA conjugates (e.g., CD71- or PD-L1-targeting) enabling precision delivery. However, off-target effects and immune activation remain key concerns due to high cell turnover in these cancers (Curreri et al. 2022; Russo and Saif 2016; Marques and Williams 2005).

In summary, hematological malignancies favor simpler delivery but demand strategies to avoid immune responses, while solid tumors require sophisticated systems to overcome physical barriers. Continued advances in siRNA design, chemical stabilization, and tumor-specific delivery will be crucial for maximizing their clinical potential (Hattab et al. 2021).

5.11 Examples of siRNA-Based Therapeutic Approaches in Various Cancers

Cancer develops as a result of multiple gene mutations within a cell. Typically, this occurs through a combination of activating mutations in oncogenes and the loss of tumor suppressor genes, leading to uncontrolled cell growth and the suppression of normal apoptosis. The identification of key mutations in "driver genes," which are responsible for initiating cancer, has significantly advanced our understanding of the disease. It is clear that siRNA therapies may be useful in the treatment of cancer. One of the key advantages of using siRNA in cancer therapy is its ability to selectively silence genes associated with cancer, regardless of whether their protein products are druggable. This opens the possibility of targeting previously "undruggable" genes. Moreover, a diverse range of therapeutic siRNA molecules can be developed to target genes involved in the various signaling pathways that are often abnormally activated in tumors (McLeod 2013; Suva et al. 2013; Walther and Schlag 2013; Bora et al. 2012) (Table 5.2).

5.12 Lung Cancer

Lung cancer is the primary cause of cancer-related fatalities globally, with non-small cell lung cancer (NSCLC) making up approximately 85% of instances. Even with advancements in traditional therapies, metastasis and resistance to drugs lead to unsatisfactory results. Therapies using siRNA present a hopeful option by specifically silencing oncogenes and can be administered either intrapulmonarily or systemically, with the latter increasing siRNA stability and reducing side effects (Molina et al. 2008). NSCLC often features EGFR mutations, particularly deletions in exon 19 that promote tumor development. SiRNAs specific to alleles that target mutant EGFR have demonstrated effectiveness in decreasing tumor mass and promoting apoptosis, while leaving wild-type alleles unharmed (Xue et al. 2014). To overcome resistance associated with KRAS mutations and p53 deletion, a mix of KRAS siRNA and a miR-34a mimic administered through nanoparticles notably reduced tumor size and enhanced survival in a mouse model. Additional targets encompass RPN2, which facilitates chemoresistance and stabilizes the mutant p53.

The knockdown of PnkRNA™ decreased tumor growth in A549 xenografts. In the same way, focusing on MDM2—a p53 suppressor—through pH-sensitive siRNA nanoparticles hindered tumor development in p53-mutant NSCLC models (Kurashige et al. 2012; Fujita et al. 2013).

5.13 Prostate Cancer

Prostate cancer is the most prevalent cancer among men, with existing treatments frequently resulting in considerable side effects. The loss of the tumor suppressor PTEN, which is a critical regulator of the PI3K pathway, occurs frequently; however, PI3K inhibitors have demonstrated restricted effectiveness, leading to a focus on other targets like oncogenes associated with metastasis and growth (Guo et al. 2013).

Myc is commonly overexpressed in prostate cancers. Silencing Myc using siRNA diminished stem-like characteristics and tumor proliferation in both in vitro and in vivo studies, resulting in nearly total tumor suppression in mice. In a similar manner, using siRNA nanocapsules to target CK2, a kinase that controls vital oncogenic pathways, resulted in tumors that were approximately 50% smaller in xenograft models (Xi et al. 2004; Jin et al. 2013; Wang et al. 2010a).

Other potential targets consist of KLK4, which influences AR and PI3K signaling. siRNA targeting KLK4 resulted in approximately 90% tumor reduction in mice (Bosch et al. 2010; Liu et al. 2014; Xu et al. 2013). In prostate cancer models, the silencing of Notch1, which is known to promote invasion, led to decreased tumor growth and enhanced apoptosis (Zhang et al. 2014).

5.14 Breast Cancer

Breast cancer is the most diagnosed cancer in women and a leading cause of cancer-related deaths. It is classified into subtypes based on receptor expression—ER, PR, and HER2—which guide treatment strategies (Howell and Dowsett 2004; Bouclier et al. 2008). siRNA therapies have shown promise in targeting key molecules in different subtypes. In ER-α-positive breast cancer (~70% of cases), siRNA against ER-α reduced tumor growth in MCF-7 xenografts when delivered via stealth nanocapsules (Bosch et al. 2010). For triple-negative breast cancer (TNBC), which lacks ER, PR, and HER2 expression and is resistant to hormone therapies, CDK1 siRNA delivered with lipid nanoparticles suppressed tumor growth in c-Myc-overexpressing models without toxicity. Targeting Myc or ATM, a DNA damage response kinase, also reduced tumor burden using nanoparticle delivery systems (Liu et al. 2014; Xu et al. 2013).

Epigenetic regulators like DNMTs are another target; DNMT-specific siRNA combined with anti-HER2 delivery restored tumor suppressor gene expression and suppressed tumor growth in HER2+ models (Wang et al. 2010b, 2017). Additionally, OPN-targeting siRNA reduced tumor growth in metastatic models, supporting its potential as a therapeutic and prognostic marker (Minai-Tehrani et al. 2012).

5.15 Advantages of siRNA in Cancer Therapy

Small interfering RNAs (siRNAs) have emerged as a revolutionary tool in molecular medicine, particularly in cancer treatment. By precisely targeting and silencing genes involved in tumor initiation, progression, drug resistance, and metastasis, siRNAs present a promising therapeutic strategy. Compared to traditional methods, siRNA-based therapies offer advantages such as enhanced specificity, flexibility, and the potential for combination treatments.

5.16 High Specificity and Reduced Off-Target Effects

RNA interference (RNAi) consists of shRNA, miRNA, and siRNA, with synthetic siRNAs being the most promising for therapeutic applications because of their accuracy and safety. siRNAs provide high specificity by base-pairing with target mRNA, reducing off-target effects. They function within the cytoplasm without incorporating into the genome, thus minimizing safety issues. siRNAs are also potent at minimal doses and can be tailored to aim at nearly any gene, owing to progress in genomic databases and sequencing technologies (Singh et al. 2018; Hu et al. 2020; Saw and Song 2020).

5.17 Potential for Personalized Medicine

Normal and cancer cells differ significantly, with cancer exhibiting 14 key hallmarks such as uncontrolled proliferation, metastasis, immune evasion, and altered metabolism. These traits offer numerous siRNA-targetable pathways, including those involved in the cell cycle (e.g., CDKs, Plks), proliferation (e.g., AKT, EphA2), metastasis (e.g., TGFβ, CCR2), angiogenesis (e.g., VEGF, FAK), apoptosis resistance (e.g., Bcl-2, Survivin), DNA repair (e.g., MGMT, PARP), immune escape (e.g., PD-1, CTLA4), metabolism (e.g., GLUTs, HIFs), and oncogenes (e.g., Myc, RAS). Ideal siRNA targets are highly expressed in tumors, essential for cancer progression, and broadly applicable across tumor types, ensuring both efficacy and minimal toxicity (Sasaki et al. 2010; Valastyan and Weinberg 2011).

In addition, there are cancer tyrosine kinase-related signaling pathways, such as epidermal growth factor receptor (EGFR), insulin-like growth factor 1 receptor (IGF-1R), Ras family proteins, BCR-ABL tyrosine kinases, and cancer cell multi-drug resistance genes (MDRs) (Assaraf et al. 2019; Bukowski et al. 2020). Given the numerous potential targets for cancer therapy, careful selection is essential, guided by the following criteria: (1) Preferential expression in tumor cells, with minimal to no expression in normal cells; (2) A significant role in promoting carcinogenesis; (3) Ideally, the target should be a broad-spectrum gene that is applicable to multiple tumor models. These criteria ensure the safety and efficacy of the treatment approach, and they encompass factors such as the Therapeutic Index and Therapeutic Window (Reddy and Kaelin Jr. 2002; Kaelin Jr. 2005). The crucial involvement of target genes in tumor development enhances the chances of successful treatment, while their broad-spectrum applicability allows for wider therapeutic use and increases market potential. Additionally, the differential expression of these genes in tumor versus normal cells helps minimize harm to healthy tissues.

5.18 Challenges in siRNA-Based Cancer Therapy

Despite the ongoing excitement in the academic community, to treat the diseases like cancer and HIV, siRNA-based cancer therapies continue to encounter challenges and hurdles in clinical advancement (Wang et al. 2017).

5.18.1 Delivery Issues

One of the major obstacles that scientists and researchers encounter in creating siRNA formulations is addressing its negative physicochemical characteristics. siRNA shows low stability, possesses a negative charge, and has a comparatively large molecular size and weight (Wang et al. 2010b). Moreover, its hydrophilic characteristics make passive diffusion for cellular delivery more challenging, leading to poor cellular absorption. Naked siRNAs are quickly broken down by serum endonucleases, passed through the glomeruli, and eliminated by the kidneys, leading to a brief half-life of 15 min to 1 h. Moreover, plasma proteins attach to nanocarriers in the bloodstream, resulting in their elimination by the reticuloendothelial system (RES). These constraints can be managed via chemical modifications, including 2′-fluoro or 2′-amine substitutions, changes to the sugar ring, phosphorothioate linkages, or PEGylation of the delivery vehicle. 2′-Modifications of sugar have been extensively studied and shown to be the most effective technique for improving the efficacy of siRNAs (Paul et al. 2022). Various 2′-modifications of ribose, including 2′-O-methyl, 2′-fluoro, and 2′-methoxyethyl, enhance thc thermodynamic stability of siRNA duplexes. Besides the intracellular and extracellular barriers, the biggest challenge for Alzheimer's disease (AD) therapy is the blood–brain barrier. The advancement of RNA-focused treatments necessitates the refinement of positional changes for siRNA studies to achieve optimal in vivo outcomes (Paul et al. 2022; Alshaer et al. 2021). The role of innate immunity is to recognize pathogens, eliminate them, and support adaptive immunity. It was previously believed that siRNA less than 30 nucleotides in length were sufficiently small to bypass the immune system and prevent nonspecific activation of the interferon response. Nonetheless, later experiments involving short synthetic siRNAs indicated that siRNA could stimulate the immune response and initiate cytokine production both in vivo and in vitro (Judge et al. 2005). This natural immune response can be activated by siRNA or by the carriers, such as cationic lipids utilized to administer siRNA in vivo. Previous research indicates that siRNAs play a role in activating monocytes in the innate immune system and proinflammatory cytokines in dendritic cells through TLR8. Additionally, certain siRNA can promote the synthesis of interferon-α (IFNα) in plasmacytoid dendritic cells through TLR7. This immune response activation can be reduced by chemically altering the ribose siRNA backbone through either 2′-O-methylation (Robbins et al. 2009; Judge and Maclachlan 2008). or phosphorothioate. Recently, it has been shown that modifying the siRNA structure with a combination of DNA analogues, 2-F-modified RNAs, and Locked Nucleic Acids can enhance silencing effectiveness while reducing immunostimulatory properties in human blood cells (Deleavey et al. 2010).

5.18.2 *Off-Target Effects (OTE)*

The siRNA must be specific to the target, and its specificity depends on the location of the nucleotides in the short siRNA design. Research has shown that a sequence homology of 8–9 nucleotides can provide partial similarity or complementarity with genes that are not targeted. Nonspecific gene silencing may occasionally lead to toxicity and provoke different interferon responses. Off-target effects (OTEs) can occur due to various reasons, such as inadequate siRNA sequence design, incorrect identification of the guide and passenger strands by the RNA-induced silencing complex (RISC), dysfunction or oversaturation of the RNAi machinery, and the application of overly high siRNA concentrations (Jackson et al. 2003; Persengiev et al. 2004). These OTEs can be reduced by refining siRNA sequences and following certain design guidelines, including suitable guanine-cytosine (GC) content, ideal sequence length, prevention of internal repeats, favored nucleotide bases at critical positions of the sense and antisense strands, and reducing secondary structures (Ryther et al. 2005).

5.19 Cellular Internalization

Despite several adjustments made to address the challenges of siRNA delivery and target precision, there remains uncertainty about whether siRNA will be delivered to the precise location. Its primary action is initiated only once it is successfully delivered into the cell's cytoplasm, while circumventing the endosomal pathway. Receptor-mediated endocytosis (pinocytosis) allows for siRNA uptake through the endosomal route. Inside endosomes, siRNA encounters acidic environments and is ultimately moved to lysosomes for additional acidification and breakdown by nucleases. To address this, specific polymers exhibiting "proton sponge" characteristics enhance the entry of protons and chloride ions into endosomes, leading to osmotic swelling and rupture, and consequently releasing siRNA into the cytosol (Oliveira et al. 2007; Ming 2011).

5.19.1 Effective Delivery of siRNA

Choosing the proper systemic delivery route is crucial for effective siRNA-driven gene silencing. Optimal delivery systems ought to be biodegradable, biocompatible, non-immunogenic, and able to effectively target cells while evading hepatic clearance (Juliano et al. 2008). Nanocarriers with sizes less than 200 nm are particularly efficient, providing benefits like increased stability, diminished clearance by the renal and reticuloendothelial system (RES), and better specificity when attached to targeting ligands. These characteristics position them as a viable option for secure and effective siRNA delivery (Ozpolat et al. 2010; Uchida et al. 2020).

In addition to standard methods, cutting-edge technologies have enhanced siRNA delivery even more. For example, Arrowhead Pharmaceuticals created the Targeted RNAi Molecule (TRiM™) platform—an efficient and robust ligand-mediated system that improves target specificity. Its dimer-conjugation architecture enables the concurrent silencing of two genes using one drug and facilitates the enhancement of pharmacokinetic characteristics, providing a flexible and efficient siRNA delivery approach. The various delivery strategies of siRNA in cancer therapy are discussed below.

Lipid-based systems, which include liposomes, solid lipid nanoparticles, micelles, and emulsions, are frequently utilized for siRNA delivery (Saw and Song 2020; Oh and Park 2009). Cationic liposomes are notably efficient because of their stability, minimal toxicity, resistance to degradation, and elevated transfection effectiveness. Polymer-mediated delivery uses biodegradable polymeric nanoparticles, which act as colloidal drug carriers (Yonezawa et al. 2020). Conjugation with siRNA enhances cellular uptake, stability, and delivery. Common polymers include PLGA and PEG, widely used in medical applications. Delivery systems based on conjugation associate siRNA with particular ligands to improve targeting and address the drawbacks of lipid and polymer systems, including size, clearance, and lack of specificity (Shukla et al. 2014). In contrast to unconjugated siRNA, bioconjugated variants provide enhanced specificity, extended half-life, and effective delivery while maintaining activity. Chemical alterations improve siRNA stability, resistance to nucleases, and duration in circulation (Uchida et al. 2020). Important modification sites comprise the phosphate bond, 2′-position, nucleobase, and ribose. Combinations such as 2′-OMe, 2′-F, and phosphorothioate have demonstrated efficacy, evidenced by approved treatments like Onpattro™ and Inclisiran (Gaglione and Messere 2010; Smith and Zain 2019).

Inorganic nanoparticles show potential as carriers for siRNA delivery because of their large surface area for encapsulation and their easily adjustable surface chemistry, allowing effective applications both in vitro and in vivo (Jiang et al. 2016).

5.20 Current and Future Clinical Applications of siRNAs in Cancer

RNA therapeutics utilize noncoding RNAs like small interfering RNAs (siRNA) and coding RNAs such as mRNA to focus on mRNA, aptamers, ribozymes, and clustered regularly interspaced short palindromic repeats CRISPR-associated (CRISPR/Cas) endonuclease to target proteins and DNA (Feng et al. 2021). Because of their varied targeting capabilities and advancements in RNA modification and delivery methods, RNA-based formulations have become viable treatment alternatives for numerous diseases (Lin et al. 2023).

5.21 SiRNA Therapeutics in Cancer and Clinical Trials

The metastasis are significant processes that contribute to the spread of cancers. Prolonged use of numerous cancer medications can cause undesirable side effects; so SiRNA therapies are currently under evaluation for the prevention or management of multiple types of cancer (Mahmoodi Chalbatani et al. 2019). KRAS mutation is the most recognized change in multiple cancers. The research (NCT01188785) on siG12D-LODER, a biodegradable polymer matrix containing siRNA against KRASG12D, showed that the polymer effectively targeted the tumor and decreased its progression. Additional research (NCT01676259) is being conducted to evaluate the effectiveness of siG12D-LODER when used alongside chemotherapy drugs like gemcitabine and nab-paclitaxel in patients with locally advanced pancreas (Golan et al. 2015; Asimgil et al. 2022). Plk, a group of five serine/threonine kinases, is frequently overexpressed in cancerous cells, and inhibiting it decreases the proliferation of cancer cells. TKM-080301, a lipid nanoparticle (LNP) that delivers siRNA aimed at PLK1 mRNA, demonstrated acceptable tolerability following intravenous treatment, yet exhibited rest-ricted antitumor effectiveness and no survival advantage in advanced hepatocellular carcinoma cancer (Feng et al. 2021). Polo-like kinases (Plks) play a role in regulating the cell cycle and promoting cell growth. Atu027, a protein kinase N3 (PKN3), effector in the phosphoinositide-3-kinase (PI3K) pathway and siRNA formulation in liposomes, exhibited no safety concerns in individuals with advanced solid tumors and achieved disease stabilization in 41% of the participants by the treatment's conclusion (Schultheis et al. 2020). The Myc oncoprotein is crucial in various cancers, and its suppression diminishes tumor growth and proliferation. DCR-MYC, a synthetic siRNA formulated with lipids that target MYC, demonstrated good tolerability along with encouraging early clinical and metabolic responses in patients suffering from solid and hepatocellular cancers (Hattab et al. 2021). Ephrin type-A receptor 2 (EPHA2), a receptor tyrosine kinase, is predominantly found in epithelial cells, and its increased expression fosters malignant transformation and tumorigenicity in mammary epithelial cells. Targeting RRM2 using siRNA may offer therapeutic benefits for melanoma patients. A therapeutic formulation, CALAA-01, was created by encapsulating RRM2 siRNA within cyclodextrin-adamantane polyethylene glycol (AD-PEG)-human transferrin protein (hTf). The systemic delivery of CALAA-01 not only diminished levels of RRM2 mRNA and protein but also effectively localized within the tumor (Hattab et al. 2021; Casacuberta-Serra et al. 2024; Goyal et al. 2022; Rahman et al. 2012).

5.22 Personalized siRNA Therapies

Customizing siRNA therapies for specific cancer profiles includes adjusting siRNA sequences and delivery techniques to address the distinct genetic and molecular features of a person's tumor. This tailored strategy boosts therapeutic accuracy, increases effectiveness, reduces off-target impacts, and aids in creating more efficient, targeted cancer treatments (Hu et al. 2020; Kato et al. 2018). siRNA-mediated gene silencing offers a promising therapeutic approach for targeting diseases at the molecular level. Recently, HDL-mimicking peptide-phospholipid nano-scaffold (HPPS) nanoparticles have been developed for efficient siRNA delivery to SCARB1-expressing cancer cells, enhancing specificity and uptake (Sousa and Videira 2025).

Elevated KIF11 expression has been recognized as an independent prognostic indicator in lung large cell carcinoma (LCC) and squamous cell carcinoma (SqCC). The targeted administration of siRNA targeting KIF11 through HPPS nanoparticles successfully lowered KIF11 expression and greatly restrained tumor growth in vivo. siRNA has become a promising therapeutic option in hormone-related cancers by allowing targeted gene silencing. In cancers like breast and ovarian, siRNAs have been created to focus on genes like HER2, PLK1, VEGF, and cyclin E. These approaches have demonstrated promise in both in vitro and in vivo experiments. Nevertheless, improvements in delivery systems and active clinical trials keep propelling advancements in siRNA-based therapies for hormone-associated cancers (Sousa and Videira 2025; Babu et al. 2016; Lee et al. 2016).

5.22.1 *siRNAs in Combination Therapies*

The emergence of cancer frequently coincides with alterations in various key proteins that play roles in tumor growth, metastasis, and survival. Therefore, alongside chemical drugs, siRNAs can serve a complementary function in making cancer cells more sensitive or blocking multidrug-resistance genes (Creixell and Peppas 2012). White and coworkers recently studied siRNA therapeutics and chemo in lung cancer models and found that downregulation of multiple genes made these lung cancer cells more responsive to a paclitaxel, and siRNA have a notable response (Lee et al. 2016).

Significantly, various recent studies suggest that the pairing of chemical drugs with siRNA provides enhanced anticancer benefits compared to using either chemical drugs or siRNA separately (Tabernero et al. 2013). The reduction of B-cell lymphoma 2 (Bcl-2) expression, a key antiapoptotic protein, has been demonstrated to hinder tumor growth. Zheng et al. (Sharma and Allison 2015) discovered that the pairing of Dtxl with Bcl-2 siRNA noticeably decreased Bcl-2 levels and improved antitumor effects, leading to marked suppression of tumor growth in a murine xenograft model of MCF-7 breast cancer, in contrast to treatment with Dtxl or siRNA alone (Sousa and Videira 2025; Chen et al. 2009; Akar et al. 2008). This discovery suggests that merging chemotherapy with RNAi technology might be an effective approach to significantly enhance chemotherapy's therapeutic effectiveness via synergistic effects (Kumar et al. 2022).

Although siRNA provides an effective means for accurate gene silencing, its complete capabilities are constrained by issues like low stability, unintended target effects, immune responses, and restricted tissue infiltration. Improvements in delivery mechanisms—such as lipid nanoparticles and tri-GalNAc conjugates—have enhanced targeting and cellular absorption. Authorized treatments like Patisiran and Inclisiran demonstrate the potential of siRNA, especially for uncommon genetic diseases and issues such as hypercholesterolemia. While only a limited number of siRNA drugs have received approval so far, continued studies on delivery methods, target choice, and production indicate an encouraging future for siRNA therapies for various illnesses (Ebenezer et al. 2025).

The application of siRNA alongside anticancer medications signifies an innovative approach in cancer treatment, providing improved effectiveness, decreased toxicity, and the possibility to surpass drug resistance (Sousa and Videira 2025; Zhang et al. 2020). By specifically suppressing genes that are either overactive or uniquely active in cancerous tissues, siRNA offers a precise approach to interfere with essential oncogenic pathways, hinder tumor development, and make tumors more responsive to current treatments. Downregulating EGFR and c-MET disrupts critical signaling pathways involved in cell growth and metastasis, and targeting HER2, MUC1, and Bcl-2 sensitizes tumors to targeted therapies and chemotherapy (Chen et al. 2009; Ngamcherdtrakul et al. 2022). To completely realize the potential of siRNA therapies, upcoming initiatives must target enhancements in delivery systems, incorporation of personalized medicine, integration with immunotherapies, refinement of dosing strategies, discovery of predictive biomarkers, and simplification of regulatory processes. Personalized medicine can significantly boost the effectiveness of siRNA therapies. Next-generation sequencing and transcriptomic analyses for genomic profiling of tumors can uncover individual genetic weaknesses, allowing for personalized treatment strategies (Sousa and Videira 2025; Ho et al. 2020).

Clinical trials need to concentrate on identifying the best dosing strategies for siRNA and concurrent medications. Assessing if sequential or simultaneous delivery produces superior results is crucial for optimizing therapeutic effectiveness. The regulatory framework for siRNA-based treatments must adapt to tackle distinct challenges, including dual-drug delivery methods and nanoparticle formulations. By tackling these critical aspects, siRNA-based treatments can surmount existing obstacles and establish themselves as a fundamental element of precision oncology (Friedrich and Aigner 2022).

5.22.2 *Purpose and Scope of This Chapter*

This chapter explores the role of small interfering RNAs (siRNAs) in gene silencing and their impact on key biological pathways involved in cancer development. The aim is to provide a comprehensive understanding of how siRNAs influence cancer cell responses to stress. This chapter also covers the progress of RNA-based therapies, focusing on the current status of siRNA technologies and their application in cancer treatment while addressing the challenges they face in clinical settings. The scope of this chapter extends to examining therapeutic potential of small interfering RNAs (siRNAs) in cancer treatment, focusing on the various strategies developed to improve their delivery, stability, and specificity. It also addresses the key challenges hindering their clinical effectiveness, such as immunogenicity, off-target effects, and the complexities of efficient delivery mechanisms. Additionally, this chapter highlights the growing interest in RNA-based therapies, aiming to provide a comprehensive overview of the advancements, current obstacles, and promising future directions for siRNA therapeutics in cancer therapy.

Funding Statement No funding received.

Conflict of Interest The authors declare no conflict of interest.

Ethical Approval This chapter does not contain any studies with human participants or animals.

References

Adams D, Gonzalez-Duarte A, O'Riordan WD et al (2018) Patisiran, an RNAi therapeutic, for hereditary transthyretin amyloidosis. N Engl J Med 379(1):11–21

Agrawal N, Dasaradhi PV, Mohmmed A, Malhotra P, Bhatnagar RK, Mukherjee SK (2003) RNA interference: biology, mechanism, and applications. Microbiol Mol Biol Rev 67(4):657–685. https://doi.org/10.1128/MMBR.67.4.657-685.2003. PMID: 14665679; PMCID: PMC309050

Akar U, Chaves-Reyez A, Barria M, Tari A, Sanguino A, Kondo Y, Kondo S, Arun B, Lopez-Berestein G, Ozpolat B (2008) Silencing of Bcl-2 expression by small interfering RNA induces autophagic cell death in MCF-7 breast cancer cells. Autophagy 4(5):669–679. https://doi.org/10.4161/auto.6083. Epub 2008 Apr 10. PMID: 18424910

Ali Zaidi S, Fatima F, Ali Zaidi S et al (2023) Engineering siRNA therapeutics: challenges and strategies. J Nanobiotechnol 21:381

Alshaer W, Zureigat H, Al Karaki A, Al-Kadash A, Gharaibeh L, Hatmal MMM et al (2021) siRNA: mechanism of action, challenges, and therapeutic approaches. Eur J Pharmacol 905:174178

Ambrosini G, Adida C, Altieri DC (1997) A novel anti-apoptosis gene, survivin, expressed in cancer and lymphoma. Nat Med 3(8):917–921

Anastasiadou E, Jacob LS, Slack FJ (2018) Non-coding RNA networks in cancer. Nat Rev Cancer 18:5–18

Asimgil H, Ertetik U, Çevik NC, Ekizce M, Doğruöz A, Gökalp M, Arık-Sever E, Istvanffy R, Friess H, Ceyhan GO, Demir IE (2022) Targeting the undruggable oncogenic KRAS: the dawn of hope. JCI Insight 7(1):e153688. https://doi.org/10.1172/jci.insight.153688. PMID: 35014625; PMCID: PMC8765045

Assaraf YG, Brozovic A, Goncalves AC et al (2019) The multi-factorial nature of clinical multidrug resistance in cancer. Drug Resist Updat 46:100645

Babu A, Muralidharan R, Amreddy N, Mehta M, Munshi A, Ramesh R (2016) Nanoparticles for siRNA-based gene silencing in tumor therapy. IEEE Trans Nanobioscience 15(8):849–863. https://doi.org/10.1109/TNB.2016.2621730. PMID: 28092499; PMCID: PMC6198667

Bai Y, Gao J, Yan Y, Zhao X (2024) The significance of long chain non-coding RNA signature genes in the diagnosis and management of sepsis patients, and the development of a prediction model. Front Immunol 15:1450014

Beaucage SL (2008) Solid-phase synthesis of siRNA oligonucleotides. Curr Opin Drug Discov Devel 11(2):203–216

Bennett CF (2019) Therapeutic antisense oligonucleotides are coming of age. Annu Rev Med 70:307–321

Bernstein E, Caudy AA, Hammond SM, Hannon GJ (2001) Role for a bidentate ribonuclease in the initiation step of RNA interference. Nature 409(6818):363–6

Bhattacharjee S, Roche B, Martienssen RA (2019) RNA-induced initiation of transcriptional silencing (RITS) complex structure and function. RNA Biol 16(9):1133–1146. https://doi.org/10.1080/15476286.2019.1621624. Epub 2019 Jun 18. PMID: 31213126; PMCID: PMC6693537

Bollag G, Hirth P, Tsai J et al (2010) Clinical efficacy of a RAF inhibitor needs broad target blockade in BRAF-mutant melanoma. Nature 467(7315):596–599

Bora RS, Gupta D, Mukkur TK, Saini KS (2012) RNA interference therapeutics for cancer: challenges and opportunities (review). Mol Med Rep 6:9–15

Bosch A, Eroles P, Zaragoza R, Vina JR, Lluch A (2010) Triple-negative breast cancer: molecular features, pathogenesis, treatment and current lines of research. Cancer Treat Rev 36:206–215

Bouclier C, Moine L, Hillaireau H, Marsaud V, Connault E, Opolon P, Couvreur P, Fattal E, Renoir JM (2008) Physicochemical characteristics and preliminary in vivo biological evaluation of nanocapsules loaded with siRNA targeting estrogen receptor alpha. Biomacromolecules 9:2881–2890

Bramsen JB, Kjems J (2013) Engineering small interfering RNAs by strategic chemical modification. Methods Mol Biol 942:87–109

Bukowski K, Kciuk M, Kontek R (2020) Mechanisms of multidrug resistance in cancer chemotherapy. Int J Mol Sci 21(9):3233

Bumcrot D, Manoharan M, Koteliansky V, Sah DW (2006) RNAi therapeutics: a potential new class of pharmaceutical drugs. Nat Chem Biol 2(12):711–719

Burnett JC, Rossi JJ (2012) RNA-based therapeutics: current progress and future prospects. Chem Biol 19(1):60–71. https://doi.org/10.1016/j.chembiol.2011.12.008. PMID: 22284355; PMCID: PMC3269031

Carpenter G (2000) The EGF receptor: a nexus for trafficking and signaling. BioEssays 22(8):697–707

Carthew RW, Sontheimer EJ (2009) Origins and Mechanisms of miRNAs and siRNAs. Cell 136(4):642–655

Casacuberta-Serra S, González-Larreategui Í, Capitán-Leo D et al (2024) MYC and KRAS cooperation: from historical challenges to therapeutic opportunities in cancer. Sig Transduct Target Ther 9:205

Castel SE, Martienssen RA (2013) RNA interference in the nucleus: roles for small RNAs in transcription, epigenetics and beyond. Nat Rev Genet 14:100–112

Chappell WH, Lasky M (2014) Targeting MYC in cancer therapy. Nat Rev Cancer 14(1):23–35

Chen AM, Zhang M, Wei D, Stueber D, Taratula O, Minko T, He H (2009) Co-delivery of doxorubicin and Bcl-2 SiRNA by mesoporous silica nanoparticles enhances the efficacy of chemotherapy in multidrug-resistant cancer cells. Small 5:2673

Corey DR (2007) Chemical modification: the key to clinical application of RNA interference? J Clin Invest 117(12):3615–3622

Creixell M, Peppas NA (2012) Co-delivery of SiRNA and therapeutic agents using Nanocarriers to overcome cancer resistance. Nano Today **7:367**

Curreri A, Sankholkar D, Mitragotri S, Zhao Z (2022) RNA therapeutics in the clinic. Bioeng Transl Med 8(1):e10374

Dana H, Chalbatani GM, Mahmoodzadeh H, Karimloo R, Rezaiean O, Moradzadeh A, Mehmandoost N, Moazzen F, Mazraeh A, Marmari V, Ebrahimi M, Rashno MM, Abadi SJ, Gharagouzlo E (2017) Molecular mechanisms and biological functions of siRNA. Int J Biomed Sci 13(2):48–57. PMID: 28824341; PMCID: PMC5542916

De La OJP, McDonald PC (2011) Targeting KRAS in cancer therapy: from laboratory to clinical trials. Mol Ther 19(5):881–888

de la Torre BG, Albericio F (2019) The Pharmaceutical Industry in 2018. An analysis of FDA drug approvals from the perspective of molecules. Molecules 24:809

Deleavey GF, Watts JK, Alain T, Robert F, Kalota A, Aishwarya V, Pelletier J, Gewirtz AM, Sonenberg N, Damha MJ (2010) Synergistic effects between analogs of DNA and RNA improve the potency of siRNA-mediated gene silencing. Nucleic Acids Res 38(13):4547–4557. https://doi.org/10.1093/nar/gkq181. Epub 2010 Apr 22. PMID: 20413581; PMCID: PMC2910058

Dietlein F, Thelen L, Reinhardt HC (2014) Cancer-specific defects in DNA repair pathways as targets for personalized therapeutic approaches. Trends Genet 30(8):326–339

Dong Y, Siegwart DJ, Anderson DG (2019) Strategies, design, and chemistry in siRNA delivery systems. Adv Drug Deliv Rev 144:133–147

Ebenezer O, Oyebamiji AK, Olanlokun JO, Tuszynski JA, Wong GK (2025) Recent update on siRNA therapeutics. Int J Mol Sci 26(8):3456. https://doi.org/10.3390/ijms26083456. PMID: 40331977; PMCID: PMC12026779

Eisenstein M (2019) Pharma's roller-coaster relationship with RNA therapies. Nature 574(7778):S4

ENCODE Project Consortium (2012) An integrated encyclopedia of DNA elements in the human genome. Nature 489:57–74

Endoh T, Ohtsuki T (2009) Cellular siRNA delivery using cell-penetrating peptides modified for endosomal escape. Adv Drug Deliv Rev 61:704–709

Fedorov Y, Anderson EM, Birmingham A et al (2006) Off-target effects by siRNA can induce toxic phenotype. RNA 12(7):1188–1196

Feng R, Patil S, Zhao X, Miao Z, Qian A (2021) RNA therapeutics - research and clinical advancements. Front Mol Biosci 8:710738. https://doi.org/10.3389/fmolb.2021.710738. PMID: 34631795; PMCID: PMC8492966

Fire A, Xu S, Montgomery MK, Kostas SA et al (1998) Potent and specific genetic interference by double-stranded RNA in *Caenorhabditis elegans*. Nature 391:806–811

Fitzgerald K, White S, Borodovsky A et al (2017) A highly durable RNAi therapeutic inhibitor of PCSK9. N Engl J Med 376(1):41–51

Friedrich M, Aigner A (2022) Therapeutic siRNA: State-of-the-Art and Future Perspectives. BioDrugs 36(5):549–571. https://doi.org/10.1007/s40259-022-00549-3. Epub 2022 Aug 23. PMID: 35997897; PMCID: PMC9396607

Fu X-D (2014) Non-coding RNA: a new frontier in regulatory biology. Natl Sci Rev 1(2):190–204

Fujita Y, Takeshita F, Mizutani T, Ohgi T, Kuwano K, Ochiya T (2013) A novel platform to enable inhaled naked RNAi medicine for lung cancer. Sci Rep 3:3325

Gaglione M, Messere A (2010) Recent progress in chemically modified siRNAs. Mini Rev Med Chem 10(7):578–595

Ganesh K, Massague J (2021) Targeting metastatic cancer. Nat Med 27(1):34–44

Gao Lf X, Dq WL, Xy Z, Yt S, Xj Z (2005) Inhibition of STAT3 expression by siRNA suppresses growth and induces apoptosis in laryngeal cancer cells. Acta Pharmacol Sin 26(3):377–383

Garber K (2018) Alnylam launches era of RNAi drugs. Nat Biotechnol 36(9):777–778

Golan T, Khvalevsky EZ, Hubert A, Gabai RM, Hen N, Segal A, Domb A, Harari G, David EB, Raskin S, Goldes Y, Goldin E, Eliakim R, Lahav M, Kopleman Y, Dancour A, Shemi A, Galun E (2015) RNAi therapy targeting KRAS in combination with chemotherapy for locally advanced pancreatic cancer patients. Oncotarget 6(27):24560–24570. https://doi.org/10.18632/oncotarget.4183. PMID: 26009994; PMCID: PMC4695206

Goyal R, Chopra H, Singh I, Dua K, Gautam RK (2022) Insights on prospects of nano-siRNA based approaches in treatment of cancer. Front Pharmacol 13:985670. https://doi.org/10.3389/fphar.2022.985670. PMID: 36091772; PMCID: PMC9452808

Gredell JA, Berger AK, Walton SP (2008) Impact of target mRNA structure on siRNA silencing efficiency: a large-scale study. Biotechnol Bioeng 100(4):744–755. https://doi.org/10.1002/bit.21798. PMID: 18306428; PMCID: PMC2658823

Grimes DR (2022) Radiofrequency radiation and cancer: a review. JAMA Oncol 8(3):456–461

Gulìa C, Signore F, Gaffi M, Gigli S, Votino R, Nucciotti R, Bertacca L, Zaami S, Baffa A, Santini E et al (2020) Y RNA: An overview of their role as potential biomarkers and molecular targets in human cancers. Cancer 12:1238

Guo J, Evans JC, O'Driscoll CM (2013) Delivering RNAi therapeutics with non-viral technology: a promising strategy for prostate cancer? Trends Mol Med 19:250–261

Ha M, Kim VN (2014) Regulation of microRNA biogenesis. Nat Rev Mol Cell Biol 15:509–524

Hahne JC, Lampis A, Valeri N (2021) Vault RNAs: hidden gems in RNA and protein regulation. Cell Mol Life Sci 78:1487–1499

Hanahan D (2022) Hallmarks of cancer: new dimensions. Cancer Discov 12(1):31–46

Hanahan D, Weinberg R (2000) The hallmarks of cancer. Cell 100:57–70

Hanna J, Hossain GS, Kocerha J (2019) The potential for microRNA therapeutics and clinical research. Front Genet 10:478

Hannon GJ (2002) RNA interference. Nature 418(6894):244–251

Hattab D, Gazzali AM, Bakhtiar A (2021) Clinical advances of siRNA-based Nanotherapeutics for cancer treatment. Pharmaceutics. 13(7):1009. https://doi.org/10.3390/pharmaceutics13071009. PMID: 34371702; PMCID: PMC8309123

Ho D, Quake SR, McCabe ERB, Chng WJ, Chow EK, Ding X, Gelb BD, Ginsburg GS, Hassenstab J, Ho CM, Mobley WC, Nolan GP, Rosen ST, Tan P, Yen Y, Zarrinpar A (2020) Enabling Technologies for Personalized and Precision Medicine. Trends Biotechnol 38(5):497–518. https://doi.org/10.1016/j.tibtech.2019.12.021. Epub 2020 Jan 21. PMID: 31980301; PMCID: PMC7924935

Howell A, Dowsett M (2004) Endocrinology and hormone therapy in breast cancer: aromatase inhibitors versus antioestrogens. Breast Cancer Res 6:269–274

Hu X, Hipolito S, Lynn R, Abraham V, Ramos S, Wong-Staal F (2004) Relative gene-silencing efficiencies of small interfering RNAs targeting sense and antisense transcripts from the same genetic locus. Nucleic Acids Res 32(15):4609–4617. https://doi.org/10.1093/nar/gkh790. PMID: 15333693; PMCID: PMC516062

Hu B, Zhong L, Weng Y, Peng L, Huang Y, Zhao Y, Liang XJ (2020) Therapeutic siRNA: state of the art. Signal Transduct Target Ther 5(1):101

Idris A, Davis A, Supramaniam A et al (2021) A SARS-CoV-2 targeted siRNA-nanoparticle therapy for COVID-19. Mol Ther 29(7):2219–2226

Iwasaki YW, Siomi MC, Siomi H (2015) PIWI-interacting RNA: its biogenesis and functions. Annu Rev Biochem 84:405–433

Jackson AL, Bartz SR, Schelter J, Kobayashi SV, Burchard J, Mao M et al (2003) Expression profiling reveals off-target gene regulation by RNAi. Nat Biotechnol 21(6):635–637

Jameson JL, Longo DL (2015) Precision medicine—personalized, problematic, and promising. N Engl J Med 372(23):2229–2234

Ji B, Wei M, Yang B (2022) Recent advances in nanomedicines for photodynamic therapy (PDT)-driven cancer immunotherapy. Theranostics 12(1):434–458

Jiang Y, Huo S, Hardie J, Liang XJ, Rotello VM (2016) Progress and perspective of inorganic nanoparticle-based siRNA delivery systems. Expert Opin Drug Deliv 13(4):547–559. https://doi.org/10.1517/17425247.2016.1134486. Epub 2016 Jan 14. PMID: 26735861; PMCID: PMC4914043

Jin Y, Qu S, Tesikova M, Wang L, Kristian A, Maelandsmo GM, Kong H, Zhang T, Jeronimo C, Teixeira MR, Yuca E, Tekedereli I, Gorgulu K, Alpay N, Sood AK, Lopez-Berestein G, Danielsen HE, Ozpolat B, Saatcioglu F (2013) Molecular circuit involving KLK4 integrates androgen and mTOR signaling in prostate cancer. Proc Natl Acad Sci USA 110:E2572–E2581

Judge A, Maclachlan I (2008) Overcoming the innate immune response to small interfering RNA. Hum Gene Ther 19:111–124

Judge AD, Sood V, Shaw JR, Fang D, McClintock K, MacLachlan I (2005) Sequence-dependent stimulation of the mammalian innate immune response by synthetic siRNA. Nat Biotechnol 23:457–462

Juliano R, Alam MR, Dixit V, Kang H (2008) Mechanisms and strategies for effective delivery of antisense and siRNA oligonucleotides. Nucleic Acids Res 36(12):4158–4171

Kaelin WG Jr (2005) The concept of synthetic lethality in the context of anticancer therapy. Nat Rev Cancer 5(9):689–698

Kaikkonen MU, Lam MT, Glass CK (2011) Non-coding RNAs as regulators of gene expression and epigenetics. Cardiovasc Res 90(3):430–440. https://doi.org/10.1093/cvr/cvr097. Epub 2011 May 9. PMID: 21558279; PMCID: PMC3096308

Kanasty R, Dorkin JR, Vegas A, Anderson D (2013) Delivery materials for siRNA therapeutics. Nat Mater 12(11):967–977

Kato T, Lee D, Huang H, Cruz W, Ujiie H, Fujino K, Wada H, Patel P, Hu HP, Hirohashi K, Nakajima T, Sato M, Kaji M, Kaga K, Matsui Y, Chen J, Zheng G, Yasufuku K (2018) Personalized siRNA-nanoparticle systemic therapy using metastatic lymph node specimens obtained with EBUS-TBNA in lung cancer. Mol Cancer Res 16(1):47–57. https://doi.org/10.1158/1541-7786.MCR-16-0341. Epub 2017 Oct 9. PMID: 28993508

Khalil DN, Smith EL, Brentjens RJ, Wolchok JD (2016) The future of cancer treatment: immunomodulation, CARs and combination immunotherapy. Nat Rev Clin Oncol 13(5):273–290

Khraiwesh B, Zhu JK, Zhu J (2012) Role of miRNAs and siRNAs in biotic and abiotic stress responses of plants. Biochim Biophys Acta 1819(2):137–148. https://doi.org/10.1016/j.bbagrm.2011.05.001. Epub 2011 May 13. PMID: 21605713; PMCID: PMC3175014

Khvorova A, Watts JK (2017) The chemical evolution of oligonucleotide therapies of clinical utility. Nat Biotechnol 35(3):238–248

Koch L (2016) Genetic screen: a network to guide precision cancer therapy. Nat Rev Genet 17(9):504–505

Kosmas CE, Munoz Estrella A, Sourlas A et al (2018) Inclisiran: a new promising agent in the management of hypercholesterolemia. Diseases 6(3):63

Kowalski PS, Rudra A, Miao L, Anderson DG (2019) Delivering the messenger: advances in Technologies for therapeutic mRNA delivery. Mol Ther 27:710–728

Kumar K, Rani V, Mishra M, Chawla R (2022) New paradigm in combination therapy of siRNA with chemotherapeutic drugs for effective cancer therapy. Curr Res Pharmacol Drug Discov 3:100103. https://doi.org/10.1016/j.crphar.2022.100103. PMID: 35586474; PMCID: PMC9108887

Kurashige J, Watanabe M, Iwatsuki M, Kinoshita K, Saito S, Nagai Y, Ishimoto T, Baba Y, Mimori K, Baba H (2012) RPN2 expression predicts response to docetaxel in oesophageal squamous cell carcinoma. Br J Cancer 107:1233–1238

Laganà A, Veneziano D, Russo F, Pulvirenti A, Giugno R, Croce CM, Ferro A (2015) Computational design of artificial RNA molecules for gene regulation. In: RNA bioinformatics Methods in molecular biology, vol 1269. Springer, New York, pp 393–412

Lee YS, Nakahara K, Pham JW, Kim K, He Z, Sontheimer EJ, Carthew RW (2004) Distinct roles for Drosophila Dicer-1 and Dicer-2 in the siRNA/miRNA silencing pathways. Cell 117(1):69–81

Lee SJ, Kim MJ, Kwon IC, Roberts TM (2016) Delivery strategies and potential targets for siRNA in major cancer types. Adv Drug Deliv Rev 104:2–15. https://doi.org/10.1016/j.addr.2016.05.010. Epub 2016 May 31. PMID: 27259398; PMCID: PMC4958528

Li Y et al (2020) Non-coding RNA in bladder cancer. Cancer Lett 485:38–44

Li C, Ni Y-Q, Xu H, Xiang Q-Y, Zhao Y, Zhan J-K, He J-Y, Li S, Liu Y-S (2021a) Roles and mechanisms of exosomal non-coding RNAs in human health and diseases. Signal Transduct Target Ther 6:383

Li C, Brant E, Budak H, Zhang B (2021b) CRISPR/Cas: a Nobel prize award-winning precise genome editing technology for gene therapy and crop improvement. J Zhejiang Univ Sci B 22(4):253–284

Li X, Loh TJ, Lim JJ, Er SP, Liao Y (2022) Glycan-RNA: a new class of non-coding RNA. BIO Integr 3:124–131

Lim SA, Cox A, Tung M, Chung EJ (2022) Clinical progress of nanomedicine-based RNA therapies. Bioact Mater 12:203–213

Lin L, Su K, Cheng Q, Liu S (2023) Targeting materials and strategies for RNA delivery. Theranostics 13(13):4667–4693. https://doi.org/10.7150/thno.87316. PMID: 37649616; PMCID: PMC10465230

Liu Y, Zhu YH, Mao CQ, Dou S, Shen S, Tan ZB, Wang J (2014) Triple negative breast cancer therapy with CDK1 siRNA delivered by cationic lipid assisted PEG-PLA nanoparticles. J Control Release 192:114–121

Loganathan T, Doss CGP (2023) Non-coding RNAs in human health and disease: potential function as biomarkers and therapeutic targets. Funct Integr Genomics 23(1):33

Ma S, Li X, Wang X et al (2019) Current progress in CAR-T cell therapy for solid tumors. Int J Biol Sci 15(12):2548–2560

Mahmoodi Chalbatani G, Dana H, Gharagouzloo E, Grijalvo S, Eritja R, Logsdon CD, Memari F, Miri SR, Rad MR, Marmari V (2019) Small interfering RNAs (siRNAs) in cancer therapy: a nano-based approach. Int J Nanomedicine 14:3111–3128. https://doi.org/10.2147/IJN.S200253. PMID: 31118626; PMCID: PMC6504672

Manoharan M (2017) Therapeutic applications of siRNA-based drugs. Annu Rev Pharmacol Toxicol 57:313–331

Marques JT, Williams BRG (2005) Activation of the mammalian immune system by siRNAs. Nat Biotechnol 23(11):1399–1405

Mateo J, Steuten L, Aftimos P et al (2022) Delivering precision oncology to patients with cancer. Nat Med 28(4):658–665

McLeod HL (2013) Cancer pharmacogenomics: early promise, but concerted effort needed. Science 339:1563–1566

Minai-Tehrani A, Jiang HL, Kim YK, Chung YS, Yu KN, Kim JE, Shin JY, Hong SH, Lee JH, Kim HJ, Chang SH, Park S, Kang BN, Cho CS, Cho MH (2012) Suppression of tumor growth in xenograft model mice by small interfering RNA targeting osteopontin delivery using biocompatible poly (amino ester). Int J Pharm 431:197–203

Ming X (2011) Cellular delivery of siRNA and antisense oligonucleotides via receptor-mediated endocytosis. Expert Opin Drug Deliv 8(4):435–449

Mirzaei S, Gholami MH, Ang HL, Hashemi F, Zarrabi A, Zabolian A, Hushmandi K, Delfi M, Khan H, Ashrafizadeh M, Sethi G, Kumar AP (2021) Pre-clinical and clinical applications of small interfering RNAs (siRNA) and co-delivery systems for pancreatic cancer therapy. Cells 10(12):3348. https://doi.org/10.3390/cells10123348. PMID: 34943856; PMCID: PMC8699513

Miyashita T, Reed JC (1993) Bcl-2 oncoprotein blocks chemotherapy-induced apoptosis in a human leukemia cell line. Blood 81(1):151–157. PMID: 8417786

Molina JR, Yang P, Cassivi SD, Schild SE, Adjei AA (2008) Non-small cell lung cancer: epidemiology, risk factors, treatment, and survivorship. Mayo Clin Proc 83:584–594

Monga I, Qureshi A, Thakur N, Gupta AK, Kumar M (2017) aspsirna: a resource of asp-sirnas having therapeutic potential for human genetic disorders and algorithm for prediction of their inhibitory efficacy. G3: Genes Genomes Genet 7(9):2931–2943

Mukai H (2003) The structure and function of PKN, a protein kinase having a catalytic domain homologous to that of PKC. J Biochem 133(1):17–27

Neuberg P, Kichler A (2014) Recent developments in nucleic acid delivery with polyethylenimines. Adv Genet 88:263–288

Ngamcherdtrakul W, Bejan DS, Cruz-Muñoz W, Reda M, Zaidan HY, Siriwon N, Marshall S, Wang R, Nelson MA, Rehwaldt JPC et al (2022) Targeted nanoparticle for co-delivery of HER2 SiRNA and a Taxane to Mirror the standard treatment of HER2+ breast cancer: efficacy in breast tumor and brain metastasis. Small 18:2107550

Niola F, Evangelisti C, Campagnola L, Massalini S, Buè MC, Mangiola A, Masotti A, Maira G, Farace MG, Ciafrè SA (2006) A plasmid-encoded VEGF siRNA reduces glioblastoma angiogenesis and its combination with interleukin-4 blocks tumor growth in a xenograft mouse model. Cancer Biol Ther 5(2):174–179

Nussinov R, Tsai CJ, Jang H (2021) Anticancer drug resistance: an update and perspective. Drug Resist Updat 59:100796

Oh YK, Park TG (2009) siRNA delivery systems for cancer treatment. Adv Drug Deliv Rev 61(10):850–862

Oliveira S, van Rooy I, Kranenburg O, Storm G, Schiffelers RM (2007) Fusogenic peptides enhance endosomal escape improving siRNA-induced silencing of oncogenes. Int J Pharm 331(2):211–214

Ozpolat B, Sood AK, Lopez-Berestein G (2010) Nanomedicine based approaches for the delivery of siRNA in cancer. J Intern Med 267(1):44–53

Paul A, Muralidharan A, Biswas A, Kamath BV, Joseph A, Alex AT (2022) siRNA therapeutics and its challenges: recent advances in effective delivery for cancer therapy. OpenNano 7:100063

Persengiev SP, Zhu X, Green MR (2004) Nonspecific, concentration-dependent stimulation and repression of mammalian gene expression by small interfering RNAs (siRNAs). RNA 10(1):12–18

Pratt AJ, MacRae IJ (2009) The RNA-induced silencing complex: a versatile gene-silencing machine. J Biol Chem 284(27):17897–17901. https://doi.org/10.1074/jbc.R900012200. Epub 2009 Apr 1. PMID: 19342379; PMCID: PMC2709356

Rahman MA, Amin AR, Wang X, Zuckerman JE, Choi CH, Zhou B, Wang D, Nannapaneni S, Koenig L, Chen Z, Chen ZG, Yen Y, Davis ME, Shin DM (2012) Systemic delivery of siRNA nanoparticles targeting RRM2 suppresses head and neck tumor growth. J Control Release

159(3):384–392. https://doi.org/10.1016/j.jconrel.2012.01.045. Epub 2012 Feb 8. PMID: 22342644; PMCID: PMC3348392
Reddy A, Kaelin WG Jr (2002) Using cancer genetics to guide the selection of anticancer drug targets. Curr Opin Pharmacol 2(4):366–373
Robbins M, Judge A, MacLachlan I (2009) siRNA and innate immunity. Oligonucleotides 19:89–102
Russo S, Saif MW (2016) Gastrointestinal cancers symposium: update on pancreatic cancer. Ann Gastroenterol 29(2):238–240
Ryther RCC, Flynt AS, Phillips J, Patton JG (2005) siRNA therapeutics: big potential from small RNAs. Gene Ther 12(1):5–11
Sajid MI, Moazzam M, Kato S, Yeseom Cho K, Tiwari RK (2020a) Overcoming barriers for siRNA therapeutics: from bench to bedside. Pharmaceuticals 13(10):294
Sajid MI, Moazzam M, Kato S, Yeseom Cho K, Tiwari RK (2020b) Overcoming barriers for siRNA therapeutics: from bench to bedside. Pharmaceuticals (Basel) 13(10):294. https://doi.org/10.3390/ph13100294. PMID: 33036435; PMCID: PMC7600125
Sánchez Y, Huarte M (2013) Long non-coding RNAs: challenges for diagnosis and therapies. Nucleic Acid Ther 23(1):15–20. https://doi.org/10.1089/nat.2012.0414. PMID: 23391415; PMCID: PMC3569944
Sasaki T, Nakashiro K, Tanaka H et al (2010) Knockdown of Akt isoforms by RNA silencing suppresses the growth of human prostate cancer cells in vitro and in vivo. Biochem Biophys Res Commun 399(1):79–83
Saw PE, Song EW (2020) siRNA therapeutics: a clinical reality. Sci China Life Sci 63:485–500
Saw PE, Song E (2024) Advancements in clinical RNA therapeutics: present developments and prospective outlooks. Cell Rep Med 5(5):101555
Schultheis B, Strumberg D, Kuhlmann J, Wolf M, Link K, Seufferlein T, Kaufmann J, Feist M, Gebhardt F, Khan M et al (2020) Safety, efficacy and pharcacokinetics of targeted therapy with the liposomal RNA interference therapeutic Atu027 combined with gemcitabine in patients with pancreatic adenocarcinoma. A randomized phase Ib/IIa study. Cancers 12:3130
Setten RL, Rossi JJ, Han SP (2019a) The current state and future directions of RNAi-based therapeutics. Nat Rev Drug Discov 18(6):421–446
Setten RL, Rossi JJ, Han SP (2019b) The current state and future directions of RNAi-based therapeutics. Nat Rev Drug Discov 18:421–446
Shakil S, Baig MH, Tabrez S et al (2019) Molecular and enzoinformatics perspectives of targeting polo-like kinase 1 in cancer therapy. Semin Cancer Biol 56:47–55
Sharma P, Allison JP (2015) Immune checkpoint targeting in cancer therapy: toward combination strategies with curative potential. Cell 161(2):205–214
Shukla RS, Qin B, Cheng K (2014) Peptides used in the delivery of small noncoding RNA. Mol Pharm 11(10):3395–3408
Singh A, Trivedi P, Jain NK (2018) Advances in siRNA delivery in cancer therapy. Artif Cells Nanomed Biotechnol 46(2):274–283
Slack FJ, Chinnaiyan AM (2019) The role of non-coding RNAs in oncology. Cell 179(5):1033–1055
Smith CE, Zain R (2019) Therapeutic oligonucleotides: state of the art. Annu Rev Pharmacol Toxicol 59(1):605–630
Sousa C, Videira M (2025) Dual approaches in oncology: the promise of siRNA and chemotherapy combinations in cancer therapies. Onco 5:2
Subhan MA, Torchilin VP (2019) Efficient nanocarriers of siRNA therapeutics for cancer treatment. Transl Res 214:62–91
Suva ML, Riggi N, Bernstein BE (2013) Epigenetic reprogramming in cancer. Science 339:1567–1570
Tabernero J, Shapiro GI, LoRusso PM, Cervantes A, Schwartz GK, Weiss GJ, Paz-Ares L, Cho DC, Infante JR, Alsina M et al (2013) First-in-humans trial of an RNA interference therapeutic targeting VEGF and KSP in cancer patients with liver involvement. Cancer Discov 3:406

Takahashi H, Asaoka M, Yan L, Rashid OM, Oshi M, Ishikawa T, Nagahashi M, Takabe K (2020) Biologically aggressive phenotype and anti-cancer immunity counterbalance in breast cancer with high mutation rate. Sci Rep 10(1):1852

Tiemann K, Rossi JJ (2009) RNAi-based therapeutics-current status, challenges and prospects. EMBO Mol Med 1(3):142–51

Tomari Y, Zamore PD (2005) Perspective: machines for RNAi. Genes Dev 19(5):517–29

Tuschl T et al (2001) Targeted gene silencing using duplex RNA in mammalian cells. Nature 403(6766):838–842

Uchida S, Perche F, Pichon C, Cabral H (2020) Nanomedicine-based approaches for mRNA delivery. Mol Pharm 17(10):3654–3684

Valastyan S, Weinberg RA (2011) Tumor metastasis: molecular insights and evolving paradigms. Cell 147(2):275–292

Vlassov VV, An D (2012) Exosome-mediated RNA delivery. RNA Biol 9(5):799–808

Walther W, Schlag PM (2013) Current status of gene therapy for cancer. Curr Opin Oncol 25:659–664

Wang KC, Chang HY (2011) Molecular mechanisms of long noncoding RNAs. Mol Cell 43(6):904–914

Wang Z, Li Y, Banerjee S, Kong D, Ahmad A, Nogueira V, Hay N, Sarkar FH (2010a) Down-regulation of notch-1 and jagged-1 inhibits prostate cancer cell growth, migration and invasion, and induces apoptosis via inactivation of Akt, mTOR, and NF-kappaB signaling pathways. J Cell Biochem 109:726–736

Wang J, Lu Z, Wientjes MG, Au JLS (2010b) Delivery of siRNA therapeutics: barriers and carriers. AAPS J 12:492–503

Wang T, Shigdar S, Shamaileh HA, Gantier MP, Yin W, Xiang D, Wang L, Zhou SF, Hou Y, Wang P, Zhang W, Pu C, Duan W (2017) Challenges and opportunities for siRNA-based cancer treatment. Cancer Lett 387:77–83. https://doi.org/10.1016/j.canlet.2016.03.045. Epub 2016 Apr 1. PMID: 27045474

Wei C, Xu Y, Shen Q, Li R, Xiao X, Saw PE, Xu X (2023) Role of long non-coding RNAs in cancer: from subcellular localization to nanoparticle-mediated targeted regulation. Mol Ther Nucleic Acids 33:774–793

Whitehead KA, Langer R (2009) Knocking down cancer with RNAi. Nat Rev Drug Discov 8(1):57–63

Winkle M, El-Daly SM, Fabbri M, Calin GA (2021) Noncoding RNA therapeutics — challenges and potential solutions. Nat Rev Drug Discov 20:629

Won JE, Byeon Y, Wi TI, Lee CM, Lee JH, Kang TH, Lee JW, Lee Y, Park YM, Han HD (2022) Immune checkpoint silencing using RNAi-incorporated nanoparticles enhances antitumor immunity and therapeutic efficacy compared with antibody-based approaches. J Immunother Cancer 10(2):e002938

Wong SC, Klein JJ, Hamilton HL et al (2012) Co-injection of a targeted, reversibly masked endosomolytic polymer dramatically improves the efficacy of cholesterol-conjugated small interfering RNAs in vivo. Nucleic Acid Ther 22(6):380–390

Xi Z, Klokk TI, Korkmaz K, Kurys P, Elbi C, Risberg B, Danielsen H, Loda M, Saatcioglu F (2004) Kallikrein 4 is a predominantly nuclear protein and is overexpressed in prostate cancer. Cancer Res 64:2365–2370

Xu R, Huang Y, Mai J, Zhang G, Guo X, Xia X, Koay EJ, Qin G, Erm DR, Li Q, Liu X, Ferrari M, Shen H (2013) Multistage vectored siRNA targeting ataxia-telangiectasia mutated for breast cancer therapy. Small 9:1799–1808

Xue W, Dahlman JE, Tammela T, Khan OF, Sood S, Dave A, Cai W, Chirino LM, Yang GR, Bronson R, Crowley DG, Sahay G, Schroeder A, Langer R, Anderson DG, Jacks T (2014) Small RNA combination therapy for lung cancer. Proc Natl Acad Sci USA 111:E3553–E3561

Yan Z, Zou H, Tian F, Grandis JR, Mixson AJ, Lu PY, Li LY (2008) Human rhomboid family-1 gene silencing causes apoptosis or autophagy to epithelial cancer cells and inhibits xenograft tumor growth. Mol Cancer Ther 7(6):1355–1364

Yeung M, Bennasser Y, Le S et al (2005) siRNA, miRNA and HIV: promises and challenges. Cell Res 15:935–946. https://doi.org/10.1038/sj.cr.7290371

Yonezawa S, Koide H, Asai T (2020) Recent advances in siRNA delivery mediated by lipid-based nanoparticles. Adv Drug Deliv Rev 154:64–78

Yoon J, Shin M, Lee JY, Lee SN, Choi JH, Choi JW (2022) RNA interference (RNAi)-based plasmonic nanomaterials for cancer diagnosis and therapy. J Control Release 342:228–240

Zhang J, Li X, Huang L (2014) Non-viral nanocarriers for siRNA delivery in breast cancer. JControl Release 190:440–450

Zhang C, Zhao Y, Zhang E, Jiang M, Zhi D, Chen H, Cui S, Zhen Y, Cui J, Zhang S (2020) Co-delivery of paclitaxel and anti-VEGF SiRNA by tripeptide lipid nanoparticle to enhance the anti-tumor activity for lung cancer therapy. Drug Deliv 27:1397

Zhang W, Liu Y, Min Z, Liang G, Mo J, Ju Z, Zeng B, Guan W, Zhang Y, Chen J, Zhang Q, Li H, Zeng C, Wei Y, Chan G-F (2021a) Circ- mine: a comprehensive database to integrate, analyze and visualize human disease-related circRNA transcriptome. Nucleic Acids Res 50(D1):D83–D92

Zhang MM, Bahal R, Rasmussen TP, Manautou JE, Zhong XB (2021b) The growth of siRNA-based therapeutics: updated clinical studies. Biochem Pharmacol 189:114432. https://doi.org/10.1016/j.bcp.2021.114432. Epub 2021 Jan 26. PMID: 33513339; PMCID: PMC8187268

Zhang Y, Vu T, Palmer DC et al (2022) A T cell resilience model associated with response to immunotherapy in multiple tumor types. Nat Med 28(10):2219–2219

Zhou M, Zou X, Cheng K et al (2022) The role of cell-penetrating peptides in potential anti-cancer therapy. Clin Transl Med 12(5):e822

Zhu Y, Zhu L, Wang X et al (2022) RNA-based therapeutics: an overview and prospectus. Cell Death Dis 13:644

Ziemer L, Kurrle R (2017) Delivery systems for RNA interference-based cancer therapy. Cancer Gene Ther 24(11):475–484

Zogg H, Singh R, Ro S (2022) Current advances in RNA therapeutics for human diseases. Int J Mol Sci 23(5):2736

Zuo Z, Liu J, Sun Z et al (2023) ERK and c-Myc signaling in host-derived tumor endothelial cells is essential for solid tumor growth. Proc Natl Acad Sci USA 120(1):e2211927120

Chapter 6
Exploring the Role of Long Noncoding RNAs (lncRNAs) as Biomarkers in Cancers

Shahid Bashir and Mohammad Uzair

Contents

Abstract The whole human genome sequence analysis has identified a large proportion of transcripts that lack protein-encoding function known as noncoding RNAs (ncRNAs). Long ncRNAs (lncRNAs) are transcripts consisting of more than 200 nucleotides. LncRNAs are identified as key regulators for diverse biological processes, including gene expressions, epigenetic modulations, transcriptional and

---

S. Bashir (✉)
Neuroscience Center, King Fahad Specialist Hospital Dammam, Dammam, Saudi Arabia

M. Uzair
Department of Bioengineering, King Fahd University of Petroleum and Minerals, Dhahran, Saudi Arabia

M. F. Ullah, A. Ahmad (eds.), *Non-coding RNAs*, Experientia Supplementum 115,
https://doi.org/10.1007/978-3-032-06948-1_6

translational regulation. LncRNAs-based mechanisms control gene activation, and cellular processes, while the dysregulated lncRNAs are associated with various disorders, including cancer. Moreover, lncRNAs have potential in diagnosis and prognosis of cancer. Several lncRNAs contribute to cancer development, metastatic cascade, and their molecular regulations in cancer development. The aberrant expression of lncRNAs regulates tumorigenesis, proliferation, metastasis, cancer stage, tumor size, and overall survival rates. Several lncRNAs are highly expressed or downregulated in several human tumors, behaving as oncogenes or tumor suppressors, respectively. The unique expression level of oncogenic lncRNAs provides a possibility to utilize them as a promising predictive biomarker of cancer.

Keywords Metastasis · Oncogenic lncRNAs · Tumor-suppressor lncRNAs · Therapeutic targets · Biomarkers

Abbreviations

EMT	Epithelial-to-mesenchymal transition
GA-5	Growth arrest-specific 5
lncRNAs	Long noncoding RNAs
ncRNAs	Noncoding RNAs
Pc2	Polycomb 2 protein
Xist	X-inactivate specific transcript

6.1 Introduction

Seventy-five percent of the human genome transcribed into RNA and only 2% of transcribed RNA has the potential for protein-coding (Djebali et al. 2012). The remaining portion is ineffective at encoding a functional protein, designated as non-coding RNAs (ncRNAs). Based on the length of nucleotides, ncRNAs are of two types: short ncRNAs (sncRNAs) that consist of ≤200 bp and long noncoding RNAs (lncRNAs) that are >200 bp long (Iyer et al. 2015). Due to their genomic location in the gene, lncRNAs are further categorized into: Intergenic lncRNAs—between two exons of protein-coding regions: Bidirectional lncRNA—transcribed in the opposite strand in an anti-parallel direction from the similar protein-coding promoter. Sense-overlapping lncRNA—overlap with introns or exons in the coding strand of RNA of different protein-coding genes, and Antisense lncRNA—transcribed from the protein-coding gene's template strand (Li et al. 2019; Ponting et al. 2009) (Fig. 6.1).

Until recently, lncRNAs were considered as nonfunctional junk DNA. However, lncRNAs play an important role in regulatory functions, making our understanding of genomic regulation significantly more complex (Batista and Chang 2013;

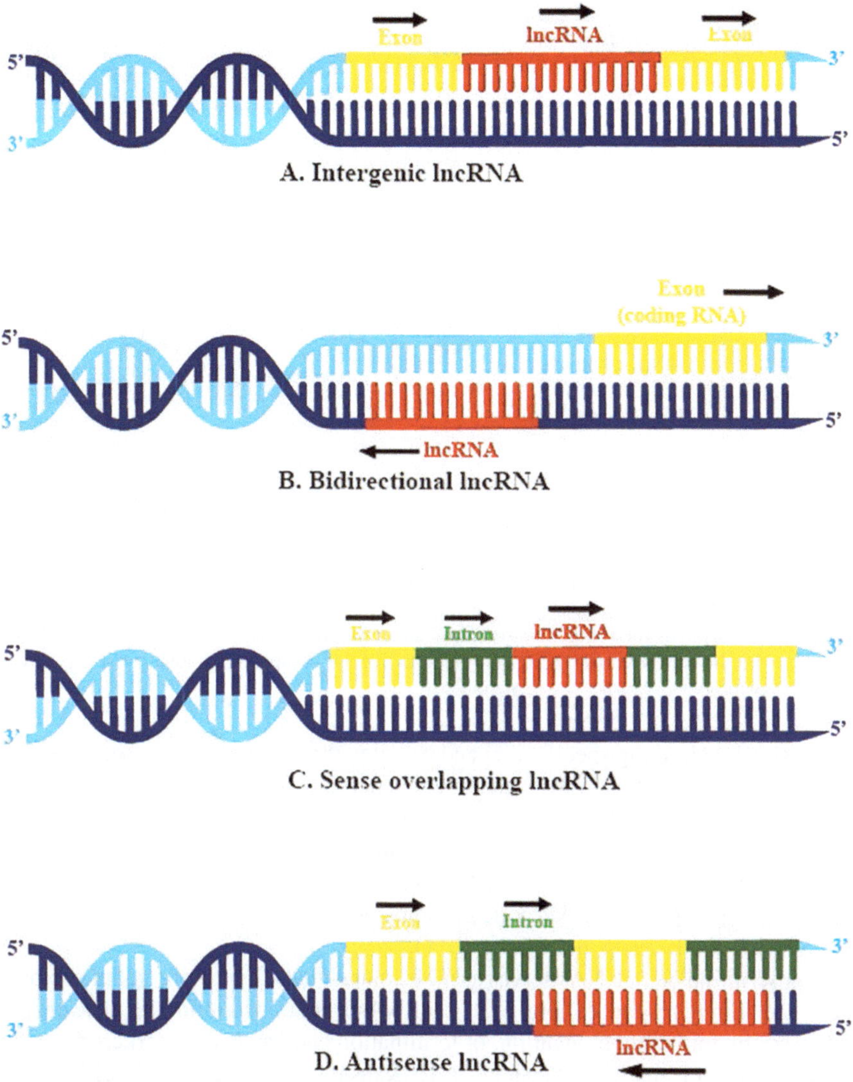

Fig. 6.1 LncRNAs classification based on genomic location. (**a**) Intergenic lncRNAs—localized between exonic regions. (**b**) Bidirectional lncRNAs—transcribed in the opposite strand in an antiparallel direction from the same protein-coding promoter. (**c**) Sense-overlapping lncRNA—overlap with introns or exons in the sense strand of RNA. (**d**) Antisense lncRNA—originated from an antisense RNA strand

Bergmann and Spector 2014). LncRNAs contribute to various biological functions as scaffolds, signals, guides, and decoys (Moran et al. 2012) (Fig. 6.2).

LncRNAs, as dynamic molecular scaffolds for cofactors, facilitate polypeptide assembly and protein backbone conformation (Fig. 6.2). They regulate genes

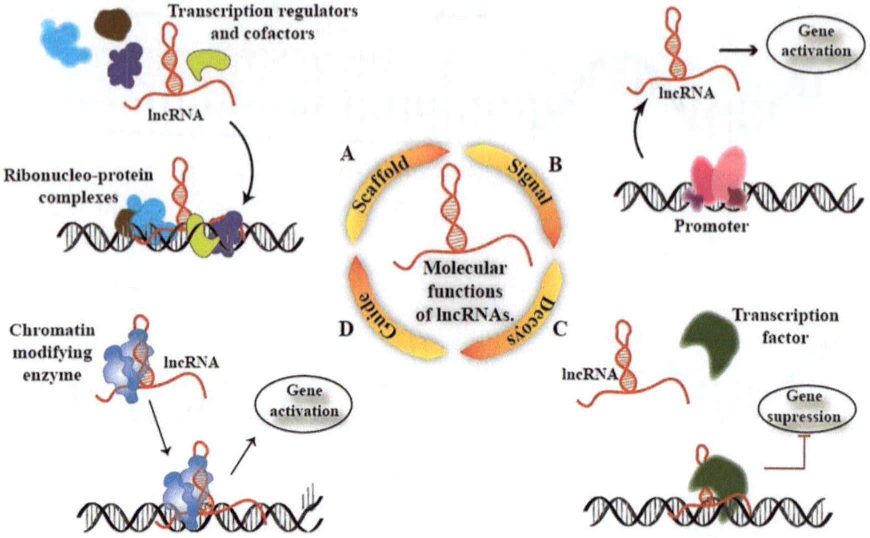

Fig. 6.2 General mechanism for lncRNA classification on basis of molecular functions. (**a**) Scaffold lncRNAs—function in assembly and protein backbone conformation. (**b**) lncRNAs—act as signaling molecules for cellular expressions and stimuli response. (**c**) lncRNAs—seize the target by binding to miRNAs. (**d**) lncRNA—guide molecular complexes to localize at a particular genomic position

expression by targeting specific loci and promote the catalytic activity of enzymatic complexes (Ribeiro et al. 2018; Mathy and Chen 2017). HOTAIR, TUG1, ANRIL, and many other lncRNAs are molecular scaffolds that modify proteins. Telomerase RNA (TERC) is a telomeric lncRNA that functions as a template for telomere replication (reverse transcriptase). TERC assembles telomerase and sustains telomere ends (5′-TTAGGG-3) of eukaryotic chromosomes to avoid genes loss· (Bergmann and Spector 2014; Wang and Chang 2011).

Transcription of lncRNAs serves as molecular signals for specific cellular expressions and stimuli responses (Fig. 6.2). LncRNAs from transcription regulate cellular processes by initiation, elongation, or termination and can act as markers for cellular functions (Wang and Chang 2011; Kansara et al. 2020). LncRNA Xist (X-inactivate specific transcript) is involved in X-chromosomal inactivation (or Iyonization) to equalize transcript levels between males and females. Xist RNA is found in the centromere of X-chromosomes and, in particular, is expressed from a Barr body (i.e., inactive X-chromosome). Xist wraps the X-chromosome and activates gene repression. Hence, X-chromosome transcribed lncRNA-Xist specifies active silencing at its genomic location (Pontier and Gribnau 2011; Loda and Heard 2019). Other lncRNAs that act as signals include KCNQ1OT1, HOTAIR, COOLAIR, linc-p21, and PANDA.

Decoy lncRNAs, such as MALAT1, PANDA, and NF-YA, negatively regulate gene expression by binding to chromatin modification proteins, transcription, microRNA (miRNA), or regulation factors, thus diverting them from their target proteins (Moran et al. 2012) (Fig. 6.2). Decoy lncRNAs sequestrate miRNAs, ribonucleoprotein complexes, and RNA-binding proteins (RBPs) to regulate gene expression (Wang and Chang 2011). LncRNA GA-5 (Growth arrest-specific 5) is known for its glucocorticoid resistance regulatory mechanism. GA5 transcribes RNA motif imitating DNA motifs of the glucocorticoid responsive gene. Thus, GA5 acts as a molecular decoy, by efficiently binding and inhibiting glucocorticoid receptors causing glucocorticoid resistance (Kino et al. 2010; Pickard and Williams 2015).

Overview

A molecular decoy is a molecule, often a type of RNA or protein, that mimics the binding site of a target molecule to sequester regulatory factors such as transcription factors, microRNAs, or proteins. By competitively binding these factors, molecular decoys prevent them from interacting with their intended targets, thereby altering cellular processes.

In contrast to decoys, some lncRNAs interact with protein assemblies or complexes and direct them at particular genes or loci (Mathy and Chen 2017) (Fig. 6.2). Such lncRNAs are classed as *guides* and required for the proper localization of regulatory protein factors on genomic loci, either in a *Trans* or *cis* manner for modulating the expression of the gene. CCND1, Xist, HOTTIP, COLDAIR, rDNA transcript, and Air are guides in *cis* while Jpx, HOTAIR, linc-p21, and other PRC2-bound RNAs are guides in *trans (*Wang and Chang 2011*;* Kansara et al. 2020*)*. Firre transcribed from the inactive X-chromosome targets chromosomes in a three-dimensional order. Firre sustains the Barr body (i.e., inactive X-chromosome) position during XCI in the *cis* configuration. In *trans*, Firre interacts with hnRNPU and forms nuclear bodies to localize genomic loci (Yang et al. 2015).

Furthermore, lncRNAs can modulate gene expression by binding to other nucleic acids and proteins. lncRNAs can potentially regulate various cellular processes, such as cell division, translation, posttranslation, transcription regulation, cell metabolism, protein trafficking, and epigenetic modifications (Jin et al. 2020a). Dysregulation and dysfunction of lncRNAs affect these processes and have a direct impact on tissue and organ' physiology causing several disorders. Recent studies have indicated that nonprotein-coding parts of the human genome are involved in numerous diseases, most notably carcinogenesis and metastasis (Olivero and Dimitrova 2020; Balas and Johnson 2018). Currently, the potential role of lncRNAs in cancer pathogenesis has got significant interest, and lncRNAs show potential in diagnosis and prognosis. LncRNAs have diverse roles in apoptosis, migration, and

Overview
A biomarker is a measurable biological indicator that helps detect or monitor a physiological state, disease, or response to treatment. Found in molecules like proteins, RNA, or metabolites, biomarkers can significantly facilitate early diagnosis, disease prognosis, and personalized therapy. In cancer, they provide insights into tumor behavior, aiding in tailored treatments and better outcomes.

invasion of cancer cells during tumorigenesis and metastasis (Bergmann and Spector 2014; Aprile et al. 2020). This chapter explains the mechanisms of lncRNAs in metastasis and tumor metastasis, especially focusing on their potential as a biomarker for cancer diagnosis and prognosis.

6.2 LncRNAs in Carcinogenesis and Tumor Progression

LncRNAs can function as regulators for normal cellular processes or, as is often the case, differentially regulate cancer development and are significantly associated with major types of cancer. Cancer, in addition to uncontrolled, unrestricted expansion of abnormal cells, involves various biological events such as metabolic disorders, DNA damage, cell stemness, and drug resistance. LncRNAs have potential roles in translation, transcription, epigenetic regulation, and gene expression, as well as cancer initiation and metastasis. Aberrant expressions of lncRNAs are associated with cancer development, metastasis, migration, and patient survival rate (Jiang et al. 2019; Jin et al. 2020a; Olivero and Dimitrova 2020). So far many lncRNAs, including HOTAIR, HOTTIP, HULC, ANRIL, BCAR4, CTBP-AS, TUG1, UCA1, XIST, LSINCT5, MALAT1, NEAT1, HEIH, KCNQ1OT1, PCGEM1, PVT1, H19, SCHLAP1, and PCAT1/5/18, have to potentially act as oncogenes and can regulate cancer development through different mechanisms (Balas and Johnson 2018; Aprile et al. 2020; Fu et al. 2019) (Fig. 6.3, Table 6.1). LncRNAs are released in the blood stream of cancer patients, due to their differential expression in cancerous states, potentially serving as prognostic molecular markers. Since lncRNAs are present in almost all body fluids and tissues, they may represent a reliable and effective approach for the diagnosis of cancer. For example, elevated levels of HOTAIR, BC200, TUG1, PANDAR, MALAT1, UCA1, CCAT1, CCAT2, and BCAR4, are associated with proliferation, invasion, differentiation, and metastasis. However, other lncRNAs such as MEG3 and GAS5 act as tumor suppressors and are found to be downregulated. LncRNAs also positively correlate with tumor size, stage, and overall patient survival rates. Thus, lncRNAs could be the future of diagnosis and prognosis of the cancers and tumor cells. Therefore,

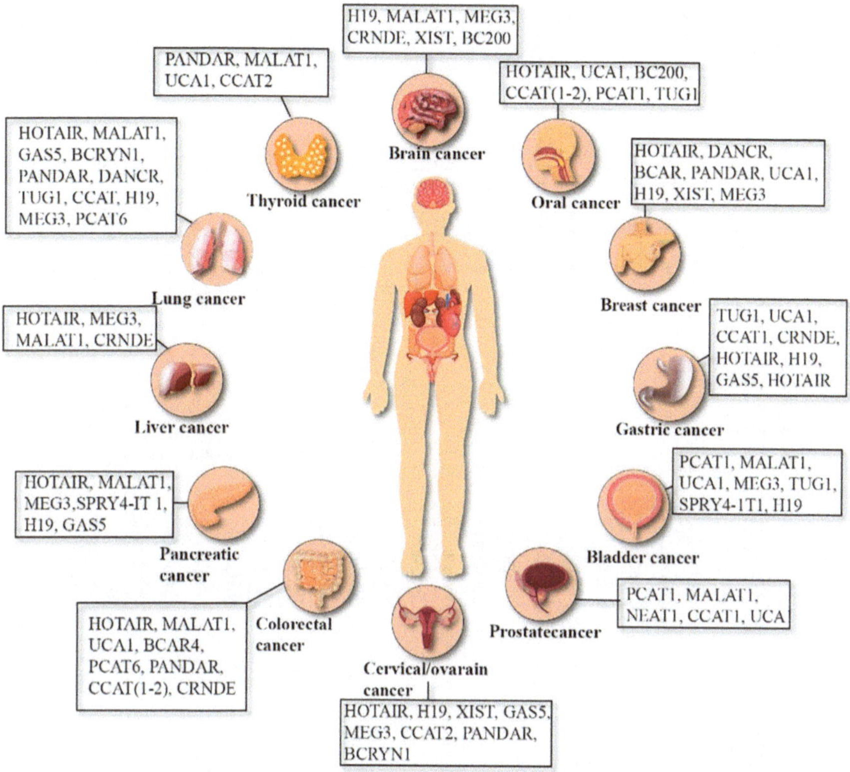

Fig. 6.3 Differentially expressed lncRNA associated with different types of cancers

lncRNAs could be potential diagnostic and therapeutic strategies for early cancer detection.

In the following sections, a selection of the commonly expressed lncRNAs in cancerous tissue will be discussed to examine their role and contribution to cancer development and the diagnostic value of lncRNA biomarkers in cancer.

6.2.1 HOTAIR

HOX Transcript Antisense Intergenic RNA (HOTAIR), a *trans*-acting lncRNA, significantly regulates gene silencing and metastasis through its dual interaction with poly-comb Repressive Complex 2 (PRC2) and Lysine-Specific Demethylase 1 (LSD1). Acting as a modular scaffold, HOTAIR recruits PRC2 and mediates the trimethylation of histone H3 at lysine 27 (H3K27me3), establishing a repressive chromatin state. Simultaneously, it directs LSD1 to remove active histone marks by demethylating histone H3 at lysine 4 (H3K4me1/2), further reinforcing

Table 6.1 Differentially expressed lncRNAs associated with different types of cancers

LncRNAs	Full name	Genomic location	Exon count	Transcript length (Base pairs)	Cancer type	Regulation/ expression	References
HOTAIR	HOX transcript antisense RNA	12q13.13	7	2370	Lung	↑	Ren et al. (2020)
					Gastric		Zhang et al. (2020a)
					Colorectal		Chen et al. (2020)
					Liver		Yuan et al. (2020)
					Breast		Mozdarani et al. (2020)
BC200 (BCYRN1)	Brain cytoplasmic 200	2p16	1	200	Lung	↑	Hu and Lu (2015)
					Breast		De Leeneer and Claes (2015)
					Ovaries		Peng et al. (2018)
					Cervical		Shin et al. (2018)
PANDAR	P21 associate d ncRNA DNA damage activated	6p21.2	1	1506	Breast	↑	Sang et al. (2016)
					Ovarian		Wang et al. (2018a)
					NSCL	↓	Han et al. (2019)
TUG1	Taurine upregulated 1	22q12.2	4	5742	Bladder	↑	Sarfi et al. (2021)
					Colorectal		Wang et al. (2016)
					Cervical	↓	Zhu et al. (2017)
					Lung		Zhang et al. (2014b)
					NSCL		
MALAT1	Metastasis-associated lung adenocarcinoma transcript 1	11q13.1	2	8779	NSCL	↑	Pan et al. (2020)
					Colorectal		Siddiqui et al. (2019)
					Pancreatic		Lu et al. (2020)
					Breast	↓	Yang et al. (2016)

UCA1	Human urothelial cancer-associated 1	19p13.12	3	2314	Bladder	↑	Li et al. (2017b)
					Thyroid		Lu and Liu (2018)
					Breast		Wu and Luo (2016)
					Lung		Wu and Zhou (2018)
					Ovaries		Qiu et al. (2017)
CCAT1	Cancer-associated transcript 1	8q24.21	2	2759	Colorectal	↑	He et al. (2014)
					Breast		Zhang et al. (2015)
					Ovaries		Lai and Cheng (2018)
					Prostate		Chen et al. (2018)
CCAT2	Cancer-associated transcript 2	8q24.21	1	1752	Skin	↑	Dehcheshmeh et al. (2018)
					Kidney		Huang et al. (2017)
					Thyroid		Xin and Ye (2020)
					Colorectal		Ozawa et al. (2017)
					Lung		Wang et al. (2020c)
MEG3	Maternally expressed gene 3	14q32.3	13	1785	Brain	↓	Zhou et al. (2012), Talebi et al. (2020), He et al. (2017), Peng et al. (2015a)
					Liver		
					Lung		
					Breast		
					Cervix		
BCAR4	Breast cancer antiestrogen resistance 4	16p13.3	5	728	Breast	↑	Peng et al. (2021)
					Prostate		Cai et al. (2018)
					Gastric		Wang et al. (2017)
					Bladder		Zou et al. (2018b)
					Lung		Gong et al. (2017)

(continued)

Table 6.1 (continued)

LncRNAs	Full name	Genomic location	Exon count	Transcript length (Base pairs)	Cancer type	Regulation/ expression	References
GAS5	Growth arrest-specific 5	1q25.1	13	656	Gastric	↓	Liu et al. (2016b)
					Bone		Wang and Kong (2018)
					NSCL		Mei et al. (2017)
					Colorectal		Liu et al. (2018)
PCAT6	Prostate cancer-associated transcript 6	1q32.1	2	764	Prostate	↑	Du et al. (2013)
					Bone		Guan et al. (2020)
					Lung		Cui et al. (2018)
					Colorectal		Wu et al. (2019a)
CRNDE	Colorectal neoplasia differentially expressed	16q12.2	7	1059	Colorectal	↑	Ellis et al. (2012)
					Cervical		Zhang and Fan (2019)
					Lung		Zhang et al. (2018b)
					Breast		Huan et al. (2017)
H19	H19 imprinted maternally expressed transcript	11p15.5	6	2362	Breast	↑	Zhang et al. (2016)
					Ovaries		Wu et al. (2019b)
					Hepatocellular	↓	Ardelt and Pachmayr (2017)
					Adrenocortical		Gao et al. (2002)
SPRY4-IT 1	Sprouty RTK signaling antagonist 4-intronic transcript 1	5q31.3	1	703	Breast	↑	Shi et al. (2015)
					Liver		Yu et al. (2017b)
					Pancreas		Guo et al. (2018)
					Bladder		Zhao et al. (2015b)
					Gastric		Peng et al. (2015b)

↑: High expression level. ↓: Low expression level

transcriptional repression (Statello et al. 2021). This coordinated mechanism silences tumor suppressor and anti-metastatic genes, facilitating cancer progression and metastasis. The differential expression of HOTAIR significantly contributes to poor prognosis in various cancers (Yuan et al. 2020), including oral (Lu et al. 2017a), lung (Ren et al. 2020), esophageal (Ma et al. 2016), colorectal (Chen et al. 2020), gastric (Zhang et al. 2020a), cervical (Zhou et al. 2020), liver (Yuan et al. 2020), and breast (Mozdarani et al. 2020). HOTAIR could be a potential biomarker for detecting such cancers. Studies indicate that high expression of HOTAIR lncRNA regulates the promotion of tumor growth, metastasis, cancer invasion, aggression, and malignancy (Zhang et al. 2014a; Liu et al. 2020). Furthermore, HOTAIR interacts with epithelial-to-mesenchymal transition (EMT) to contribute to cancer stem cells development. Therefore, HOTAIR triggers EMT and cancer cell stemness acquisition via the control of multiple signaling mechanisms during tumorigenesis (Pádua Alves et al. 2013). The overexpression of HOTAIR lncRNA in cancerous epithelial cells is linked with metastasis and poor prognosis outcomes in various tumors.

HOTAIR is significantly involved in chemotherapy resistance by promoting multidrug resistance (MDR), leading to decreased drug sensitivity and increased resistance across various cancers. HOTAIR resists to a wide range of anticancer agents, including anthracyclines (e.g., doxorubicin, epirubicin), platinum-based drugs (e.g., cisplatin, oxaliplatin), fluoropyrimidines (e.g., 5-fluorouracil), taxanes (e.g., paclitaxel, docetaxel), and tyrosine kinase inhibitors (e.g., imatinib, gefitinib, crizotinib, sunitinib, sorafenib). Furthermore, it affects responsiveness to hormonal agents (e.g., progesterone, dexamethasone), alkylating agents (e.g., temozolomide, mitomycin), and antibody-based therapies (e.g., trastuzumab). HOTAIR-mediated chemoresistance is further modulated by its interaction with various miRNAs, such as miR-34a, miR-126, miR-125, miR-145, etc., which regulate key molecular pathways. The Wnt/β-catenin and PI3K/Akt signaling pathways are central mediators of HOTAIR-induced resistance (Zhu et al. 2022; Liu et al. 2016a; Cantile et al. 2024), highlighting HOTAIR's role as a regulator of chemotherapy efficacy and a potential target to overcome drug resistance in cancer.

6.2.2 BC200 (BCYRN1)

Brain cytoplasmic RNA 1 (or BC200) is the shortest oncogenic factor lncRNA, first discovered in primate brain (Watson and Sutcliffe 1987). BC200 is predominantly expressed in neurons, where it regulates local protein synthesis by modulating translation (Peng et al. 2020). It exhibits RHAU (for RNA helicase associated with AU rich element) helicase activity by binding to RHAU, enabling it to unwind cytosine-rich tetra-strand structures at its 3′ end, a function essential for its regulatory interactions. BC200 binds to various proteins, including poly(A)-binding protein, and synaptic cytoplasmic interaction proteins, further modulating translational and synaptic processes (Jang et al. 2020; Tan et al. 2020; Booy et al. 2016).

BC200 has been linked to various human diseases, specifically Alzheimer's disease, and numerous types of cancer through alterations in RNA expression. BC200 lncRNA is detected in cancerous masses of esophageal, breast, oral, ovarian, colon, and lung cancers and proposed a possible link between BC200 and tumorigenesis. Various studies upregulated and overexpressed BC200 in breast (De Leeneer and Claes 2015), stomach (Ren et al. 2018), esophageal, lung, tongue, parotid, ovarian, and cervical cancers (Shin et al. 2018; Peng et al. 2018). Consequently, a higher expression of BC200 lncRNA predicts poor prognosis and can be used as a molecular marker for cancer management (Zhao et al. 2016). In non-small-cell lung cancer (NSCLC), c-MYC binds to the BC200, with a higher expression observed in MYC-related cancers (Hu and Lu 2015). In the hematologic malignancies, i.e., myelodysplastic syndrome (MDS), BC200, as a competing endogenous RNA (ceRNA), sponges miR-150-5p and upregulates MYB expression, a key driver of malignant proliferation. This BC200/miR-150-5p/MYB feedback loop has been identified as a critical pathway promoting the progression of MDS (Liu et al. 2022), positioning BC200 as a potential biomarker and therapeutic target.

Overview
ceRNA is a type of RNA that regulates gene expression by competing for shared microRNAs (miRNAs). These RNAs act as "sponges" to bind and sequester miRNAs, preventing them from targeting other RNA molecules.

6.2.3 PANDAR

P21-associated ncRNA DNA damage activated (PANDAR) regulates the repair of DNA damage (Guttman and Rinn 2012). During DNA damage, PANDAR induced by tumor suppressor p53 interacts with nuclear transcription factor Y subunit alpha (NF-YA) to limit proapoptotic genes expression, thereby contributing to cell survival (Hung et al. 2011). PANDAR functions as a transcriptional coactivator and interacts with scaffold attachment factor A (SAFA) and Bmi1 to modulate gene expression (Puvvula et al. 2014). Deregulation of the lncRNA PANDAR gene is associated with tumor development and serve as a potential biomarker for cancer identification (Zhan et al. 2016; Zou et al. 2018a), with high expression and upregulation observed in thyroid (Li et al. 2017a), gastric (Ma et al. 2016), bladder (Zhan et al. 2016), breast (Sang et al. 2016), ovarian (Wang et al. 2018a), and colorectal (Lu et al. 2017b) carcinomas. However, the low expression of PANDAR in NSCLC and high expression in colorectal cancer are associated with poor prognosis and survival rate (Liz and Esteller 2016; Han et al. 2019). In oral squamous cell carcinoma (OSCC), PANDAR is highly expressed, and facilitates proliferation and reduces apoptosis. PANDAR binds to serine and arginine-rich splicing factor 7

(SRSF7), facilitating SRSF7-dependent upregulation of serine/threonine-protein kinase pim-1 (PIM1), which promotes OSCC progression (Jia et al. 2021). These findings highlight PANDAR's role as a critical regulator in cancer biology, presenting it as a potential therapeutic target. Aberrant expression and dysregulation of the PANDAR gene, in various malignancies, involves migration, proliferation, invasion, and anti-apoptosis (Li et al. 2017a).

6.2.4 *TUG1*

Taurine upregulated gene 1 (TUG1) functions as an epigenetic regulatory factor for transcription. TUG1 develops retinal cells in response to taurine and promotes the development of photoreceptors in the retina (Balas and Johnson 2018; Young et al. 2005). TUG1 regulates the cell cycle when induced by p53 and interacts with polycomb repressor complex1 and 2 (PRC1-PRC2) (Khalil et al. 2009). TUG1 functions as a ceRNA by sponging miRNA. TUG1 negatively correlates with miRNAs, particularly miR-143-5p. TUG1 and miR-143-5p are in the same RISC complex. However, TUG1 expression is not influenced by up- or downregulation of miR-143-5p (Yu et al. 2019). Several studies reported the aberrant expression and dysregulation of TUG1 in numerous human cancers. TUG1 is a poor prognostic factor in esophageal (Jin et al. 2020b), bladder (Sarfi et al. 2021), colorectal (Wang et al. 2016), cervical (Zhu et al. 2017), ovarian (Li et al. 2018a), and hepatocellular (Li et al. 2021) carcinomas, and it shows upregulated expression. Conversely, the lower expression of TUG1 in lung cancer and NSCLC is associated with larger tumor size and cancer progression (Zhang et al. 2014b). TUG1 affects tumorigenesis by cell proliferation, invasion, migration, and resisting apoptosis (Zhu et al. 2017; Dai et al. 2020).

6.2.5 *MALAT1*

Metastasis-associated lung adenocarcinoma transcript 1 (MALAT 1), also known as NEAT2 (noncoding nuclear-enriched abundant transcript 2), is a spliced lncRNA highly abundant in the mammalian nucleus (Hutchinson et al. 2007). MALAT1 has diverse regulatory functions in alternative splicing, gene expression, autophagy, and cancer progression. MALAT1 acts as a scaffold and regulates the localization and function of Polycomb 2 protein (Pc2) by facilitating interactions between unmethylated Pc2, E2F transcription factor, activation-associated histone markers, and transcriptional coactivator complexes during serum stimulation, thereby promoting gene expression (Yang et al. 2011; Zhang et al. 2017a). MALAT1 also modulates alternative pre-mRNA splicing by regulating the phosphorylation status and nuclear speckle localization of SRSF1/2/3, and others, affecting the distribution of these splicing factors and influencing mRNA splicing (Arun et al. 2020; Tripathi et al. 2010).

MALAT1 plays a critical regulatory role in the development and progression of several cancers (Li et al. 2018b). For instance, MALAT1 is upregulated in the lung (Pan et al. 2020), colorectal (CRC) (Siddiqui et al. 2019), bladder (Cao et al. 2020), oral (Xiao et al. 2020), and thyroid (Peng et al. 2020). However, MALAT1 negative regulation regulates breast cancer progression, via suppression of CCND1 gene expression (Yang et al. 2016). MALAT1 also contributes to chemotherapy resistance and immune evasion (Hou et al. 2023). It competitively binds to miR-23b-3p, reducing its inhibitory effect on ATG12, thereby upregulating ATG12 expression and promoting autophagy-mediated chemotherapy resistance in gastric cancer (YiRen et al. 2017). MALAT1 also enhances immune escape and immunosuppression, leading to poor immunotherapy responses and unfavorable prognoses (Hou et al. 2023). Furthermore, it contributes to drug resistance by augmenting DNA repair through homologous recombination (HR) and nonhomologous end joining (NHEJ) pathways, as well as by altering drug distribution via the regulation of ABC transport proteins (Li and Heyer 2008; Chang et al. 2017; Hosoya and Miyagawa 2014). For instance, in glioblastoma, silencing MALAT1 (si-MALAT1) decreases the expression of LRP1, MDR1, MRP5, and ZEB1, enhancing Temozolomide (TMZ) sensitivity, whereas its overexpression induces TMZ resistance (Hou et al. 2023; Wang et al. 2020a; Kim et al. 2018). Thus, the abnormal expression of MALAT1 facilitates cancer progression, drug resistance, and immune evasion, making it a potential biomarker in cancer.

6.2.6 UCA1

Human urothelial cancer-associated 1 (UCA1) functions as gene expression during the postfertilization primary phase and plays a significant role throughout embryogenesis. UCA1 is overexpressed in bladder cancer and increases cancer cell proliferation (Wang et al. 2006, 2008). UCA1 was first found in urine in 2006 and has proven to be a sensitive and precise prognostic marker for bladder cancer (Wang et al. 2006; Cui et al. 2017). UCA1 acts as a ceRNA in bladder cancer and performs an important role in mitochondrial function. UCA1 contributes to enhancing mitochondrial function by upregulating ARL2 and inhibiting the miR-195 signaling pathway in bladder tumorigenesis (Li et al. 2017b). Moreover, UCA1 abnormal expressions have been evaluated in various malignancies, with most studies demonstrating a high expression in cancer. For example, elevated expression of UCA1 has been found in thyroid (Lu and Liu 2018), prostate (Ghiam et al. 2017), breast (Wu and Luo 2016), bladder (Fang et al. 2016), lung (Wu and Zhou 2018), and ovarian (Qiu et al. 2017) tumors. UCA1 exerts its oncogenic effects through interactions with multiple cancer-associated signaling pathways, including AKT, mTOR, Hippo, and Wnt/β pathways (Huang et al. 2021; Liu et al. 2021; Yang et al. 2021; Guan et al. 2022; Wang et al. 2021a). These interactions regulate critical cancer-related processes, particularly EMT, a key mechanism in tumor development and progression (Liu et al. 2021). UCA1 silencing has been shown to significantly increase the expression of E-cadherin, a hallmark of epithelial phenotype, while reducing the expression of mesenchymal markers such as N-cadherin, Snail, and Vimentin. In breast cancer, UCA1 suppression also upregulates GSK-3β and p-GSK-3β, leading to reduced levels of cyclin D1, β-catenin, and MMP-7, thereby inhibiting Wnt-catenin pathway (Jiao et al. 2016; Ghafouri-Fard and Taheri 2019). Through these regulatory mechanisms, UCA1 promotes EMT and contributes to cancer progression, highlighting it as a possible therapeutic target in malignancies.

6.2.7 CCAT1 and CCAT2

Colon cancer associated transcript 1 (CCAT1) is an RNA gene in humans, affiliated with the oncogenic lncRNA class. 2.6 kb long CCAT1 was initially identified in cancer, where it was upregulated and enhanced cancer progression (Alaiyan et al. 2013). CCAT1 has two isoforms: CCAT1-L, characterized by an extended second exon, and 5L-CCAT1, featuring an extended 5′UTR. CCAT1-L, located upstream of the MYC oncogene, regulates chromatin looping at the MYC locus (Amjadi-Moheb et al. 2021), enhancing its transcriptional activity. Abnormal expressions of CCAT1 are associated with various carcinomas by contributing to tumorigenesis, metastasis, tumor growth, size, stage, and survival rate (Guo and Hua 2017; Zhang et al. 2018a; Xin et al. 2016). CCAT1 has been found to be upregulated and promote the pathogenesis and progression of colorectal cancer (He et al. 2014), esophagus (Zhang et al. 2017b), breast (Zhang et al. 2015), and hepatocellular (Deng et al. 2015), and lung adenocarcinomas (Lin et al. 2018), ovarian (Lai and Cheng 2018), prostate (Chen et al. 2018), and gastric cancers (Zhang et al. 2018a). In gastric cancer (GC), CCAT1 significantly contributes to cancer development via the PTBP1/PKM2/glycolysis pathway. By interacting with PTBP1, CCAT1 stabilizes this critical splicing factor by preventing its ubiquitin-mediated degradation. PTBP1, in turn, facilitates the splicing transition from PKM1 to PKM2, augmenting glycolytic activity and promoting Warburg effect (Peng et al. 2024), thus, supporting the metabolic reprogramming of GC cells. These mechanisms highlight CCAT1's multifaceted role in cancer biology, making it a promising biomarker and therapeutic target in both AML and GC (Zhang et al. 2023). Moreover, as a ceRNA, CCAT1 sequesters miR-155, thereby regulating its activity and promoting the expression of downstream targets such as c-Myc. This ceRNA activity inhibits myeloid cell differentiation, promotes proliferation, and drives acute myeloid leukemia progression (Bohrer et al. 2024; Chen et al. 2016a).

Colon cancer-associated transcript 2 (CCAT2) is a 1.7 kb long oncogenic lncRNA, located at a recurrently amplified region 8q24 in colon cancer and other carcinomas. This transcript upregulates the proto-oncogene, MYC, to enhance tumor growth, metastasis, and chromosomal instability (Ling et al. 2013; Ozawa et al. 2017). The overexpression of CCAT2 has been identified as key activator in the initiation and progression of many different of cancers, including skin (Dehcheshmeh et al. 2018), kidney (Huang et al. 2017), thyroid (Xin and Ye 2020), colorectal (Ozawa et al. 2017), myeloid (Shah et al. 2018), and lung. Overexpression of CCAT2 suppresses mature miR-145 expression but enhances pre-miR-145 expression. CCAT2 regulates the miR-145 maturation process, since expression of CCAT2 negatively associated with mature miR-145 and positively with pre-miR-145 expression (Yu et al. 2017a). In hepatocellular carcinoma, CCAT2 enhances cellular proliferation and metastasis by upregulating the expression of NDRG1 through increased promoter activity (Liu et al. 2019a). In colorectal cancer, CCAT2 is overexpressed and promtes cell proliferation and metastasis by interacting with TAF15, thereby transcriptionally activating RAB14 and the AKT/GSK3β signaling pathway (Wang et al. 2021b). In esophageal cancer, CCAT2 negatively regulates miR-145 and p70 ribosomal protein S6 kinase 1 (p70S6K1), modulating key signaling pathways. It inhibits the Akt, ERK, and p70S6K1 phosphorylation while enhancing radiotherapy resistance by reducing apoptosis and altering the Bax/Bcl2 and active-caspase 3/caspase 3 ratios. CCAT2 drives resistance through the miR-145/p70S6K1 and p53 signaling pathways, underscoring its role in reducing EC radiosensitivity (Wang et al. 2020b). Therefore, CCAT2 regulates cancer cell proliferation, and therapy resistance, making it a promising therapeutic target for enhancing treatment efficacy across various malignancies.

6.2.8 MEG3

Maternally expressed gene 3 (MEG3) is a maternally expressed, imprinted, and polyadenylated lncRNA. Multiple isoform transcripts of MEG3 are transcribed from this gene via alternative splicing of 10 exons (Zhang et al. 2010; Zhang et al. 2003). This gene is highly expressed in the normal human pituitary, including normal gonadotroph cells; however, loss of MEG3 expression has been reported in cancers (Al-Rugeebah et al. 2019) and contributes to tumor suppression and cellular regulation. As a ceRNA, MEG3 sponges miRNAs such as miR-21, miR-181a, and miR-421, sequestering them away from their target genes to modulate gene expression (Zhang et al. 2022).

Downregulated MEG3 levels are associated with the larger size of tumor, advanced stage, invasion, and poor prognosis. MEG3 downregulation may be due to hypermethylation of the gene promoter region, causing the c-AMP-binding site to be blocked (Zhang et al. 2017c; Zhao et al. 2005). MEG3 is also a critical regulator of p53, enhancing its transcriptional activity and promoting posttranslational modifications, which increase p53 expression and exert negative control over the cell cycle (Zhu et al. 2015). MEG-3, as a tumor suppressor, activates p53, increasing the p53 protein level. MEG3 overexpression inhibits proliferation and increases apoptosis of cancer (Zhou et al. 2012). In various cancers, MEG3 is downregulated, including brain, bladder, liver, prostate, breast, colorectal, lung, gastrointestinal tract, and cervix cancers (Talebi et al. 2020; He et al. 2017; Peng et al. 2015a).

MEG3 regulates various cancer-associated signaling pathways, including TGF-β, PTEN/PI3K/Akt, PI3K/Akt/mTOR, JAK2/STAT3, and p53/PI3K/Akt pathways (Baazaoui et al. 2025). Through these pathways, MEG3 influences processes such as apoptosis, proliferation, and tumor progression. MEG3 inhibits oncogenic signaling by PI3K/Akt/mTOR pathway, while its regulation of the JAK2/STAT3 pathway suppresses inflammation-driven tumorigenesis. These multifaceted roles position MEG3 as a potential tumor suppressor and a potential therapeutic target in various cancers (Baazaoui et al. 2025).

6.2.9 BCAR4

Breast cancer antiestrogen resistance 4(BCAR4), a spliced oncogenic lncRNA, promotes progression and metastasis of breast cancer. BCAR4 is associated with antiestrogen tamoxifen-resistance phenotype (Meijer et al. 2006; Dorssers and Veldscholte 1997) and is only expressed in human placenta and oocytes (Godinho et al. 2011). In breast cancer, elevated expression of BCAR4 involves lymph node metastasis, cancer cell proliferation, and reduced survival rate (Tu et al. 2019). BCAR4 drives metastasis and invasion by regulating noncanonical Hedgehog pathway (Xing et al. 2015). BCAR4 and the regulating transcriptional factor Glioma-associated oncogene homolog 1, 2 (GLI 1/2) are physically correlated with each other. GLI2-dependent gene transcription promotes tumor cell progression and metastasis. BCAR4 activates GLI2 transcriptional regulation via direct binding to smad nuclear-interacting protein 1 (SNIP 1) and serine/threonine-protein phosphatase 1 regulatory subunit 10 (PPP1R10, also known as PNUTS) (Xing et al. 2014; Chen et al. 2016b; ten Haaf et al. 2009). In response to the release of chemokine CCL21, BCAR4 interacts with transcription factors SNIP1 and PNUTS, initiating a cascade that activates GLI-dependent transcription. BCAR4 binding relieves SNIP1's inhibition of p300-dependent histone acetylation, facilitating chromatin remodeling. Simultaneously, BCAR4-recruited PNUTS binds to acetylated H3K18 (H3K18ac), which activates PP1 phosphatase, leading to the derepression of RNA polymerase II (Xing et al. 2014, 2015). This coordinated mechanism drives the transcription of a noncanonical Hedgehog/GLI2 program, ultimately promoting enhanced cancer metastasis. Upregulation of BCAR4 expression promotes cancer progression not only in breast cancer (Peng et al. 2021), but also prostate (Cai et al. 2018), colorectal (Li et al. 2016a), gastrointestinal tract (Wang et al. 2017), cervical (Zou et al. 2018b), bladder (Zhang et al. 2020b), and lung cancers (Gong et al. 2017). Thus, BCAR4 is a potential biomarker and therapeutic target for tumor progression.

6.2.10 GAS5

Growth arrest-specific 5 (GAS5) is a tumor suppressing lncRNA that has characteristics of an upstream oligopyrimidine tract of the 5′-terminal (Smith and Steitz 1998). GAS5 acts as a decoy for the glucocorticoid response element (GRE) by suppressing its transcriptional regulation effect. GAS5 competes with DNA GREs through bonding to the DNA-binding domain (DBD) of the ligand-activated glucocorticoid receptor (GR), thus, inhibiting transcription factor GR from targeting GRE's gene. GAS5 also represses the transcription of some steroids hormone receptors, including, the progesterone, mineralocorticoid, and androgen receptors, by binding to their DBDs (Kino et al. 2010; Lucafo et al. 2015). GAS5 is also associated with cell growth arrest and apoptosis. GAS5, as a tumor suppressor, is downregulated in various cancers (Ji et al. 2019). GAS5 acts as a ceRNA and functions as miRNA sponges to regulate cell metabolism programs and signaling pathways (Ji et al. 2019). In tumorigenesis, GAS5 suppresses cell proliferation and metastasis by negatively regulating the expression of miR-21 (Zhang et al. 2013), miR-23 (Liu et al. 2016b), miR-23a (Mei et al. 2017), miR-103 (Guo et al. 2015), miR-135b (Xue et al. 2017), miR-203a (Wang and Kong 2018), miR-221 (Liu et al. 2018), miR-222 (Zhao et al. 2015a), and many others. GAS5 downregulation is significantly associated with larger size, distant metastasis, and advanced tumor stage (Guo et al. 2015). This evidence supports the role of GAS5 as a mediator of oncogenesis.

6.2.11 PCAT6

Prostate cancer-associated transcript 6 (PCAT6) is expressed exclusively in the testes and related to spermatogenesis (Chen et al. 2019; Du et al. 2013). LncRNA PCAT6 is upregulated in prostate cancer and significantly associated with tumor development, cell proliferation, and metastasis (Lv et al. 2019). Aberrant expression of PCAT6 also negatively correlates with prognosis and cancer patient survival rates in other cancers, including bone (Guan et al. 2020), lung (Cui et al. 2018), gastric (Xu et al. 2018), and colorectal (Wu et al. 2019a). The level of PCAT6 increases in cancerous tissues. It also promotes the progression, invasion, and cancer TNM (tumor, node, and metastasis) stages. In NSCLC, PCAT6 serves as a ceRNA by binding directly and targeting miR-330-5p. Research shows that the levels of PCAT6 and miR-330-5p have a reciprocal correlation with each other. The increase in PCAT6 level decreases the expression level of miR-330-5p, thus proving the negative correlation of both PCAT6 and miR-330-5p (Cui et al. 2018). Thus, PCAT6 could be a potential biomarker for diagnosis and prognosis of cancer.

6.2.12 CRNDE

Colorectal neoplasia differentially expressed (CRNDE) lncRNA is expressed highly in colorectal cancer (Dai et al. 2018). CRNDE has tissue-specific expression levels with almost no expression in normal cells and tissues, for instance, white blood cells (WBCs), liver, and colorectal mucosa. In contrast, the parotid gland, testes, breast, skin, and bronchial epithelium have elevated expression of CNDRE (Ellis et al. 2012, 2014). CRNDE may function as a scaffold for regulatory protein complexes, such as PRC2 and chromatin-modifying complexes, for the gene activation by epigenetic modifications (Khalil et al. 2009; Ellis et al. 2012). Ellis et al. demonstrated that the role of CRNDE in CRC was regulated by insulin and insulin-like growth factors (IGFs) inducing glycolytic metabolism or the Warburg effect of cancerous cells through various metabolic changes (Ellis et al. 2014). Moreover, the elevated expression of CRNDE in various human carcinomas, including cervical (Zhang and Fan 2019), lung (Zhang et al. 2018b), breast (Huan et al. 2017), and ovarian cancers (Szafron et al. 2015), is correlated with larger tumor size, advanced stage, metastasis, and poor survival rate (Graham et al. 2011; Zhou et al. 2019; Liu et al. 2016c).

6.2.13 H19

H19 imprinted maternally expressed transcript is a long noncoding oncogenic fetal-specific RNA, mapping to an imprinted region in close proximity to the insulin-like growth factor 2 (IGF2) gene (Brunkow and Tilghman 1991; Pope et al. 2017). H19 expresses itself exclusively when inherited from the maternal chromosome, whereas IGF2 is a paternally expressed gene. Both H19 and IGF2 genes' imprinting show an inverse association; in general, IGF2 is expressed in the absence of H19 while expression of IGF2 suppresses in the presence of H19 (Leighton et al. 1995; Nordin et al. 2014). Dysregulation of the H19 gene has been implicated in Beckwith–Wiedemann syndrome (BWS) and associated with genetic or epigenetic malformations and Wilms tumor development in the kidney (Weksberg et al. 2001). The methylation level of H19 is associated with male fecundity. In the semen of sterile males, H19 gene methylation is lower compared to fertile males. Therefore, the aberrant H19 methylation is associated with male infertility, possibly serving as a potential biomarker for deficiency in spermatogenesis (Li et al. 2016b; Rotondo et al. 2013). Moreover, aberrantly elevated H19 expression is involved in the pathogenesis of various human cancers (Ghafouri-Fard et al. 2020), with overexpression in cancer patients associated with poor prognosis and poor survival rate. H19 is significantly highly expressed in various human cancers, including gastric (Chen et al. 2016c), breast (Zhang et al. 2016), bladder (Wang et al. 2018b), ovarian (Wu et al. 2019b), and lung cancers (Liu et al. 2019b). In contrast to other cancers, H19 expression in hepatocellular carcinoma (Ardelt and Pachmayr 2017) and adrenocortical neoplasms (Gao et al. 2002) is lower and is coupled with elevated IGF2 expression. This evidence proves the potential of H19 as a biomarker for cancer recurrence.

6.2.14 SPRY4-Intronic Transcript 1 (SPRY4-IT4)

SPRY4-IT4 encodes from the second intronic region of Sprouty RTK signaling antagonist 4 (SPRY4) gene, localized at chromosome 5q31.3. The unique secondary structures of SPRY4-IT1 contain hairpins (Khaitan et al. 2011; Li et al. 2017c). This transcript plays a novel oncogenic role by inhibiting melanoma cell invasion, proliferation, and promotes apoptosis, thus regulating carcinogenesis (Mazar et al. 2014). Emerging evidence shows that dysregulated SPRY4-IT1 (high expression) is present in many cancers, including breast (Shi et al. 2015), liver (Yu et al. 2017b), pancreatic (Guo et al. 2018), bladder (Zhao et al. 2015b), and gastric cancers (Peng et al. 2015b). In lung cancer, SPRY4-IT1 expression decreases more than in normal tissues. Furthermore, a lower expression of SPRY4-IT1 is associated with a poor prognosis (Sun et al. 2014). In contrast, the serum levels of SPRY4-IT1 significantly increases in NSCLC (Guo et al. 2018; Hu et al. 2016). Hence, SPRY4-IT1 has the potential to be a predictive molecular marker for diagnosing human cancers.

SPRY4-IT1 facilitates cancer progression by regulating multiple molecular signaling pathways. It binds to the RNA-binding protein STAU1, facilitating STAU1 recruitment to the 3′-UTR of TCEB1 mRNA (Zhao et al. 2021). This interaction promotes TCEB1 mRNA degradation through STAU1-mediated mRNA decay (SMD), upregulating hypoxia-inducible factor 1α (HIF-1α), which drives cancer metastasis. In addition, NF-κB/p65 transactivates SPRY4-IT1, further suppressing TCEB1 expression and amplifying HIF-1α levels (Zhao et al. 2021).

Moreover, SPRY4-IT1 sponges miR-101-3p and upregulates EZH2 and induces VEGFA expression, which promotes cell proliferation and angiogenesis in glioma. In hypoxia in prostate cancer, SPRY4-IT1 expression increases, which in turn enhances S-phase entry and elevates the levels of cell cycle regulators CDK2 and cyclin D1 (Wang et al. 2023). Furthermore, SPRY4-IT1 activates the AKT signaling pathway through increased AKT phosphorylation, contributing to tumor growth and progression (Sang et al. 2023). These multifaceted roles position SPRY4-IT1 as a significant regulator of cancer metastasis, and angiogenesis highlighting its potential as a therapeutic target.

6.3 Conclusions

LncRNAs are a diverse group of RNAs that have no potential to encode a proper or functional protein. LncRNAs emerged as an important biological molecule that significantly affects various physiological and pathological processes, from regulating gene expression to translational, transcriptional, epigenetic modulations, and other cellular functions. LncRNAs significantly complicated the understanding of transcriptomics and genomics. LncRNAs can also express as a scaffold, decoys, ceRNA, signals, and guides for specific regulations and modifications. Many lncRNAs have been identified as oncogenic ncRNAs that are associated with tumor progression, distant metastasis, invasion, apoptosis, and poorer survival rate. Aberrant expression and dysregulation of lncRNAs mediate cancer phenotypes and particularly regulate metastasis via cell proliferation, cell stemness, and lymph node metastasis. Cancer-related lncRNAs, either freely circulating in body fluids or existing in tumors, function through significant expression levels and several cellular mechanisms. The unique aberrant expression of lncRNAs serves as a valuable predictive biomarker for the diagnosis and prognosis of cancers. Thus, lncRNAs have a significant role in tumorigenesis, providing a new paradigm for developing clinical approaches for cancer diagnosis and treatment. However, more studies are required to understand the role of lncRNAs in oncogenesis, facilitating the development of lncRNA therapeutic targets with the potential for novel cancer treatments.

References

Alaiyan B et al (2013) Differential expression of colon cancer associated transcript1 (CCAT1) along the colonic adenoma-carcinoma sequence. BMC Cancer 13(1):196

Al-Rugeebah A, Alanazi M, Parine NR (2019) MEG3: an oncogenic long non-coding RNA in different cancers. Pathol Oncol Res 25(3):859–874

Amjadi-Moheb F, Paniri A, Akhavan-Niaki H (2021) Insights into the links between MYC and 3D chromatin structure and epigenetics regulation: implications for cancer therapy. Cancer Res 81(8):1925–1936

Aprile M et al (2020) LncRNAs in cancer: from garbage to junk. Cancer 12(11):3220

Ardelt MA, Pachmayr J (2017) The long non-coding RNA H19-a new player in hepatocellular carcinoma. Cell Stress 1(1):4

Arun G, Aggarwal D, Spector DL (2020) MALAT1 long non-coding RNA: functional implications. Noncoding RNA 6(2):22

Baazaoui N et al (2025) Potential role of long noncoding RNA maternally expressed gene 3 (MEG3) in the process of neurodegeneration. Neuroscience 565:487–498

Balas MM, Johnson AM (2018) Exploring the mechanisms behind long noncoding RNAs and cancer. Non-coding RNA Res 3(3):108–117

Batista PJ, Chang HY (2013) Long noncoding RNAs: cellular address codes in development and disease. Cell 152(6):1298–1307

Bergmann JH, Spector DL (2014) Long non-coding RNAs: modulators of nuclear structure and function. Curr Opin Cell Biol 26:10–18

Bohrer C et al (2024) CCAT1 lncRNA is chromatin-retained and post-transcriptionally spliced. Histochem Cell Biol 162(1):91–107

Booy EP et al (2016) RNA helicase associated with AU-rich element (RHAU/DHX36) interacts with the 3'-tail of the long non-coding RNA BC200 (BCYRN1). J Biol Chem 291(10):5355–5372

Brunkow ME, Tilghman S (1991) Ectopic expression of the H19 gene in mice causes prenatal lethality. Genes Dev 5(6):1092–1101

Cai Z et al (2018) BCAR4 activates GLI2 signaling in prostate cancer to contribute to castration resistance. Aging 10(12):3702

Cantile M et al (2024) The role of HOTAIR in the modulation of resistance to anticancer therapy. Front Mol Biosci 11:11

Cao Y et al (2020) Long non-coding RNA in bladder cancer. Clin Chim Acta 503:113–121
Chang HHY et al (2017) Non-homologous DNA end joining and alternative pathways to double-strand break repair. Nat Rev Mol Cell Biol 18(8):495–506
Chen L et al (2016a) Long non-coding RNA CCAT1 acts as a competing endogenous RNA to regulate cell growth and differentiation in acute myeloid leukemia. Mol Cells 39(4):330–336
Chen F, Mo J, Zhang L (2016b) Long noncoding RNA BCAR4 promotes osteosarcoma progression through activating GLI2-dependent gene transcription. Tumour Biol 37(10):13403–13412
Chen J et al (2016c) H19 serves as a diagnostic biomarker and up-regulation of H19 expression contributes to poor prognosis in patients with gastric cancer. Neoplasma 63(2):223–230
Chen H et al (2018) Long non-coding RNA CCAT1 promotes the migration and invasion of prostate cancer PC-3 cells. Eur Rev Med Pharmacol Sci 22(10):2991–2996
Chen S et al (2019) Gene amplification derived a cancer-testis long noncoding RNA PCAT6 regulates cell proliferation and migration in hepatocellular carcinoma. Cancer Med 8(6):3017–3025
Chen S, Zhang C, Feng M (2020) Prognostic value of lncRNA HOTAIR in colorectal cancer: a meta-analysis. Open Med 15(1):76–83
Cui X et al (2017) Accuracy of the urine UCA1 for diagnosis of bladder cancer: a meta-analysis. Oncotarget 8(21):35222
Cui LH et al (2018) lncRNA PCAT6 promotes non-small cell lung cancer cell proliferation, migration and invasion through regulating miR-330-5p. Onco Targets Ther 11:7715
Dai M, Li S, Qin X (2018) Colorectal neoplasia differentially expressed: a long noncoding RNA with an imperative role in cancer. Onco Targets Ther 11:3755
Dai Q et al (2020) Long non-coding RNA TUG1 promotes cell progression in hepatocellular carcinoma via regulating miR-216b-5p/DLX2 axis. Cancer Cell Int 20(1):1–13
De Leeneer K, Claes K (2015) Non coding RNA molecules as potential biomarkers in breast cancer. In: Advances in cancer biomarkers. Springer, pp 263–275
Dehcheshmeh IS, Karimi MM, Jafarisani M (2018) Increased expression of CCAT2 LncRNA in non-melanoma skin cancer. Int J Health Stud 4(1):1–4
Deng L et al (2015) Long noncoding RNA CCAT1 promotes hepatocellular carcinoma progression by functioning as let-7 sponge. J Exp Clin Cancer Res 34(1):18
Djebali S et al (2012) Landscape of transcription in human cells. Nat Genet 489(7414):101–108
Dorssers LC, Veldscholte J (1997) Identification of a novel breast-cancer-anti-estrogen-resistance (BCAR2) locus by cell-fusion-mediated gene transfer in human breast-cancer cells. Int J Cancer 72(4):700–705
Du Z et al (2013) Integrative genomic analyses reveal clinically relevant long noncoding RNAs in human cancer. Nat Struct Mol Biol 20(7):908
Ellis BC, Molloy PL, Graham LD (2012) CRNDE: a long non-coding RNA involved in cancer, neurobiology, and development. Front Genet 3:270
Ellis BC, Graham LD, Molloy PL (2014) CRNDE, a long non-coding RNA responsive to insulin/IGF signaling, regulates genes involved in central metabolism. BBA-Mol Cell Res 1843(2):372–386
Fang Q, Chen X, Zhi X (2016) Long non-coding RNA (LncRNA) urothelial carcinoma associated 1 (UCA1) increases multi-drug resistance of gastric cancer via downregulating miR-27b. Med Sci Monit 22:3506
Fu P-F et al (2019) Role of cytoplasmic lncRNAs in regulating cancer signaling pathways. J Zhejiang Univ-Sci B 20(1):1–8
Gao ZH et al (2002) Association of H19 promoter methylation with the expression of H19 and IGF-II genes in adrenocortical tumors. J Clin Endocrinol Metab 87(3):1170–1176
Ghafouri-Fard S, Taheri M (2019) UCA1 long non-coding RNA: an update on its roles in malignant behavior of cancers. Biomed Pharmacother 120:109459
Ghafouri-Fard S, Esmaeili M, Taheri M (2020) H19 lncRNA: roles in tumorigenesis. Biomed Pharmacother 123:109774
Ghiam AF et al (2017) Long non-coding RNA urothelial carcinoma associated 1 (UCA1) mediates radiation response in prostate cancer. Oncotarget 8(3):4668

Godinho M et al (2011) Characterization of BCAR4, a novel oncogene causing endocrine resistance in human breast cancer cells. J Cell Physiol 226(7):1741–1749
Gong J et al (2017) Increased expression of long non-coding RNA BCAR4 is predictive of poor prognosis in patients with non-small cell lung cancer. Tohoku J Exp Med 241(1):29–34
Graham LD et al (2011) Colorectal neoplasia differentially expressed (CRNDE), a novel gene with elevated expression in colorectal adenomas and adenocarcinomas. Genes Cancer 2(8):829–840
Guan X, Xu Y, Zheng J (2020) Long non-coding RNA PCAT6 promotes the development of osteosarcoma by increasing MDM2 expression. Oncol Rep 44(6):2465–2474
Guan J et al (2022) LncRNA UCA1 accelerates osteosarcoma progression via miR-145 and Wnt/β-catenin pathway. Am J Transl Res 14(9):6029
Guo X, Hua Y (2017) CCAT1: an oncogenic long noncoding RNA in human cancers. J Cancer Res Clin Oncol 143(4):555–562
Guo C et al (2015) LncRNA-GAS5 induces PTEN expression through inhibiting miR-103 in endometrial cancer cells. J Biomed Sci 22(1):1–9
Guo W et al (2018) Long non-coding RNA SPRY4-IT1 promotes cell proliferation and invasion by regulation of Cdc20 in pancreatic cancer cells. PLoS One 13(2):e0193483
Guttman M, Rinn JL (2012) Modular regulatory principles of large non-coding RNAs. Nature 482(7385):339–346
Han L et al (2019) Prognostic and Clinicopathological significance of long non-coding RNA PANDAR expression in cancer patients: a meta-analysis. Front Oncol 9:1337
He X et al (2014) C-Myc-activated long noncoding RNA CCAT1 promotes colon cancer cell proliferation and invasion. Tumour Biol 35(12):12181–12188
He Y et al (2017) Potential applications of MEG3 in cancer diagnosis and prognosis. Oncotarget 8(42):73282
Hosoya N, Miyagawa K (2014) Targeting DNA damage response in cancer therapy. Cancer Sci 105(4):370–388
Hou J et al (2023) Functions and mechanisms of lncRNA MALAT1 in cancer chemotherapy resistance. Biomark Res 11(1):23
Hu T, Lu Y-R (2015) BCYRN1, a c-MYC-activated long non-coding RNA, regulates cell metastasis of non-small-cell lung cancer. Cancer Cell Int 15(1):1–8
Hu X et al (2016) The plasma lncRNA acting as fingerprint in non-small-cell lung cancer. Tumour Biol 37(3):3497–3504
Huan J et al (2017) Long noncoding RNA CRNDE activates Wnt/β-catenin signaling pathway through acting as a molecular sponge of microRNA-136 in human breast cancer. Am J Transl Res 9(4):1977
Huang J-L et al (2017) Long non-coding RNA CCAT2 promotes cell proliferation and invasion through regulating Wnt/β-catenin signaling pathway in clear cell renal cell carcinoma. Tumour Biol 39(7):1010428317711314
Huang G et al (2021) Upregulated UCA1 contributes to oxaliplatin resistance of hepatocellular carcinoma through inhibition of miR-138-5p and activation of AKT/mTOR signaling pathway. Pharmacol Res Perspect 9(1):e00720
Hung T et al (2011) Extensive and coordinated transcription of noncoding RNAs within cell-cycle promoters. Nat Genet 43(7):621–629
Hutchinson JN et al (2007) A screen for nuclear transcripts identifies two linked noncoding RNAs associated with SC35 splicing domains. BMC Genomics 8:39
Iyer MK et al (2015) The landscape of long noncoding RNAs in the human transcriptome. Nat Genet 47(3):199–208
Jang S, Shin H, Lee Y (2020) Functional analysis of RNA motifs essential for BC200 RNA-mediated translational regulation. BMB Rep 53(2):94–99
Ji J et al (2019) The role of long non-coding RNA GAS5 in cancers. Cancer Manag Res 11:2729
Jia T et al (2021) Knockdown of LncRNA PANDAR by CRISPR-dCas9 decreases proliferation and increases apoptosis in Oral squamous cell carcinoma. Front Mol Biosci 8:653787

Jiang M-C et al (2019) Emerging roles of lncRNA in cancer and therapeutic opportunities. Am J Cancer Res 9(7):1354

Jiao C et al (2016) lncRNA-UCA1 enhances cell proliferation through functioning as a ceRNA of Sox4 in esophageal cancer. Oncol Rep 36(5):2960–2966

Jin K-T et al (2020a) Roles of lncRNAs in cancer: focusing on angiogenesis. Life Sci 252:117647

Jin G et al (2020b) LncRNA TUG1 promotes tumor growth and metastasis of esophageal squamous cell carcinoma by regulating XBP1 via competitively binding to miR-498. Neoplasma 67(4):751–761

Kansara S et al (2020) Mechanistic involvement of long non-coding RNAs in oncotherapeutics resistance in triple-negative breast cancer. Cells 9(6):1511

Khaitan D et al (2011) The melanoma-upregulated long noncoding RNA SPRY4-IT1 modulates apoptosis and invasion. Cancer Res 71(11):3852–3862

Khalil AM et al (2009) Many human large intergenic noncoding RNAs associate with chromatin-modifying complexes and affect gene expression. Proc Natl Acad Sci USA 106(28):11667–11672

Kim S-S et al (2018) Targeted nanocomplex carrying siRNA against MALAT1 sensitizes glioblastoma to temozolomide. Nucleic Acids Res 46(3):1424–1440

Kino T et al (2010) Noncoding RNA gas5 is a growth arrest–and starvation-associated repressor of the glucocorticoid receptor. Sci Signal 3(107):ra8

Lai XJ, Cheng HF (2018) LncRNA colon cancer-associated transcript 1 (CCAT1) promotes proliferation and metastasis of ovarian cancer via miR-1290. Eur Rev Med Pharmacol Sci 22(2):322–328

Leighton PA et al (1995) An enhancer deletion affects both H19 and Igf2 expression. Genes Dev 9(17):2079–2089

Li X, Heyer W-D (2008) Homologous recombination in DNA repair and DNA damage tolerance. Cell Res 18(1):99–113

Li Q et al (2016a) Differentially expressed long non-coding RNAs and the prognostic potential in colorectal cancer. Neoplasma 63(6):977–983

Li XP et al (2016b) H19 gene methylation status is associated with male infertility. Exp Ther Med 12(1):451–456

Li J et al (2017a) PANDAR: a pivotal cancer-related long non-coding RNA in human cancers. Mol BioSyst 13(11):2195–2201

Li HJ et al (2017b) LncRNA UCA1 promotes mitochondrial function of bladder cancer via the MiR-195/ARL2 signaling pathway. Cell Physiol Biochem 43(6):2548–2561

Li J et al (2017c) SPRY4-IT1: a novel oncogenic long non-coding RNA in human cancers. Tumour Biol 39(6):1010428317711406

Li T et al (2018a) LncRNA TUG1 promotes cells proliferation and inhibits cells apoptosis through regulating AURKA in epithelial ovarian cancer cells. Medicine (Baltimore) 97(36):e12131

Li Z-X et al (2018b) MALAT1: a potential biomarker in cancer. Cancer Manag Res 10:6757

Li Z et al (2019) The role of long noncoding RNAs in gene expression regulation, in gene expression profiling in cancer. IntechOpen, Rijeka

Li W et al (2021) LncRNA TUG1 promotes hepatocellular carcinoma migration and invasion via targeting miR-137/AKT2 Axis. Cancer Biother Radiopharm 36(10):850–862

Lin H et al (2018) Overexpression of the long noncoding RNA CCAT1 promotes metastasis via epithelial-to-mesenchymal transition in lung adenocarcinoma. Oncol Lett 16(2):1809–1814

Ling H et al (2013) CCAT2, a novel noncoding RNA mapping to 8q24, underlies metastatic progression and chromosomal instability in colon cancer. Genome Res 23(9):1446–1461

Liu M-Y et al (2016a) Elevated HOTAIR expression associated with cisplatin resistance in non-small cell lung cancer patients. J Thorac Dis 8(11):2016

Liu X et al (2016b) Long non-coding RNA GAS5 acts as a molecular sponge to regulate miR-23a in gastric cancer. Int J Exp Pathol 9:11412–11419

Liu T et al (2016c) Increased expression of the long noncoding RNA CRNDE-h indicates a poor prognosis in colorectal cancer, and is positively correlated with IRX5 mRNA expression. Onco Targets Ther 9:1437
Liu L et al (2018) Prognostic and predictive value of long non-coding RNA GAS5 and mircoRNA-221 in colorectal cancer and their effects on colorectal cancer cell proliferation, migration and invasion. Cancer Biomark 22(2):283–299
Liu Y et al (2019a) Long noncoding RNA CCAT2 promotes hepatocellular carcinoma proliferation and metastasis through up-regulation of NDRG1. Exp Cell Res 379(1):19–29
Liu L, Liu L, Lu S (2019b) lncRNA H19 promotes viability and epithelial-mesenchymal transition of lung adenocarcinoma cells by targeting miR-29b-3p and modifying STAT3. Int J Oncol 54(3):929–941
Liu L, Zhang Y, Lu J (2020) The roles of long noncoding RNAs in breast cancer metastasis. Cell Death Dis 11(9):1–14
Liu Z et al (2021) Regulatory role of long non-coding RNA UCA1 in signaling pathways and its clinical applications. Oncol Lett 21(5):1–14
Liu Z et al (2022) LncRNA BC200/miR-150-5p/MYB positive feedback loop promotes the malignant proliferation of myelodysplastic syndrome. Cell Death Dis 13(2):126
Liz J, Esteller M (2016) lncRNAs and microRNAs with a role in cancer development. Biochim Biophys Acta 1859(1):169–176
Loda A, Heard E (2019) Xist RNA in action: past, present, and future. PLoS Genet 15(9):e1008333
Lu H, Liu X (2018) UCA1 promotes papillary thyroid carcinoma development by stimulating cell proliferation via Wnt pathway. Eur Rev Med Pharmacol Sci 22(17):5576–5582
Lu M-Y et al (2017a) Targeting LncRNA HOTAIR suppresses cancer stemness and metastasis in oral carcinomas stem cells through modulation of EMT. Oncotarget 8(58):98542
Lu M et al (2017b) The high expression of long non-coding RNA PANDAR indicates a poor prognosis for colorectal cancer and promotes metastasis by EMT pathway. J Cancer Res Clin Oncol 143(1):71–81
Lu H et al (2020) MALAT1 promotes epithelial-mesenchymal transition of pancreatic cancer cells through miR-141-5p/TGFBR1/TGFBR2 axis. Eur Cytokine Netw 35(3):28–37
Lucafo M et al (2015) Long noncoding RNA GAS5: a novel marker involved in glucocorticoid response. Curr Mol Med 15(1):94–99
Lv XJ et al (2019) Long noncoding RNA PCAT6 regulates cell growth and metastasis via Wnt/β-catenin pathway and is a prognosis marker in cervical cancer. Eur Rev Med Pharmacol Sci 23(5):1947–1956
Ma P et al (2016) Increased expression of LncRNA PANDAR predicts a poor prognosis in gastric cancer. Biomed Pharmacother 78:172–176
Mathy NW, Chen X-M (2017) Long non-coding RNAs (lncRNAs) and their transcriptional control of inflammatory responses. J Biol Chem 292(30):12375–12382
Mazar J et al (2014) The functional characterization of long noncoding RNA SPRY4-IT1 in human melanoma cells. Oncotarget 5(19):8959
Mei Y et al (2017) Long noncoding RNA GAS5 suppresses tumorigenesis by inhibiting miR-23a expression in non-small cell lung cancer. Oncol Res 25(6):1027–1037
Meijer D et al (2006) Functional screen for genes responsible for tamoxifen resistance in human breast cancer cells. Mol Cancer Res 4(6):379–386
Moran VA, Perera RJ, Khalil AM (2012) Emerging functional and mechanistic paradigms of mammalian long non-coding RNAs. Nucleic Acids Res 40(14):6391–6400
Mozdarani H, Ezzatizadeh V, Rahbar Parvaneh R (2020) The emerging role of the long non-coding RNA HOTAIR in breast cancer development and treatment. J Transl Med 18:1–15
Nordin M et al (2014) Epigenetic regulation of the Igf2/H19 gene cluster. Cell Prolif 47(3):189–199
Olivero CE, Dimitrova N (2020) Identification and characterization of functional long noncoding RNAs in cancer. FASEB J 34(12):15630–15646

Ozawa T et al (2017) CCAT1 and CCAT2 long noncoding RNAs, located within the 8q. 24.21 'gene desert', serve as important prognostic biomarkers in colorectal cancer. Ann Oncol 28(8):1882–1888

Pádua Alves C et al (2013) Brief report: the lincRNA Hotair is required for epithelial-to-mesenchymal transition and stemness maintenance of cancer cell lines. Stem Cells 31(12):2827–2832

Pan J et al (2020) Long noncoding RNA MALAT1 as a candidate serological biomarker for the diagnosis of non-small cell lung cancer: a meta-analysis. Thorac Cancer 11(2):329–335

Peng W et al (2015a) Long non-coding RNA MEG3 functions as a competing endogenous RNA to regulate gastric cancer progression. J Exp Clin Cancer Res 34(1):1–10

Peng W et al (2015b) Long noncoding RNA SPRY4-IT1 predicts poor patient prognosis and promotes tumorigenesis in gastric cancer. Tumour Biol 36(9):6751–6758

Peng J et al (2018) Long non-coding RNA BCYRN1 promotes the proliferation and metastasis of cervical cancer via targeting microRNA-138 in vitro and in vivo. Oncol Lett 15(4):5809–5818

Peng Y et al (2020) MiR-146b-5p regulates the expression of long noncoding RNA MALAT1 and its effect on the invasion and proliferation of papillary thyroid cancer, vol 36. Cancer Biother Radiopharm

Peng R et al (2021) Variant in BCAR4 gene correlated with the breast cancer susceptibility and mRNA expression of lncRNA BCAR4 in Chinese Han population. Breast Cancer 28(2):424–433

Peng L et al (2024) PRMT1 promotes Warburg effect by regulating the PKM2/PKM1 ratio in non-small cell lung cancer. Cell Death Dis 15(7):504

Pickard MR, Williams GT (2015) Molecular and cellular mechanisms of action of tumour suppressor GAS5 LncRNA. Genes 6(3):484–499

Pontier DB, Gribnau J (2011) Xist regulation and function explored. Hum Genet 130(2):223–236

Ponting CP, Oliver PL, Reik W (2009) Evolution and functions of long noncoding RNAs. Cell Death Dis 136(4):629–641

Pope C et al (2017) Targeting H19, an imprinted long non-coding RNA, in hepatic functions and liver diseases. Diseases 5(1):11

Puvvula PK et al (2014) Long noncoding RNA PANDA and scaffold-attachment-factor SAFA control senescence entry and exit. Nat Commun 5:5323

Qiu YR et al (2017) Expression of IncRNA UCA1 in ovarian cancer and its clinical significance. Eur J Gynaecol Oncol 38(2):191–195

Ren H et al (2018) Upregulation of LncRNA BCYRN1 promotes tumor progression and enhances EpCAM expression in gastric carcinoma. Oncotarget 9(4):4851

Ren MM et al (2020) Roles of HOTAIR in lung cancer susceptibility and prognosis. Mol Genet Genomic Med 8:e1299

Ribeiro DM et al (2018) Protein complex scaffolding predicted as a prevalent function of long non-coding RNAs. Nucleic Acids Res 46(2):917–928

Rotondo JC et al (2013) Methylation loss at H19 imprinted gene correlates with methylenetetrahydrofolate reductase gene promoter hypermethylation in semen samples from infertile males. Epigenetics 8(9):990–997

Sang Y et al (2016) LncRNA PANDAR regulates the G1/S transition of breast cancer cells by suppressing p16 INK4A expression. Sci Rep 6(1):1–10

Sang W et al (2023) LncRNA SPRY4-IT1 is upregulated and promotes the proliferation of prostate cancer cells under hypoxia in vitro. Oncol Lett 25(4):138

Sarfi M, Abbastabar M, Khalili EJ (2021) Increased expression of urinary Exosomal LnCRNA TUG-1 in early bladder cancer. Gene Rep 22:101010

Shah MY et al (2018) Cancer-associated rs6983267 SNP and its accompanying long noncoding RNA CCAT2 induce myeloid malignancies via unique SNP-specific RNA mutations. Genome Res 28(4):432–447

Shi Y et al (2015) The long noncoding RNA SPRY4-IT1 increases the proliferation of human breast cancer cells by upregulating ZNF703 expression. Mol Cancer 14(1):1–13

Shin H et al (2018) BC200 RNA: an emerging therapeutic target and diagnostic marker for human cancer. Mol Cells 41(12):993

Siddiqui H et al (2019) Roles of long non-coding RNAs in colorectal cancer tumorigenesis: a review. Mol Clin Oncol 11(2):167–172

Smith CM, Steitz JA (1998) Classification of gas5 as a multi-small-nucleolar-RNA (snoRNA) host gene and a member of the 5′-terminal oligopyrimidine gene family reveals common features of snoRNA host genes. Mol Cell Biol 18(12):6897–6909

Statello L et al (2021) Gene regulation by long non-coding RNAs and its biological functions. Nat Rev Mol Cell Biol 22(2):1–23

Sun M et al (2014) EZH2-mediated epigenetic suppression of long noncoding RNA SPRY4-IT1 promote s NSCLC cell proliferation and metastasis by affecting the epithelial–mesenchymal transition. Cell Death Dis 5(6):e1298–e1298

Szafron LM et al (2015) The putative oncogene, CRNDE, is a negative prognostic factor in ovarian cancer patients. Oncotarget 6(41):43897

Talebi A et al (2020) The relevance of long noncoding RNAs in colorectal cancer biology and clinical settings. J Cancer Res Ther 16(Suppl 1):S22–S33

Tan N et al (2020) Effect of lncRNA-BC200 on proliferation and migration of liver cancer cells in vitro and in vivo. Oncol Rep 43(2):461–470

ten Haaf A et al (2009) Expression of the glioma-associated oncogene homolog (GLI) 1in human breast cancer is associated with unfavourable overall survival. BMC Cancer 9(1):298

Tripathi V et al (2010) The nuclear-retained noncoding RNA MALAT1 regulates alternative splicing by modulating SR splicing factor phosphorylation. Mol Cell 39(6):925–938

Tu C et al (2019) The value of LncRNA BCAR4 as a prognostic biomarker on clinical outcomes in human cancers. J Cancer 10(24):5992

Wang KC, Chang HY (2011) Molecular mechanisms of long noncoding RNAs. Mol Cell 43(6):904–914

Wang Y, Kong D (2018) LncRNA GAS5 represses osteosarcoma cells growth and metastasis via sponging MiR-203a. Cell Physiol Biochem 45(2):844–855

Wang X-S et al (2006) Rapid identification of UCA1 as a very sensitive and specific unique marker for human bladder carcinoma. Clin Cancer Res 12(16):4851–4858

Wang F et al (2008) UCA1, a non-protein-coding RNA up-regulated in bladder carcinoma and embryo, influencing cell growth and promoting invasion. FEBS Lett 582(13):1919–1927

Wang L et al (2016) Long non-coding RNA TUG1 promotes colorectal cancer metastasis via EMT pathway. Oncotarget 7(32):51713–51719

Wang L et al (2017) BCAR4 increase cisplatin resistance and predicted poor survival in gastric cancer patients. Eur Rev Med Pharmacol Sci 21(18):4064–4070

Wang H et al (2018a) The cisplatin-induced lncRNA PANDAR dictates the chemoresistance of ovarian cancer via regulating SFRS2-mediated p53 phosphorylation. Cell Death Dis 9(11):1–15

Wang J et al (2018b) Determination of serum exosomal H19 as a noninvasive biomarker for bladder cancer diagnosis and prognosis. Med Sci Monit 24:9307

Wang J et al (2020a) Overexpression of ABCB1 transporter confers resistance to mTOR inhibitor WYE-354 in cancer cells. Int J Mol Sci 21(4):1387

Wang M et al (2020b) lncRNA CCAT2 promotes radiotherapy resistance for human esophageal carcinoma cells via the miR-145/p70S6K1 and p53 pathway. Int J Oncol 56(1):327–336

Wang Z et al (2020c) Long non coding RNA CCAT2 enhances proliferation, invasion and drug-resistances of non-small lung cancer via activating the miR-204-3p-suppresed IGFBP2/AKT/Bcl2 pathway. https://doi.org/10.21203/rs.3.rs-25814/v2

Wang M et al (2021a) Advances in understanding the LncRNA-mediated regulation of the hippo pathway in cancer. Onco Targets Ther Volume 14:2397–2415

Wang D, Li Z, Yin H (2021b) Long non-coding RNA CCAT2 a in colorectal cancer. Front Oncol 11:11

Wang J et al (2023) LncRNA SPRY4-IT1 facilitates cell proliferation and angiogenesis of glioma via the miR-101 3p/EZH2/VEGFA signaling axis. Cancer Med 12(6):7309–7326

Watson JB, Sutcliffe JG (1987) Primate brain-specific cytoplasmic transcript of the Alu repeat family. Mol Cell Biol 7(9):3324–3327

Weksberg R et al (2001) Tumor development in the Beckwith–Wiedemann syndrome is associated with a variety of constitutional molecular 11p15 alterations including imprinting defects of KCNQ1OT1. Hum Mol Genet 10(26):2989–3000

Wu C, Luo J (2016) Long non-coding RNA (lncRNA) urothelial carcinoma-associated 1 (UCA1) enhances tamoxifen resistance in breast cancer cells via inhibiting mTOR signaling pathway. Med Sci Monit 22:3860

Wu H, Zhou C (2018) Long non-coding RNA UCA1 promotes lung cancer cell proliferation and migration via microRNA-193a/HMGB1 axis. Biochem Biophys Res Commun 496(2):738–745

Wu H et al (2019a) Long non-coding RNA PCAT6 targets miR-204 to modulate the chemoresistance of colorectal cancer cells to 5-fluorouracil-based treatment through HMGA2 signaling. Cancer Med 8(5):2484–2495

Wu Y et al (2019b) Long non-coding RNA H19 mediates ovarian cancer cell cisplatin-resistance and migration during EMT. Int J Clin Exp Pathol 12(7):2506

Xiao L et al (2020) lncRNA MALAT1 promotes cell proliferation and invasion by regulating the miR-101/EZH2 axis in oral squamous cell carcinoma. Oncol Lett 20(5):1–1

Xin S, Ye X (2020) Knockdown of long non-coding RNA CCAT2 suppresses the progression of thyroid cancer by inhibiting the Wnt/β-catenin pathway. Int J Mol Med 46(6):2047–2056

Xin Y et al (2016) CCAT1: a pivotal oncogenic long non-coding RNA in human cancers. Cell Prolif 49(3):255–260

Xing Z et al (2014) lncRNA directs cooperative epigenetic regulation downstream of chemokine signals. Cell 159(5):1110–1125

Xing Z et al (2015) LncRNA BCAR4 wires up signaling transduction in breast cancer. RNA Biol 12(7):681–689

Xu Y et al (2018) PCAT6 participates in the development of gastric cancer through endogenously competition with microRNA-30. Eur Rev Med Pharmacol Sci 22(16):5206–5213

Xue Y et al (2017) Long noncoding RNA GAS5 inhibits tumorigenesis and enhances radiosensitivity by suppressing miR-135b expression in non-small cell lung cancer. Oncol Res 25(8):1305–1316

Yang L et al (2011) ncRNA- and Pc2 methylation-dependent gene relocation between nuclear structures mediates gene activation programs. Cell 147(4):773–788

Yang F et al (2015) The lncRNA firre anchors the inactive X chromosome to the nucleolus by binding CTCF and maintains H3K27me3 methylation. Genome Biol 16(1):52

Yang Z et al (2016) Downregulation of long non-coding RNA MALAT1 induces tumor progression of human breast cancer through regulating CCND1 expression. Open Life Sci 11(1):232–236

Yang D et al (2021) Silencing of lncRNA UCA1 inhibited the pathological progression in PCOS mice through the regulation of PI3K/AKT signaling pathway. J Ovarian Res 14:1–9

YiRen H et al (2017) Long noncoding RNA MALAT1 regulates autophagy associated chemoresistance via miR-23b-3p sequestration in gastric cancer. Mol Cancer 16:1–12

Young TL, Matsuda T, Cepko CL (2005) The noncoding RNA taurine upregulated gene 1 is required for differentiation of the murine retina. Curr Biol 15(6):501–512

Yu Y et al (2017a) A novel mechanism of lncRNA and miRNA interaction: CCAT2 regulates miR-145 expression by suppressing its maturation process in colon cancer cells. Mol Cancer 16(1):1–11

Yu G et al (2017b) Long non-coding RNA SPRY4-IT1 promotes development of hepatic cellular carcinoma by interacting with ERRα and predicts poor prognosis. Sci Rep 7(1):1–9

Yu X et al (2019) Long non-coding RNA taurine upregulated gene 1 promotes osteosarcoma cell metastasis by mediating HIF-1α via miR-143-5p. Cell Death Dis 10(4):280

Yuan C, Ning Y, Pan Y (2020) Emerging roles of HOTAIR in human cancer. J Cell Biochem 121(5–6):3235–3247

Zhan Y et al (2016) Up-regulation of long non-coding RNA PANDAR is associated with poor prognosis and promotes tumorigenesis in bladder cancer. J Exp Clin Cancer Res 35(1):83

Zhang J-J, Fan L-P (2019) Long non-coding RNA CRNDE enhances cervical cancer progression by suppressing PUMA expression. Biomed Pharmacother 117:108726

Zhang X et al (2003) A pituitary-derived MEG3 isoform functions as a growth suppressor in tumor cells. J Clin Endocrinol Metabol 88(11):5119–5126

Zhang X et al (2010) Maternally expressed gene 3 (MEG3) noncoding ribonucleic acid: isoform structure, expression, and functions. Endocrinology 151(3):939–947

Zhang Z et al (2013) Negative regulation of lncRNA GAS5 by miR-21. Cell Death Differ 20(11):1558–1568

Zhang J et al (2014a) Long non-coding RNA HOTAIR in carcinogenesis and metastasis. Acta Biochim Biophys Sin 46(1):1–5

Zhang EB et al (2014b) P53-regulated long non-coding RNA TUG1 affects cell proliferation in human non-small cell lung cancer, partly through epigenetically regulating HOXB7 expression. Cell Death Dis 5(5):e1243

Zhang XF et al (2015) Overexpression of long non-coding RNA CCAT1 is a novel biomarker of poor prognosis in patients with breast cancer. Int J Clin Exp Pathol 8(8):9440–9445

Zhang K et al (2016) Circulating lncRNA H19 in plasma as a novel biomarker for breast cancer. Cancer Biomark 17(2):187–194

Zhang X, Hamblin MH, Yin KJ (2017a) The long noncoding RNA Malat1: its physiological and pathophysiological functions. RNA Biol 14(12):1705–1714

Zhang E et al (2017b) H3K27 acetylation activated-long non-coding RNA CCAT1 affects cell proliferation and migration by regulating SPRY4 and HOXB13 expression in esophageal squamous cell carcinoma. Nucleic Acids Res 45(6):3086–3101

Zhang J et al (2017c) Downregulation of long noncoding RNA MEG3 is associated with poor prognosis and promoter hypermethylation in cervical cancer. J Exp Clin Cancer Res 36(1):1–9

Zhang Z et al (2018a) Long non-coding RNA CCAT1 as a diagnostic and prognostic molecular marker in various cancers: a meta-analysis. Oncotarget 9(34):23695

Zhang M et al (2018b) Long noncoding RNA CRNDE/PRC2 participated in the radiotherapy resistance of human lung adenocarcinoma through targeting p21 expression. Oncol Res 26(8):1245–1255

Zhang J et al (2020a) HOTAIR contributes to the carcinogenesis of gastric cancer via modulating cellular and exosomal miRNAs level. Cell Death Dis 11(9):1–15

Zhang R et al (2020b) lncRNA BCAR4 sponges miR-370-3p to promote bladder cancer progression via Wnt signaling. Int J Mol Med 45(2):578–588

Zhang L et al (2022) The biological roles and molecular mechanisms of long non-coding RNA MEG3 in the hallmarks of cancer. Cancers (Basel) 14(24):6032

Zhang C et al (2023) LncRNA CCAT1 facilitates the progression of gastric cancer via PTBP1-mediated glycolysis enhancement. J Exp Clin Cancer Res 42(1):246

Zhao J et al (2005) Hypermethylation of the promoter region is associated with the loss of MEG3 gene expression in human pituitary tumors. J Clin Endocrinol Metabol 90(4):2179–2186

Zhao X et al (2015a) Gas5 exerts tumor-suppressive functions in human glioma cells by targeting miR-222. Mol Ther 23(12):1899–1911

Zhao X-L et al (2015b) Increased expression of SPRY4-IT1 predicts poor prognosis and promotes tumor growth and metastasis in bladder cancer. Int J Clin Exp Pathol 8(2):1954

Zhao R-H et al (2016) BC200 LncRNA a potential predictive marker of poor prognosis in esophageal squamous cell carcinoma patients. Onco Targets Ther 9:2221

Zhao L et al (2021) NF-κB-activated SPRY4-IT1 promotes cancer cell metastasis by downregulating TCEB1 mRNA via Staufen1-mediated mRNA decay. Oncogene 40(30):4919–4929

Zhou Y, Zhang X, Klibanski A (2012) MEG3 noncoding RNA: a tumor suppressor. J Cancer 48(3):R45–R53

Zhou Y et al (2019) Prognostic value of long noncoding RNA CRNDE as a novel biomarker in solid cancers: an updated systematic review and meta-analysis. J Cancer 10(11):2386

Zhou Y-H et al (2020) Long non-coding RNA HOTAIR in cervical cancer: molecular marker, mechanistic insight, and therapeutic target. In: Advances in clinical chemistry. Elsevier, Amsterdam, pp 117–140

Zhu J et al (2015) Long noncoding RNA MEG3 interacts with p53 protein and regulates partial p53 target genes in Hepatoma cells. PLoS One 10(10):e0139790

Zhu J et al (2017) Long non-coding RNA TUG1 promotes cervical cancer progression by regulating the miR-138-5p-SIRT1 axis. Oncotarget 8(39):65253–65264

Zhu C et al (2022) Functions and underlying mechanisms of lncRNA HOTAIR in cancer chemotherapy resistance. Cell Death Discov 8(1):383

Zou Y et al (2018a) Long non-coding PANDAR as a novel biomarker in human cancer: a systematic review. Cell Prolif 51(1):e12422

Zou R et al (2018b) Up-regulated BCAR4 contributes to proliferation and migration of cervical cancer cells. Surg Oncol 27(2):306–313

Chapter 7
MicroRNAs as Biomarkers for the Diagnosis and Prognosis of Different Diseases and Cancer

Rashid Mir, Ajaz Ahmad Waza, and Mohammad Muzaffar Mir

Contents

Abstract Small, endogenous noncoding RNAs known as miRNAs are widely distributed and use posttranscriptional control to suppress the expression of target genes. miRNAs have crucial roles in the development of disease and carcinogenesis, and new research suggests that miRNAs might be reliable diseases biomarkers. The identification of miRNAs in bodily fluids, in particular, has created a compelling possibility for the creation of noninvasive biomarkers for disease diagnosis, prognosis, and response prediction to treatment. Deregulation of the expression of microRNAs, which are essential for the regulation of several biological processes, has been connected to the emergence of metabolic, cardiovascular, neurodegenera-

R. Mir (✉)
Prince Fahd Bin Sultan Research Chair for Biomedical Research, University of Tabuk, Tabuk, Kingdom of Saudi Arabia
e-mail: rashid@ut.edu.sa

A. A. Waza
Multidisciplinary Research Unit (MRU), Government Medical College (GMC), Srinagar, Jammu & Kashmir, India

M. M. Mir
Department of Clinical Biochemistry, College of Medicine, University of Bisha, Bisha, Asir Region, Kingdom of Saudi Arabia

M. F. Ullah, A. Ahmad (eds.), *Non-coding RNAs*, Experientia Supplementum 115,
https://doi.org/10.1007/978-3-032-06948-1_7

tive, and cancerous disorders. An exhaustive, up-to-date, and comprehensive investigation of the role of miRNAs in disease is presented in this chapter. Its goal is to stimulate further study in this field since small miRNAs may be utilized for effective disease detection, prognosis, and therapy.

Keywords MicroRNAs · Diagnostics markers · Cancer · Therapy · Cardiovascular disease

Abbreviations

ARDS	Acute respiratory distress syndrome
HCC	Hepatocellular carcinoma
HDAC5	Histone deacetylase 5
MS	Multiple sclerosis
ncRNAs	Noncoding RNAs
PND	Perioperative neurocognitive disorders

7.1 Introduction

A family of noncoding RNAs known as miRNAs is involved in the posttranslational control of gene expression and has been linked to a number of diseases, including CAD, T2D, cancer, tumor progression, and metastasis. Furthermore, miRNAs have potential as metastatic, prognostic, and diagnostic biomarkers. MiRNAs primarily attach to particular locations within the target messenger RNA (mRNA) molecules' 3′ untranslated region (UTR), which prevents translation or causes the mRNA to degrade. As a result, the target gene's ability to produce proteins is suppressed. Numerous eukaryotic, prokaryotic, and viral species have shown to have thousands of miRNAs that carry out diverse tasks at different periods. It is believed that miRNAs govern around 30% of the mammalian protein-coding genome, and these molecules are phylogenetically conserved in a number of species, including humans and dogs.

Numerous biological processes, including the cell cycle, proliferation, differentiation, and death, are significantly influenced by miRNAs. All body fluids and tissues, including blood, serum, plasma, urine, saliva, semen, pleura, and ascitic effusion, contain miRNAs. Because of their important involvement in the formation of cancer, miRNAs are one of the most researched subclasses of noncoding RNAs (ncRNAs) in pathological circumstances, namely, in human malignancies. The identification of miRNAs in bodily fluids has created a compelling possibility for the development of noninvasive biomarkers for disease diagnosis, prognosis, and response prediction to treatment.

7.2 Role of miR-142-3p

miR-142-3p is widely expressed across human tissues and organs and has been identified as a key player in cancer biology (Liang et al. 2020b). Specifically, this miRNA functions as a tumor suppressor in colon cancer, where its expression is significantly downregulated. In colon cancer, miR-142-3p has been shown to target three critical risk genes associated with poor clinical outcomes: CD133, ABCG2, and Lgr5 (Shen et al. 2013). These genes are well-known markers of cancer stem cells (CSCs), which play a pivotal role in tumor initiation, chemoresistance, and metastasis. Studies have demonstrated a significant inverse correlation between miR-142-3p levels and the expression of these genes in colon cancer tissues. This suggests that the downregulation of miR-142-3p contributes to the upregulation of CSC markers, thereby promoting cancer aggressiveness and resistance to therapy (Shen et al. 2013).

Beyond its regulation of CSC markers, miR-142-3p has been implicated in the modulation of other critical oncogenic pathways in colorectal cancer (Liu et al. 2020). For instance, it targets the transcription factor 7 (TCF-7), a component of the Wnt/β-catenin signaling pathway, which is frequently dysregulated in colorectal cancer. By inhibiting TCF-7, miR-142-3p can disrupt Wnt-driven transcriptional programs that promote cell proliferation and invasion. Additionally, miR-142-3p negatively regulates fatty acid synthase (FASN), which is often overexpressed in cancer cells to support rapid growth and metastasis. The miRNA also targets the MYC oncogene, a master regulator of cell cycle progression and metabolism, further emphasizing its tumor-suppressive potential.

miR-142-3p has garnered attention as a promising invasive biomarker due to its dual role in regulating tumor progression and metastasis in several malignancies, including hepatocellular carcinoma (HCC), cervical cancer, and other cancer types (Xie et al. 2021). In HCC, miR-142-3p functions as a tumor suppressor by inhibiting the migration and invasion of cancer cells through direct regulation of genes such as HMGB1. The downregulation of miR-142-3p, often influenced by long noncoding RNA (lncRNA) TUG1, contributes to EMT. This regulatory mechanism suggests that miR-142-3p serves as an important player in controlling tumor aggressiveness. In addition, miR-142-3p modulates key pathways related to cell adhesion, migration, and invasion by targeting critical genes involved in these processes. Monitoring the expression of miR-142-3p in tumor tissues or biofluids could provide valuable insights into the metastatic potential of cancer and the likelihood of disease progression. As a result, miR-142-3p presents itself as an effective invasive biomarker for assessing the clinical status of patients, guiding prognosis, and informing therapeutic strategies aimed at halting metastasis.

Beyond its role in cancer, miR-142-3p's diagnostic and prognostic value extends to autoimmune diseases like multiple sclerosis (MS), where it has been implicated in inflammation-driven synaptic dysfunction. It exerts its effects by targeting glutamate-aspartate transporter (GLAST), a key player in glutamate homeostasis, thereby disrupting glutamate function and contributing to synaptic alterations that worsen disease progression (De Vito et al. 2022). This role of miR-142-3p in neuroinflammation highlights its potential as a biomarker for not only monitoring disease activity but also predicting patient outcomes. Furthermore, the negative correlation between miR-142-3p levels and favorable prognosis in both cancer and autoimmune diseases underscores its significance as a prognostic factor.

As miR-142-3p levels fluctuate in response to treatment and disease progression, integrating its expression profiling into clinical practice could enhance diagnostic accuracy, help assess disease severity, and enable more personalized treatment approaches. Ultimately, miR-142-3p's ability to influence immune modulation, tumor behavior, and disease pathology makes it an invaluable noninvasive tool for managing a variety of clinical conditions, from cancer to autoimmune diseases.

7.3 Role of miR-124-3p

miR-124 was first identified in mice (Lagos-Quintana et al. 2002) and is recognized as one of the most abundant microRNAs (miRNAs) in the brain (Mishima et al. 2007). Beyond its prominent presence in the nervous system, miR-124 expression has also been observed in a variety of tissues (Qin and Liu 2019). The miR-124 family comprises three members—miR-124-1, miR-124-2, and miR-124-3—that are encoded on distinct chromosomal regions. Notably, the mature form of miR-124 is highly conserved, with an identical sequence found in humans, mice, and rats. This miRNA plays a pivotal role in the nervous system, contributing to critical processes such as neuronal development, neurotransmission, and modulation of synapse morphology. Additionally, miR-124 is involved in regulating myeloid cell activity and hematopoiesis (Soreq and Wolf 2011). One of its significant molecular functions includes the downregulation of histone deacetylase 5 (HDAC5), which enhances neurite outgrowth by activating the MEF2C-GPM6A signaling axis (Gu et al. 2018). The dysregulation of miR-124 has been implicated in the pathogenesis of various conditions (Fig. 7.1). Furthermore, aberrant miR-124 expression has been associated with malignancies in several tissue types.

Distinct studies have highlighted the upregulation of miR-124 in various nonmalignant disorders, including myocardial infarction (MI), Alzheimer's disease (AD), cardiovascular diseases (CVD), spinal cord injury (SCI), and heart failure. This upregulation is often associated with repair mechanisms following injury. In SCI models, miR-124 contributes to neuronal repair. In a rat model, agomiR-124 reduced neuronal loss, inhibited astrocyte activation, and increased NF-200 levels in the dorsal horn. It altered the expression of 165 mRNAs, targeting Tal1 to enhance neuronal precursor cell proliferation and inhibit differentiation (Wang et al. 2020). miR-124 also upregulated Neat1, improving locomotion in SCI models (Cui et al. 2019).

In smokers, increased miR-124-3p expression in monocytes correlated with higher atherosclerosis risk and elevated surface proteins CD29, CD45RA, and CD206 (De Ronde et al. 2017). miR-124 plays a dual role in cardiovascular

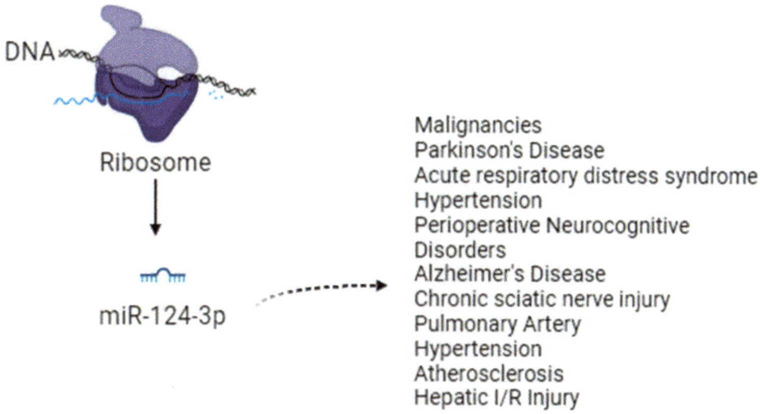

Fig. 7.1 Diverse role of miR-124-3p in various human diseases

conditions. While its upregulation intensifies angiotensin II-induced HF and inhibits angiogenesis by targeting CD151, it also increases apoptosis and reduces nitric oxide secretion, cell viability, and migration in endothelial cells (Zhao et al. 2018). Suppressing miR-124 protects human osteoblasts from H_2O_2-induced toxicity by restoring cyclin-dependent kinase 6 and ROCK1 levels (Liang et al. 2020a). In diabetic vascular complications, a high-glucose medium decreases circHIPK3, leading to miR-124 buildup in endothelial cells. Suppressing miR-124 mitigates glucose-induced damage, suggesting the circHIPK3-miR-124 axis as a therapeutic target (Cao et al. 2018). Table 7.1 summarizes studies demonstrating miR-124-3p in various human and animal disorders.

miR-124-3p has garnered significant attention due to its involvement in multiple physiological and pathological pathways. Its altered expression has been linked to numerous conditions, making it a promising invasive biomarker for both diagnostic and prognostic applications. Detection of miR-124-3p levels in biological samples such as blood, serum, or tissue biopsies can provide valuable insights into disease onset, progression, and outcomes.

The diagnostic potential of miR-124-3p has been demonstrated across a range of diseases. For example, in acute ischemic stroke (AIS), elevated levels of miR-124-3p in serum exosomes have been identified as reliable indicators of ischemic events. These findings suggest that measuring miR-124-3p levels could facilitate early diagnosis and timely intervention. Similarly, in Parkinson's disease (PD), plasma levels of miR-124-3p are significantly altered and have been associated with disease severity, including cases complicated by comorbid depression. In ARDS, the downregulation of miR-124-3p correlates with increased inflammation and lung injury, reinforcing its utility as a diagnostic marker in inflammatory pulmonary conditions.

Beyond its diagnostic utility, miR-124-3p also shows promise as a prognostic biomarker, offering insights into disease progression and therapeutic outcomes. In Alzheimer's disease (AD), dysregulated expression of miR-124-3p in cerebrospinal fluid and brain tissues has been linked to tau protein hyperphosphorylation and

Table 7.1 miR-124-3p in human and animal disorders

Type of disease	Related molecules	Function	References
Parkinson's disease (PD)	DAPK1, Caspase-3, TH, MALAT1	LncRNA MALAT1 via miR-124-3p/APK1 axis regulates cell apoptosis and motor disorders in PD	Lu et al. (2020)
Acute respiratory distress syndrome (ARDS)	P65, IL-1β, IL-6, TNF-α	miR-124-3p targeting P65 reduces inflammation and pulmonary injury in ARDS mouse model	Liang et al. (2020c)
Hypertension	EGR1	miR-124-3p inhibits EGR1 to reduce angiotensin II-dependent hypertension	Lv et al. (2021)
Perioperative neurocognitive disorders (PND)	MYL2–2	LncRNA MYL2–2 downregulates miR-1, contributing to PND after cardiac surgery	Deng et al. (2020)
Alzheimer's disease (AD)	Caveolin-1, tau, PI3K, Akt, GSK3β	miR-124-3p modulates caveolin-1/PI3K/Akt/GSK3β axis to reduce tau-induced apoptosis	Kang et al. (2017)
Chronic sciatic nerve injury (CCI)	EZH2, IL-6A, IL-1βB, TNF-α	miR-124-3p targets EZH2 to reduce neuropathic pain caused by CCI	Zhang et al. (2019b)
Pulmonary artery hypertension (PAH)	KLF5, PCNA, cyclin-A1, cyclin-D1, cyclin-E1, MALAT1	MALAT1 sponging miR-124-3p.1/KLF5 axis promotes HPASMC growth and migration	Wang et al. (2019a)
Atherosclerosis (AS)	MEKK3, PCNA, Bcl-2, Bax, p38MAPK	miR-124-3p targets MEKK3 and inhibits p38MAPK pathway to reduce macrophage proliferation in AS	Zhai et al. (2020)
Hepatic I/R injury	TNF-α, IL-β, Epha2, mmu_circRNA_005186	Silencing mmu_circRNA_005186 via miR-124-3p/Epha2 axis reduces LPS-induced inflammation in hepatic injury	Zhang et al. (2019a)

neuronal apoptosis, key factors in cognitive decline. Similarly, in atherosclerosis, miR-124-3p influences macrophage proliferation and inflammatory responses, providing a measure of disease progression. In CCI, reduced levels of miR-124-3p exacerbate neuropathic pain, highlighting its potential for monitoring treatment efficacy in pain management.

The functional roles of miR-124-3p in cellular signaling pathways underpin its biomarker potential. It regulates inflammatory cytokines such as IL-1β, IL-6, and TNF-α, which play pivotal roles in autoimmune and neuroinflammatory diseases. Additionally, miR-124-3p modulates apoptosis and autophagy by targeting proteins like Caspase-3 and p65, influencing cell survival in conditions like PD and ARDS. Dysregulation of miR-124-3p also impacts vascular integrity through pathways such as PI3K/Akt/GSK3β and NF-κB, contributing to pathologies like hypertension and stroke.

The clinical application of miR-124-3p as a biomarker is highly promising, but several challenges remain. Standardizing detection methods, such as quantitative real-time PCR or next-generation sequencing, is crucial for consistent and reproducible results. Large-scale validation studies are needed to establish reference ranges and diagnostic thresholds across diverse populations. Additionally, integrating miR-124-3p with other biomarkers could enhance specificity and sensitivity in complex diseases, paving the way for its use in precision medicine.

In conclusion, miR-124-3p holds immense potential as an invasive biomarker for diagnosing and prognosticating various diseases. Its accessibility through minimally invasive liquid biopsies and its sensitivity to pathological changes make it an attractive candidate for clinical translation. As research progresses, miR-124-3p could become a cornerstone of biomarker-based diagnostics and targeted therapeutic strategies, contributing significantly to the advancement of personalized medicine.

7.4 Role of miR-182-5p

MicroRNA-182-5p (miR-182-5p) is an important component of the microRNA regulatory network that plays a crucial role in various diseases. Recent studies have demonstrated that miR-182-5p has significant regulatory effects on several pathological conditions (Krishnan et al. 2013). Its involvement in these diverse diseases has highlighted its importance as a multifaceted molecule that can regulate key cellular functions. miR-182-5p exerts its effects by modulating the expression of genes involved in crucial cellular processes such as cell proliferation, apoptosis, and metastasis, which are fundamental to disease progression (Wu et al. 2020).

In cancer, miR-182-5p has been shown to play a vital role in tumorigenesis by targeting various oncogenes and tumor suppressor genes, impacting key processes such as EMT (Sameti et al. 2023). Additionally, miR-182-5p has been implicated in regulating immune responses, influencing inflammatory pathways that are often altered in diseases such as autoimmune conditions (Pucella et al. 2015). Recent studies also suggest that miR-182-5p may serve as a valuable biomarker for both diagnosis and prognosis, providing a noninvasive approach for early detection and monitoring of therapeutic responses. These findings underscore the growing significance of miR-182-5p in disease pathology and its potential as a therapeutic target in various clinical applications.

MicroRNA-182-5p (miR-182-5p) has emerged as a critical component of the molecular landscape, showing great promise as both a diagnostic and prognostic biomarker in various disease contexts, particularly cardiovascular and oncological disorders. In cardiovascular diseases, miR-182-5p is gaining attention as a potential noninvasive biomarker for diagnosing unprotected left main coronary artery disease (uLMCAD) (Zhu et al. 2019). Studies have demonstrated that circulating miR-182-5p, along with miR-5187-5p, can distinguish uLMCAD patients from the non-CAD population before patients undergo invasive diagnostic procedures such as coronary angiography (CAG). This ability to predict cardiovascular conditions early on could facilitate timely interventions and prevent further complications. Additionally, miR-182-5p plays a significant role in the pathophysiology of myocardial infarction (MI), where it exacerbates myocardial injury by promoting apoptosis (Pei et al. 2023). This insight suggests that interventions designed to decrease miR-182-5p levels could be critical in reducing myocardial damage and improving outcomes in MI patients. Therefore, miR-182-5p holds substantial promise not only as a diagnostic marker but also as a therapeutic target in cardiovascular disease management, providing a foundation for further exploration in clinical practice.

In the realm of cancer, miR-182-5p has demonstrated diverse and impactful roles, particularly in regulating tumor progression, metastasis, and resistance to therapy. In colorectal cancer (CRC), miR-182-5p functions as a tumor suppressor by inhibiting the metastasis and proliferation of CRC cells. It does so by directly targeting MTDH (Metadherin), a gene that is implicated in cancer progression and metastasis. The miR-182-5p/MTDH axis represents a promising therapeutic target in CRC, potentially leading to the development of treatments that hinder tumor spread and improve patient survival (Jin et al. 2019). Furthermore, miR-182-5p plays a pivotal role in non-small cell lung cancer (NSCLC), where it drives epithelial-mesenchymal transition (EMT) and promotes metastasis by targeting EPAS1, a transcription factor involved in tumor progression.

This highlights miR-182-5p's role in facilitating tumor aggressiveness and its potential utility in predicting metastatic potential. Additionally, the overexpression of miR-182-5p has been observed in breast cancer, where it serves as an oncomir, contributing to cancer development and progression. Given its broad involvement in various cancers, miR-182-5p serves as a valuable diagnostic biomarker for identifying aggressive forms of cancer and can be leveraged to predict prognosis and guide treatment decisions. As ongoing research uncovers the specific pathways and target genes regulated by miR-182-5p, it is expected that its clinical application will be refined, providing personalized treatment strategies and improving the management of various malignancies. The integration of miR-182-5p into clinical workflows promises to significantly enhance diagnostic accuracy, guide therapeutic decisions, and contribute to the development of novel cancer treatments.

7.5 Role of miR-200b

The miR-200 family, an important group of microRNAs, consists of five members: miR-200a, miR-200b, miR-200c, miR-141, and miR-429 (Fig. 7.2). Despite this minor variation in their seed sequences, the members of the miR-200 family are thought to have a high degree of functional overlap, as they can theoretically target similar mRNAs. The conservation of the seed sequences across these family members allows them to regulate shared downstream genes and participate in the same biological processes, which include the regulation of EMT, a key event in cancer metastasis (Korpal and Kang 2008).

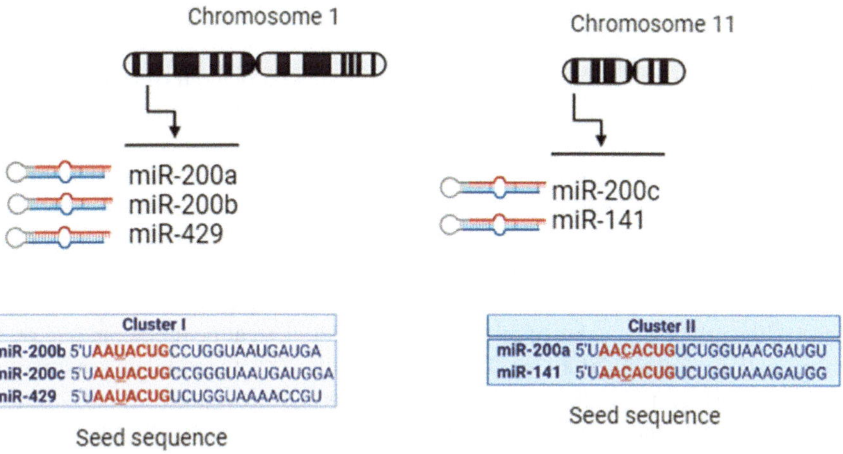

Fig. 7.2 The miR-200 family of microRNAs is encoded by two genomic clusters: miR-200b, miR-200a, and miR-429 on chromosome 1, and miR-200c and miR-141 on chromosome 11. These microRNAs share a conserved AA(U/C)ACUG seed sequence essential for mRNA target recognition and silencing, underscoring their roles in differentiation, migration, and tumor suppression

Recent research has shown that the miR-200 family plays a crucial role in various biological functions and disease processes, particularly in cancer progression. The miR-200 family is well-known for its involvement in the regulation of EMT, which are essential for cell migration, invasion, and resistance to apoptosis (Mongroo and Rustgi 2010). The miR-200 family counteracts EMT by targeting key transcription factors such as ZEB1 and ZEB2, which are critical for the induction of mesenchymal traits in cancer cells. Through this mechanism, the miR-200 family can inhibit metastasis and reduce the invasiveness of cancer cells, thus influencing tumor progression and clinical outcomes.

Moreover, due to their ability to regulate a wide range of target genes, the members of the miR-200 family are also being explored as potential diagnostic and prognostic biomarkers. Their expression levels can reflect the aggressiveness of cancers and predict responses to treatment. The therapeutic potential of miR-200 family members lies in their ability to reverse EMT and inhibit metastatic behavior, making them attractive candidates for targeted therapies in cancer treatment.

Among the miR-200 family members, miR-200b has gained significant attention as a potential diagnostic and prognostic biomarker for various cancers (Thi Chung Duong et al. 2022). Elevated levels of circulating miR-200b have been linked to different malignancies, and its expression is associated with the regulation of critical factors involved in EMT, such as ZEB1 and ZEB2, which are essential for cancer cell migration and invasion. Notably, miR-200b's role in EMT suppression and its involvement in regulating genes related to cell adhesion and migration have made it a valuable target for cancer therapy. Furthermore, miR-200b is also regulated by epigenetic modifications such as DNA methylation, with studies showing that DNA methyltransferase (DNMT3A) represses miR-200b expression in triple-negative breast cancer, highlighting the potential for miR-200b as an epigenetic biomarker.

From a diagnostic and prognostic perspective, miR-200b holds promise due to its ability to be detected in various body fluids, such as serum, plasma, and exosomes, making it a noninvasive biomarker for cancer detection. Elevated circulating miR-200b levels have been observed in the context of anovulatory women, linking its expression to potential biomarkers for hormonal or cancerous alterations (Eisenberg et al. 2017). Its downregulation in exosomes derived from cancer-associated fibroblasts (CAFs) in hypoxic environments has been shown to reduce chemotherapy sensitivity, particularly to 5-fluorouracil in colorectal cancer, underscoring the prognostic value of miR-200b in predicting treatment outcomes (Loss of exosomal micro-RNA-200b-3p). The ability of miR-200b to modulate sensitivity to treatment further highlights its role in assessing prognosis.

In colorectal cancer, miR-200b influences resistance to therapy through targeting high-mobility group box 3 (HMGB3), a protein involved in DNA damage response and repair (Chen et al. 2022). Additionally, the involvement of miR-200b in myofibroblasts transdifferentiation in precancerous oral submucous fibrosis suggests that miR-200b could serve as a biomarker for the early detection and progression of precancerous conditions. Collectively, these findings indicate that miR-200b is not only a critical player in the regulation of EMT and metastasis but also holds promise as a valuable biomarker for the diagnosis, prognosis, and therapeutic management of various cancers.

7.6 Role of microRNA-488

MicroRNA-488 (miR-488) has been implicated in various human diseases, with its functions and mechanisms offering promising clinical applications. Existing studies indicate its potential as a diagnostic and prognostic biomarker, particularly in cancer and noncancerous conditions. Numerous studies have demonstrated a significant association between miR-488 expression and the clinicopathological characteristics of various diseases (Table 7.2). In cancers such as colorectal cancer, esophageal squamous cell carcinoma (ESCC), gastric cancer, ovarian cancer, and pancreatic cancer, miR-488 expression correlates with key clinical parameters.

For CRC, miR-488 expression has been linked to tumor size, differentiation degree, infiltration depth, lymph node metastasis, TNM stage, and Duke stage across four studies. In ESCC, Yang et al. revealed that miR-488-3p expression is significantly downregulated in advanced stages compared to early stages, suggesting a correlation with tumor progression (Yang et al. 2019). Similarly, in gastric cancer, miR-488-3p expression is associated with TNM stage and metastases. Guo et al. reported that miR-488-3p levels in ovarian cancer patients are significantly correlated with FIGO stage and lymph node metastasis (Guo et al. 2020). In pancreatic cancer, data from the TCGA-PAAD database show a strong association between miR-488-5p expression and T stage, N stage, and TNM stages (Han et al. 2022).

Table 7.2 Clinical significance of miR-488 in various human diseases

Disease	Clinical application	Associated characteristics	References
Acute ischemic stroke (AIS)	Indicates poor prognosis	Correlates with NIHSS scores	Liang and Zhang (2020)
Atherosclerosis (AS)	Differentiates AS patients from healthy controls	Positively associated with carotid intima-media thickness	Li et al. (2021)
Breast cancer (BC)	Independent prognostic factor for RFS; not linked with OS	No significant link with various clinical features (e.g., age, tumor size)	Masuda et al. (2018)
Colorectal cancer (CRC)	Predicts 3-year survival; stratifies patients by tumor characteristics	Linked with larger tumors, poor differentiation, deeper infiltration, advanced Duke stages	Deng et al. (2021)
Colorectal cancer (CRC)	–	Associated with infiltration depth, differentiation, lymph node metastasis, and Duke stage	Wang et al. (2019b)
Colorectal cancer (CRC)	Predicts overall survival (OS)	Associated with TNM stage	Lv et al. (2017)
Esophageal squamous cell carcinoma (ESCC)	Predicts OS	Linked with clinical tumor stage	Yang et al. (2019)
Gastric cancer (GC)	–	Associated with TNM stage and metastasis	Yang et al. (2023)
Gastric cancer (GC)	–	Associated with TNM stage	Zhao et al. (2016)
Malignant melanoma (MM)	–	No significant correlation with features like age, ulceration, metastasis, or stage	Li et al. (2017)
Ovarian cancer (OC)	Predicts OS	Linked with FIGO stage and lymph node metastasis	Guo et al. (2020)
Pancreatic cancer (PC)	Predicts OS	Associated with T stage, N stage, and TNM stages	Han et al. (2022)

Beyond oncology, miR-488 expression also plays a role in noncancerous conditions such as AIS and AS. In AIS, elevated miR-488 levels in peripheral blood positively correlate with NIHSS scores, reflecting disease severity. In AS, serum miR-488-3p levels show a strong positive correlation with the carotid intima-media thickness (CIMT) value, a key marker of atherosclerotic burden. Distinct expression patterns of miR-488 support its diagnostic potential in several conditions. For instance, Li et al. demonstrated that serum miR-488-3p effectively differentiates AS patients from healthy controls, achieving an area under the ROC curve (AUC) value of 0.892 (Li et al. 2021). Similarly, in CRC, miR-488-3p expression is significantly lower in metastatic or recurrent cases compared to nonmetastatic cases. ROC curve analysis further underscores its ability to distinguish CRC patients based on tumor size, differentiation degree, infiltration depth, and Duke stage, highlighting its potential as a diagnostic tool (Deng et al. 2021).

miR-488 has shown prognostic value in multiple diseases, particularly cancers. For example, downregulation of miR-488-3p in CRC, ESCC, and ovarian cancer correlates with poor overall survival (OS). Similarly, low levels of miR-488-5p in pancreatic cancer patients are associated with shorter OS. In breast cancer, high serum pre-miR-488 levels independently predict poor recurrence-free survival (RFS), although no significant association with OS was observed. Deng et al. reported that higher miR-488-3p levels predicted favorable 3-year survival in CRC patients (Deng et al. 2021). Outside of cancer, Liang et al. found that elevated peripheral blood miR-488 levels correlate with poorer prognosis in AIS patients, further supporting its role as a prognostic biomarker (Liang and Zhang 2020).

Research on miR-488 has significantly advanced our understanding of its roles in regulating cellular processes, such as proliferation, apoptosis, migration, invasion, metastasis, cell cycle regulation, chemoresistance, and glycolysis, particularly in cancer. Its relevance extends to noncancerous diseases, including AMI, CIS, endometriosis, osteoarthritis (OA), osteoporosis (OP), and SCI. Despite its potential, several critical questions remain. The mechanisms underlying miR-488 dysregulation require further exploration, as promoter methylation has been proposed as a contributing factor, but relevant studies are limited. Additionally, its diverse gene targets and pathways necessitate disease-specific investigations. Bridging the gap between research and clinical application remains a challenge, with future studies needed to evaluate its utility as a biomarker and therapeutic target in specific diseases. In conclusion, miR-488 holds a significant promise for clinical applications. Addressing existing gaps in knowledge will be essential to fully harness its potential as a diagnostic and prognostic biomarker across a range of human diseases.

7.7 Conclusion

This chapter provided an overview of the miRNA signatures in different diseases and cancer with varying molecular and clinical features. We have demonstrated the potential use of circulating and cellular miRNAs as therapeutic targets and biomarkers for diagnosis and prognosis.

Funding Statement No funding received.

Conflict of Interest The authors declare no conflict of interest.

Ethical Approval This chapter does not contain any studies with human participants or animals.

References

Cao Y, Yuan G, Zhang Y, Lu R (2018) High glucose-induced circHIPK3 downregulation mediates endothelial cell injury. Biochem Biophys Res Commun 507:362–368

Chen S, Tu Y, Yuan H, Shi Z, Guo Y, Gong W, Tu S (2022) Regulatory functions of miR-200b-3p in tumor development (Review). Oncol Rep 47:96

Cui Y, Yin Y, Xiao Z, Zhao Y, Chen B, Yang B, Xu B, Song H, Zou Y, Ma X, Dai J (2019) LncRNA Neat1 mediates miR-124-induced activation of Wnt/beta-catenin signaling in spinal cord neural progenitor cells. Stem Cell Res Ther 10:400

De Ronde MWJ, Kok MGM, Moerland PD, Van Den Bossche J, Neele AE, Halliani A, Van Der Made I, De Winther MPJ, Meijers JCM, Creemers EE, Pinto-Sietsma SJ (2017) High miR-124-3p expression identifies smoking individuals susceptible to atherosclerosis. Atherosclerosis 263:377–384

De Vito F, Musella A, Fresegna D, Rizzo FR, Gentile A, Stampanoni Bassi M, Gilio L, Buttari F, Procaccini C, Colamatteo A, Bullitta S, Guadalupi L, Caioli S, Vanni V, Balletta S, Sanna K, Bruno A, Dolcetti E, Furlan R, Finardi A, Licursi V, Drulovic J, Pekmezovic T, Fusco C, Bruzzaniti S, Hornstein E, Uccelli A, Salvetti M, Matarese G, Centonze D, Mandolesi G (2022) MiR-142-3p regulates synaptopathy-driven disease progression in multiple sclerosis. Neuropathol Appl Neurobiol 48:e12765

Deng X, Chen Y, Liu Z, Xu J (2020) MiR-124-3p.1 sensitizes ovarian cancer cells to mitochondrial apoptosis induced by carboplatin. Onco Targets Ther 13:5375–5386

Deng X, Li D, Ke X, Wang Q, Yan S, Xue Y, Zheng H (2021) Mir-488 alleviates chemoresistance and glycolysis of colorectal cancer by targeting PFKFB3. J Clin Lab Anal 35:e23578

Eisenberg I, Nahmias N, Novoselsky Persky M, Greenfield C, Goldman-Wohl D, Hurwitz A, Haimov-Kochman R, Yagel S, Imbar T (2017) Elevated circulating micro-ribonucleic acid (miRNA)-200b and miRNA-429 levels in anovulatory women. Fertil Steril 107:269–275

Gu X, Fu C, Lin L, Liu S, Su X, Li A, Wu Q, Jia C, Zhang P, Chen L, Zhu X, Wang X (2018) miR-124 and miR-9 mediated downregulation of HDAC5 promotes neurite development through activating MEF2C-GPM6A pathway. J Cell Physiol 233:673–687

Guo JY, Wang XQ, Sun LF (2020) MicroRNA-488 inhibits ovarian cancer cell metastasis through regulating CCNG1 and p53 expression. Eur Rev Med Pharmacol Sci 24:2902–2910

Han D, Zhu S, Li X, Li Z, Huang H, Gao W, Liu Y, Zhu H, Yu X (2022) The NF-kappaB/miR-488/ERBB2 axis modulates pancreatic cancer cell malignancy and tumor growth through cell cycle signaling. Cancer Biol Ther 23:294–309

Jin Y, Zhang ZL, Huang Y, Zhang KN, Xiong B (2019) MiR-182-5p inhibited proliferation and metastasis of colorectal cancer by targeting MTDH. Eur Rev Med Pharmacol Sci 23:1494–1501

Kang Q, Xiang Y, Li D, Liang J, Zhang X, Zhou F, Qiao M, Nie Y, He Y, Cheng J, Dai Y, Li Y (2017) MiR-124-3p attenuates hyperphosphorylation of tau protein-induced apoptosis via caveolin-1-PI3K/Akt/GSK3beta pathway in N2a/APP695swe cells. Oncotarget 8:24314–24326

Korpal M, Kang Y (2008) The emerging role of miR-200 family of microRNAs in epithelial-mesenchymal transition and cancer metastasis. RNA Biol 5:115–119

Krishnan K, Steptoe AL, Martin HC, Wani S, Nones K, Waddell N, Mariasegaram M, Simpson PT, Lakhani SR, Gabrielli B, Vlassov A, Cloonan N, Grimmond SM (2013) MicroRNA-182-5p targets a network of genes involved in DNA repair. RNA 19:230–242

Lagos-Quintana M, Rauhut R, Yalcin A, Meyer J, Lendeckel W, Tuschl T (2002) Identification of tissue-specific microRNAs from mouse. Curr Biol 12:735–739

Li N, Ma Y, Ma L, Guan Y, Yang D (2017) MicroRNA-488-3p sensitizes malignant melanoma cells to cisplatin by targeting PRKDC. Cell Biol Int 41:622–629

Li Z, Xu C, Sun D (2021) MicroRNA-488 serves as a diagnostic marker for atherosclerosis and regulates the biological behavior of vascular smooth muscle cells. Bioengineered 12:4092–4099

Liang J, Zhang Z (2020) Higher peripheral blood MiR-488 level predicts poor prognosis of acute ischemic stroke. Clin Lab 66:1391

Liang J, Shen YC, Zhang XY, Chen C, Zhao H, Hu J (2020a) Circular RNA HIPK3 downregulation mediates hydrogen peroxide-induced cytotoxicity in human osteoblasts. Aging 12:1159–1170

Liang L, Fu J, Wang S, Cen H, Zhang L, Mandukhail SR, Du L, Wu Q, Zhang P, Yu X (2020b) MiR-142-3p enhances chemosensitivity of breast cancer cells and inhibits autophagy by targeting HMGB1. Acta Pharm Sin B 10:1036–1046

Liang Y, Xie J, Che D, Zhang C, Lin Y, Feng L, Chen J, Chen L, Wu Z (2020c) MiR-124-3p helps to protect against acute respiratory distress syndrome by targeting p65. Biosci Rep 40:BSR20192132

Liu P, Cao F, Sui J, Hong Y, Liu Q, Gao X, Gong H, Hao L, Lou Z, Zhang W (2020) MicroRNA-142-3p inhibits tumorigenesis of colorectal cancer via suppressing the activation of Wnt signaling by directly targeting to beta-catenin. Front Oncol 10:552944

Lu Y, Gong Z, Jin X, Zhao P, Zhang Y, Wang Z (2020) LncRNA MALAT1 targeting miR-124-3p regulates DAPK1 expression contributes to cell apoptosis in Parkinson's disease. J Cell Biochem 121:4838–4848

Lv Y, Shi Y, Han Q, Dai G (2017) Histone demethylase PHF8 accelerates the progression of colorectal cancer and can be regulated by miR-488 in vitro. Mol Med Rep 16:4437–4444

Lv L, Shen J, Xu J, Wu X, Zeng C, Lin L, Mao W, Wei T (2021) MiR-124-3p reduces angiotensin II-dependent hypertension by down-regulating EGR1. J Hum Hypertens 35:696–708

Masuda T, Shinden Y, Noda M, Ueo H, Hu Q, Yoshikawa Y, Tsuruda Y, Kuroda Y, Ito S, Eguchi H, Ohno S, Mimori K (2018) Circulating pre-microRNA-488 in peripheral blood is a potential biomarker for predicting recurrence in breast cancer. Anticancer Res 38:4515–4523

Mishima T, Mizuguchi Y, Kawahigashi Y, Takizawa T (2007) RT-PCR-based analysis of microRNA (miR-1 and -124) expression in mouse CNS. Brain Res 1131:37–43

Mongroo PS, Rustgi AK (2010) The role of the miR-200 family in epithelial-mesenchymal transition. Cancer Biol Ther 10:219–222

Pei G, Chen L, Wang Y, He C, Fu C, Wei Q (2023) Role of miR-182 in cardiovascular and cerebrovascular diseases. Front Cell Dev Biol 11:1181515

Pucella JN, Yen WF, Kim MV, Van Der Veeken J, Luo CT, Socci ND, Naito Y, Li MO, Iwai N, Chaudhuri J (2015) miR-182 is largely dispensable for adaptive immunity: lack of correlation between expression and function. J Immunol 194:2635–2642

Qin Z, Liu X (2019) miR-124, a potential therapeutic target in colorectal cancer. Onco Targets Ther 12:749–751

Sameti P, Tohidast M, Amini M, Bahojb Mahdavi SZ, Najafi S, Mokhtarzadeh A (2023) The emerging role of MicroRNA-182 in tumorigenesis; a promising therapeutic target. Cancer Cell Int 23:134

Shen WW, Zeng Z, Zhu WX, Fu GH (2013) MiR-142-3p functions as a tumor suppressor by targeting CD133, ABCG2, and Lgr5 in colon cancer cells. J Mol Med 91:989–1000

Soreq H, Wolf Y (2011) NeurimmiRs: microRNAs in the neuroimmune interface. Trends Mol Med 17:548–555

Thi Chung Duong T, Nguyen THN, Thi Ngoc Nguyen T, Huynh LH, Ngo HP, Thi Nguyen H (2022) Diagnostic and prognostic value of miR-200 family in breast cancer: a meta-analysis and systematic review. Cancer Epidemiol 77:102097

Wang D, Xu H, Wu B, Jiang S, Pan H, Wang R, Chen J (2019a) Long non-coding RNA MALAT1 sponges miR-124-3p.1/KLF5 to promote pulmonary vascular remodeling and cell cycle progression of pulmonary artery hypertension. Int J Mol Med 44:871–884

Wang YB, Shi Q, Li G, Zheng JH, Lin J, Qiu W (2019b) MicroRNA-488 inhibits progression of colorectal cancer via inhibition of the mitogen-activated protein kinase pathway by targeting claudin-2. Am J Physiol Cell Physiol 316:C33–C47

Wang J, Li H, Chen L, Dong J, Yang J, Gong Z, Wang B, Zhao X (2020) mRNA profiling for miR-124-mediated repair in spinal cord injury. Neuroscience 438:158–168

Wu X, Chen H, Wu M, Peng S, Zhang L (2020) Downregulation of miR-182-5p inhibits the proliferation and invasion of triple-negative breast cancer cells through regulating TLR4/NF-kappaB pathway activity by targeting FBXW7. Ann Transl Med 8:995

Xie N, Meng Q, Zhang Y, Luo Z, Xue F, Liu S, Li Y, Huang Y (2021) MicroRNA-142-3p suppresses cell proliferation, invasion and epithelial-to-mesenchymal transition via RAC1-ERK1/2 signaling in colorectal cancer. Mol Med Rep 24:568

Yang Y, Li H, He Z, Xie D, Ni J, Lin X (2019) MicroRNA-488-3p inhibits proliferation and induces apoptosis by targeting ZBTB2 in esophageal squamous cell carcinoma. J Cell Biochem 120:18702–18713

Yang D, Shi M, You Q, Zhang Y, Hu Z, Xu J, Cai Q, Zhu Z (2023) Tumor- and metastasis-promoting roles of miR-488 inhibition via HULC enhancement and EZH2-mediated p53 repression in gastric cancer. Cell Biol Toxicol 39:1341–1358

Zhai C, Cong H, Hou K, Hu Y, Zhang J, Zhang Y, Zhang H (2020) Effects of miR-124-3p regulation of the p38MAPK signaling pathway via MEKK3 on apoptosis and proliferation of macrophages in mice with coronary atherosclerosis. Adv Clin Exp Med 29:803–812

Zhang C, Liang R, Gan X, Yang X, Chen L, Jian J (2019a) MicroRNA-384-5p/Beclin-1 as potential indicators for epigallocatechin Gallate against cardiomyocytes ischemia reperfusion injury by inhibiting autophagy via PI3K/Akt pathway. Drug Des Devel Ther 13:3607–3623

Zhang Y, Cai H, Chen S, Sun D, Zhang D, He Y (2019b) Exosomal transfer of miR-124 inhibits normal fibroblasts to cancer-associated fibroblasts transition by targeting sphingosine kinase 1 in ovarian cancer. J Cell Biochem 120:13187–13201

Zhao Y, Lu G, Ke X, Lu X, Wang X, Li H, Ren M, He S (2016) miR-488 acts as a tumor suppressor gene in gastric cancer. Tumour Biol 37:8691–8698

Zhao Y, Yan M, Chen C, Gong W, Yin Z, Li H, Fan J, Zhang XA, Wang DW, Zuo H (2018) MiR-124 aggravates failing hearts by suppressing CD151-facilitated angiogenesis in heart. Oncotarget 9:14382–14396

Zhu L, Chen T, Ye W, Wang JY, Zhou JP, Li ZY, Li CC (2019) Circulating miR-182-5p and miR-5187-5p as biomarkers for the diagnosis of unprotected left main coronary artery disease. J Thorac Dis 11:1799–1808

Chapter 8
The Transcriptional Role of Circular RNAs in Drug Resistance by Modulating the miRNA/mRNA Axis

Mohammad H. Ghazimoradi, Amirmohammad Moghadam, Parastoo Akbarabadi, and Sadegh Babashah

Contents

Abstract Drug resistance hinders cancer treatment and directly affects cancer outcome and survival rate. While drug resistance is one of the worst features of cancers, there is limited knowledge about its mechanism and initiation in cancers. Noncoding RNAs are one of the most impactful epigenetic elements that are deregulated in many diseases. There is a strong correlation between cancers and noncoding RNAs. In this regard, we tried to find the connection between cancers and this epigenetic factor. In our investigation, we find an axis in which circular RNA, a dominant type of noncoding RNA, could regulate miRNA to control mRNA degradation. We encountered many deregulated pathways and tried to explain the mechanism underlying the CirRNA/miRNA/mRNA axis in drug resistance of cancer, including the most widely recognized resistance type of cancer in the body.

Keywords Drug resistance · Noncoding RNA · Circular RNA · miRNA · mRNA

M. H. Ghazimoradi · A. Moghadam · P. Akbarabadi · S. Babashah (✉)
Department of Molecular Genetics, Faculty of Biological Sciences, Tarbiat Modares University, Tehran, Iran
e-mail: babashah@modares.ac.ir

M. F. Ullah, A. Ahmad (eds.), *Non-coding RNAs*, Experientia Supplementum 115,
https://doi.org/10.1007/978-3-032-06948-1_8

Abbreviations

ACT	Adoptive cell therapy
ATG7	Autophagy related 7
ceRNA	Competing endogenous RNA
CHD4	Chromodomain helicase DNA-binding Protein 4
circRNA/CircRNA	Circular RNA
CSC	Cancer stem cell
DUSP7	Dual specificity phosphatase 7
E2F3	E2F transcription factor 3
E-CAD	E-Cadherin
EMT	Epithelial-mesenchymal transition
ERα	Estrogen receptor alpha
FOXO3	Forkhead Box O3
HIPK3	Homeodomain interacting protein kinase 3
ICI	Immune checkpoint inhibitor
ITGB1	Integrin beta 1
miRNA	MicroRNA
mRNA	Messenger RNA
MYC	Myelocytomatosis viral oncogene homolog
MyD88	Myeloid differentiation primary response 88
ncRNA	Noncoding RNA
NF-κB	Nuclear factor kappa-light-chain-enhancer of activated B cells
PI3K	Phosphoinositide 3-Kinase
PIK3CA	Phosphatidylinositol-4,5-Bisphosphate 3-Kinase Catalytic Subunit Alpha
PTX	Paclitaxel
RhoA	Ras homolog family member A
ROCK	Rho-associated coiled-coil containing Protein Kinase
Src	Proto-oncogene tyrosine-protein kinase Src
TGF-β	Transforming growth factor beta
TLR4	Toll-like receptor 4
TNBC	Triple-negative breast cancer
USP4	Ubiquitin-specific peptidase 4
UTR	Untranslated region

8.1 Introduction

Drug resistance is one of the significant challenges to cancer treatment, increasing the risk of treatment failure and death (Vasan et al. 2019). Doxorubicin (Chen et al. 2023), tamoxifen (Chang 2012), paclitaxel (Esparza-Lopez et al. 2022), and even targeted therapy, are among the drugs that can develop drug resistance in cancers (Vander Velde et al. 2020). Several mechanisms of drug resistance have been identified. These pathways facilitate cancer cells to withstand anticancer medicines. Cancer cells, for instance, may transport anticancer drugs via uncontrolled molecular pumps or inhibit therapeutic agents from reaching cancer cells (Mansoori et al. 2017; Housman et al. 2014). Furthermore, cancer cells may confer a survival advantage due to modifications to their survival pathways. This pathway also contributes to pharmacological effects by inhibiting the drug's target molecular processes, including DNA repair, apoptosis, and EMT (Hanahan 2022). It was also asserted that because of their characteristics, including reduced metabolism, self-renewal, and differentiation, CSCs may perform better than anticancer therapies (Phi et al. 2018).

The immune system is critical in defending the body against cancer. Cancer immunotherapy targets malignancies through harnessing the body's immune system. It facilitates the immune system identify and combat cancerous cells. Immune checkpoint inhibitors (ICIs), adoptive cell therapy (ACT), cancer vaccines, oncolytic virus treatment, and immune system modulators are some of the many types of cancer immunotherapy (Farkona et al. 2016). An immunosuppressive tumor microenvironment, a lack of tumor antigen and decreased immunogenicity, a lack of antigen presentation, T cell dysfunction and increased expression of suppressive immune checkpoints, and differences in immunological functions are some of the causes of drug resistance in immunotherapy (Bai et al. 2020). The function of circular RNAs (circRNAs), a unique family of noncoding RNAs that form covalently closed circular structures that contradict the linear structure of other RNA molecules, in cancer cell proliferation, migration, invasion, and, most importantly, chemoresistance by influencing the availability of associated miRNAs and targeted genes, has been the focus of recent research, despite the fact that medication resistance is a significant obstacle to cancer treatment (Almouh et al. 2022). Circular RNA dysregulation has been found in a number of cancer types, and numerous studies have shown their role in cancer development and metastasis. Originally believed to be splicing artifacts, RNA-binding protein sequestering agents, miRNA sponges, or transcriptional regulators, they have emerged as critical regulators in regulating gene expression (Kristensen et al. 2018). One of cancer's most important and uncontrolled processes is the spiking of miRNA to circular RNA's microRNA response regions (Memczak et al. 2013). Through this process, circular RNA sponges the miRNA and inhibits its function as an oncogene messenger RNA silencer. It also disrupts tumor suppressors by disrupting their native pathways and genes linked to drug resistance (Liu et al. 2022a; Ghazimoradi et al. 2023). Interestingly, microRNAs can also modulate gene expression and carry out transactional repression, thus their silencing effect is not limited to mRNA degradation (Ghazimoradi and Babashah 2022). Even subtle perturbations in the CircRNA/miRNA/mRNA system can result in profound pathological consequences because circular RNA can sponge to tens of these regulatory RNAs and miRNAs have thousands of targets through various mechanisms (Memczak et al. 2013). CircRNA/miRNA/mRNA has been demonstrated to interfere with doxorubicin (Li et al. 2020a), tamoxifen (Liang et al. 2019), paclitaxel (Hu et al. 2023a), 5-fluorouracil (Yao et al. 2021a), and a variety of other medicines, including immunotherapy

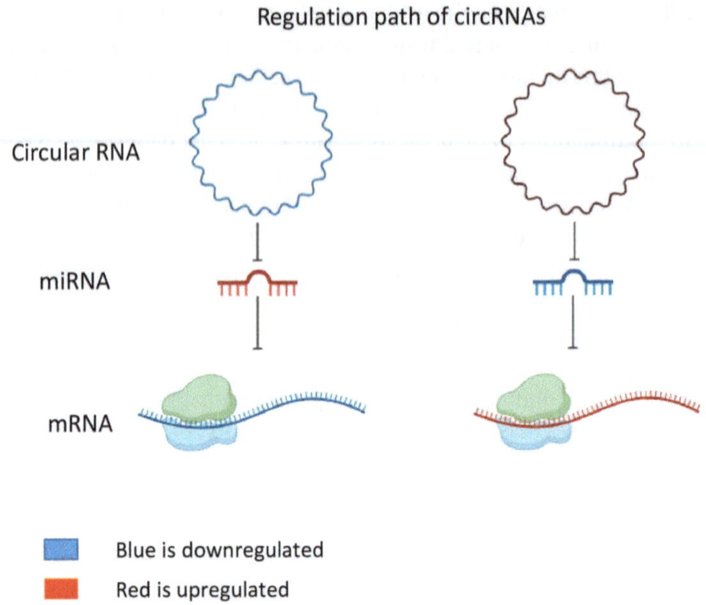

Fig. 8.1 The mechanism by which CirRNA/miRNA/mRNA transcription axis regulates drug resistance in cancers

(Fig. 8.1). Although this important route has been extensively investigated, its significance in drug resistance, particularly in breast cancer, needs to be reassessed. Understanding this process in drug resistance will provide novel treatment targets for overcoming drug resistance.

8.2 CircRNA/miRNA/mRNA Pathway in Drug-Resistant Breast Cancer

Drug-resistant cancer cell populations are a major contributor to breast cancer mortality (Huang et al. 2015). Circ_0085495, a cytoplasm-localized circular RNA, has been found to be significantly upregulated in doxorubicin-resistant breast cancer cells. Functionally, Circ_0085495 promotes chemoresistance, cellular proliferation, and metastatic potential by acting as a competitive endogenous RNA (ceRNA) for miR-873-5p. Through this sponging mechanism, Circ_0085495 downregulates miR-873-5p activity, resulting in the derepression of integrin β1 (ITGB1), a critical mediator implicated in the modulation of metastasis, proliferation, and multidrug resistance via multiple oncogenic signaling pathways (Xie and Zheng 2022). Hsa_circ_0092276 facilitates doxorubicin resistance in breast cancer cells by modulating autophagic activity through the miR-348/ATG7 regulatory axis (Wang et al. 2021a). miR-1299 has been identified as a downstream target of circ_0006528 in breast cancer. In doxorubicin-resistant breast cancer cell lines and tissues, miR-1299 expression is markedly suppressed, whereas circ_0006528 and CHD4 exhibit significant upregulation compared to drug-sensitive counterparts. circ_0006528 exerts oncogenic functions by acting as a molecular sponge for miR-1236-3p, thereby facilitating the upregulation of CHD4 a known downstream target of miR-1299 and an activator of the RhoA/ROCK signaling pathway. Additionally, circ_0006528 is implicated in mediating tamoxifen resistance in breast cancer cells (Hao et al. 2021). NCOA3 is a transcriptional coactivator possessing intrinsic acetyltransferase activity. It plays a critical role in mediating resistance to endocrine therapy, particularly in the context of HER2/neu overexpression, activating mutations in PIK3CA (PI3K pathway), and activating mutations in the proto-oncogene tyrosine-protein kinase Src (Li et al. 2022). NCOA3 has been validated as a direct target of miR-4458 through luciferase reporter assays. Circ_0001667 functions as a competing endogenous RNA (ceRNA), sequestering miR-4458. Both Circ_0001667 and NCOA3 are significantly upregulated in breast cancer tissues and cell lines, particularly in doxorubicin-resistant phenotypes, whereas miR-4458 expression is notably downregulated. Cui et al. demonstrated that enforced expression of miR-4458 or silencing of Circ_0001667 or NCOA3 enhances the sensitivity of breast cancer cells to doxorubicin. Additionally, CircRNA_104889 has been identified as another circular RNA that sponges miR-4458, promoting tumor progression in lung adenocarcinoma (Cui et al. 2022). Circ_UBE2D2, an exosome-derived circular RNA capable of sequestering multiple microRNAs, is significantly upregulated in tamoxifen-resistant breast cancer tissues and cell lines. Silencing Circ_UBE2D2 enhances the sensitivity of breast cancer cells to tamoxifen and suppresses their invasive and migratory capabilities in vitro. Conversely, exogenous delivery of exosomes enriched with Circ_UBE2D2 promotes tamoxifen resistance, as well as increases cell migration, invasion, and tumor growth in murine models. Mechanistically, Circ_UBE2D2 downregulates E-cadherin (E-CAD) and estrogen receptor alpha (ERα), while upregulating VIMENTIN expression, thereby promoting epithelial-mesenchymal transition (EMT) and resistance. Notably, miR-200a-3p downregulated in tamoxifen-resistant breast cancer has been shown to suppress Circ_UBE2D2 expression (Hu et al. 2020). Hsa_circ_0025202 functions as a competing endogenous RNA (ceRNA) by sponging miR-182-5p and miR-197-3p, thereby upregulating the expression of Forkhead box O3 (FOXO3) and Homeodomain Interacting Protein Kinase 3 (HIPK3). FOXO3, a well-established tumor suppressor, exerts its effects by inhibiting cell proliferation and metastasis while promoting apoptosis across various cancer types, including breast cancer. The overexpression of hsa_circ_0025202 has been shown to reverse tumor progression and tamoxifen resistance both in vitro and in vivo, effects that are counteracted by the introduction of miR-182-5p and miR-197-3p mimics. Furthermore, knockdown of FOXO3 and HIPK3 recapitulates the phenotype associated with low hsa_circ_0025202 expression, inducing drug resistance in MCF7 and T47D breast cancer cell lines. Of note, in glioblastoma models, CircFOXO3 enhances cell proliferation and invasion by sponging miR-138-5p and miR-432-5p, highlighting the pivotal oncogenic and regulatory roles of the FOXO3 axis in multiple cancer types (Hu et al. 2020; Li et al.

2021a). Hsa_circ_0000199 promotes chemoresistance in triple-negative breast cancer (TNBC) by modulating the PI3K/Akt/mTOR signaling pathway through sponging of miR-206 and miR-613. This regulatory mechanism contributes to reduced sensitivity of TNBC cells to multiple chemotherapeutic agents, including cisplatin, adriamycin, paclitaxel, and gemcitabine (Li et al. 2021b). Ubiquitin-specific protease 4 (USP4) is a deubiquitinating enzyme that plays crucial roles in regulating several key signaling pathways, including NF-kB, TGF-β, Wnt/β-catenin, and p53 (Luo et al. 2023; Young et al. 2019). CircBMPR2 regulates USP4 expression by acting as a sponge for miR-553. Additionally, circBMPR2 suppresses the motility of breast cancer cells, with its expression being lower in cancerous tissues compared to normal tissues. Loss- and gain-of-function experiments demonstrate that circBMPR2 plays a critical role in promoting tamoxifen resistance in breast cancer cells by inhibiting apoptosis. Moreover, overexpression of miR-553 counteracts the tumor-suppressive effects of circBMPR2, further supporting its involvement in chemoresistance (Young et al. 2019). Elevated expression of Toll-like receptor 4 (TLR4) is linked to poor overall survival in epithelial ovarian cancer patients and enhanced metastasis in breast cancer. TLR4 activation has been shown to initiate the MyD88 signaling cascade, leading to the activation and nuclear translocation of NF-κB. In breast cancer, high TLR4 expression is inversely correlated with miR-361-5p levels. MiR-361-5p targets TLR4, and the use of miR-361-5p mimics can enhance the viability of paclitaxel-resistant breast cancer cells upon paclitaxel treatment. CircGFRA1 functions as a competing endogenous RNA (ceRNA) by sponging miR-361-5p, thereby modulating the sensitivity of breast cancer cells to paclitaxel (Zheng et al. 2021). E2F Transcription Factor 3 (E2F3) is a downstream target of miR-140-5p. In breast cancer, both Circ-RNF111 and E2F3 are upregulated in resistant cell lines and cancerous tissues, while miR-140-5p expression is downregulated. MiR-140-5p directly targets the 3′ untranslated region (3′ UTR) of E2F3, and Circ-RNF111 enhances the translation of E2F3 by sponging miR-140-5p. This interaction promotes paclitaxel resistance in breast cancer cells (Zang et al. 2020). Circ-ABCB10 acts as a competing endogenous RNA (ceRNA), sponging Let-7a-5p to regulate DUSP7. Let-7 is one of the most extensively studied microRNAs, known for its inhibitory effects on various cancers by targeting MYC and other key genes involved in cell cycle regulation, proliferation, and apoptosis. Gain- and loss-of-function assays revealed that knockdown of Circ-ABCB10 and DUSP7 significantly enhances paclitaxel (PTX) sensitivity in breast cancer cells by activating Let-7a. Furthermore, overexpression of Circ-ABCB10 or DUSP7 increases PTX sensitivity, induces apoptosis, and suppresses the invasion and autophagy of paclitaxel-resistant breast cancer cells (Yang et al. 2020). Circ_0006528 promotes resistance in breast cancer cells by modulating the miR-1236-3p/CHD4 axis (Liu et al. 2020). Circ_0006528 also contributes to paclitaxel resistance in breast cancer. In paclitaxel-resistant cells, Circ_0006528 functions as a competing endogenous RNA by sponging miR-1299, thereby increasing the expression of cyclin-dependent kinase 8 (CDK8). Notably, CDK8 is upregulated in various cancers, and elevated expression of this kinase is associated with poor prognosis and reduced patient survival (Menzl et al. 2019). Circ_0006528 contributes to paclitaxel resistance in breast cancer by acting as a competing endogenous RNA (ceRNA) that sponges miR-1299, leading to an increase in the expression of cyclin-dependent kinase 8 (CDK8). CDK8 is known to be upregulated in various cancers, and its high expression is correlated with poor patient prognosis and reduced survival rates (Liu et al. 2020). Yang and colleagues identified the overexpression of CDR1as and its involvement in 5-FU resistance in breast cancer cell lines, although its expression in clinical patient samples remains to be fully understood. They demonstrated that overexpression of CDR1as results in reduced levels of miR-7 and increased expression of cyclin E1 (CCNE1). CCNE1 overexpression is associated with tumorigenesis and chromosomal instability. Additionally, Yang and colleagues showed that transfection of MCF7 and MDA-MB-231-resistant cells with si-CDR1as or miR-7 mimics resulted in a decreased IC50, indicating enhanced drug sensitivity (Yang et al. 2019). These changes lead to a reduced colony formation rate and an increased expression of BAX/BCL and cleaved-caspase-3/caspase-3, indicating enhanced apoptosis. These findings suggest that inhibiting CDR1as and overexpressing miR-7 can increase the chemosensitivity of 5-FU-resistant breast cancer cells. The direct interaction between CDR1as and miR-7 was confirmed through an immunoprecipitation assay, while the binding of miR-7 to the 3′ UTR of CCNE1 was validated via a dual-luciferase reporter assay. Additionally, this interaction was observed in an in vivo model following knockdown of CDR1as using siRNA or miR-7 overexpression (Yang et al. 2019). High Mobility Group AT-Hook 2 (HMGA2) is a downstream target of miR-216b. Elevated expression of HMGA2 and its various mutations have been observed across numerous cancers, particularly in breast cancer. Notably, overexpression of HMGA2 in breast cancer is strongly correlated with increased metastasis (Mansoori et al. 2021; Young and Narita 2007). Overexpression of CircFBXL5 and downregulation of miR-216b have been observed in breast cancer, particularly in MDA-MB-231 and MCF7-resistant cells. Knockdown of CircFBXL5 or the introduction of miR-216b mimics can restore sensitivity to 5-FU in MDA-MB-231-resistant cells. Furthermore, the direct interaction between miR-216b and HMGA2 has been demonstrated, along with the inhibitory effects of CircFBXL5 on miR-216b expression (Zhu et al. 2021). CircFAT1, through the miR-525-5p/SKA1 axis, plays a critical role in the development of oxaliplatin (OXA) resistance in breast cancer. CircFAT1 is significantly upregulated in OXA-resistant breast cancer patients, and its knockdown reduces resistance in cancer cell lines. Mechanistically, CircFAT1 acts as a sponge for miR-525, leading to increased expression of SKA1, a direct target of miR-525. SKA1 has been shown to activate the Notch and WNT signaling pathways, contributing to the resistance phenotype in various cancers (Yao et al. 2021b). Monastrol-resistant breast cancer cells have also been observed, with studies suggesting that circRNAMTO1 (hsa_circRNA-007874) is overexpressed in these cells upon monastrol treatment. This circular RNA is believed to interact with tumor necrosis factor receptor-associated factor 4 (TRAF4) and suppress the protein levels of EG5. However, the precise role of circRNAMTO1 in resistance cells is yet to be fully understood (Liu et al. 2018). Trastuzumab is a monoclonal antibody targeting HER2, commonly used in the treatment of HER2-positive breast cancer. Despite being one of the most effective therapies for

HER2-positive breast cancer and a key component of many combination treatments, a subset of patients with HER2 overexpression exhibit varying degrees of resistance to this targeted therapy (Maximiano et al. 2016). The hsa_circ_0001598/miR-1184/PD-L1 signaling pathway is suggested to play a significant role in trastuzumab resistance in breast cancer. Huang and colleagues demonstrated that circ_0001598 is overexpressed in breast cancer, particularly in trastuzumab-resistant patients. Additionally, miR-1184 was identified and validated as a target of circ_0001598, leading to the upregulation of PD-L1, a direct target of circ_0001598. Although the precise mechanism driving trastuzumab resistance in these cells remains unclear, there is evidence of immune evasion, particularly through the escape of CD8+ T cells (Huang et al. 2021a). Apart from hsa_circ_0001598, exosome-mediated transfer of circHIPK3 from trastuzumab-resistant cells can induce resistance in susceptible cells by sponging various microRNAs (Zhang et al. 2021a). Lapatinib, a kinase inhibitor, serves as another example of resistance in cells targeting HER2, as well as EGFR (HER1). It works by inhibiting the phosphorylation of HER1 and HER2 receptors (Kute et al. 2004). A significant proportion of patients treated with lapatinib for advanced-stage breast cancer develop resistance to the drug. Studies have shown that circ-MMP11 contributes to this resistance by sponging miR-153, leading to the upregulation of ANLN, a key gene involved in cell division and metastasis. Notably, circ-MMP11 can be transferred via exosomes, which may play a role in metastasis and the formation of the tumor microenvironment (Wang et al. 2019). Trastuzumab is a targeted therapy for treating HER2-positive breast cancer, yet drug resistance continues to pose a significant challenge. The circular RNA hsa_circ_0001598 contributes to trastuzumab resistance by promoting immune escape through the upregulation of programmed death-ligand 1 (PD-L1). This is achieved by sponging miR-1184 in breast cancer cells (Huang et al. 2021a).

8.3 Impact of Circular RNA/miRNA/mRNA Axis on Drug Resistance of Gastrointestinal (GI) Tract

Knockdown of circ_0000338 has been shown to enhance the sensitivity of 5-fluorouracil (5-FU)-resistant colorectal cancer (CRC) cells to 5-FU by facilitating apoptosis and suppressing cell proliferation through the modulation of miR-217 and miR-485-3p (Zhao et al. 2021). Exosome-derived circ-FBXW7 restores oxaliplatin (OX) sensitivity in resistant cells by promoting OX-induced apoptosis, inhibiting epithelial-mesenchymal transition (EMT), and reducing drug efflux. This effect is mediated through its function as a sponge for miR-18b-5p (Zang et al. 2021). Conversely, hsa_circ_0005963 is delivered via exosomes from oxaliplatin (OX)-resistant colorectal cancer (CRC) cells to OX-sensitive cells, promoting glycolytic activity and drug resistance. This occurs through the upregulation of PKM2 expression via suppression of miR-122 (Amelimojarad et al. 2022). Recent findings indicate that circ_0020095 contributes to cisplatin resistance by functioning as a sponge for miR-487a-3p, thereby upregulating the expression of SOX9 (Sun et al. 2021). In doxorubicin (DOX)-resistant colorectal cancer (CRC) cells, silencing of circCSPP1 enhances sensitivity to DOX, suppresses malignant phenotypes, and promotes apoptosis via modulation of the miR-944/FZD7 signaling axis (Xi et al. 2021). In esophageal cancer cells, cisplatin resistance is promoted by exosomal circ_0000337, which modulates the miR-377-3p/JAK2 signaling pathway (Zang et al. 2021). Furthermore, silencing of circ_0006168 or its downstream effector JMJD1C significantly suppresses cell proliferation, invasion, and migration while enhancing apoptosis, thereby increasing the sensitivity of esophageal squamous cell carcinoma (ESCC) cells to taxol in vitro (Qu et al. 2021). This mechanism may lead to elevated expression of MDR-1 protein, thereby enhancing the resistance of MGC-803/MTA gastric cancer cells to Pemetrexed through regulation of the miR-194-5p/JMJD1C axis (Qu et al. 2021). CircFN1 promotes cisplatin resistance in gastric cancer by acting as a molecular sponge for miR-182-5p, thereby attenuating its tumor-suppressive effects (Huang et al. 2021b). Exosome-derived circ-PVT1 modulates invasion, autophagy, and apoptosis, thereby facilitating cisplatin resistance in gastric cancer (GC) cells through the miR-30a-5p/YAP1 signaling axis, highlighting its potential as a promising therapeutic target in GC (Yao et al. 2021c). Furthermore, silencing of circ_ASAP2 or hsa_circ_0081143 enhances cisplatin sensitivity in gastric cancer (GC) by acting as competing endogenous RNAs (ceRNAs), thereby suppressing the progression of GC (Xue et al. 2019). It has been found that circAKT3 plays a role in cisplatin (CDDP) resistance in gastric cancer (GC) by promoting DNA damage repair mechanisms and inhibiting apoptosis in GC cells (Huang et al. 2019). CircCCDC66 and circDONSON have been shown to contribute to cisplatin (CDDP) resistance in gastric cancer (GC) by regulating specific miRNAs and their downstream target genes (Liu et al. 2022a). All the aforementioned studies highlight that certain circRNAs are overexpressed in cisplatin (CDDP)-resistant gastric cancer (GC), potentially serving as biomarkers for poor prognosis. In contrast, Zhang et al. demonstrated that the upregulation of circ_0001017 inhibited malignant biological behaviors in GC and enhanced CDDP sensitivity in resistant GC cells, partially through the miR-543/PHLPP2 signaling axis (Hao et al. 2021). Similarly, circCUL2 functions as a tumor suppressor and increases cisplatin (CDDP) sensitivity in gastric cancer by regulating miR-142-3p/ROCK2-mediated autophagy (Peng et al. 2020). Exosomal circ_0032821 promotes oxaliplatin (OXA) resistance in oxaliplatin-sensitive gastric cancer (GC) cells by sponging miR-515-5p, leading to the upregulation of SOX9 expression (Zhong et al. 2021a). Gastric cancer presents a major clinical challenge, and despite the introduction of anti-PD-1 therapy, most patients do not achieve favorable outcomes. In a study by Chen et al., upregulation of circDLG1 was found to be significantly correlated with an aggressive tumor phenotype and poor prognosis in gastric cancer patients undergoing anti-PD-1 therapy. Mechanistically, circDLG1 upregulates CXCL12 expression by sponging miR-141-3p. The increased CXCL12 levels contribute to the expansion of myeloid-derived suppressor cells (MDSCs) and reduced infiltration of CD8+ T cells into the tumor microenvironment (TME). Notably, knocking down CXCL12 enhanced the sensitivity of gastric cancer to anti-PD-1 therapy, highlighting the critical role of circDLG1 in mediating immunotherapeutic resistance (Chen et al. 2021a). In colorectal cancer (CRC) patients, upregulation of circQSOX1 has been shown to contribute to resistance against anti-CTLA-4 treatment. Mechanistically, circQSOX1 acts as a sponge for miR-326 and miR-330-5p, leading to the upregulation of phosphoglycerate mutase 1 (PGAM1), which promotes glycolysis in CRC cells. Further studies indicate that knockdown of circQSOX1 reduces the infiltration of immunosuppressive Treg cells, thus remodeling the tumor microenvironment (TME) and driving immunotherapy resistance (Liu et al. 2022b).

8.4 CircRNAs Deregulation Regulates HCC and Beyond

CircPTGR1 and ABCC1 are significantly overexpressed in hepatocellular carcinoma (HCC) cells, with circPTGR1 playing a key role in modulating 5-fluorouracil (5-FU) resistance in HCC via the miR-129-5p/ABCC1 axis. Additionally, knockdown of circPTGR1 and circARNT2 has been shown to induce apoptosis in HCC cells and inhibit resistance to 5-FU and cisplatin (CDDP), respectively (Liu et al. 2022a). Moreover, circRNA, acting as a competing endogenous RNA (ceRNA), suppresses the activity of miR-326, leading to the upregulation of RUNX2 expression. This, in turn, enhances cisplatin (CDDP) resistance in cancer cells (Guan et al. 2020). Two studies have explored the mechanism of circRNA-SORE in sorafenib resistance in hepatocellular carcinoma (HCC) from different perspectives. One study reveals that circRNA-SORE inhibits PRP19-mediated YBX1 degradation, thereby impacting the expression of downstream gene targets of YBX1. These targets include key signaling molecules such as ERK, AKT, c-Myc, Raf1, and TGF-α1, all of which are involved in regulating cell survival, proliferation, and drug resistance (Xu et al. 2020a). Another study demonstrates that circRNA-SORE acts as a microRNA sponge by interacting with miR-103a-2-5p and miR-660-3p. This interaction subsequently activates the Wnt/β-catenin signaling pathway, contributing to the development of sorafenib resistance in hepatocellular carcinoma (HCC) cells (Xu et al. 2020b). Huang et al. further demonstrate that the overexpression of circMET facilitates the progression of hepatocellular carcinoma (HCC) and promotes immune tolerance via the miR-30-5p/Snail/DPP4/CXCL10 signaling axis. This pathway is involved in regulating tumor growth and immune evasion, contributing to the aggressive behavior of HCC cells (Huang et al. 2020a). Another example is circMTO1, which suppresses liver cancer progression by acting as a sponge for miR-9. This interaction leads to the upregulation of the tumor suppressor p21, contributing to the inhibition of liver cancer cell growth (Liu and Zhang 2018). Liver cancer, a major global malignancy, is primarily represented by hepatocellular carcinoma (HCC), accounting for approximately 90% of all liver cancer cases. HCC is often diagnosed at advanced stages, and while current therapies provide some survival benefits, the persistent challenge of immunotherapy resistance remains due to the immunosuppressive tumor microenvironment (TME) (Gu et al. 2024; Xie et al. 2022). In hepatocellular carcinoma (HCC) cells, elevated expression of circMET is associated with increased invasion, metastasis, and resistance to immunotherapy. Mechanistically, circMET sponges miR-30-5p, leading to the upregulation of Snail and its downstream target DPP4. This interaction promotes immunosuppression through the miR-30-5p/Snail/DPP4 axis, resulting in the degradation of CXCL10. Notably, the DPP4 inhibitor Sitagliptin enhances the efficacy of anti-PD-1 therapy, suggesting a potential therapeutic strategy for overcoming immunotherapy resistance in HCC (Huang et al. 2020a). In hepatocellular carcinoma (HCC), resistance to anti-PD-1 treatment can be triggered through interactions between the tumor and macrophages. Exosomal circTMEM181 in the tumor microenvironment (TME) upregulates the expression of CD39 in macrophages, which activates the ATP-adenosine pathway and induces immunosuppression. Targeting the adenosine pathway can potentially enhance the effectiveness of immune checkpoint blockade (ICB) therapy, offering a strategy to overcome resistance in HCC (Lu et al. 2021). Similarly, exosome-delivered circUHRF1 induces immunosuppressive effects in hepatocellular carcinoma (HCC) by upregulating TIM3 expression in natural killer (NK) cells. This upregulation leads to NK cell exhaustion and suppresses the secretion of key cytokines like IFN-α and TNF-α, ultimately contributing to HCC resistance to anti-PD-1 therapy. The molecular mechanism behind this involves circUHRF1-mediated degradation of miR-449c-5p, which results in increased TIM3 expression in NK cells, thereby promoting immunosuppression and resistance to immunotherapy (Xu et al. 2023). Exosome-derived circCCAR1 targets miR-127-5p to upregulate WTAP in activated T cells, leading to CD8+ T cell dysfunction. This dysfunction promotes hepatocellular carcinoma (HCC) resistance to anti-PD-1 therapy, suggesting that circCCAR1 plays a critical role in mediating immune evasion and therapy resistance through the regulation of miR-127-5p and WTAP in T cells (Hu et al. 2023b). High levels of CircRHBDD1 are linked to poor prognosis and decreased sensitivity to anti-PD-1 therapy in patients. Inhibition of CircRHBDD1 enhances the effectiveness of anti-PD-1 treatment by interacting with YTHDF1, which promotes the translation of PIK3R1, suggesting that CircRHBDD1 plays a role in modulating immune response and resistance to immunotherapy through the regulation of the PIK3R1 pathway (Cai et al. 2022). CircSOD2, which is highly expressed in HCC tissues, is associated with reduced CD8+ T cell activity in the tumor microenvironment (TME). It contributes to resistance to anti-PD-1 therapy by sponging miR-497-5p, leading to the upregulation of ANXA11. This interaction ultimately impairs the effectiveness of anti-PD-1 treatment by modulating immune responses in the TME (Ye et al. 2023).

8.5 Glioma as a Target of Circular/miRNA/mRNA Axis

CircRNA and Glioma Drug Resistance: Glioma represents the most prevalent type of aggressive brain tumor, known for its high mortality rate, propensity for metastasis, and substantial challenges related to drug resistance (Rahnama et al. 2021). In this study, we summarize the role of circRNAs in chemoresistance in glioma patients. It has been demonstrated that exosomal hsa_circ_0042003, transported via heparanase, transfers from temozolomide (TMZ)-resistant glioma cells to drug-sensitive cells, thereby contributing to TMZ resistance in glioma (Si et al. 2021). In glioma, exosomal circ _0072083 regulates NANOG and ALKBH5 by sponging miR-1252-5p and inducing demethylation, thereby contributing to temozolomide (TMZ) resistance (Ding et al. 2021). Hsa _circ _0110757 and circ_0005198 promote temozolomide (TMZ) resistance and suppress glioma cell apoptosis by modulating the miR-1298-5p/ITGA1 and miR-198/TRIM14 axes, respectively (Li et al. 2021c; Deng et al. 2020). In line with these findings, the downregulation of circ-VPS18, hsa_circ_0000936, circHIPK3, and circCEP128 enhances temozolomide (TMZ) sensitivity and inhibits glioma progression (Li et al. 2021d; Hua et al. 2019). The key finding is that the downregulation of circHIPK3 inhibits the PI3K/AKT signaling pathway, partially through the miR-524-5p/KIF2A axis (Yin and Cui 2021). Similarly, in TMZ-resistant glioma, exosomal circ-HIPK3 can regulate the miR-421/ZIC5 axis, promoting cell progression and contributing to TMZ resistance (Han et al. 2021). Overexpression of CircASAP1 enhances glioblastoma cell proliferation and temozolomide (TMZ) resistance through the circASAP1/miR-502-5p/NRAS regulatory network, suggesting that CircASAP1 could serve as a potential therapeutic target for TMZ-resistant glioblastoma (Wei et al. 2021a). Downregulation of circ_0008344 inhibits glioma growth and enhances radiosensitivity by regulating the miR-433-3p/RNF2 axis (Di et al. 2021). Downregulation of hsa_circ_0076248 or upregulation of its target miR-181a not only inhibited cell proliferation and migration but also significantly enhanced the sensitivity of glioma cells to TMZ. Additionally, hsa_circ_0110757 was found to be highly expressed in TMZ-resistant glioma cells, and its knockdown improved chemosensitivity to TMZ by targeting the miR-1298-5p/ITGA1 axis (Li et al. 2021c; Lei et al. 2019). CircATP8B4 and circCPA4 are upregulated in glioma and decrease radiation sensitivity by acting as sponges for miR-766 and miR-760, respectively. Studies have also demonstrated that circ_VCAN is more highly expressed in radioresistant glioma cells than in sensitive ones. Moreover, overexpression of circ_VCAN promotes carcinogenesis and diminishes the radiosensitivity of gliomas by regulating miR-1183 (Katsushima et al. 2023). Exosomal circWDR62 contributes to temozolomide resistance and malignant progression in glioma by regulating the miR-370-3p/MGMT axis (Geng et al. 2022).

8.6 CircRNAs and Osteosarcoma Drug Resistance

Knockdown of Circ_0081001 enhances methotrexate (MTX) sensitivity in osteosarcoma cells by suppressing miR-494-3p, which in turn reduces the upregulation of TGM2 (Wei et al. 2021b). Similarly, hsa_circ_0000073 enhances malignant behaviors such as proliferation, invasion, and migration and contributes to methotrexate (MTX) resistance in osteosarcoma cells by sponging miR-145-5p and miR-151-3p, thereby upregulating NRAS expression (Li et al. 2020b). DOX resistance in osteosarcoma is promoted by hsa_circ_0004674 through the Wnt/β-catenin pathway, which is regulated by the miR-342-3p/FBN1 axis (Bai et al. 2021). Additionally, several circRNAs, including circ_0001721, circSAMD4A, circ_0002060, and circ_0003496, enhance DOX resistance and promote the progression of osteosarcomas (Ji et al. 2023; Xie et al. 2020). Zhu et al. reported that the resistance to DOX and CDDP in osteosarcoma cells can be reduced by inhibiting circPVT1 expression (Kun-Peng et al. 2018). It has been reported that the knockdown of circ-CHI3L1.2 promotes apoptosis and reduces resistance in CDDP-resistant osteosarcoma cells (Zhang et al. 2021b). Additionally, exosomal hsa_circ_103801 can enhance the role of exosomes in promoting chemoresistance of osteosarcoma cells to CDDP (Pan et al. 2021). Similarly, circTADA2A can sponge miR-129-5p, which competes with TRPS1 and YAP1, thereby influencing osteosarcoma cell proliferation and resistance to CDDP (Zhang et al. 2020a).

8.7 Ovarian Urinary System Drug Resistance and the Role of Circular/miRNA Axis

Ovarian cancer is a prevalent gynecologic malignancy, and traditional treatment options are often hindered by chemoresistance (Stewart et al. 2019). Both circ_C20orf11 and exosomal circFoxp1 contribute to CDDP resistance in ovarian cancer cells (Yin et al. 2021; Luo and Gui 2020). In contrast, Cdr1as enhances sensitivity to CDDP in ovarian cancer by regulating the miR-1270/SCAI axis (Zhao et al. 2019). The downregulation of circRNA_0000735 and upregulation of circRNA_0006404 can inhibit the expression of p-GP, leading to tolerance to DTX treatment (Chen and Tai 2022). Two studies have shown that circ_CELSR1 is upregulated in paclitaxel (PTX)-resistant ovarian cancer cells, and silencing circ_CELSR1 alleviated PTX resistance in ovarian cancer in vivo (Wei et al. 2021c; Zhang et al. 2020b). One study also found that inhibiting circCELSR1 led to G0/G1 phase cell cycle arrest and promoted apoptosis in ovarian cancer cells (Zhang et al. 2020b). Similarly, circTNPO3 enhances PTX resistance in ovarian cancer cells by sponging miR-1299, which upregulates NEK2 expression (Xia et al. 2020). Research on circRNAs in drug resistance of urinary system tumors is still emerging. One study suggests that the AR-mediated ADAR2/circFNTA/miR-370-3p/FNTA pathway influences KRAS, which in turn alters bladder cancer cell invasion and chemosensitivity to CDDP (Chen et al. 2020a). Hypoxia-induced overexpression of circELP3 enhances CDDP resistance in bladder cancer cells (Su et al. 2019). Circ0008399 enhances the formation of the WTAP/METTL3/METTL14 m6A methyltransferase complex, leading to increased TNFAIP3 expression in an m6A-dependent manner. Moreover, elevated levels of circ0008399 and WTAP are associated with poor prognosis in bladder cancer patients (Wei et al. 2021d). A study indicates that hsa_circ_0000285 is expressed at significantly lower levels in bladder cancer tissues and in CDDP-resistant patients. However, the specific mechanism by which hsa_circ_0000285 influences chemosensitivity requires further investigation (Chi et al. 2019). Additionally, hsa_circ_0035483 plays a role in GEM-induced autophagy and enhances resistance of renal clear cell carcinoma (RCC) to GEM by regulating the hsa-miR-335/CCNB1 axis (Yan et al. 2019). A recent study reveals that the downregulation of circFAM13B expression in bladder cancer (BCa) correlates with a poorer prognosis and resistance to anti-PD-1 treatment. Mechanistically, circFAM13B interacts with IGF2BP1 via its K homology 3–4 (KH3–4) domains to inhibit the stability of PKM2 mRNA, thereby suppressing glycolysis, a process involving PKM2. When circFAM13B is decreased, glycolysis is not properly inhibited, leading to an immunosuppressive tumor microenvironment (TME). In HuNOG mouse models with BCa tumors, overexpression of circFAM13B enhances CD8+ T cell activity and improves sensitivity to anti-PD-1 therapy (Lv et al. 2023). The downregulation of circMGA reduces the therapeutic sensitivity to anti-PD-1 treatment. CircMGA forms an RNA-protein complex by binding with HNRNPL, which leads to the upregulation of CCL5. This increase in CCL5 enhances the infiltration of CD8+ T cells, thereby improving the immunotherapeutic effect of anti-PD-1 (Sun et al. 2023).

8.8 Lung Cancer as the Most Affected Cancer Drug Resistance by Circular RNAs

According to the global cancer statistics of 2020, lung cancer continues to be one of the leading causes of cancer-related deaths worldwide (Sung et al. 2021). CDDP (cisplatin) is one of the most effective and widely used anticancer drugs in clinical practice; however, the development of resistance to CDDP significantly impedes its therapeutic efficacy in cancer treatment (Shen et al. 2012). The involvement of circRNAs in drug resistance in lung cancer is summarized in Table 8.1. Increasing evidence suggests that circRNAs play a significant role in CDDP resistance in lung cancer. For instance, the knockdown or silencing of specific circRNAs, including circ-RNF121, circ_0076305, circRNA_103615, circ_0008928, hsa_circRNA_103809, circ-PRMT5, and circ_0007385, has been shown to suppress CDDP resistance in non-small cell lung cancer (NSCLC) (Liu et al. 2022c; Liang et al. 2021). Additionally, circAKT3 disrupts the balance of glycolysis and enhances CDDP resistance, whereas circRNA-FOXO3 accelerates glycolysis and promotes CDDP sensitivity in lung cancer cells (Xu et al. 2020c; Zhang et al. 2021c). Both hsa_circ_0014235 and hsa_circ_0096157 overexpression enhances CDDP resistance and promotes the malignancy of NSCLC cells, including increased proliferation, migration, and invasion (Xu et al. 2020d; Lu et al. 2020). CircRNA_100565 is associated with poor prognosis in NSCLC, and its deletion reduces CDDP resistance. This effect can be reversed by inhibiting miR-377-3p or overexpressing ADAM28 (Zhong et al. 2021b). Circ_PIP5K1A modulates CDDP sensitivity in NSCLC through the miR-493-5p/ROCK1 axis, while circ_0085131 contributes to

Table 8.1 Summary of circular RNA/miRNA/mRNA axis in drug resistance

Circular RNA	miRNA	mRNA
1. Breast cancer		
circ_0085495	miR-873-5p	integrin β1
hsa_circ_0092276	miR-348	ATG7
circ_0006528	miR-1299/miR-1236-3p	CHD4/CDK8
circ_0001667	miR-4458	NCOA3
circUBE2D2	miR-200a-3p (indirect)	E-CAD/Erα/VIMENTIN
hsa_circ_0025202	miR-182-5p/miR-197-3p	FOXO3/HIPK3
circBMPR2	miR-553	USP4
circGFRA1	miR-361-5p	TLR4
circ-RNF111	miR-140-5p	E2F3
circ-ABCB10	let-7a-5p	DUSP7
CDR1as	miR-7	CCNE1
circFBXL5	miR-216b	HMGA2
circFAT1	miR-525-5p	SKA1
circ_0001598	miR-1184	PD-L1
circ-MMP11	miR-153	ANLN
2. Gastrointestinal cancers Colorectal cancer (CRC):		
circ_0000338	miR-217/miR-485-3p	AEG-1/TPX2 (putative)
Circ-FBXW7	miR-18b-5p	mTOR/ZEB1 (suggested)
circ_0005963	miR-122	PKM2
circ_0020095	miR-487a-3p	SOX9
circCSPP1	miR-944	FZD7
Esophageal Cancer: Esophageal cancer:		
circ_0000337	miR-377-3p	JAK2
circ_0006168	miR-194-5p	JMJD1C
Gastric cancer (GC):		
circFN1	miR-182-5p	BCL2/PTEN
circ-PVT1	miR-30a-5p	YAP1
circ_ASAP2 hsa_circ_0081143	Unspecified miRNAs	CDK6
circAKT3	miR-198	PIK3R1/DNA damage repair (CDDP resistance)
circCCDC66/ circDONSON	Unspecified miRNAs(miR-618-suggested)	BCL2
circ_0001017	miR-543	PHLPP2
circCUL2	miR-142-3p	ROCK2
circ_0032821	miR-515-5p	SOX9
circDLG1	miR-141-3p	CXCL12
circQSOX1	miR-326/miR-330-5p	PGAM1
3. Hepatocellular carcinoma (HCC):		
circPTGR1	miR-129-5p	ABCC1

(continued)

Table 8.1 (continued)

Circular RNA	miRNA	mRNA
circARNT2	–	–
circRNA-SORE	miR-103a-2-5p /miR-660-3p	Wnt/β-catenin
circMET	miR-30-5p	Snail/DPP4/CXCL10
circMTO1	miR-9	p21
circTMEM181	miR-488-3p	CD39 ATP-adenosine pathway (CD39 in macrophages)
circUHRF1	miR-449c-5p	TIM3 (NK cells)
circCCAR1	miR-127-5p	WTAP
CircRHBDD1	YTHDF1	PIK3R1
CircSOD2	miR-497-5p	ANXA11
4. Glioma		
circ_0042003	Exosomal, miR unspecified	–
circ_0072083	miR-1252-5p	NANOG/ALKBH5
circ_0110757	miR-1298-5p	ITGA1
circ_0005198	miR-198	TRIM14
circ-VPS18/ circ_0000936 circHIPK3/ circ_CEP128	miR-1229-3p/ miR-124, miR-654/ miR-199a-5p	BCAT1 CCND2, IGF2BP3 CDK4
circASAP1	miR-502-5p	NRAS
circ_0008344	miR-433-3p	RNF2
circATP8B4/ circCPA4	miR-766/miR-760/	Nestin
circ_VCAN	miR-1183 let-7 miRNA	PD-L1
circWDR62	miR-370-3p	MGMT
5. Osteosarcoma		
circ_0081001	miR-494-3p	TGM2
hsa_circ_0000073	miR-145-5p/miR-151-3p	NRAS
hsa_circ_0004674	miR-342-3p	FBN1
circ_0001721 circSAMD4A circ_0002060 circ_0003496	Various miRs)miR-372-3p، miR-569، miR-599, miR-1244، miR-342-3p, miR-198, miR-370)	MDM2، FZD7 ABCB1 KLF12
circPVT1	miR-423-5p، miR-526b، miR-205-5p، miR-137	Wnt5a/Ror2، FOXC2، c-FLIP، TRIAP1
circ-CHI3L1.2	miR-340-5p	LPAATβ
circ_103801	Unspecified	Unspecified
circTADA2A	miR-129-5p	TRPS1/YAP1
6. Ovarian and urinary system cancers		
circ_C20orf11/ circFoxp1	miR-22 ¸ miR-150-3p miR-527	CEBPG, FMNL3,YWHAZ
Cdr1as	miR-1270	SCAI

(continued)

Table 8.1 (continued)

Circular RNA	miRNA	mRNA
circRNA_0000735 circRNA_0006404	miR-526b, miR-346	DKK4 (p-GP related) DKK3
circ_CELSR1	miR-149-5p PTX resistance	SIK2 (salt-inducible kinase 2)
circTNPO3	miR-1299	NEK2
circFNTA	miR-370-3p	FNTA
circELP3	Unspecified	Unspecified
circ0008399	Unspecified	TNFAIP3
hsa_circ_0035483	miR-335	CCNB1
circFAM13B	miR-212	PKM2 mRNA stability E2F5
circMGA	HNRNPL	CCL5
7. Lung cancer:		
circ-RNF121 circ_0076305 circRNA_103615 circ_0008928 hsa_circRNA_103809 circ-PRMT5 circ_0007385	miR-646 miR-296-5p miR-338-3p miR-4302 , miR-377-3p miR-377, miR-382, miR-498 , miR-4458 miR-493-3p	SOX4 STAT3 RAB22A EZH2, REV3L ZNF121, GOT1 ABCB1
circAKT3	–	Glycolysis (CDDP resistance)
circ_0001821 circ_ZFR	miR-101-3p	CUL4B GRK5 KPNA4
circ_0011292	miR-379-5p	TRIM65
circ_0002874	miR-1273f	MDM2/P53
circ_0030998 circ_0002483	miR-558 miR-182-5p/ miR-125a-3p	MMP1, MMP17 CCL4
circ_0014130	miR-545-3p	YAP1
circ_0003998	miR-136-5p	CORO1C
circ_0005909	miR-338-3p	SOX4
circASK1	–	ASK1-dependent apoptosis
circSETD3	miR-520 h	ABCG2
circRNA_102481	miR-30a-5p	ROR1
circFGFR1 circIGF2BP3 circCELF1 circASCC3 circZNF451	miR-381-3p miR-328-3p/miR-3173-5p miR-636 miR-432-5p	Immune checkpoints

drug resistance by targeting miR-654-5p and upregulating ATG7 expression (Feng et al. 2021; Kong 2020). The sensitivity of CDDP in lung cancer cells can be enhanced by the knockdown of circABCB10 (Wu et al. 2020). However, overexpression of circ_0000079 disrupts the formation of the FXR1/PRCKI complex by regulating FXR1, thereby inhibiting cell invasion and reducing CDDP resistance in NSCLC (Chen et al. 2020b). Paclitaxel (PTX), a widely utilized chemotherapeutic agent for various solid tumors including lung cancer, exhibits increased efficacy in non-small cell lung cancer (NSCLC) upon silencing of circ_0001821 and circ_ZFR, which suppress cellular proliferation and metastasis while enhancing PTX sensitivity and apoptosis through the downregulation of GRK5 and KPNA4 via miRNA sponging mechanisms (Liu et al. 2021a; Li et al. 2021e). Similarly, circ_0011292 enhances paclitaxel resistance and promotes tumor progression in NSCLC by modulating the miR-379-5p/TRIM65 regulatory pathway (Guo et al. 2022). Downregulation of hsa_circ_0002874 has been shown to overcome paclitaxel resistance in NSCLC and promote apoptosis through modulation of the miR-1273f/MDM2/p53 signaling axis (Xu et al. 2021). Unlike the aforementioned circRNAs, overexpression of circ_0030998 and circ_0002483 suppresses the malignant phenotype of lung cancer cells and increases their paclitaxel sensitivity by targeting miR-558 and miR-182-5p, respectively (Li et al. 2019, 2021f). Du et al. investigated a distinct circ_0014130/miR-545-3p/YAP1 signaling axis involved in regulating chemoresistance and malignant progression in docetaxel (DTX)-resistant non-small cell lung cancer (NSCLC) cells (Du et al. 2021). Knockdown of circ_0003998 inhibits colony formation and enhances apoptosis and DTX sensitivity by acting as a sponge for miR-136-5p, thereby regulating the expression of CORO1C in docetaxel (DTX)-resistant non-small cell lung cancer (NSCLC) cells (Zhang et al. 2021d). Knockdown of circ_0005909 has been shown to inhibit the proliferation, metastasis, and drug resistance of non-small cell lung cancer (NSCLC) cells (Song et al. 2021). Additionally, a novel protein encoded by circASK1 activates ASK1-dependent apoptosis, which facilitates to alleviate gefitinib resistance in lung adenocarcinoma (LUAD) (Wang et al. 2021b). Studies have demonstrated that circSETD3 overexpression diminishes gefitinib sensitivity by interacting with miR-520 h and ABCG2, ultimately reducing the intracellular concentration of gefitinib (Huang et al. 2020b). Exosomes are small extracellular vesicles that play crucial roles in immunity, signal transduction, tumor progression, and the development of drug resistance (Pegtel and Gould 2019). Yang et al. proposed that tumor-derived exosomal circRNA_102481 contributes to EGFR-TKI resistance in NSCLC by acting as a sponge for microRNA-30a-5p, thereby regulating ROR1 (Yang et al. 2021). In the context of immune checkpoint inhibitors, hsa_circ_0003222 and circFGFR1 have been shown to enhance resistance of NSCLC to anti-PD-L1 and anti-PD-1 therapies, respectively (Li et al. 2021g; Zhang et al. 2019). Additionally, the inhibition of circIGF2BP3/PKP3 enhances the therapeutic efficacy of anti-PD-1 treatment in a Lewis LADC mouse model (Liu et al. 2021b). CircRNAs play a crucial role in reducing immunotherapy sensitivity in lung cancer by engaging in complex molecular mechanisms. A key example is CircIGF2BP3 in NSCLC, which is linked to poor prognosis. It reduces CD8+ T cell infiltration and increases PKP3 expression by sponging miR-328-3p and miR-3173-5p. This action inhibits PD-L1 degradation, thereby promoting immune evasion and enhancing resistance to immunotherapy (Liu et al. 2021b). CircCELF1 impedes T cell recruitment to the tumor microenvironment (TME) and hinders the effectiveness of anti-PD-1 therapy by sponging miR-491-5p. This interaction results in the upregulation of EGFR expression, contributing to immune evasion and reduced therapeutic sensitivity (Ge et al. 2021). CircASCC3 activates C5a, which in turn remodels the immunosuppressive tumor microenvironment (TME), leading to the development of resistance to anti-PD-1 therapy. This process contributes to immune evasion and reduces the effectiveness of immunotherapy (Gao et al. 2022a). CircFGFR1 exerts immunosuppressive effects by sponging miR-381-3p, which leads to a reduction in CD8+ T cell infiltration in the tumor microenvironment. This contributes to immune evasion and hinders the effectiveness of immune-based therapies (Zhang et al. 2019), Hsa_circ_0003222, hsa_circ_0020714, and hsa_circ_0000190 promote proliferation, enhance stemness, and contribute to resistance against anti-PD-1 therapy in NSCLC by modulating key molecular pathways within the tumor microenvironment (Li et al. 2021g; Luo et al. 2021; Wu et al. 2022). Exosomal CircUSP7 impairs CD8+ T cell function, thereby promoting resistance to immunotherapy by disrupting immune cell activation and function within the tumor microenvironment (Chen et al. 2021b). In lung adenocarcinoma (LUAD), circRUNX1 modulates antigen presentation by acting as a sponge for miR-4739, potentially altering the immune response and contributing to the tumor's ability to evade immune surveillance (Zhang et al. 2021e), Circ_0004140 diminishes the effectiveness of anti-PD-1 therapy by sponging miR-1184, which leads to the upregulation of CCL22 expression and subsequent recruitment of Treg cells, contributing to immune suppression in the tumor microenvironment (Liu et al. 2022d). Exosomal circZNF451 plays a critical role in reshaping the tumor-immune microenvironment by modulating macrophage polarization via the FXR1-ELF4-IRF4 axis. This makes it a potential biomarker for predicting the response to PD-1 blockade in lung adenocarcinoma (LUAD) (Gao et al. 2022b). Additionally, circHMGB2 impairs the efficacy of PD-1 blockade by upregulating CARM1 through its interaction with miR-181a-5p. This process inhibits the type 1 IFN response, ultimately desensitizing tumor cells to immune attacks mediated by cytotoxic T cells (Zhang et al. 2022).

Acknowledgments We would like to thank all authors responsible for the insights that we attempted to summarize.

Author Contributions MHG is responsible for conceiving the idea, and writing the chapter. AM and PA contributed to editing and revising the chapter. SB supervised the process.

Funding This chapter did not receive any specific grant from funding agencies in the public, commercial, or not-for-profit sectors.

Data Availability All of the data has been presented in the manuscript.

Competing Interests The authors declare that they have no competing interests related to the content of this chapter.

Ethics Approval This chapter does not involve any studies with human participants or animals performed by any of the authors. Therefore, ethics approval was not required.

References

Almouh M, Razmara E, Bitaraf A, Ghazimoradi MH, Hassan ZM, Babashah S (2022) Circular RNAs play roles in regulatory networks of cell signaling pathways in human cancers. Life Sci [Internet] 309:120975. Available from: https://www.sciencedirect.com/science/article/pii/S0024320522006750

Amelimojarad M, AmeliMojarad M, Nazemalhosseini-Mojarad E (2022) Exosomal noncoding RNAs in colorectal cancer: an overview of functions, challenges, opportunities, and clinical applications. Pathol Res Pract 238:154133

Bai R, Chen N, Li L, Du N, Bai L, Lv Z et al (2020) Mechanisms of cancer resistance to immunotherapy. Front Oncol. Frontiers Media S.A. 10:1290

Bai Y, Li Y, Bai J, Zhang Y (2021) Hsa_circ_0004674 promotes osteosarcoma doxorubicin resistance by regulating the miR-342-3p/FBN1 axis. J Orthop Surg Res 16:1–12

Cai J, Chen Z, Zhang Y, Wang J, Wang J, Zhang Z et al (2022) Circrhbdd1 augments M6a-dependent metabolic reprogramming and restricts anti-PD-1 therapy in hepatocellular carcinoma. Mol Ther Oncolytics 24:755–771

Chang M (2012) Tamoxifen resistance in breast cancer. Biomol Ther 20:256–267

Chen YY, Tai YC (2022) Hsa_circ_0006404 and hsa_circ_0000735 regulated ovarian cancer response to docetaxel treatment via regulating p-GP expression. Biochem Genet 60(1):395–414

Chen J, Sun Y, Ou Z, Yeh S, Huang C, You B et al (2020a) Androgen receptor-regulated circ FNTA activates KRAS signaling to promote bladder cancer invasion. EMBO Rep 21(4):e48467

Chen C, Zhang M, Zhang Y (2020b) Circ_0000079 decoys the RNA-binding protein FXR1 to interrupt formation of the FXR1/PRCKI complex and decline their mediated cell invasion and drug resistance in NSCLC. Cell Transplant 29:0963689720961070

Chen DL, Sheng H, Zhang DS, Jin Y, Zhao BT, Chen N et al (2021a) The circular RNA circDLG1 promotes gastric cancer progression and anti-PD-1 resistance through the regulation of CXCL12 by sponging miR-141-3p. Mol Cancer 20(1):166

Chen SW, Zhu SQ, Pei X, Qiu BQ, Xiong D, Long X et al (2021b) Cancer cell-derived exosomal circUSP7 induces CD8+ T cell dysfunction and anti-PD1 resistance by regulating the miR-934/SHP2 axis in NSCLC. Mol Cancer 20:1–18

Chen NN, Di MX, Miao Z, Zhang XM, Han BY, Almaamari AA et al (2023) Doxorubicin resistance in breast cancer is mediated via the activation of FABP5/PPARγ and CaMKII signaling pathway. Front Pharmacol 14:1150861

Chi BJ, Zhao DM, Liu L, Yin XZ, Wang FF, Bi S et al (2019) Downregulation of hsa_circ_0000285 serves as a prognostic biomarker for bladder cancer and is involved in cisplatin resistance. Neoplasma 66(2):197

Cui Y, Fan J, Shi W, Zhou Z (2022) Circ_0001667 knockdown blocks cancer progression and attenuates adriamycin resistance by depleting NCOA3 via releasing miR-4458 in breast cancer. Drug Dev Res 83(1):75–87

Deng Y, Zhu H, Xiao L, Liu C, Meng X (2020) Circ_0005198 enhances temozolomide resistance of glioma cells through miR-198/TRIM14 axis. Aging (Albany NY) 13(2):2198

Di L, Zhao X, Ding J (2021) Knockdown of circ_0008344 contributes to radiosensitization in glioma via miR-433-3p/RNF2 axis. J Biosci 46:1–13

Ding C, Yi X, Chen X, Wu Z, You H, Chen X et al (2021) Warburg effect-promoted exosomal circ_0072083 releasing up-regulates NANGO expression through multiple pathways and enhances temozolomide resistance in glioma. J Exp Clin Cancer Res 40(1):164

Du D, Cao X, Duan X, Zhang X (2021) Blocking circ_0014130 suppressed drug resistance and malignant behaviors of docetaxel resistance-acquired NSCLC cells via regulating miR-545-3p-YAP1 axis. Cytotechnology 73(4):571–584

Esparza-Lopez J, Longoria O, De La Cruz-Escobar EN, Garibay-Diaz JC, Leon-Rodriguez E, De Jesús I-SM (2022) Paclitaxel resistance is mediated by NF-κB on mesenchymal primary breast cancer cells. Oncol Lett 23(2):50

Farkona S, Diamandis EP, Blasutig IM (2016) Cancer immunotherapy: the beginning of the end of cancer? BMC Med. BioMed Central Ltd 14:73

Feng N, Guo Z, Wu X, Tian Y, Li Y, Geng Y et al (2021) Circ_PIP5K1A regulates cisplatin resistance and malignant progression in non-small cell lung cancer cells and xenograft murine model via depending on miR-493-5p/ROCK1 axis. Respir Res 22:1–12

Gao J, Zhang LX, Ao YQ, Jin C, Zhang PF, Wang HK et al (2022a) Elevated circASCC3 limits antitumor immunity by sponging miR-432–5p to upregulate C5a in non-small cell lung cancer. Cancer Lett 543:215774

Gao J, Ao YQ, Zhang LX, Deng J, Wang S, Wang HK et al (2022b) Exosomal circZNF451 restrains anti-PD1 treatment in lung adenocarcinoma via polarizing macrophages by complexing with TRIM56 and FXR1. J Exp Clin Cancer Res 41(1):295

Ge W, Chi H, Tang H, Xu J, Wang J, Cai W et al (2021) Circular RNA CELF1 drives immunosuppression and anti-PD1 therapy resistance in non-small cell lung cancer via the miR-491-5p/EGFR axis. Aging (Albany NY) 13(22):24560

Geng X, Zhang Y, Lin X, Zeng Z, Hu J, Hao L et al (2022) Exosomal circWDR62 promotes temozolomide resistance and malignant progression through regulation of the miR-370-3p/MGMT axis in glioma. Cell Death Dis 13(7):596

Ghazimoradi MH, Babashah S (2022) The role of CircRNA/miRNA/mRNA axis in breast cancer drug resistance. Front Oncol. Frontiers Media S.A. 12:966083

Ghazimoradi MH, Karimpour-Fard N, Babashah S (2023) The promising role of non-coding RNAs as biomarkers and therapeutic targets for leukemia. Genes (Basel) 14(1):131

Gu XY, Huo JL, Yu ZY, Jiang JC, Xu YX, Zhao LJ (2024) Immunotherapy in hepatocellular carcinoma: an overview of immune checkpoint inhibitors, drug resistance, and adverse effects. Oncologie 26(1):9–25

Guan Y, Zhang Y, Hao L, Nie Z (2020) CircRNA_102272 promotes cisplatin-resistance in hepatocellular carcinoma by decreasing MiR-326 targeting of RUNX2. Cancer Manag Res 12:12527–12534

Guo C, Wang H, Jiang H, Qiao L, Wang X (2022) Circ_0011292 enhances paclitaxel resistance in non-small cell lung cancer by regulating miR-379-5p/TRIM65 axis. Cancer Biother Radiopharm 37(2):84–95

Han C, Wang S, Wang H, Zhang J (2021) Exosomal circ-HIPK3 facilitates tumor progression and temozolomide resistance by regulating miR-421/ZIC5 axis in glioma. Cancer Biother Radiopharm 36(7):537–548

Hanahan D (2022) Hallmarks of cancer: new dimensions. Cancer Discov. American Association for Cancer Research Inc 12:31–46

Hao J, Du X, Lv F, Shi Q (2021) Knockdown of circ_0006528 suppresses cell proliferation, migration, invasion, and Adriamycin Chemoresistance via regulating the miR-1236-3p/CHD4 Axis in breast cancer. J Surg Res [Internet] 260:104–115. Available from: https://www.sciencedirect.com/science/article/pii/S0022480420308118

Housman G, Byler S, Heerboth S, Lapinska K, Longacre M, Snyder N et al (2014) Drug resistance in cancer: an overview. Cancers. MDPI AG 6:1769–1792

Hu K, Liu X, Li Y, Li Q, Xu Y, Zeng W et al (2020) Exosomes mediated transfer of Circ_UBE2D2 enhances the resistance of breast cancer to tamoxifen by binding to MiR-200a-3p. Med Sci Monit 26:e922253–e922251

Hu S, Zhang Q, Sun J, Xue J, Wang C (2023a) Circular RNA circ_0000376 promotes paclitaxel resistance and tumorigenesis of non-small cell lung cancer via positively modulating KPNA4 by sponging miR-1298-5p. Thorac Cancer 14(22):2116–2126

Hu Z, Chen G, Zhao Y, Gao H, Li L, Yin Y et al (2023b) Exosome-derived circCCAR1 promotes CD8+ T-cell dysfunction and anti-PD1 resistance in hepatocellular carcinoma. Mol Cancer 22(1):55

Hua L, Huang L, Zhang X, Feng H, Shen B (2019) Knockdown of circular RNA CEP128 suppresses proliferation and improves cytotoxic efficacy of temozolomide in glioma cells by regulating miR-145-5p. Neuroreport 30(18):1231–1238

Huang J, Li H, Ren G (2015) Epithelial-mesenchymal transition and drug resistance in breast cancer (review). Int J Oncol. Spandidos Publications 47:840–848

Huang X, Li Z, Zhang Q, Wang W, Li B, Wang L et al (2019) Circular RNA AKT3 upregulates PIK3R1 to enhance cisplatin resistance in gastric cancer via miR-198 suppression. Mol Cancer 18:1–20

Huang XY, Zhang PF, Wei CY, Peng R, Lu JC, Gao C et al (2020a) Circular RNA circMET drives immunosuppression and anti-PD1 therapy resistance in hepatocellular carcinoma via the miR-30-5p/snail/DPP4 axis. Mol Cancer 19:1–18

Huang Y, Dai Y, Wen C, He S, Shi J, Zhao D et al (2020b) circSETD3 contributes to acquired resistance to gefitinib in non-small-cell lung cancer by targeting the miR-520h/ABCG2 pathway. Mol Ther Nucleic Acids 21:885–899

Huang L, Ma J, Cui M (2021a) Circular RNA hsa_circ_0001598 promotes programmed death-ligand-1-mediated immune escape and trastuzumab resistance via sponging miR-1184 in breast cancer cells. Immunol Res 69(6):558–567

Huang X, Zhang Q, Hu H, Jin Y, Zeng A, Xia Y et al (2021b) A novel circular RNA circFN1 enhances cisplatin resistance in gastric cancer via sponging miR-182-5p. J Cell Biochem 122(9):1009–1020

Ji Y, Liu J, Zhu W, Ji J (2023) circ_0002060 enhances doxorubicin resistance in osteosarcoma by regulating the miR-198/ABCB1 axis. Cancer Biother Radiopharm 38(9):585–595

Katsushima K, Joshi K, Perera RJ (2023) Diagnostic and therapeutic potential of circular RNA in brain tumors. Neurooncol Adv 5(1):vdad063

Kong R (2020) Circular RNA hsa_circ_0085131 is involved in cisplatin-resistance of non-small-cell lung cancer cells by regulating autophagy. Cell Biol Int 44(9):1945–1956

Kristensen LS, Hansen TB, Venø MT, Kjems J (2018) Circular RNAs in cancer: opportunities and challenges in the field. Oncogene. Nature Publishing Group 37:555–565

Kun-Peng Z, Xiao-Long M, Chun-Lin Z (2018) Overexpressed circPVT1, a potential new circular RNA biomarker, contributes to doxorubicin and cisplatin resistance of osteosarcoma cells by regulating ABCB1. Int J Biol Sci 14(3):321

Kute T, Lack CM, Willingham M, Bishwokama B, Williams H, Barrett K et al (2004) Development of Herceptin resistance in breast cancer cells. Cytometry A 57(2):86–93

Lei B, Huang Y, Zhou Z, Zhao Y, Thapa AJ, Li W et al (2019) Circular RNA hsa_circ_0076248 promotes oncogenesis of glioma by sponging miR-181a to modulate SIRT1 expression. J Cell Biochem 120(4):6698–6708

Li X, Yang B, Ren H, Xiao T, Zhang L, Li L et al (2019) Hsa_circ_0002483 inhibited the progression and enhanced the Taxol sensitivity of non-small cell lung cancer by targeting miR-182-5p. Cell Death Dis 10(12):953

Li X, He J, Ren X, Zhao H, Zhao H (2020a) Circ_0003998 enhances doxorubicin resistance in hepatocellular carcinoma by regulating miR-218-5p/EIF5A2 pathway. Diagn Pathol 15(1):141

Li X, Liu Y, Zhang X, Shen J, Xu R, Liu Y et al (2020b) Circular RNA hsa_circ_0000073 contributes to osteosarcoma cell proliferation, migration, invasion and methotrexate resistance by sponging miR-145-5p and miR-151-3p and upregulating NRAS. Aging (Albany NY) 12(14):14157

Li H, Li Q, He S (2021a) Hsa_circ_0025202 suppresses cell tumorigenesis and tamoxifen resistance via miR-197-3p/HIPK3 axis in breast cancer. World J Surg Oncol 19:1–12

Li H, Xu W, Xia Z, Liu W, Pan G, Ding J et al (2021b) Hsa_circ_0000199 facilitates chemotolerance of triple-negative breast cancer by interfering with miR-206/613-led PI3K/Akt/mTOR signaling. Aging (Albany NY) 13(3):4522

Li H, Liu Q, Chen Z, Wu M, Zhang C, Su J et al (2021c) Hsa_circ_0110757 upregulates ITGA1 to facilitate temozolomide resistance in glioma by suppressing hsa-miR-1298-5p. Cell Death Dis 12(3):252

Li W, Ma Q, Liu Q, Yan P, Wang X, Jia X (2021d) Circ-VPS18 knockdown enhances TMZ sensitivity and inhibits glioma progression by MiR-370/RUNX1 Axis. J Mol Neurosci 71(6):1234–1244

Li J, Fan R, Xiao H (2021e) Circ_ZFR contributes to the paclitaxel resistance and progression of non-small cell lung cancer by upregulating KPNA4 through sponging miR-195-5p. Cancer Cell Int 21:1–15

Li X, Feng Y, Yang B, Xiao T, Ren H, Yu X et al (2021f) A novel circular RNA, hsa_circ_0030998 suppresses lung cancer tumorigenesis and Taxol resistance by sponging miR-558. Mol Oncol 15(8):2235–2248

Li C, Zhang J, Yang X, Hu C, Chu T, Zhong R et al (2021g) hsa_circ_0003222 accelerates stemness and progression of non-small cell lung cancer by sponging miR-527. Cell Death Dis 12(9):807

Li Y, Liang J, Dang H, Zhang R, Chen P, Shao Y (2022) NCOA3 is a critical oncogene in thyroid cancer via the modulation of major signaling pathways. Endocrine 75:149–158

Liang Y, Song X, Li Y, Ma T, Su P, Guo R et al (2019) Targeting the circBMPR2/miR-553/USP4 Axis as a potent therapeutic approach for breast cancer. Mol Ther Nucleic Acids 17:347–361

Liang H, Lin Z, Lin H, Zhao L, Huang W (2021) circRNA_103615 contributes to tumor progression and cisplatin resistance in NSCLC by regulating ABCB1. Exp Ther Med 22(3):934

Liu J, Zhang N (2018) Circular RNA shifts the miRNA binding panel and takes a role in hepatocellular carcinoma progression. Transl Cancer Res 7(Suppl 1):S77

Liu Y, Dong Y, Zhao L, Su L, Luo J (2018) Circular RNA-MTO1 suppresses breast cancer cell viability and reverses monastrol resistance through regulating the TRAF4/Eg5 axis. Int J Oncol 53(4):1752–1762

Liu G, Zhang Z, Song Q, Guo Y, Bao P, Shui H (2020) Circ_0006528 contributes to paclitaxel resistance of breast cancer cells by regulating miR-1299/CDK8 axis. Onco Targets Ther 13:9497–9511

Liu Y, Li C, Liu H, Wang J (2021a) Circ_0001821 knockdown suppresses growth, metastasis, and TAX resistance of non-small-cell lung cancer cells by regulating the miR-526b-5p/GRK5 axis. Pharmacol Res Perspect 9(4):e00812

Liu Z, Wang T, She Y, Wu K, Gu S, Li L et al (2021b) N 6-methyladenosine-modified circIGF2BP3 inhibits CD8+ T-cell responses to facilitate tumor immune evasion by promoting the deubiquitination of PD-L1 in non-small cell lung cancer. Mol Cancer 20:1–25

Liu XY, Zhang Q, Guo J, Zhang P, Liu H, Bin TZ et al (2022a) The role of circular RNAs in the drug resistance of cancers. Front Oncol. Frontiers Media S.A. 11:790589

Liu Z, Zheng N, Li J, Li C, Zheng D, Jiang X et al (2022b) N6-methyladenosine-modified circular RNA QSOX1 promotes colorectal cancer resistance to anti-CTLA-4 therapy through induction of intratumoral regulatory T cells. Drug Resist Updat 65:100886

Liu Y, Zhai R, Hu S, Liu J (2022c) Circular RNA circ-RNF121 contributes to cisplatin (DDP) resistance of non-small-cell lung cancer cells by regulating the miR-646/SOX4 axis. Anti-Cancer Drugs 33(1):e186–e197

Liu Y, Zhang H, Zhang W, Xiang L, Yin Z, Xu H et al (2022d) circ_0004140 promotes LUAD tumor progression and immune resistance through circ_0004140/miR-1184/CCL22 axis. Cell Death Discov 8(1):181

Lu H, Xie X, Wang K, Chen Q, Cai S, Liu D et al (2020) Circular RNA hsa_circ_0096157 contributes to cisplatin resistance by proliferation, cell cycle progression, and suppressing apoptosis of non-small-cell lung carcinoma cells. Mol Cell Biochem 475(1):63–77

Lu JC, Zhang PF, Huang XY, Guo XJ, Gao C, Zeng HY et al (2021) Amplification of spatially isolated adenosine pathway by tumor–macrophage interaction induces anti-PD1 resistance in hepatocellular carcinoma. J Hematol Oncol 14:1–20

Luo Y, Gui R (2020) Circulating exosomal circFoxp1 confers cisplatin resistance in epithelial ovarian cancer cells. J Gynecol Oncol 31(5):e75

Luo YH, Yang YP, Chien CS, Yarmishyn AA, Adekunle Ishola A, Chien Y et al (2021) Circular RNA hsa_circ_0000190 facilitates the tumorigenesis and immune evasion by upregulating the expression of soluble PD-L1 in non-small-cell lung cancer. Int J Mol Sci 23(1):64

Luo W, Zhang G, Wang Z, Wu Y, Xiong Y (2023) Ubiquitin-specific proteases: vital regulatory molecules in bone and bone-related diseases. Int Immunopharmacol 118:110075

Lv J, Li K, Yu H, Han J, Zhuang J, Yu R et al (2023) HNRNPL induced circFAM13B increased bladder cancer immunotherapy sensitivity via inhibiting glycolysis through IGF2BP1/PKM2 pathway. J Exp Clin Cancer Res [Internet] 42(1):41. Available from: https://doi.org/https://doi.org/10.1186/s13046-023-02614-3

Mansoori B, Mohammadi A, Davudian S, Shirjang S, Baradaran B (2017) The different mechanisms of cancer drug resistance: a brief review. Adv Pharm Bull. Tabriz University of Medical Sciences 7:339–348

Mansoori B, Mohammadi A, Ditzel HJ, Duijf PHG, Khaze V, Gjerstorff MF et al (2021) HMGA2 as a critical regulator in cancer development. Genes (Basel) 12(2):269

Maximiano S, Magalhaes P, Guerreiro MP, Morgado M (2016) Trastuzumab in the treatment of breast cancer. BioDrugs 30:75–86

Memczak S, Jens M, Elefsinioti A, Torti F, Krueger J, Rybak A et al (2013) Circular RNAs are a large class of animal RNAs with regulatory potency. Nature 495(7441):333–338

Menzl I, Witalisz-Siepracka A, Sexl V (2019) CDK8-novel therapeutic opportunities. Pharmaceuticals 12(2):92

Pan Y, Lin Y, Mi C (2021) Cisplatin-resistant osteosarcoma cell-derived exosomes confer cisplatin resistance to recipient cells in an exosomal circ_103801-dependent manner. Cell Biol Int 45(4):858–868

Pegtel DM, Gould SJ (2019) Exosomes. Annu Rev Biochem 88:487–514

Peng L, Sang H, Wei S, Li Y, Jin D, Zhu X et al (2020) circCUL2 regulates gastric cancer malignant transformation and cisplatin resistance by modulating autophagy activation via miR-142-3p/ROCK2. Mol Cancer 19:1–19

Phi LTH, Sari IN, Yang YG, Lee SH, Jun N, Kim KS et al (2018) Cancer stem cells (CSCs) in drug resistance and their therapeutic implications in cancer treatment. Stem Cells Int. Hindawi Limited 2018:5416923

Qu F, Wang L, Wang C, Yu L, Zhao K, Zhong H (2021) Circular RNA circ_0006168 enhances Taxol resistance in esophageal squamous cell carcinoma by regulating miR-194-5p/JMJD1C axis. Cancer Cell Int 21(1):273

Rahnama S, Bakhshinejad B, Farzam F, Bitaraf A, Ghazimoradi MH, Babashah S (2021) Identification of dysregulated competing endogenous RNA networks in glioblastoma: a way toward improved therapeutic opportunities. Life Sci [Internet]. 277:119488. Available from: https://www.sciencedirect.com/science/article/pii/S0024320521004732

Shen DW, Pouliot LM, Hall MD, Gottesman MM (2012) Cisplatin resistance: a cellular self-defense mechanism resulting from multiple epigenetic and genetic changes. Pharmacol Rev 64(3):706–721

Si J, Li W, Li X, Cao L, Chen Z, Jiang Z (2021) Heparanase confers temozolomide resistance by regulation of exosome secretion and circular RNA composition in glioma. Cancer Sci 112(9):3491–3506

Song HM, Meng D, Wang JP, Zhang XY (2021) circRNA hsa_circ_0005909 predicts poor prognosis and promotes the growth, metastasis, and drug resistance of non-small-cell lung cancer via the miRNA-338-3p/SOX4 pathway. Dis Markers 2021(1):8388512

Stewart C, Ralyea C, Lockwood S (2019) Ovarian cancer: an integrated review. Semin Oncol Nurs. Elsevier 35:151–156

Su Y, Yang W, Jiang N, Shi J, Chen L, Zhong G et al (2019) Hypoxia-elevated circELP3 contributes to bladder cancer progression and cisplatin resistance. Int J Biol Sci 15(2):441

Sun Y, Cao Z, Shan J, Gao Y, Liu X, Ma D et al (2021) Hsa_circ_0020095 promotes oncogenesis and cisplatin resistance in colon cancer by sponging miR-487a-3p and modulating SOX9. Front Cell Dev Biol 8:604869

Sun J, Zhang H, Wei W, Xiao X, Huang C, Wang L et al (2023) Regulation of CD8+ T cells infiltration and immunotherapy by circMGA/HNRNPL complex in bladder cancer. Oncogene [Internet] 42(15):1247–1262. Available from: https://doi.org/https://doi.org/10.1038/s41388-023-02637-2

Sung H, Ferlay J, Siegel RL, Laversanne M, Soerjomataram I, Jemal A et al (2021) Global cancer statistics 2020: GLOBOCAN estimates of incidence and mortality worldwide for 36 cancers in 185 countries. CA Cancer J Clin 71(3):209–249

Vander Velde R, Yoon N, Marusyk V, Durmaz A, Dhawan A, Miroshnychenko D et al (2020) Resistance to targeted therapies as a multifactorial, gradual adaptation to inhibitor specific selective pressures. Nat Commun 11(1):2393

Vasan N, Baselga J, Hyman DM (2019) A view on drug resistance in cancer. Nature. Nature Publishing Group 575:299–309

Wang X, Zhang H, Chen X (2019) Drug resistance and combating drug resistance in cancer. Cancer Drug Resist 2(2):141

Wang Q, Liang D, Shen P, Yu Y, Yan Y, You W (2021a) Hsa_circ_0092276 promotes doxorubicin resistance in breast cancer cells by regulating autophagy via miR-348/ATG7 axis. Transl Oncol 14(8):101045

Wang T, Liu Z, She Y, Deng J, Zhong Y, Zhao M et al (2021b) A novel protein encoded by circASK1 ameliorates gefitinib resistance in lung adenocarcinoma by competitively activating ASK1-dependent apoptosis. Cancer Lett 520:321–331

Wei Y, Lu C, Zhou P, Zhao L, Lyu X, Yin J et al (2021a) EIF4A3-induced circular RNA ASAP1 promotes tumorigenesis and temozolomide resistance of glioblastoma via NRAS/MEK1/ERK1–2 signaling. Neuro Oncol 23(4):611–624

Wei W, Ji L, Duan W, Zhu J (2021b) Circular RNA circ_0081001 knockdown enhances methotrexate sensitivity in osteosarcoma cells by regulating miR-494-3p/TGM2 axis. J Orthop Surg Res 16:1–12

Wei S, Qi L, Wang L (2021c) Overexpression of circ_CELSR1 facilitates paclitaxel resistance of ovarian cancer by regulating miR-149-5p/SIK2 axis. Anti-Cancer Drugs 32(5):496–507

Wei W, Sun J, Zhang H, Xiao X, Huang C, Wang L et al (2021d) Circ0008399 interaction with WTAP promotes assembly and activity of the m6A methyltransferase complex and promotes cisplatin resistance in bladder cancer. Cancer Res 81(24):6142–6156

Wu Z, Gong Q, Yu Y, Zhu J, Li W (2020) Knockdown of circ-ABCB10 promotes sensitivity of lung cancer cells to cisplatin via miR-556-3p/AK4 axis. BMC Pulm Med 20(1):10

Wu J, Zhu MX, Li KS, Peng L, Zhang PF (2022) Circular RNA drives resistance to anti-PD-1 immunotherapy by regulating the miR-30a-5p/SOX4 axis in non-small cell lung cancer. Cancer Drug Resist 5(2):261

Xi L, Liu Q, Zhang W, Luo L, Song J, Liu R et al (2021) Circular RNA circCSPP1 knockdown attenuates doxorubicin resistance and suppresses tumor progression of colorectal cancer via miR-944/FZD7 axis. Cancer Cell Int 21:1–13

Xia B, Zhao Z, Wu Y, Wang Y, Zhao Y, Wang J (2020) Circular RNA circTNPO3 regulates paclitaxel resistance of ovarian cancer cells by miR-1299/NEK2 signaling pathway. Mol Ther Nucleic Acids 21:780–791

Xie H, Zheng R (2022) Circ_0085495 knockdown reduces adriamycin resistance in breast cancer through miR-873-5p/integrin β1 axis. Anticancer Drugs [Internet] 33(1):e166. Available from: https://journals.lww.com/anti-cancerdrugs/fulltext/2022/01000/circ_0085495_knockdown_reduces_adriamycin.34.aspx

Xie C, Liang G, Xu Y, Lin E (2020) Circular RNA hsa_circ_0003496 contributes to tumorigenesis and chemoresistance in osteosarcoma through targeting (microRNA) miR-370/Krüppel-like factor 12 Axis. Cancer Manag Res 12:8229–8240

Xie Q, Zhang P, Wang Y, Mei W, Zeng C (2022) Overcoming resistance to immune checkpoint inhibitors in hepatocellular carcinoma: challenges and opportunities. Front Oncol 12:958720

Xu J, Ji L, Liang Y, Wan Z, Zheng W, Song X et al (2020a) CircRNA-SORE mediates sorafenib resistance in hepatocellular carcinoma by stabilizing YBX1. Signal Transduct Target Ther 5(1):298

Xu J, Wan Z, Tang M, Lin Z, Jiang S, Ji L et al (2020b) N 6-methyladenosine-modified CircRNA-SORE sustains sorafenib resistance in hepatocellular carcinoma by regulating β-catenin signaling. Mol Cancer 19:1–16

Xu Y, Jiang T, Wu C, Zhang Y (2020c) CircAKT3 inhibits glycolysis balance in lung cancer cells by regulating miR-516b-5p/STAT3 to inhibit cisplatin sensitivity. Biotechnol Lett 42(7):1123–1135

Xu X, Tao R, Sun L, Ji X (2020d) Exosome-transferred hsa_circ_0014235 promotes DDP chemoresistance and deteriorates the development of non-small cell lung cancer by mediating the miR-520a-5p/CDK4 pathway. Cancer Cell Int 20:1–15

Xu J, Ni L, Zhao F, Dai X, Tao J, Pan J et al (2021) Overexpression of hsa_circ_0002874 promotes resistance of non-small cell lung cancer to paclitaxel by modulating miR-1273f/MDM2/p53 pathway. Aging (Albany NY) 13(4):5986

Xu HZ, Lin XY, Xu YX, Xue HB, Lin S, Xu TW (2023) An emerging research: the role of hepatocellular carcinoma-derived exosomal circRNAs in the immune microenvironment. Front Immunol 14:1227150

Xue M, Li G, Fang X, Wang L, Jin Y, Zhou Q (2019) Hsa-circ-0081143 promotes cisplatin resistance in gastric cancer by targeting miR-646/CDK6 pathway. Cancer Cell Int 19(1):25

Yan L, Liu G, Cao H, Zhang H, Shao F (2019) Hsa_circ_0035483 sponges hsa-miR-335 to promote the gemcitabine-resistance of human renal cancer cells by autophagy regulation. Biochem Biophys Res Commun 519(1):172–178

Yang W, Gu J, Wang X, Wang Y, Feng M, Zhou D et al (2019) Inhibition of circular RNA CDR1as increases chemosensitivity of 5-FU-resistant BC cells through up-regulating miR-7. J Cell Mol Med 23(5):3166–3177

Yang W, Gong P, Yang Y, Yang C, Yang B, Ren L (2020) Circ-ABCB10 contributes to paclitaxel resistance in breast cancer through let-7a-5p/DUSP7 axis. Cancer Manag Res 12:2327–2337

Yang B, Teng F, Chang L, Wang J, Liu DL, Cui YS et al (2021) Tumor-derived exosomal circRNA_102481 contributes to EGFR-TKIs resistance via the miR-30a-5p/ROR1 axis in non-small cell lung cancer. Aging (Albany NY) 13(9):13264

Yao W, Wang J, Meng F, Zhu Z, Jia X, Xu L et al (2021a) Circular RNA CircPVT1 inhibits 5-fluorouracil chemosensitivity by regulating ferroptosis through MiR-30a-5p/FZD3 Axis in esophageal cancer cells. Front Oncol 11:780938

Yao Y, Li X, Cheng L, Wu X, Wu B (2021b) Circular RNA FAT atypical cadherin 1 (circFAT1)/microRNA-525-5p/spindle and kinetochore-associated complex subunit 1 (SKA1) axis regulates oxaliplatin resistance in breast cancer by activating the notch and Wnt signaling pathway. Bioengineered 12(1):4032–4043

Yao W, Guo P, Mu Q, Wang Y (2021c) Exosome-derived Circ-PVT1 contributes to cisplatin resistance by regulating autophagy, invasion, and apoptosis via miR-30a-5p/YAP1 axis in gastric cancer cells. Cancer Biother Radiopharm 36(4):347–359

Ye R, Lu X, Liu J, Duan Q, Xiao J, Duan X et al (2023) CircSOD2 contributes to tumor progression, immune evasion and anti-PD-1 resistance in hepatocellular carcinoma by targeting miR-497-5p/ANXA11 axis. Biochem Genet 61(2):597–614

Yin H, Cui X (2021) Knockdown of circHIPK3 facilitates temozolomide sensitivity in glioma by regulating cellular behaviors through miR-524-5p/KIF2A-mediated PI3K/AKT pathway. Cancer Biother Radiopharm 36(7):556–567

Yin J, Huang HY, Long Y, Ma Y, Kamalibaike M, Dawuti R et al (2021) circ_C20orf11 enhances DDP resistance by inhibiting miR-527/YWHAZ through the promotion of extracellular vesicle-mediated macrophage M2 polarization in ovarian cancer. Cancer Biol Ther 22(7–9):440–454

Young ARJ, Narita M (2007) Oncogenic HMGA2: short or small? Genes Dev 21(9):1005–1009

Young MJ, Hsu KC, Lin TE, Chang WC, Hung JJ (2019) The role of ubiquitin-specific peptidases in cancer progression. J Biomed Sci 26:1–14

Zang H, Li Y, Zhang X, Huang G (2020) Circ-RNF111 contributes to paclitaxel resistance in breast cancer by elevating E2F3 expression via miR-140-5p. Thorac Cancer 11(7):1891–1903

Zang R, Qiu X, Song Y, Wang Y (2021) Exosomes mediated transfer of Circ_0000337 contributes to cisplatin (CDDP) resistance of esophageal cancer by regulating JAK2 via miR-377-3p. Front Cell Dev Biol 9:673237

Zhang PF, Pei X, Li KS, Jin LN, Wang F, Wu J et al (2019) Circular RNA circFGFR1 promotes progression and anti-PD-1 resistance by sponging miR-381-3p in non-small cell lung cancer cells. Mol Cancer 18:1–13

Zhang J, Ma X, Zhou R, Zhou Y (2020a) TRPS1 and YAP1 regulate cell proliferation and drug resistance of osteosarcoma via competitively binding to the target of circTADA2A–miR-129-5p. Onco Targets Ther Volume 13:12397–12407

Zhang S, Cheng J, Quan C, Wen H, Feng Z, Hu Q et al (2020b) circCELSR1 (hsa_circ_0063809) contributes to paclitaxel resistance of ovarian cancer cells by regulating FOXR2 expression via miR-1252. Mol Ther Nucleic Acids 19:718–730

Zhang H, Yan C, Wang Y (2021a) Exosome-mediated transfer of circHIPK3 promotes trastuzumab chemoresistance in breast cancer. J Drug Target 29(9):1004–1015

Zhang Z, Zhou Q, Luo F, Zhou R, Xu J, Xiao J et al (2021b) Circular RNA circ-CHI3L1. 2 modulates cisplatin resistance of osteosarcoma cells via the miR-340-5p/LPAATβ axis. Hum Cell 34(5):1558–1568

Zhang Y, Ge P, Zhou D, Xing R, Bai L (2021c) Circular RNA FOXO3 accelerates glycolysis and improves cisplatin sensitivity in lung cancer cells via the miR-543/Foxo3 axis. Oncol Lett 22(6):839

Zhang W, Song C, Ren X (2021d) Circ_0003998 regulates the progression and docetaxel sensitivity of DTX-resistant non-small cell lung cancer cells by the miR-136-5p/CORO1C axis. Technol Cancer Res Treat 20:1533033821990040

Zhang PF, Xu YF, Zhang LX, Wu J, Fan Y (2021e) CircRUNX1 drives immune evasion and anti-PD1 immunotherapy resistance in lung adenocarcinoma by the miR-4739/PCSK9/MHC I Axis. https://doi.org/10.21203/rs.3.rs-835938/v1

Zhang LX, Gao J, Long X, Zhang PF, Yang X, Zhu SQ et al (2022) The circular RNA circHMGB2 drives immunosuppression and anti-PD-1 resistance in lung adenocarcinomas and squamous cell carcinomas via the miR-181a-5p/CARM1 axis. Mol Cancer 21(1):110

Zhao Z, Ji M, Wang Q, He N, Li Y (2019) RETRACTED: circular RNA Cdr1as upregulates SCAI to suppress cisplatin resistance in ovarian cancer via miR-1270 suppression. Mol Ther Nucleic Acids 18:24–33

Zhao K, Cheng X, Ye Z, Li Y, Peng W, Wu Y et al (2021) Exosome-mediated transfer of circ_0000338 enhances 5-fluorouracil resistance in colorectal cancer through regulating MicroRNA 217 (miR-217) and miR-485-3p. Mol Cell Biol 41(5):e00517–e00520

Zheng SR, Huang QD, Zheng ZH, Zhang ZT, Guo GL (2021) circGFRA1 affects the sensitivity of triple-negative breast cancer cells to paclitaxel via the miR-361-5p/TLR4 pathway. J Biochem 169(5):601–611

Zhong Y, Wang D, Ding Y, Tian G, Jiang B (2021a) Circular RNA circ_0032821 contributes to oxaliplatin (OXA) resistance of gastric cancer cells by regulating SOX9 via miR-515-5p. Biotechnol Lett 43:339–351

Zhong Y, Lin H, Li Q, Liu C, Shen J (2021b) CircRNA_100565 contributes to cisplatin resistance of NSCLC cells by regulating proliferation, apoptosis and autophagy via miR-337-3p/ADAM28 axis. Cancer Biomark 30(2):261–273

Zhu M, Wang Y, Wang F, Li L, Qiu X (2021) CircFBXL5 promotes the 5-FU resistance of breast cancer via modulating miR-216b/HMGA2 axis. Cancer Cell Int 21:1–12 [illegible] (2019) Circular RNA [illegible] resistance [illegible] lung cancer cells. Mol Cancer [illegible]

[illegible]

Chapter 9
Silhouette of Probiotics in the Regulation of miRNAs and lncRNAs in Carcinogenesis and Metastasis: Is It a Silver Lining or a Cross to Bear

Rukset Attar, Asmaa Hamid, Laiba Shamshad Dar, Nodirali Normakhamatov, and Ammad Ahmad Farooqi

Contents

Abstract Cancer is a genomically complex and multifaceted disease. Wealth of information distilled through decades of high-throughput research has revealed a broad spectrum of oncogenic signaling cascades, immunological evasion, and drug

---

R. Attar
Department of Obstetrics and Gynecology, Faculty of Medicine, Yeditepe University Hospital, Yeditepe University, Istanbul, Turkey

A. Hamid
Department of Nutrition and Health Promotion, University of Home Economics, Lahore, Pakistan

L. S. Dar
Shifa College of Pharmaceutical Sciences, Shifa Tameer-e-Millat University, Islamabad, Pakistan

N. Normakhamatov
Tashkent Pharmaceutical Institute, Ministry of the Health of Uzbekistan, Tashkent, Uzbekistan

A. A. Farooqi (✉)
Institute of Biomedical and Genetic Engineering (IBGE), Islamabad, Pakistan

M. F. Ullah, A. Ahmad (eds.), *Non-coding RNAs*, Experientia Supplementum 115,
https://doi.org/10.1007/978-3-032-06948-1_9

resistance, which play a steering role in carcinogenesis and metastasis. Noncoding RNAs have also emerged as key players in the regulation of multiple stages of cancer progression and metastatic spread of cancer cells to secondary sites. The concept of probiotics has started to gain limelight due to its ability to pharmacologically modulate the host microbiome and the immunological responses. The genomics era has provided impetus for the discovery and characterization of bacterial probiotic effector molecules that stimulate specific responses. We have witnessed an exponential increase in the seminal studies which provided proof-of-concept about the mechanistic regulation of cell signaling pathways and noncoding RNAs by probiotics. These exciting and groundbreaking studies ignited an outburst of data generated using several "omics" technologies. In this chapter, we have provided a summary of seminal studies associated with the anticancer and antimetastatic role of probiotics in animal models. However, circumstantial evidence has also underlined tumor-promoting role of probiotics in animal model studies. Therefore, there is a need to scrupulously reinterpret the existing pieces of evidence related to conflicting data about pro-tumorigenic and tumor-inhibitory roles of probiotics. We also critically summarized how probiotics modulated noncoding RNAs to prevent/inhibit cancer progression. Surprisingly, probiotics-mediated regulation of noncoding RNAs has not been comprehensively explored in different cancers. In accordance with this approach, in-depth analysis of target long noncoding RNAs and circular RNAs by probiotics will allow the researchers to develop near-to-complete signaling landscape to reap the full benefits of the medicinal significance of probiotics.

Keywords Probiotics · Cancer · Metastasis · Noncoding RNA · Circular RNA

Abbreviations

circRNAs	Circular RNAs
lncRNAs	Long non-coding RNAs
m6A	Methyladenosine
miRNAs	microRNAs

9.1 Introduction

Probiotics were brought from shadow to the limelight because of their significant pharmacological effects in different diseases. Consumption of over-the-counter probiotics for the promotion of health and well-being has grown exponentially in recent years. Clinical treatment of drug-resistant and advanced cancers is highly challenging and complicated because of tumor heterogeneity. Deregulated signaling pathways centrally drive the drug resistance, loss of apoptosis, epithelial-to-mesenchymal transition, and progression of the cancers. Metastasis is the spread of cancer cells from a primary site to distant organs. A wealth of information has unveiled complexities of immune escape mechanisms used throughout the metastatic cascades to promote dissemination, invasion, colonization, and outgrowth of cancer cells at secondary sites. In this chapter, we have attempted to analyze the anticancer and antimetastatic role of probiotics. However, we have also included important evidence linked to the pro-tumorigenic role of probiotics. In the later stages of the chapter, we have summarized how probiotics modulated microRNAs and long non-coding RNAs in cancers. In the upcoming section, we will set the spotlight on the role of probiotics in the inhibition of metastatic dissemination in tumor-bearing animal models.

9.2 Role of Probiotics in the Inhibition of Metastasis in Animal Models Studies

In this perspective, we highlight major breakthroughs, challenges, and limitations in the unbiased interpretation of the large amount of data regarding the antimetastatic and pro-tumorigenic role of probiotics.

N6-methyladenosine (m6A) is an RNA modification in coding and noncoding RNAs, catalyzed by the METTL3–METTL14 methyltransferase complexes. METTL3 overexpression facilitated the proliferation, migration, and invasion of colorectal cancer cells. *Clostridium butyricum* was found to induce shrinkage of the palpable tumors in xenografted mice. There was a significant reduction in the expression levels of METTL3 in the tumor tissues (Zhang et al. 2023).

Clostridium butyricum significantly inhibited high-fat diet-mediated development of intestinal tumors in $Apc^{min/+}$ mice (Chen et al. 2020).

VSL#3 (Ferring Pharmaceuticals, Italy) is a commercially available probiotic mixture containing eight strains of probiotic bacteria. VSL#3 caused a significant reduction in the pulmonary metastasis in mice intravenously injected with B16F10 cells. Short-chain fatty acids (SCFAs) primarily propionate and butyrate, inhibit the metastatic dissemination of melanoma cells. Propionate and butyrate promoted the chemotactic recruitment of Th17 cells in the lungs. CCL20 expressed in pulmonary endothelial cells enhanced the recruitment of Th17 cells to the lungs. Consequently, recruitment of Th17 cells in the lungs inhibited pulmonary metastasis of melanoma cells. Whereas, CCL20 knockdown not only reduced the percentage of Th17 cells in the lungs but also triggered an increase in the number of tumor foci (Chen et al. 2021).

9.3 Engineered Probiotics

Bioengineered probiotics have offered new opportunities to improve colorectal cancer screening, treatment and prevention. *Escherichia coli* Nissle 1917 (EcN), a probiotic strain has been reported to be safe for use. A bioluminescent labeling system is used in different studies to track the bacteria in cell cultures and animal models.

In vivo imaging of mice administered with bioluminescent bacteria provided evidence about retention of the bacteria in the ApcMin/+ mouse gut for a duration of 7 weeks. Enrichment of bioluminescent bacteria in neoplasia was studied through ex vivo imaging of intestinal tissues, where bioluminescence co-localized with visible macroadenomas. Importantly, greater intensity of bioluminescence was detected in the distal small intestine with a high adenoma burden. Engineered *Escherichia coli* Nissle 1917 was used to produce immuno-therapeutics for the regression of tumor and reshaping of the tumor-immune microenvironment (Gurbatri et al. 2024).

Intra-tumorally injected probiotic neoantigen delivery vector invoked stronger antitumor efficacy with a complete response observed in BALB/c mice xenografted with CT26 cancer cells. Likewise, the cancer inhibition was significant in xenografted mice intravenously injected with a probiotic neoantigen delivery vector. Importantly, there was a considerable reduction in progressive lung metastases and the number of pulmonary metastatic foci in mice intravenously injected with CT26 cancer cells. Probiotic neoantigen delivery vectors efficiently enhanced the population of tumor-infiltrating $FOXP3^{-}CD4^{+}$, $CD8^{+}$ T cells, and NK cells (Redenti et al. 2024).

9.4 Darker Side of the Probiotics

Germ-free mice first received transplantation of fecal microbiota from the donor stool of a complete responder patient to anti-PD-1 blockade. Afterwards, mice were orally gavaged with one of two probiotics (*Lactobacillus rhamnosus* GG or *Bifidobacterium longum*–based). Mice xenotransplanted with murine melanoma tumors were treated with anti-PD-L1 (anti-PD-1 ligand-1) therapy. It was noted that mice that received probiotics failed to elicit antitumor responses to treatment with anti-PD-L1 with an evident increase in the size of the palpable tumors. Cellular analysis of tumor-infiltrating immune subsets from anti-PD-L1-treated mice demonstrated a significant reduction in the frequency of IFN-γ (interferon-γ) positive CD8+ T cells in tumor tissues of mice treated with probiotics. This study suggested a reduction in the population of cytotoxic T cells within the tumor microenvironment of animal models treated with probiotics. These findings are in concordance with previous studies that suggested a palpable increase in colorectal carcinoma in probiotic-treated mice. Melanoma-bearing C57BL/6 mice were given a standard fiber-rich whole grain diet and treated with anti-PD-1 therapeutic treatment. Principally, mice receiving a fiber-rich diet demonstrated significant regression of the tumor xenografts as compared to mice that received a fiber-poor diet during treatment with anti-PD-1 (Spencer et al. 2021).

VSL#3, a commercially available probiotic cocktail, did not provide protection against tumorigenesis in the chemically induced (AOM/$Il10^{-/-}$) model of colitis-associated colorectal cancer (Arthur et al. 2013).

9.5 Regulation of Non-coding RNAs by Probiotics

Advancements in high-throughput sequencing technologies have allowed researchers to explore that the genome is pervasively transcribed into RNAs, seemingly not translated into proteins. The discovery of noncoding RNAs has transformed the field of molecular oncology. Different types of noncoding RNAs particularly microRNAs (miRNAs), long noncoding RNAs (lncRNAs), and circular RNAs (circRNASs) have been structurally and functionally characterized. In this section, we have critically summarized available scientific evidence related to the role of prebiotics in the regulation of miRNAs and lncRNAs.

9.6 Role of Probiotics in Regulation of miRNAs

Lactobacillus casei conditioned media were used to treat HT-29 cells. *L. casei* condition medium markedly induced apoptotic death in a time- and dose-dependent manner in HT-29 cells. It has been indicated that micoRNA-21 negatively regulated PTEN (Phosphatase and tensin homolog) and PDCD4 (programmed cell death 4) in colorectal cancer cells (shown in Fig. 9.1). *L. casei* condition medium significantly downregulated miR-21 in cancer cells (Hosseinzadeh et al. 2023).

Extracts of *Lactobacillus acidophilus* and *Saccharomyces cerevisiae* standard probiotics were used for the evaluation of anticancer effects against HT29 and SW480 cancer cells. Probiotics extract induced upregulation of BAX, caspase-9 and caspase-3 and activated the apoptotic pathway. BAX is a proapoptotic protein and is involved in the activation of the intrinsic apoptotic pathway. Caspase-9 is activated as a consequence of signaling through the release of proteins from mitochondria. Subsequently, caspase-9 activates caspase-3 and induces apoptosis in the cancer cells. However, these extracts significantly reduced the levels of Bcl-2, MMP2, and MMP9. MicroRNAs were also found to be altered in the cancer cells treated with probiotic extracts. There was an increase in the expression of miR-34, miR-135, miR-25, miR-16, miR-195, miR-27, miR-98, and let7. Levels of miR-9, miR-106b, miR-17, miR-21, miR-155, and miR-221 were found to be significantly reduced (Saffar et al. 2024).

The expression level of tumor suppressor lncRNA, including LincRNA-P21, was decreased significantly in rectal cancer patients. Expression levels of lincRNA-P21 were found to increase significantly following the consumption of probiotics and were substantially higher in the probiotic users. Importantly, levels of tumor suppressor miRs, including miR-34a, miR-133b, miR-378a, miR-548 ac, and miR-601 were found to be significantly reduced in the rectal cancer patients. However, there was a considerable increase in these tumor suppressor miRNAs after the consumption of probiotics (Khodaii et al. 2022).

Chemically induced colon cancer animal models were given 1.5 g of powders of *Lactobacillus acidophilus* and *Bifidobacterium bifidum* in drinking water. Expression

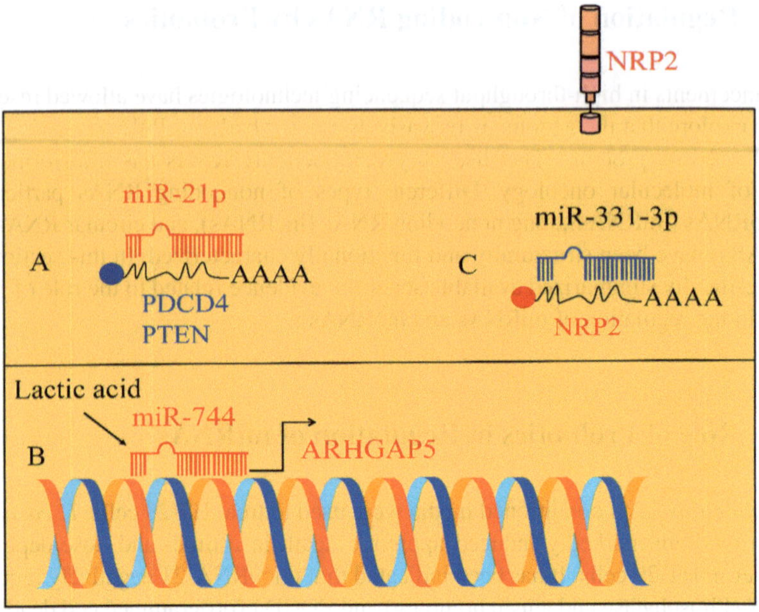

Fig. 9.1 (**a**) Targeting of PDCD4 and PTEN by oncogenic miR-21. (**b**) miR-744 stimulates the expression of ARHGAP5. miR-744 interacts with the promoter region of ARHGAP5 and upregulates its expression at the transcriptional level. (**c**) miR-331-3p acts as a tumor suppressor and inhibits the expression of NRP2

levels of miR-135b and miR-155 were noticed to be upregulated in both the plasma and the colon tissue samples in the colon cancer models. However, probiotics consumption exerted repressive effects on their expression. Importantly, miR-18a and miR-26b levels were reduced in the animal models, but the consumption of the probiotics stimulated their expression (Heydari et al. 2019).

Lactic acid is usually produced by Lactobacillus species. Probiotics *Lactobacillus crispatus* and *Lactobacillus gasseri* have a central role in cervical cancer progression. Lactic acid has a pivotal role in enhancing cellular motility that fuels metastatic spread. Oncogenic MiR-744 is upregulated in lactic acid-treated SiHa cells. Mechanistically, miR-744 was found to interact with the promoter region of ARHGAP5 and increased the promoter activity of the target mRNA (shown in Fig). ARHGAP5 inhibition led to dramatic reduction in the invasion and migration abilities and clearly impaired miR-744-mediated increase in the migratory and invasive properties of SiHa cells (Li et al. 2019).

Administration of the probiotics in tumor-bearing mice led to a significant increase in p-c-Jun and p53 levels and a simultaneous reduction in Bcl-2 levels. miR-331-3p is a tumor suppressor microRNA involved in the regression of cancer. Expression levels of miR-331-3p were found to be considerably enhanced in HCT116 and SW480 cells treated with probiotics. Neuropilin-2 (NRP2) is a non-tyrosine kinase transmembrane glycoprotein. miR-331-3p negatively regulated NRP2 in cancer cells and inhibited cancer progression (shown in Fig. 9.1). It was shown that probiotic treatment remarkably reduced NRP2 expression in HCT116 and SW480 cells. Probiotic administration induced shrinkage of the palpable tumors in mice primarily through upregulation of miR-331-3p and downregulation of NRP2 levels (Li and Chen 2021).

9.7 Role of Probiotics in Regulation of LncRNAs

Clostridium butyricum (CB), a probiotic induced apoptosis in SW480 cancer cells. *Clostridium butyricum* probiotic stimulated upregulation of 50 lncRNAs by more than fivefold and exerted repressive effects on 152 lncRNAs by more than fivefold. BNIP3L is a mitochondrial protein that belongs to the BH3-only protein of the BCL2 family. Probiotics stimulated the levels of caspase-1, tumor necrosis factor (TNF) and BNIP3L, while the BCL family members (BCL11B and BCL2L11) were found to be considerably reduced (Yang and Wei 2025).

To date, just a small fraction of the mechanistic role of probiotics in the regulation of lncRNAs has been studied, and what we already know might just be the tip of the iceberg.

9.8 Concluding Remarks

Probiotics-mediated regulation of cell signaling pathways and noncoding RNAs has started to provide previously unprecedented scientific information. Increasingly, it is being realized that crosstalk between signaling cascades further complicates the cellular responses, thus challenging the development of effective therapeutic strategies. Although experimental evidence related to probiotics-mediated regulation of miRNAs has started to emerge, we still have an insufficient and incomplete understanding of the regulation of lncRNAs and circular RNAs. Detailed investigation related to the regulation of lncRNAs by probiotics will enable researchers to gain insights about sponging effects/competitive endogenous RNAs and epigenetic regulation of different subsets of genes in different cancers. Identification of tumor suppressor lncRNAs will open avenues for a combinatorial approach consisting of lncRNA mimics and probiotics in tumor-bearing animal models. The development of probiotics as an efficient therapeutic strategy needs to be guided by a comprehensive understanding of oncogenic signaling cascades and noncoding RNAs, leading to the identification of key nodes within these networks that can be exploited pharmacologically.

Conflict of Interest The authors declare no conflict of interest.

Ethical Approval This chapter does not contain any studies with human participants or animals.

Funding No funding received.

References

Arthur JC, Gharaibeh RZ, Uronis JM, Perez-Chanona E, Sha W, Tomkovich S, Mühlbauer M, Fodor AA, Jobin C (2013) VSL#3 probiotic modifies mucosal microbial composition but does not reduce colitis-associated colorectal cancer. Sci Rep 3:2868. https://doi.org/10.1038/srep02868

Chen D, Jin D, Huang S, Wu J, Xu M, Liu T, Dong W, Liu X, Wang S, Zhong W, Liu Y, Jiang R, Piao M, Wang B, Cao H (2020) Clostridium butyricum, a butyrate-producing probiotic, inhibits intestinal tumor development through modulating Wnt signaling and gut microbiota. Cancer Lett 469:456–467. https://doi.org/10.1016/j.canlet.2019.11.019

Chen L, Zhou X, Wang Y, Wang D, Ke Y, Zeng X (2021) Propionate and butyrate produced by gut microbiota after probiotic supplementation attenuate lung metastasis of melanoma cells in mice. Mol Nutr Food Res 65(15):e2100096. https://doi.org/10.1002/mnfr.202100096

Gurbatri CR, Radford GA, Vrbanac L, Im J, Thomas EM, Coker C, Taylor SR, Jang Y, Sivan A, Rhee K, Saleh AA, Chien T, Zandkarimi F, Lia I, Lannagan TRM, Wang T, Wright JA, Kobayashi H, Ng JQ, Lawrence M, Sammour T, Thomas M, Lewis M, Papanicolas L, Perry J, Fitzsimmons T, Kaazan P, Lim A, Stavropoulos AM, Gouskos DA, Marker J, Ostroff C, Rogers G, Arpaia N, Worthley DL, Woods SL, Danino T (2024) Engineering tumor-colonizing E. coli Nissle 1917 for detection and treatment of colorectal neoplasia. Nat Commun 15(1):646. https://doi.org/10.1038/s41467-024-44776-4

Heydari Z, Rahaie M, Alizadeh AM, Agah S, Khalighfard S, Bahmani S (2019) Effects of lactobacillus acidophilus and Bifidobacterium bifidum probiotics on the expression of MicroRNAs 135b, 26b, 18a and 155, and their involving genes in mice colon cancer. Probiotics Antimicrob Proteins 11(4):1155–1162. https://doi.org/10.1007/s12602-018-9478-8

Hosseinzadeh R, Bahadori A, Ghorbani M, Mohammadimehr M (2023) Lactobacillus casei condition medium downregulates miR-21 relative expression in HT-29 colorectal cancer cell line. FEMS Microbiol Lett 370:fnad089. https://doi.org/10.1093/femsle/fnad089

Khodaii Z, Mehrabani Natanzi M, Khalighfard S, Ghandian Zanjan M, Gharghi M, Khori V, Amiriani T, Rahimkhani M, Alizadeh AM (2022) Novel targets in rectal cancer by considering lncRNA-miRNA-mRNA network in response to lactobacillus acidophilus consumption: a randomized clinical trial. Sci Rep 12(1):9168. https://doi.org/10.1038/s41598-022-13297-9

Li S, Chen T (2021) Colorectal carcinoma growth inhibition by dietary care combined with probiotic intervention through targeting NRP2 expression. Dokl Biochem Biophys 500(1):354–359. https://doi.org/10.1134/S1607672921050161

Li C, Jia L, Yu Y, Jin L (2019) Lactic acid induced microRNA-744 enhances motility of SiHa cervical cancer cells through targeting ARHGAP5. Chem Biol Interact 298:86–95. https://doi.org/10.1016/j.cbi.2018.10.027

Redenti A, Im J, Redenti B, Li F, Rouanne M, Sheng Z, Sun W, Gurbatri CR, Huang S, Komaranchath M, Jang Y, Hahn J, Ballister ER, Vincent RL, Vardoshivilli A, Danino T, Arpaia N (2024) Probiotic neoantigen delivery vectors for precision cancer immunotherapy. Nature 635(8038):453–461. https://doi.org/10.1038/s41586-024-08033-4

Saffar KN, Larypoor M, Torbati MB (2024) Analyzing of colorectal cancerrelated genes and microRNAs expression profiles in response to probiotics lactobacillus acidophilus and Saccharomyces cerevisiae in colon cancer cell lines. Mol Biol Rep 51(1):122. https://doi.org/10.1007/s11033-023-09008-w

Spencer CN, McQuade JL, Gopalakrishnan V, McCulloch JA, Vetizou M, Cogdill AP, Khan MAW, Zhang X, White MG, Peterson CB, Wong MC, Morad G, Rodgers T, Badger JH, Helmink BA, Andrews MC, Rodrigues RR, Morgun A, Kim YS, Roszik J, Hoffman KL, Zheng J, Zhou Y, Medik YB, Kahn LM, Johnson S, Hudgens CW, Wani K, Gaudreau PO, Harris AL, Jamal MA, Baruch EN, Perez-Guijarro E, Day CP, Merlino G, Pazdrak B, Lochmann BS, Szczepaniak-Sloane RA, Arora R, Anderson J, Zobniw CM, Posada E, Sirmans E, Simon J, Haydu LE, Burton EM, Wang L, Dang M, Clise-Dwyer K, Schneider S, Chapman T, Anang NAS, Duncan S, Toker J, Malke JC, Glitza IC, Amaria RN, Tawbi HA, Diab A, Wong MK, Patel

SP, Woodman SE, Davies MA, Ross MI, Gershenwald JE, Lee JE, Hwu P, Jensen V, Samuels Y, Straussman R, Ajami NJ, Nelson KC, Nezi L, Petrosino JF, Futreal PA, Lazar AJ, Hu J, Jenq RR, Tetzlaff MT, Yan Y, Garrett WS, Huttenhower C, Sharma P, Watowich SS, Allison JP, Cohen L, Trinchieri G, Daniel CR, Wargo JA (2021) Dietary fiber and probiotics influence the gut microbiome and melanoma immunotherapy response. Science 374(6575):1632–1640. https://doi.org/10.1126/science.aaz7015

Yang X, Wei L (2025) Analysis of long non-coding RNA and mRNA expression in clostridium butyricum-induced apoptosis in SW480 colon cancer cells. Gene 940:149208. https://doi.org/10.1016/j.gene.2024.149208

Zhang K, Dong Y, Li M, Zhang W, Ding Y, Wang X, Chen D, Liu T, Wang B, Cao H, Zhong W (2023) Clostridium butyricum inhibits epithelial-mesenchymal transition of intestinal carcinogenesis through downregulating METTL3. Cancer Sci 114(8):3114–3127. https://doi.org/10.1111/cas.15839

Chapter 10
Noncoding RNAs in Metabolic Reprogramming of Breast Cancer: Dietary Opportunities and Translational Implications

Yahya Mukhlis, Rajendran Amalraj, Jagadish Natesh, and Syed Musthapa Meeran

Contents

Abstract Breast cancer remains one of the leading causes of cancer-related death among women worldwide, characterized by significant molecular and metabolic heterogeneity. Metabolic reprogramming has been shown to enable tumor cells to

Y. Mukhlis · R. Amalraj · J. Natesh · S. M. Meeran (✉)
Department of Biochemistry, CSIR-Central Food Technological Research Institute, Mysuru 570020, Karnataka, India

Academy of Scientific and Innovative Research (AcSIR), Ghaziabad 201002, India
e-mail: s.musthapa.cftri@csir.res.in

M. F. Ullah, A. Ahmad (eds.), *Non-coding RNAs*, Experientia Supplementum 115,
https://doi.org/10.1007/978-3-032-06948-1_10

adapt to the dynamic microenvironment, supporting uncontrolled proliferation and survival. Emerging evidence highlights the critical roles of noncoding RNAs (ncRNAs), including microRNAs, long noncoding RNAs, and circular RNAs, in coordinating the complex regulatory networks underlying metabolic reprogramming in breast cancer. These ncRNAs influence key metabolic pathways such as glycolysis, lipid metabolism, and amino acid metabolism by targeting transcription factors, enzymes, and signaling cascades. Evidence suggests that targeting dysregulated ncRNAs holds significant potential for modulating cancer cell metabolism and offers novel strategies for breast cancer management. Furthermore, bioactive compounds derived from dietary sources have demonstrated the ability to modulate ncRNA expression and function, presenting exciting prospects for dietary or nutritional interventions in breast cancer therapy. This chapter compiles the intricate relationship between ncRNAs and metabolic reprogramming in breast cancer, with a focus on innovative techniques to target ncRNAs and the potential of dietary strategies to influence these regulatory pathways.

Keywords Breast cancer · Dietary bioactives · LncRNAs · miRNAs · Cancer metabolism

Abbreviations

CCND1	Cyclin D1
circRNAs	Circular RNAs
CYP1B1	Cytochrome P450 1B1
lncRNAs	Long noncoding RNAs
METTL14	Methyltransferase-like 14
miRNAs	microRNAs
ncRNA	noncoding RNA

10.1 Introduction

Breast cancer is the common cause of cancer-related mortality among women globally (Wang et al. 2020a; Bray et al. 2024). There has been significant development over the past years in the field of breast cancer diagnosis and treatment; however, challenges such as the recurrence of tumors and metastasis are persisting (Riggio et al. 2021). Research has shown numerous mechanisms involved in the progression, metastasis, and recurrence of breast cancer; among these mechanisms, metabolic reprogramming in the cancer cells has gained significant attention in recent times (Wang et al. 2020a).

The term “metabolic reprogramming” refers to the modulation in cellular metabolism that enables cancer cells to meet the increased energy demands and biosynthetic requirements that result in the rapid proliferation of cancer cells (DeBerardinis et al. 2008; Schiliro and Firestein 2021). In metabolic reprogramming, the *Warburg effect* is the hallmark in cancer cells, wherein cancer cells preferentially undergo aerobic glycolysis over oxidative phosphorylation regardless of sufficient oxygen supply (Hsu and Sabatini 2008). This switch is important since the changeover from the use of fuels to glucose enables the active conversion of glucose into metabolites required for nucleotide, amino acids, and lipid synthesis (Pavlova and Thompson 2016). This phenomenon aids in further active cellular growth and division. Apart from glucose metabolism, cancer cells also demonstrate changes in lipid and amino acid metabolism, further contributing to their malignant phenotype (Pavlova and Thompson 2016). This altered cancer cell metabolism plays a crucial role in tumor progression and survival by regulating key signaling pathways. It has a significant impact on antitumor responses by modulating the expression of oncogenes and tumor suppressor genes (Safi et al. 2023).

Metabolic reprogramming in cancer cells is a complex process regulated by various mechanisms such as genetic and epigenetic alterations, dysregulations in signaling pathways, the influence of the tumor microenvironment, and noncoding RNA (ncRNA)-mediated gene regulations (Tarrado-Castellarnau et al. 2016; Xu et al. 2023; Aden et al. 2025). Among these, ncRNAs are centrally involved in the regulation of cancer cell metabolism (Lin et al. 2021). NcRNAs are a diverse class of RNA molecules that are not translated into proteins. However, they play a pivotal role in the regulation of gene expression at multiple levels, including chromatin remodeling, transcription, post-transcriptional modifications, and signal transduction (Anastasiadou et al. 2018; Safi et al. 2023). Various studies have indicated that ncRNAs play a vital role in regulating various metabolic processes, including glucose metabolism, glycolysis, the pentose phosphate pathway (PPP), the tricarboxylic acid (TCA) cycle, amino acid metabolism, and fatty acid metabolism (Lin et al. 2021).

NcRNAs can be broadly classified into several categories; among them, microRNAs (miRNAs), long noncoding RNAs (lncRNAs), and circular RNAs (circRNAs) are the most studied and have been associated with gene regulation involved in metabolic reprogramming (Xu et al. 2024b). Among them, miRNAs are short, approximately 21-23 nucleotide-long RNAs that are the most studied for their functionalities. miRNAs mainly function by binding to the 3′ untranslated regions (UTR) of target mRNAs, leading to their degradation or translational repression (Krol et al. 2010; O'Brien et al. 2018). In contrast, lncRNAs are transcripts that are longer than 200 nucleotides and have been shown to exert their regulatory functions through various mechanisms, including chromatin remodeling, transcriptional control, and post-transcriptional modulation (Statello et al. 2021; Natesh et al. 2022). Additionally, circRNAs are characterized by their covalently closed loop structures and function by sponging miRNA, modulating splicing and transcription, and modifying parental genes (Pamudurti et al. 2017).

Studies have shown an intricate interplay between ncRNAs and metabolic reprogramming in breast cancer cells; thus, targeting these regulatory networks provides an excellent opportunity for therapeutic interventions. Interestingly, bioactives from dietary sources have been shown to modulate the expression and function of ncRNAs, thereby influencing cancer cell metabolism. Additionally, various bioactives are under different stages of clinical trials for their therapeutic efficacy against breast cancer. Therefore, understanding the relationship between ncRNA regulation and metabolic reprogramming and targeting it with dietary bioactives may offer promising translational implications. Thus, this chapter is focused on compiling the role of ncRNA in the metabolic reprogramming of breast cancer cells and exploring the potential of dietary bioactives in influencing metabolic reprogramming by regulating ncRNAs.

10.2 Role of Noncoding RNAs in Breast Cancer Progression and Metastasis

10.2.1 miRNAs Regulate Breast Cancer Cell Proliferation

miRNAs are small, ncRNA molecules that are critical in regulating gene expression at the post-transcriptional level. Typically, miRNAs downregulate gene expression by binding to specific mRNAs, resulting in either mRNA degradation or the inhibition of translation, depending on their complementarity to the target mRNA sequences. Dysregulated expression of miRNAs has been implicated in the development and progression of various human diseases, including breast cancer. Some of the majorly dysregulated miRNAs and their function in BC progression are compiled in Table 10.1.

Uncontrolled cell proliferation is a key characteristic of cancer cells and serves as the primary driver of breast tumor development (Fouad and Aanei 2017). The progression of the cell cycle is crucial for maintaining a fine balance between facilitating cell proliferation and cell cycle arrest. Healthy cells have a limited capacity for cell division. Once they reach finite cell division in the tissue, the cells are arrested at the G_0 phase of the cell cycle and remain quiescent. This mechanism of physiologically adaptive cell cycle arrest becomes dysregulated in cancer cells (White et al. 2010). Cyclin E1, a key regulator of the G_1–S cell cycle transition, is frequently targeted by tumor-suppressive miRNAs (tsmiRs) such as miR-497, miR-16, and miR-30c-2-3p, which are found to be downregulated in breast cancer (Luo et al. 2013a; Guo et al., 2015; Shukla et al. 2015a). Overexpression of these miRNAs inhibits cell proliferation and disrupts cell cycle progression. Similarly, miR-483-3p is also downregulated in breast cancer, which directly targets cyclin E1. Its overexpression reduces cyclin E1 levels, inhibits the G_1-S transition, blocks DNA synthesis initiation by downstream target p-NPAT, and prevents the formation of the cyclin E1–CDK2 complex, thereby suppressing breast cancer cell proliferation and progression (Huang and Lyu 2018). The ectopic expression of miR-26a-5p inhibited breast cancer cell proliferation by significantly downregulating cyclin D1,

Table 10.1 miRNAs and their regulatory roles in breast cancer progression

Sl. no.	NcRNAs	Targets	Role	Significance	Ref
1	miR-497	Cyclin E1	Inhibits cell proliferation and G_1–S transition	Downregulated in breast cancer	Luo et al. (2013b)
2	miR-16	Cyclin E1	Inhibits cell proliferation and G_1–S transition	Downregulated in breast cancer	Guo et al. (2015)
3	miR-30c-2-3p	Cyclin E1	Inhibits cell proliferation and G_1–S transition	Downregulated in breast cancer	Shukla et al. (2015b)
4	miR-483-3p	Cyclin E1	Reduces cyclin E1 levels, inhibits G_1–S transition, prevents DNA synthesis	Downregulated in breast cancer	Huang and Lyu (2018)
5	miR-26a-5p	Cyclin D1, CDK4, and CDK6	Inhibits proliferative growth, enhances tumor suppressor proteins	Downregulated in breast cancer cells	Huang et al. (2019)
6	miR-200c/141	SerpinB2	Increases metastatic potential, promotes lung, and lymph node metastasis	Overexpressed in metastatic breast cancer	Jin et al. (2017)
7	miR-200c/141	FOXP3 and KAT2B	Regulates EMT and metastasis, potential biomarker for breast cancer	Elevated in plasma during metastasis	Zhang et al. (2017)
8	miR-7-5p	REGγ	Suppresses proliferation, induces apoptosis by targeting REGγ	Regulates apoptosis in breast cancer	Shi et al. (2015)
9	miR-15a/16	BMI1	Induces apoptosis, activates mitochondrial apoptotic pathways	Regulates apoptosis in breast cancer	Patel et al. (2016)
10	miR-17-5p	STAT3	Enhances apoptosis, sensitizes to paclitaxel-induced apoptosis	Regulates apoptosis in breast cancer	Liao et al. (2017)

CDK4, and CDK6 expressions while upregulated the expression levels of tumor suppressor proteins p21, p27, and p53 (Huang et al. 2019).

Epithelial–mesenchymal transition (EMT) is crucial for breast cancer metastasis, providing cancer cells stem-like traits and enhancing migration and invasion capabilities (Kalluri and Weinberg 2009). EMT involves E-cadherin loss, reducing cell–cell contact and potentially activating the Wnt/β-catenin pathway (Jeanes et al. 2008; Tian et al. 2011). Overexpression of vimentin and the E-cadherin-to-N-cadherin switch further promotes the metastatic phenotype of cancer cells (Liu et al. 2015). miRNAs, such as the miR-200c/141 cluster, play a significant role in the metastatic and invasive potential of breast cancer cells by regulating EMT and genes involved in cell motility and invasion. In a study, Jin et al. demonstrated that miR-200c/141 overexpression increased the metastatic potential of breast cancer cells through upregulation of SerpinB2 expression, promoting lung and lymph node metastasis in a xenograft model. SerpinB2 overexpression was linked to higher metastasis risk in triple-negative breast cancer (TNBC) (Jin et al. 2017). Other studies showed that miR-200c/141 expression is induced by FOXP3 and KAT2B, with increased plasma levels of miR-200c/141 observed in metastatic breast cancer, suggesting its potential as a biomarker for metastasis (Zhang et al. 2017).

Apoptosis, a coordinated process essential for maintaining cellular balance, is often disrupted in cancer, leading to malignant cell survival, tumor progression, metastasis, and resistance to therapies (Wong 2011). Several miRNAs regulate apoptosis in breast cancer cells by targeting key molecules. For instance, miR-7-5p suppresses breast cancer cell proliferation and induces apoptosis by downregulating proteasome activator subunit 3 (REGγ), which modulates proteolysis of p21 and p53 (Li et al. 2007; Shi et al. 2015). Similarly, ectopic expression of miR-15a and miR-16 downregulated BMI1, elevated ROS level in mitochondria, and subsequently led to impaired mitochondrial membrane potential, resulting in mitoptosis in breast cancer cells (Patel et al. 2016). Furthermore, miR-17-5p enhances apoptosis by inhibiting STAT3 and upregulating p53, thereby sensitizing breast cancer cells to paclitaxel-induced apoptosis (Liao et al. 2017). These miRNAs collectively highlight their therapeutic potential in restoring apoptotic pathways in breast cancer.

10.2.2 lncRNAs Regulate Breast Cancer Cell Proliferation

lncRNAs are a type of ncRNA that has more than 200 nucleotides and does not encode proteins. Studies have shown the critical roles of lncRNAs in various stages of cancer development and progression. Dysregulated expression of lncRNAs has been linked to breast cancer progression. Functionally, lncRNAs can act as either enhancers or suppressors of breast cancer cell invasion and metastatic behavior (Liu et al. 2020a). Several studies have highlighted the crucial roles of lncRNAs in breast cancer cell proliferation, apoptosis, metastasis, invasion, EMT, and resistance to drugs, as compiled in Table 10.2 (Wang et al. 2019; Bao et al. 2020; Du et al. 2020a). LncRNA MIR210HG is overexpressed in breast cancer tissues, and its downregulation has been shown to inhibit cell proliferation in MCF-7 and MDA-MB-231 cells by acting as a ceRNA, sequestering miR-1226-3p to repress mucin1 (MUC1-C). Further, MIR210HG has also been shown to promote tumor progression by enhancing the Warburg effect (Li et al. 2019b; Du et al. 2020c).

Another oncogenic LncRNA, H19, plays an important role in breast cancer progression by regulating the H19/miR-138/SOX4 axis. Transient knockdown of H19 inhibits cell proliferation by attenuating miR-138-mediated suppression of SOX4, highlighting its significance in tumor development (Yan et al. 2020; Chen et al. 2021). Similarly, LINC00673 has been identified as a critical oncogenic lncRNA. Silencing LINC00673 represses tumor growth and cell viability in vitro and in vivo (Qiao et al. 2019). Mechanistic studies revealed that it functions by sponging miR-515-5p, leading to the upregulation of MARK4 and subsequent inactivation of the Hippo signaling pathway, thereby promoting cancer progression (Qiao et al. 2019). Likewise, oncogenic lncRNA Titin antisense1 (TTN-AS1) also

Table 10.2 lncRNAs and their regulatory roles in breast cancer progression

Sl. no.	NcRNAs	Targeting molecules	Role	Significance	Ref
1	MIR210HG	miR-1226-3p	Promotes breast cancer cell proliferation by regulating MUC1-C	Knockdown inhibits cell proliferation in MDA-MB-231 and MCF-7 cells	Li et al. (2019b)
2	H19	miR-138 and SOX4	Regulates H19/ miR-138/SOX4 axis, enhancing tumor progression	Knockdown inhibits cell proliferation	Si et al. (2019)
3	LINC00673	miR-515-5p and MARK4	Acts as an oncogenic lncRNA by regulating the hippo signaling pathway	Silencing represses tumor growth in vitro and in vivo	Qiao et al. (2019)
4	TTN-AS1	miR-524-5p and RRM2	Enhances breast cancer cell proliferation by targeting miR-524-5p	Promotes tumor growth	Feng et al. (2020); Zheng et al. (2021)
5	HOTAIR	miR-601 and ZEB1	Sponges miR-601, regulates ZEB1 expression to promote tumorigenesis	Inhibition suppresses cell growth in vivo and in vitro	Wang et al. (2020c)
6	SOX21-AS1	PI3K/AKT pathway	Upregulated in breast cancer, promotes cell growth *via* PI3K/ AKT	Silencing reduces cell growth in MCF-7 and BT-20 cells	Sheng et al. (2020)
7	NEAT1	miR-410-3p and CCND1	Promotes cell proliferation by sponging miR-410-3p	Inhibition reduces cell proliferation in SKBR3 cells	Liu et al. (2020b)
8	GAS5	miR-196a-5p	Suppresses cell growth by downregulating miR-196a-5p	Tumor suppressor role	Li et al. (2018)
9	MT1JP	miR-24-3p	Suppresses cell proliferation in breast cancer	Downregulated in breast cancer	Zhu et al. (2019a)
10	HAND2-AS1	miR-3118 and miR-1275	Inhibits cell growth by modulating miR-3118 and miR-1275	Tumor suppressor role	Yang et al. (2019a); Dong et al. (2020)
11	LINC01189	miR-586 and Wnt/β-catenin pathway	Impedes tumor growth by deactivating the Wnt/β-catenin pathway	Tumor suppressor role	Zhang et al. (2021a)

(continued)

Table 10.2 (continued)

Sl. no.	NcRNAs	Targeting molecules	Role	Significance	Ref
12	MICAL2-1	miR-25 and DKK3	Inhibits cell proliferation by regulating the miR-25/DKK3 axis	Tumor suppressor role	Yao et al. (2022)
13	IGF2-AS	IGF2 and PI3K/AKT/mTOR pathway	Reduces cell growth by repressing IGF2 and the PI3K/AKT/mTOR pathway	Downregulated in breast cancer	Zhang et al. (2021b)
14	ZFAS1	STAT3 pathway	Limits cell proliferation by activating the STAT3 pathway	Downregulated in breast cancer	Sharma et al. (2021)
15	LINC00641	miR-194-5p	Blocks tumor growth by sponging miR-194-5p	Downregulated in breast cancer	Han and Zhang (2021)
16	LNC942	METTL14	Inhibits apoptosis by stabilizing CXCR4 and CYP1B1	Enhances mRNA stability	Sun et al. (2020)
17	DLX6-AS1	miR-505-3p and RUNX2	Reduces apoptosis by upregulating RUNX2	Promotes survival in cancer cells	Zhao et al. (2019)
18	CERS6-AS1	CERS6	Inhibits apoptosis by modulating apoptosis-related proteins	Promotes cancer cell survival	Bao et al. (2020)
19	HOXC-AS3	YBX1, TK1	Promotes apoptosis by activating YBX1/TK1 axis	Induces apoptosis	Zhang et al. (2018); Su et al. (2022)
20	Linc-ROR	miR-194-3p	Act as ceRNA sponge for miR-194-3p, which targets MECP2	Promotes the cell proliferation, colony formation, cell migration, and invasion by decreasing the sensitivity of breast cancer cells to rapamycin	Zhou et al. (2020)
21	Linc-ROR	TIMP3	Recruit transmethylase MLL1, thereby promotes H3K4 trimethylation in TIMP3 promoter region	Promotes tumor progression in BC cells	Hu et al. (2021)

plays a crucial role in breast cancer cell proliferation. It acts as a competitive sponge for miR-524-5p, upregulating RRM2, and also targets the miR-139-5p/ZEB1 axis to drive tumor growth (Feng et al. 2020; Zheng et al. 2021).

Previous research has shown that HOTAIR is significantly overexpressed in breast cancer, and its inhibition suppresses cell growth in both in vivo and in vitro models. Mechanistically, HOTAIR exerts its oncogenic effects by sponging miR-601 and subsequently regulating ZEB1 expression (Wang et al. 2020c). Similarly, lncRNA SRY-Box 21 antisense RNA 1 (SOX21-AS1) is markedly upregulated in breast cancer tissues, and its silencing in MCF-7 and BT-20 cells reduces cell growth through the PI3K/AKT pathway Sheng et al. (2020). Furthermore, the lncRNA NEAT1 is upregulated in breast cancer tissues, and its inhibition in SKBR3 cells reduces cell proliferation by sponging miR-410-3p, leading to the downregulation of cyclin D1 (CCND1) (Liu et al. 2020d).

In contrast, several lncRNAs act as growth suppressors in breast cancer. GAS5, for instance, reduces cell growth by downregulating miR-196a-5p and inhibiting the PI3K/Akt pathway (Li et al. 2018). Similarly, MT1JP, which is downregulated in breast cancer, suppresses cell proliferation by targeting miR-24-3p (Zhu et al. 2019a). HAND2-AS1 has been observed to inhibit cell growth by modulating miR-3118 and miR-1275, while LINC01189 impedes tumor growth by sponging miR-586 and deactivating the Wnt/β-catenin pathway (Yang et al. 2019a; Dong et al. 2020; Zhang et al. 2021a). Furthermore, MICAL2-1, whose expression is downregulated in breast cancer, curbs cell proliferation by regulating the miR-25/Dickkopf 3 (DKK3) axis (Yao et al. 2022). Similar to these, many other lncRNAs also demonstrate tumor-suppressive functions. IGF2-AS, with reduced levels in breast cancer specimens, inhibits cell growth by repressing IGF2 and the PI3K/Akt/mTOR pathway (Zhang et al. 2021b). ZFAS1, downregulated in breast cancer, limits cell proliferation by activating the STAT3 pathway when silenced (Han and Zhang 2021; Sharma et al. 2021). LINC00641 is also downregulated, effectively blocking tumor growth by sponging miR-194-5p (Han and Zhang 2021). These findings collectively highlight the pivotal roles of lncRNAs in regulating breast cancer cell proliferation, suggesting that targeting specific lncRNAs could offer promising therapeutic strategies to combat breast cancer.

Long noncoding RNAs (lncRNAs) play crucial roles in regulating apoptosis in breast cancer. LNC942 enhances mRNA stability of C-X-C Motif Chemokine Receptor 4 (CXCR4) and cytochrome P450 1B1 (CYP1B1) via N6-methyladenine (m6A)-related protein methyltransferase-like 14 (METTL14), inhibiting apoptosis, while DLX6-AS1 sponges miR-505-3p to upregulate RUNX2, reducing apoptosis (Sun et al. 2020). Similarly, CERS6-AS1 and LINC00673 inhibit apoptosis by upregulating CERS6 and modulating apoptosis-related proteins, respectively (Qiao et al. 2019). HOXC-AS3 and LINC01121 silencing promotes apoptosis through the Y-box binding protein 1 (YBX1)/TK1 axis and HMGA2 repression (Zhang et al. 2018; Su et al. 2022; Xie et al. 2023). SOX21-AS1 depletion induces proapoptotic proteins, and ROR knockdown alters Bcl-2 and Bax levels, facilitating apoptosis (Li et al. 2017a; Peng et al. 2020; Liu et al. 2020c; Zhou et al. 2020; Hu et al. 2021). Conversely, some lncRNAs promote apoptosis. IGF2-AS upregulation enhances apoptotic death, FGF14-AS2 induces apoptosis by regulating miR-205-5p, and PTENP1 triggers apoptosis via the miR-20a/PTEN/PI3K/AKT axis. These findings underscore the dual role of lncRNAs in breast cancer apoptosis regulation (Gao et al. 2019; Yang et al. 2019b; Zhang et al. 2021b).

10.3 Mechanisms of ncRNA-Mediated Regulation of Metabolic Enzymes

Metabolic reprogramming allows tumor cells to adapt to energy demands and sustain uncontrolled proliferation. In breast cancer, alterations in glucose, lipid, and amino acid metabolism provide critical survival advantages, supporting rapid growth, metastasis, and resistance to therapy. NcRNAs, including miRNAs, lncRNAs, and circRNAs, have emerged as key regulators of metabolic pathways by targeting enzymes, transporters, and signaling molecules involved in these processes (Zhang et al. 2021b; Xu et al. 2024a). Dysregulation of ncRNAs can disrupt metabolic homeostasis, driving oncogenesis by enhancing glycolysis, lipid synthesis, and amino acid uptake while inhibiting metabolic checkpoints such as ferroptosis (Safi et al. 2023; Xiang et al. 2024). This section explores the intricate role of ncRNAs in modulating breast cancer metabolism, focusing on their regulation of glucose, lipid, and amino acid metabolic pathways, and highlights their potential as therapeutic targets (Table 10.3).

10.3.1 Metabolic Reprogramming of Glucose by ncRNA in Breast Cancer

Dysregulation in glucose metabolism is one of the leading causes of cancer development. Cancer cells usually acquire energy through aerobic glycolysis. The dysregulation of ncRNAs involved in cancer glucose metabolism leads to mitochondrial dysfunction, activation of key metabolic enzymes, changes in isozyme profiles, and disruptions in glucose metabolic signaling pathways (Shankaraiah et al. 2018; Liu et al. 2020b; Huang et al. 2022; Kansara et al. 2023). Recent studies have highlighted ncRNAs as key regulators in the reprogramming of glucose metabolism in breast cancer cells, where they play crucial roles in promoting cancer cell proliferation and metastasis (Fig. 10.1 and Table 10.3).

Glucose transporters (GLUTs) are essential for facilitating glucose transport across cell membranes and regulating glucose metabolism (Zhao and Zhang 2016). GLUTs are often overexpressed in tumor tissues, increasing glucose metabolism in cancer cells. Certain miRNAs have been found to modulate GLUTs, thereby inhibiting tumor growth and invasion. Genomic analyses revealed that miR-140-5p is downregulated in breast cancer, which directly targets GLUT1 to exert antitumor

Table 10.3 ncRNAs and their role in regulating energy metabolism and their enzymes in breast cancer

Sl. no.	NcRNA	Status in BC cells	Targeting enzyme/ protein	Role	Ref.
1.	Circ_007299	Upregulated	GLUT1	Increase the effectiveness of glycolysis and glucose absorption by increasing GLUT1 protein levels in breast cancer	Qi et al. (2020)
2	miR-155	Upregulated	HK2	Upregulates the expression of HK2 and promotes glycolysis	Jiang et al. (2012)
3	miR-143	Downregulated	HK2	Directly suppresses HK2 expression in breast cancer	Jiang et al. (2012)
4	mir-195	Downregulated	FASN, ACACA	Inhibits proliferation, invasion, and metastasis in breast cancer cells by targeting FASN and ACACA	Singh et al. (2015)
5	mir-335	Downregulated	PKM2/HK2	By targeting PKM2, miR-335 can reduce the conversion of phosphoenolpyruvate to pyruvate, thereby slowing down glycolytic flux and decreasing cancer cell proliferation. Through its regulation of HK2, miR-335 can minimize glucose uptake and the initial step of glycolysis, further inhibiting the glycolytic pathway	Huang et al. (2022)
6	circ_0102273	Upregulated	PFKFB3	Upregulates the expression of PFKFB3 and promotes breast cancer proliferation and metastasis by increasing glycolysis	Yu et al. (2022)
7	HOTAIR	Upregulated	GLUT1	HOTAIR enhances GLUT1 expression through activating the mTOR pathway or through interacting with GLUT1 directly	Liu et al. (2019)
8	LINC00473	Upregulated	PI3K/AKT/ mTOR	Activates the PI3K/AKT/ mTOR pathway, which is central to stimulating glycolysis in cancer cells. This activation leads to an increase in glycolytic enzyme levels, enhancing the Warburg effect in hepatocellular carcinomas	Song et al. (2020)

(continued)

Table 10.3 (continued)

Sl. no.	NcRNA	Status in BC cells	Targeting enzyme/ protein	Role	Ref.
9	UCA1	Upregulated	HK2	Upregulation of HK2 expression and increased the aerobic glycolysis in breast cancer	Mota et al. (2018)
10	MEG3	Downregulated	PI3K/Akt	In breast cancer, MEG3 acts as a tumor suppressor and directly inhibits miR-21 expression via the PI3K/Akt pathway, slowing the Warburg effect in tumors	Zhu et al. (2019b)
11	LINC00346	Upregulated	GLUT1	Regulates glycolysis by modulation of glucose transporter 1 in breast cancer cells	Li et al. (2020b)
12	miR-5096	Downregulated	SLC7A11	Downregulates SLC7A11 to induce ferroptosis	Yadav et al. (2021)
13	Circ-BGN	Upregulated	SLC7A11	Enhances SLC7A11 stability to inhibit ferroptosis	Wang et al. (2022b)
14	miR-876-3p	Downregulated	SLC1A5	Inhibits SLC1A5, reducing glutamine transport	Yuan et al. (2022)
15	Circ_0062558	Upregulated	SLC1A5	Promotes glutamine transport through miR-876-3p/SLC1A5 axis	Yuan et al. (2022)
16	miR-149-5p	Downregulated	SLC1A5	Regulates glutamine uptake and induces apoptosis	Wang et al. (2021b)
17	circSEPT9	Upregulated	SLC1A5	Sponging miR-149-5p, thereby enhances SLC1A5 expression	Wang et al. (2021a)
18	tRiMetF31	Downregulated	PFKFB3	Profoundly suppresses angiogenesis by silencing PFKFB3	Wang et al. (2022a)
19	Circ_KIF4A	Upregulated	PKM2	Promotes liver metastasis in TNBC	Huang et al. (2022)
20	HOXC-AS3	Upregulated	PFK1, GLUT, HK2, LDHA, PDK1, and SIRT6	Regulates the metabolic reprogramming of breast cancer cells under metabolic stress	Zhu et al. (2022)
21	mir-16-1-3p	Downregulated	PGK1	Suppress BC tumor progression by targeting 3'-UTR of *PGK1*	Ye et al. (2020)

(continued)

Table 10.3 (continued)

Sl. no.	NcRNA	Status in BC cells	Targeting enzyme/ protein	Role	Ref.
22	CircARL8B	Upregulated	FASN, ACC1, FABP5	Enhanced BC cell growth, metastasis and fatty acid metabolism via modulating miR-653-5p/HMGA2 axis	Wu et al. (2021)
23	mir-33a	Downregulated	FASN	Inhibits fatty acid synthase metabolism	Ozfiliz-Kilbas et al. (2021)
24	Lnc030	Upregulated	SQLE	Increases cholesterol synthesis, activates PI3K/Akt pathway that governs breast cancer stem cells	Qin et al. (2021)
25	mir-513c/ miR-3163	Downregulated	Glutaminase (GLS)	Regulates glutamine metabolism by targeting GLS	Delgir et al. (2021)

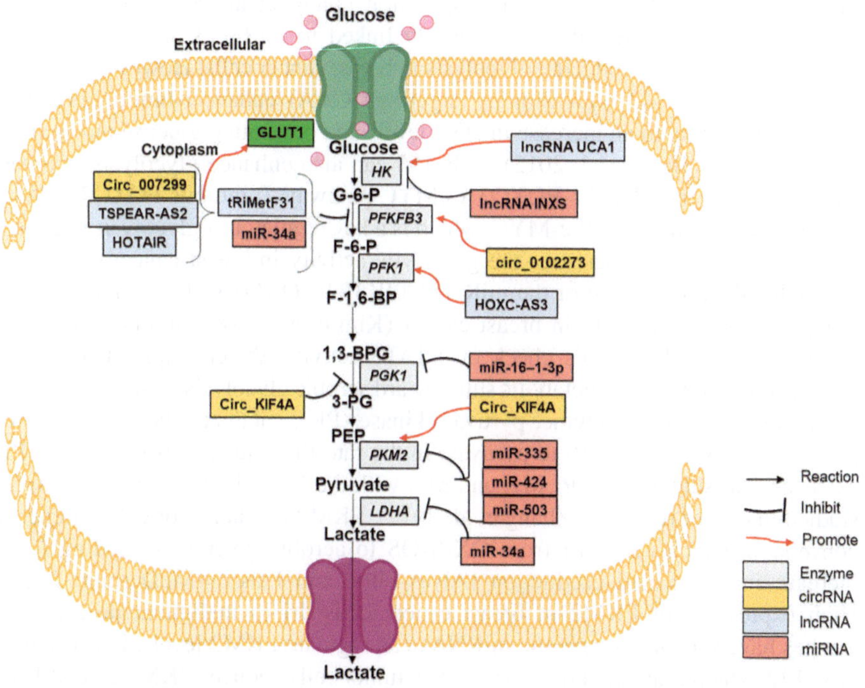

Fig. 10.1 Regulation of glucose metabolism by ncRNAs in breast cancer. ncRNAs regulate glucose uptake and glycolytic flux by modulating GLUTs and glycolytic enzymes. In metabolic reprogrammed breast cancer cells, the uptake of glucose is facilitated by GLUTs and converted into pyruvate through glycolysis by the key enzymes, even under normoxic conditions. Various ncRNAs play a crucial role in regulating this metabolic shift by influencing the expression of GLUT1 and essential glycolytic enzymes, including HK2, PFKFB3, PGK1, PKM2, and LDHA. The regulatory role of ncRNA in these metabolic steps is indicated using different symbols

effects (He et al. 2019). Additionally, miR-Let-7a-5p, a tumor suppressor, targets 3′-UTR of GLUT12 and downregulates its expression in TNBC, thereby suppressing GLUT12-mediated glycolysis in TNBC, making let-7a-5p/GLUT12 axis a therapeutic target for TNBC (Shi et al. 2020). Moreover, oncogenic ncRNAs like Circ_007299 play a pivotal role in promoting aerobic glycolysis in breast cancer via miR-149-5p/*SHMT2* axis (Qi et al. 2020). Similarly, lncRNA TSPEAR-AS2 stabilizes GLUT1 mRNA through an insulin-like growth factor 2 mRNA-binding protein 2 (IGF2BP2)-dependent pathway reprograms glucose metabolism in breast cancer. In addition, GLUT1 expression was repressed by the knockdown of lncRNA LINC00346 in BC cells, and LINC00346 was found to be oncogenic by targeting miR-148a/b, thereby promoting GLUT1 expression (Li et al., 2020c).

In the glycolytic pathway, hexokinase (HK), the initial rate-limiting enzyme, plays a crucial role in cancer metabolism. The upregulation of HK2 is strongly associated with increased malignancy in cancer, and this expression is influenced by various ncRNAs (Li et al. 2014). In breast cancer cells, the overexpression of HK2 enhances tumor progression and metastasis by increasing glycolysis, contributing to a poor prognosis. HK2 has emerged as a significant target for cancer therapy, with growing research indicating that targeting HK2 can effectively regulate glucose metabolism and influence cancer progression (Jiang et al. 2012). miR-155, an inflammation-mediated miRNA, is closely linked to miR-143. miR-143 directly inhibits HK2, thereby reducing glycolysis in breast cancer cells. However, miR-155 can suppress miR-143 expression by targeting CCAAT enhancer-binding protein β (C/EBPβ), leading to an increase in HK2 levels and promoting glucose metabolism in cancer cells (Jiang et al. 2012). miR-155 can also enhance glycolysis in breast cancer through the PIK3R1-FOXO3A-cMYC pathway. It inhibits FOXO3A, leading to the destabilization of c-MYC, which is a key regulator of glycolysis in cancer cells. Which, in turn, increases the glycolytic activity in breast cancer cells. This highlights the critical role of the miR-155/PIK3R1-FOXO3A-cMYC axis in regulating glucose metabolism in breast cancer (Kim et al. 2018). Similarly, lncRNA UCA1 disengaged Merlin's check on STAT3 activity, thereby upregulated HK2 expression, resulting in a metabolic shift toward aerobic glycolysis (Mota et al. 2018).

Another important enzyme, pyruvate kinase (PK), catalyzes the conversion of phosphoenolpyruvate (PEP) to pyruvate in the rate-limiting step of glycolysis. PK exists as four subtypes: L, R, M1, and M2, with the M2 subtype being extensively studied in cancer research (Zhang et al. 2019). PKM2 is crucial for cellular energy conversion, which can shift from OXPHOS to aerobic glycolysis, leading to an increase in lactic acid production (Alves-Filho and Pålsson-McDermott 2016). High expression of PKM2 in breast cancer cells drives malignant progression. Studies have shown that specific miRNAs can regulate PKM2 levels. For instance, miR-122, which is abundantly expressed by tumor cells, reduces PKM2 and GLUT1 expressions and suppresses glucose metabolism in breast cancer cells (Fong et al. 2015). miR-Let-7a-5p is a known tumor-suppressive miRNA in various tissues; recently, a study has shown miR-Let-7a-5p targets signal transducer and activator of transcription 3 (Stat3), which in turn downregulates expression of heterogeneous nuclear ribonucleoprotein (hnRNP-A1), thereby regulating PKM2 expression to modulate glucose metabolism of breast cancer cells. Whereas Stat3 was upregulated, hnRNP-A1 blocked miR-Let-7a-5p biogenesis in breast cancer cells, indicating a functional feedback loop in existence among miR-Let-7a-5p, Stat3, and hnRNP-A1 for regulating PKM2 expression. Thus, this miR-Let-7a-5p/Stat3/hnRNP-A1/PKM2 axis serves as a potential target for breast cancer treatment (Yao et al. 2019).

Lactate dehydrogenase A (LDHA) plays a crucial role in converting pyruvate to lactic acid, catalyzing the final step of glycolysis. It is a key enzyme regulating aerobic glycolysis in cancer cells and is central to modulating the Warburg effect in tumors. Targeting LDHA is considered an effective strategy for disrupting tumor metabolism (Shim et al. 1997). Certain miRNAs act as tumor suppressors by inhibiting LDHA expression, thereby reducing the Warburg effect and suppressing breast cancer growth and metastasis. For instance, miR-34a and miR-30a-5p directly target LDHA, lowering its levels and decreasing glycolysis, glucose uptake, and lactate production in cancer cells (Xiao et al. 2016; Li et al. 2017b). MEG3 acts as a tumor suppressor by sponging miR-21, thereby reversing miR-21-mediated activation of the PI3K/Akt pathway and glycolytic activity. On the other hand, H19 upregulates LDHA by sponging miR-519d-3p, promoting aerobic glycolysis and enhancing energy production, thus supporting tumor progression. These ncRNAs highlight the intricate regulatory mechanisms influencing cancer metabolism (Zhu et al. 2019b; Sun et al. 2021b).

Additionally, research has found that monocarboxylate transporter 1 (MCT1) is also upregulated in breast cancer and facilitates lactate uptake, promoting OXPHOS and angiogenesis. In TNBC, miR-342-3p negatively regulates MCT1, and its downregulation increases MCT1, thereby altering lactate and glucose metabolism in TNBC and disrupting the metabolic balance. Targeting MCT1 presents a potential therapeutic approach, with the miR-342-3p/MCT1 relationship offering a new treatment strategy for breast cancer (Romero-Cordoba et al. 2018; Wang et al. 2021a).

10.3.2 Metabolic Reprogramming of Lipids by ncRNA in Breast Cancer

Fatty acids (FAs) are the main building blocks of several lipid species, including phospholipids, sphingolipids, and triglycerides, which are essential for the rapid proliferation of breast cancer cells by sustaining membrane biosynthesis and providing energy sources under metabolic stress (Koundouros and Poulogiannis 2020). Some of the major ncRNA regulations of lipid metabolism in BC are shown in Fig. 10.2 and Table 10.3. In breast cancer, it was found that circARL8B regulates lipid metabolism by influencing enzymes like acetyl CoA carboxylase 1 (ACC1), fatty acid synthase (FASN), and fatty acid binding protein 5 (FABP5). Silencing circARL8B reduces phospholipids, triglycerides, and lipid-metabolizing enzymes, inhibiting tumor growth and metastasis. This effect was mediated through the

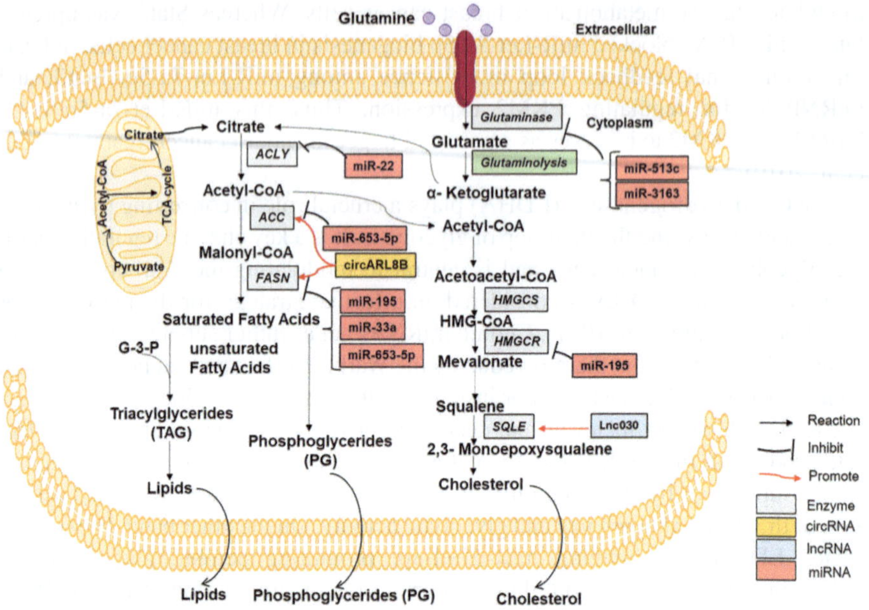

Fig. 10.2 Regulation of lipogenesis, cholesterol, and glutamine metabolism by ncRNAs in breast cancer. ncRNAs modulate lipids and cholesterol by regulating the intermediate metabolism through various key enzymes, including ACLY, ACC, HMGCS, HMGCR, and SQLE. Additionally, ncRNA plays a crucial role in modulating the conversion of glutamine to glutamate by suppressing the activity of glutaminase. The regulatory role of ncRNA in these metabolic steps is indicated using different symbols

miR-653-5p/HMGA2 axis, suggesting circARL8B can be a therapeutic target for disrupting lipid metabolism in breast cancer (Wu et al. 2021). FASN is a crucial enzyme in the endogenous lipogenic pathway, facilitating the conversion of acetyl and malonyl coenzyme A into long-chain saturated fatty acids, with NADPH serving as the primary reducing agent. In a study, it was found that miR-195 inhibits proliferation, invasion, and metastasis by targeting lipid biosynthesis enzymes FASN and ACC1, showcasing its role in altering lipid metabolism (Singh et al. 2015). CircZFAND6 functioned as a competing endogenous RNA (ceRNA) by binding to miR-647, leading to an increase in FASN expression, which in turn promoted the proliferation and metastasis of breast cancer (Huang et al. 2023).

Cholesterol synthesis occurs in three stages: conversion of acetyl-CoA to HMG-CoA, transformation of HMG-CoA into mevalonate (MVA) by HMG-CoA reductase (HMGCR), and the subsequent formation of cholesterol from squalene through multiple reactions. HMGCR is a key enzyme in this pathway. Overexpression of miR-195 downregulated HMGCR, ACC1, and FASN expression, leading to reduced cholesterol and triglyceride levels. This reduction mitigated the epithelial-mesenchymal transition in breast cancer cells, suggesting that miR-195 could target genes involved in adipogenesis and cholesterol synthesis, offering a potential therapeutic approach for breast cancer treatment (Singh et al. 2015).

Additionally, squalene epoxidase (SQLE) is the enzyme responsible for the initial oxygenation step in the transformation of squalene into 2,3(S)-monoxysqualene (MOS) during the third phase of cholesterol synthesis (Sun et al. 2021a). SQLE contributes to oncogenesis in several types of cancer, including breast cancer. Elevated SQLE expression is commonly seen in various cancers and is linked to tumor metastasis and patient prognosis (Helms et al. 2008). In a study, authors demonstrated that lnc030 stabilized SQLE mRNA by interacting with poly(rC)-binding protein 2 (PCBP2), which in turn increased cholesterol production. This triggered the PI3K/Akt signaling pathway via the SQLE/lnc030 axis. This pathway preserved the stemness of breast cancer stem cells (BCSCs) and facilitated breast tumorigenesis and growth (Qin et al. 2021).

10.3.3 Metabolic Reprogramming of Amino Acids by ncRNA in Breast Cancer

Amino acids are essential for the rapid proliferation of cancer cells, acting as building blocks for protein synthesis. They are classified into two types: essential amino acids, which must be obtained through the diet, and nonessential amino acids, which the body can produce. In tumor cells, the loss of expression of enzymes that are responsible for synthesizing non-essential amino acids via mutation or silencing impairs their utilization and affects metabolic processes (Vettore et al. 2020). Some of the major ncRNA regulation of amino acid metabolism in BC are shown in Fig. 10.2 and Table 10.3.

Cancer cells exploit elevated blood glutamine levels as a source of carbon and nitrogen to sustain biosynthesis, energy production, and cellular homeostasis, fueling tumor growth. Glutamine uptake occurs via transporters like solute carrier family 1 (SCL1) and amino acid transporter member 5 (SLC1A5). Additionally, intracellular glutamine-derived glutamate can be exchanged for extracellular cystine through the SLC7A11 transporter (Altman et al. 2016). The circSEPT9 promotes breast cancer progression through the miR-149-5p/SLC1A5 pathway. A reduction in SLC1A5 expression impaired glutamine uptake, suppressed cell proliferation, and triggered apoptosis in breast cancer cells (Wang et al. 2021a; Yuan et al. 2022). Similarly, silencing of circ_0062558, a circular RNA that promotes the growth, migration, and glutamine metabolism of TNBC, decreased glutamine consumption, α-ketoglutarate, and glutamate production, while increased levels of miR-876-3p markedly suppressed SLC1A5 expression (Yuan et al. 2022).

The cystine/glutamate antiporter SLC7A11 supports tumor growth by facilitating cystine uptake, which is crucial for glutathione production and antioxidant defense. Additionally, it inhibits ferroptosis, a form of iron-dependent cell death marked by excessive iron accumulation and lipid peroxidation (Li et al. 2020a).

Overexpression of SLC7A11 is linked to poor survival in breast cancer patients. miR-5096 targets the 3'UTR of SLC7A11, downregulating its expression and inducing cell death through ferroptosis, involving changes in ROS, hydroxyl radicals, and lipid peroxides in breast cancer cells (Yadav et al. 2021). Additionally, Circ-BGN binds to SLC7A11 and OTUB1, enhancing OTUB1-mediated deubiquitination and stabilizing SLC7A11 protein levels, thereby preventing ferroptosis. These findings highlight multiple mechanisms regulating ferroptosis in breast cancer (Wang et al. 2022b).

Overall, these findings highlight the critical role of ncRNAs in regulating glucose, lipid, and amino acid metabolism in cancer cells, underscoring their potential as therapeutic targets in breast cancer.

10.4 Therapeutic Targeting of ncRNA in Cancer Metabolism

In the previous sections, we have presented evidence on the role of ncRNAs in regulating energy metabolism in cancer. Targeting tumor metabolic pathways is an emerging strategy for inhibiting tumor growth and progression (Xiao et al. 2023). Consequently, therapeutic interventions aimed at modulating ncRNAs involved in tumor metabolism offer promising, precise approaches to cancer treatment. There are two well-established strategies for targeting ncRNAs therapeutically, such as restoring the activity or expression of tumor-suppressor ncRNAs in cancer cells and directly inhibiting the activity or expression of oncogenic ncRNAs (Ji et al. 2017; Abd-Aziz et al. 2020). This section explores various molecular approaches developed to modulate ncRNA expression, including miRNA mimics, antisense oligonucleotides (ASOs), miRNA sponges, short hairpin RNAs (shRNA), small interfering RNAs (siRNA), therapeutic circular RNAs (circRNAs), antisense miRs (anti-miRs), and CRISPR–Cas9 gene-editing technology.

10.4.1 miRNA Mimics

miRNA mimics are synthetic oligonucleotide duplexes designed to restore the expression of tumor-suppressor miRNAs that are downregulated in cancer (Winkle et al. 2021). For instance, miR-128 is a well-established tumor suppressor known for its ability to inhibit tumor growth and promote apoptosis (Fang et al. 2022). A cohort study found that low miR-128 expression was associated with shorter overall survival in both TNBC and non-TNBC patients, while shorter disease-free survival was observed only in TNBC patients. However, ectopic expression of miR-128 inhibited glucose metabolism, mitochondrial respiration, and TNBC cell proliferation, suggesting its potential as a therapeutic target for TNBC treatment (Xiao et al.

2018). Several miRNA mimics, such as MRX34 and the miR-16 mimic MesomiR-1, are currently being evaluated in clinical trials for cancer treatment (ClinicalTrials.gov identifier: NCT01829971, NCT02369198).

10.4.2 Antisense miRs (Anti-miRs)

Antisense miRs (anti-miRs) are chemically modified, single-stranded oligonucleotides that selectively inhibit oncogenic miRNAs (oncomiRs). They work by competitively binding to target miRNAs, preventing their interaction with mRNAs (Stenvang et al. 2012; Winkle et al. 2021). Du et al. demonstrated that overexpression of the oncomiR miR-210-3p in TNBC cells enhanced aerobic glycolysis. miR-210-3p stabilized HIF-1α and repressed p53 activity by targeting glycerol-3-phosphate dehydrogenase 1 and cytoglobin, respectively, thereby promoting TNBC cell proliferation. These findings suggest that miR-210-3p could be a promising therapeutic target for TNBC (Du et al. 2020b). A similar strategy has been explored for tRNA-derived small RNA (tsRNA) regulation. Using small RNA inhibitors, Zhu et al. constructed a knockdown model for sncRNA $tRF^{Lys-CTT-010}$, which led to reduced G6PC protein levels and decreased lactate production. Silencing $tRF^{Lys-CTT-010}$ significantly inhibited TNBC cell growth, movement, and invasion (Zhu et al. 2021). This evidence indicates that the $tRF^{Lys-CTT-010}$/G6PC axis can be manipulated as a therapeutic target in TNBC treatment. Despite their therapeutic potential, miRNA mimics and anti-miRs face challenges due to their physiochemical limitations, including poor cell membrane permeability, immunotoxicity, off-target effects, and rapid degradation by nucleases (Segal and Slack 2020). Chemical modifications such as 2′-O-methylation and locked nucleic acids (LNA) have been introduced to improve their stability and specificity. Additionally, lipid-based nanocarriers are commonly used for in vivo RNA drug delivery, while emerging peptide- and polymer-based systems offer promising alternatives (Winkle et al. 2021).

10.4.3 miRNA Sponges

miRNA sponges are engineered RNA molecules designed to sequester multiple oncogenic miRNAs, thereby preventing their interaction with target mRNAs. These sponges contain multiple miRNA-binding sites, enabling them to inhibit entire miRNA clusters effectively (Esau 2008). The ability of miRNA sponges to effectively target the miR-17 ~ 92 cluster, including miR-17, miR-18a, miR-19, and miR-92a has been demonstrated to successfully silence all the miRNAs within the cluster simultaneously. Additionally, in vitro studies revealed that antisense oligonucleotides (ASOs) aimed at a single miRNA were less effective than miRNA sponges targeting the entire miR-17 ~ 92 cluster in inhibiting the growth of B-cell lymphoma cells (Kluiver et al. 2012).

10.4.4 Circular RNAs (circRNAs) as miRNA Sponges

The therapeutic potential of miRNA inhibition is further expanded by the discovery of circular long noncoding RNAs (circRNAs) that act as natural miRNA sponges. ciRS-7, the first identified circRNA, contains over 70 conserved miR-7 binding sites and is highly expressed in the human brain (Hansen et al. 2013). CircRNAs are generally downregulated in cancerous tissues, suggesting their potential role in tumor suppression (Kristensen et al. 2019). Both endogenous and synthetic circRNAs can serve as stable and potent miRNA sponge therapies. However, foreign circRNAs have been shown to activate the PAMP receptor RIG1, triggering an immune response (Chen et al. 2017). Modifications such as N6-methyladenosine (m6A) alterations, which are found in endogenous circRNAs, may help mitigate this immunogenic effect (Chen et al. 2019). Thus, further evaluation of synthetic circRNAs is required to harness their full potential as miRNA sponge therapies.

10.5 Dietary Bioactives Regulating ncRNAs in Breast Cancer Metabolism

Dietary bioactives are well known for their diverse biological functions, including maintaining overall metabolic homeostasis. Dietary bioactives play a crucial role in modulating breast cancer metabolism by influencing key signaling pathways involved in energy balance, oxidative stress, and cellular proliferation. These bioactives, derived from fruits, vegetables, and medicinal plants, can regulate glucose and lipid metabolism, inhibit aerobic glycolysis (Warburg effect), and restore mitochondrial function in cancer cells. They also impact metabolic enzymes and oncogenic pathways by modulating ncRNA expression, thereby altering tumor progression. Additionally, bioactives such as polyphenols, flavonoids, and alkaloids exhibit anti-inflammatory and antioxidant properties, reducing metabolic stress and enhancing therapeutic efficacy in breast cancer. They exert their effects through various mechanisms and have shown promising applications in cancer therapy. A growing body of evidence suggests that dietary bioactives possess antitumor potential across multiple cancer types (Mierziak et al. 2021). Recent studies indicate that bioactives can directly interact with mature ncRNAs, influencing metabolic pathways (Baselga-Escudero et al. 2014). Emerging evidence suggests that plant-based ncRNAs and dietary bioactives like curcumin, resveratrol, and genistein can modulate ncRNA expression and activity. Preclinical studies indicate that these bioactives either directly influence the activity of metabolic enzymes or influence endogenous ncRNAs that have an impact on metabolic enzymes, thereby regulating key metabolic pathways in breast cancer (Table 10.4).

Table 10.4 Preclinical research on dietary bioactives on ncRNA-mediated breast cancer metabolism

Bioactives	NcRNA	Metabolic targets	Pathway	Ref.
Resveratrol	miR-34a	LDHA	Glycolysis	Xiao et al. (2016); Otsuka et al. (2018)
Quercetin	↓ miR-22	ACL	Fatty acid metabolism	Koufaris et al. (2016); Wang et al. (2020b)
	↓ miR-206	PFKFB3	Glycolysis	Ge et al. (2015); Zhang et al. (2020)
	↓ UCA1	HK2	Glycolysis	Mota et al. (2018); Rezaie et al. (2021)
Sulforaphane	↓ miR-23b, miR-92b, miR-381, and miR-382	HK2 and PKM2	Cell cycle, clonogenicity, and migration	Lewinska et al. (2017)
Diindolylmethane	↑ miR-212/132	GLUT1, HK2, and LDHA	Inhibits proliferation, migration, and invasion	Hanieh (2015); Dai et al. (2017); Zhang et al. (2020)

10.5.1 Dietary Bioactive on ncRNA-Mediated Breast Cancer Metabolism

Resveratrol, a phytoalexin from grapes, berries, and peanuts, has been shown to induce tumor-suppressive miRNAs such as miR-34a, miR-424, and miR-503, thereby inhibiting breast cancer cell proliferation by suppressing heterogeneous nuclear ribonucleoprotein A1 (HNRNPA1). Moreover, the study also found that resveratrol exerts broad effects on the HNRNPA1-related pre-mRNA splicing pathway (Otsuka et al. 2018). Another study showed that hnRNPI, hnRNPA1, and hnRNPA2 mediate the alternative splicing of PKM. The overexpression of these proteins facilitates the inclusion of exon 10 of PKM2, promoting aerobic glycolysis in various cancers. Additionally, miR-34a-mediated repression of LDHA inhibited aerobic glycolysis and cell proliferation in breast cancer cells in vitro (David et al. 2010; Xiao et al. 2016). These studies provide novel insights into the regulatory roles of resveratrol on energy metabolism in breast cancer.

Quercetin, a flavonoid found in berries, apples, and onions, has been shown to modulate miR-22 expression in HOK cells and mitigate LPS-induced injury (Wang et al. 2020b). miR-22 functions as both a tumor suppressor and an oncogene, depending on the cellular microenvironment. Koufaris et al. demonstrated that miR-22 regulates tumor metabolism by repressing de novo fatty acid synthesis and elongation through targeting ATP-citrate lyase (ACLY) and long-chain fatty acid elongase (ELOVL6) (Koufaris et al. 2016). Similarly, miR-206 can act as both a tumor promoter and a suppressor. Quercetin promotes osteogenesis in bone marrow mesenchymal stem cells (BMSCs) by modulating the miR-206/Cx43 pathway. It also directly interacts with the 3'-UTR of 6-phosphofructo-2-kinase (*PFKFB3*), thereby inhibiting glycolysis, reducing lactate production, and suppressing proliferation and migration in breast cancer cells (Ge et al. 2015; Zhang et al. 2020). Additionally, lncRNA Urothelial Cancer-Associated 1 (UCA1) has been implicated in promoting cancer cell proliferation, migration, and invasion by sponging miR-126. UCA1 is upregulated in multiple cancers through the cooperative action of the Hippo and TGF-β transcription effectors. Its upregulation disrupts Merlin's inhibition of STAT3, leading to HK2 upregulation and a shift toward aerobic glycolysis in breast cancer cells (Mota et al. 2018; Sun et al. 2018). Interestingly, quercetin has been shown to induce cell death and arrest the cell cycle in MCF-7 breast cancer cells by upregulating lncRNA INXS and downregulating UCA1 (Rezaie et al. 2021).

Sulforaphane (SFN), an isothiocyanate from cruciferous vegetables, has been shown to significantly reduce the levels of miR-23b, miR-92b, miR-381, and miR-382 in MCF-7, MDA-MB-231, and SK-BR-3 breast cancer cells. Elevated mRNA levels of *TGFB2* and *CDKN1A* were linked to a large decrease in miR-23b, miR-92b, and miR-381 in SK-BR-3 cells, which could explain p21-mediated cell cycle arrest. In addition, onco-miRNA miR-382 regulated the RERG/Ras/ERK signaling axis, thereby promoting the viability, clonogenicity, survival, migration, invasion, and tumorigenesis/metastasis of breast cancer cells. The expression of miR-382 is negatively correlated with tumor suppressor RERG expression. SFN also reduces HK2 and PKM2 expression in breast cancer cells, further suppressing aerobic glycolysis (Lewinska et al. 2017).

Another potent bioactive from cruciferous vegetables, 3,3′-diindolylmethane (DIM), has been reported to induce metabolic stress in triple-negative breast cancer (TNBC) by downregulating key glycolytic mediators such as PKM2, GLUT1, and HIF-1α. A recent study found that DIM upregulates miR-212/132 in TNBC cells, leading to SOX4 inhibition and reduced invasion and metastasis of MDA-MB-231 cells (Hanieh 2015; Penta et al. 2022). Another independent study has reported that SOX4 promoted key glycolytic genes, including GLUT1, HK2, and LDHA, through activation of AKT8 in rodent T-cell lymphoma (AKT) signaling pathways (Dai et al. 2017). These findings highlight bioactives from cruciferous vegetables as potent inhibitors of aerobic glycolysis in breast cancer.

10.5.2 Dietary ncRNA-Mediated Regulation on Breast Cancer Metabolism

In recent times, plant-derived miRNAs have gained significant attention as they may participate in cross-kingdom gene regulation, influencing mammalian gene expression and physiological processes. A pioneering study by Zhang and colleagues detected plant-derived miRNAs in human and animal serum, suggesting that dietary ncRNAs, especially plant miRNAs, could regulate genes across species. This finding has sparked significant research interest in the potential role of plant-derived miRNAs in health and disease (Chen et al. 2008; Mitchell et al. 2008; Zhang et al. 2012). Plant miRNAs differ from animal miRNAs due to distinct biochemical characteristics, notably the 2′-O-methylation at the ribose of the 3′ end nucleotide. This specific modification, unique to plant miRNAs, has been utilized to differentiate them from endogenous miRNAs. Researchers suggest that this alteration enhances the stability of these dietary miRNAs during biological processes (Yu et al. 2005). This structural advantage has been pivotal in their detection in serum and tissues following dietary consumption, as demonstrated with rice-derived miR168a. This miRNA, shown to target LDL receptor adaptor protein 1 (LDLRAP1) in mouse liver, exemplifies the regulatory potential of dietary ncRNAs on host transcriptomes (Zhang et al. 2012).

The presence of serum miR159 is negatively correlated with breast cancer incidence and progression in the Western population. These miR159 have predominantly been detected in serum inside extracellular vesicles with 2′-O-methylation on the 3′ terminal ribose, suggesting a plant origin. miR159 mimics targeted 3′ UTR of TCF7, a transcription factor in the Wnt signaling pathway, inhibited breast cancer cell proliferation. Further, oral administration of miR159 significantly inhibited in vivo tumor growth in xenograft mice (Chin et al. 2016). Another study has shown that miR159 constitutes 60% of broccoli-derived miRNAs (Del Pozo-Acebo et al. 2022). Similarly, miR156a, another dietary ncRNA abundant in broccoli, a miR156a mimic repressed EMT transition in human nasopharyngeal cancer cells by targeting 3′ UTR of junctional adhesion molecule A (Tian et al. 2016). An independent study showed miR156a interferes in the Wnt/β-catenin pathway by targeting Wnt10b (Li et al. 2019a). These studies open a potential avenue for targeting breast cancer metabolic pathways using plant-derived ncRNAs.

Even though there are plenty of proven biological effects of plant miRNAs in cross-kingdom regulation, there are some major hurdles to overcome. Low copy number in serum, bioavailability, and safety concerns are major concerns regarding the use of dietary ncRNAs as therapeutic agents (Zhang et al. 2012; Dickinson et al. 2013; Sundaram 2019). With the advancement in research to overcome these challenges, dietary ncRNAs could provide an alternative approach to target various pathological conditions, including breast cancer.

10.6 Clinical Translation of Dietary Bioactive in Breast Cancer Therapy

Emerging research suggests that numerous dietary bioactives, including polyphenols, flavonoids, and indoles, can modulate noncoding RNA (ncRNA) expression, thereby influencing cancer progression and metastasis through metabolic regulation in preclinical models. However, the clinical translation of these dietary interventions remains in its early stages, with ongoing studies evaluating their efficacy, bioavailability, and potential integration into breast cancer therapy. While several clinical trials have explored the role of dietary bioactives in breast cancer prevention and treatment, no clinical studies have specifically examined their effects on ncRNAs in breast cancer.

Genistein, an isoflavonoid phytoestrogen, is mainly available from soybeans. Epidemiological correlations have exhibited that the soy-rich diet and cancer chemoprevention were linked to the presence of genistein in soy-based foods (Li et al. 2013). Its chemopreventive effects have been attributed to multiple mechanisms, including epigenetic modifications (Tuli et al. 2019). A phase I clinical trial in postmenopausal women evaluated high-dose soy isoflavones (900 mg/day) and found no significant changes in DNA damage or apoptosis in peripheral lymphocytes, though estrogenic effects were observed. Nonetheless, unconjugated soy isoflavones appeared to be safe and well-tolerated at high doses (Pop et al. 2008). Similarly, a phase IIB trial in women at high risk for breast cancer assessed six months of soy isoflavone supplementation (PTIG-2535) but found no significant reduction in breast tissue density or cell proliferation (Khan et al. 2012). Additionally, an ongoing phase II clinical trial is investigating the effect of a combination therapy involving genistein and gemcitabine in stage IV breast cancer patients (ClinicalTrials.gov identifier: NCT00244933).

Quercetin is also extensively studied in many clinical investigations. A phase II open-label, single-arm study is currently evaluating the safety and efficacy of a combination therapy involving quercetin and Dasatinib alongside chemotherapy in metastatic TNBC. Additionally, an early phase I trial is examining a quadruple therapy comprising quercetin, zinc, metformin, and epigallocatechin gallate (EGCG) as an adjuvant treatment for TNBC (ClinicalTrials.gov identifier: NCT06355037 and NCT05680662). Furthermore, a randomized, double-blinded, placebo-controlled phase II clinical trial evaluated the effects of green tea extract (GTE), rich in EGCG, supplementation over 12 months (four decaffeinated capsules per day, totaling 1315 mg catechins, including 843 mg EGCG) on mammographic density in postmenopausal women. While no significant overall changes were observed, a reduction in mammographic density was noted in younger women (aged 50–55 years), suggesting an age-dependent effect similar to tamoxifen (Samavat et al. 2017).

Sulforaphane (SFN), a bioactive compound from cruciferous vegetables, has also been clinically investigated. SFN is a known dietary epigenetic modulator and has been shown to inhibit the progression of TNBC by various mechanisms, including epigenetic modifications (Meeran et al. 2010; Sinha et al. 2015). A triple-blinded, randomized controlled trial assessed the effects of glucoraphanin supplementation (a precursor to SFN) in patients with atypical mammograms. Results indicated a downregulation of Ki-67 and HDAC3 expression in benign breast tissue, along with reduced HDAC activity in peripheral blood mononuclear cells (Atwell et al. 2015). Another study found an inverse correlation between cruciferous vegetable consumption and Ki-67 expression in ductal carcinoma in situ (DCIS), though no significant associations were observed with other biomarkers (Zhang et al. 2016). Additionally, an ongoing phase II trial is evaluating SFN co-administration in patients undergoing doxorubicin treatment to mitigate doxorubicin-induced cardiac dysfunction by modulating SIRT1 and Nrf2 target genes (ClinicalTrials.gov identifier: NCT03934905).

Resveratrol, another promising dietary bioactive, has demonstrated potential in regulating carcinogenesis-related enzymes. A clinical study found that a daily intake of 1 g resveratrol for four weeks modulated the expression of phase I enzymes, responsible for carcinogen activation, and phase II enzymes, involved in detoxification, supporting its chemopreventive potential (Chow et al. 2010). A summary of ongoing clinical trials evaluating dietary bioactives, including curcumin, DIM, SFN, quercetin, and EGCG, in breast cancer prevention and therapy is given in Table 10.5.

The discrepancy between the strong preclinical efficacy of these bioactives and their clinical outcomes is likely due to poor systemic bioavailability and rapid first-pass metabolism, which necessitates sustained high-dose intake for effective therapeutic benefits (Anand et al. 2007; Ciska et al. 2009; Yang et al. 2012; Springer and Moco 2019; Williams 2021). While preclinical evidence strongly supports their chemopreventive and therapeutic potential, some studies have yielded contradictory results, emphasizing the need to prioritize clinical validation before further expanding preclinical research (Mondal et al. 2022). Additionally, the absence of clinical studies specifically correlating these bioactives with ncRNA-mediated metabolic reprogramming in breast cancer highlights an important avenue for future research.

10.7 Clinical Challenges and Future Perspective

Metabolic reprogramming is a hallmark of breast cancer, enabling tumor cells to adapt to nutrient fluctuations, sustain proliferation, and evade immune surveillance. Targeting these metabolic alterations is a promising approach to cancer prevention and therapy. While current metabolism-targeting strategies—such as FDA-approved glycolysis inhibitors (2-deoxy-D-glucose), BCAT1 inhibitors (gabapentin), and AMPK activators (metformin), along with experimental agents like CPI-613 (PDH inhibitor) and IACS-010759 (OXPHOS inhibitor)—show potential, their efficacy is often limited by toxicity, resistance, and lack of specificity (Lukey et al. 2017; Pujalte-Martin et al. 2024). This underscores the need for novel, more precise regulatory mechanisms.

NcRNAs, including miRNAs, lncRNAs, and circRNAs, offer a promising approach by selectively modulating key metabolic pathways at the post-transcriptional level. As metabolic modulators, ncRNAs have shown promising potential in vitro and in vivo, yet their clinical translation remains challenging due to the heterogeneity of breast cancer subtypes, and the dynamic nature of ncRNA expression complicates their use as reliable biomarkers or therapeutic targets. Additionally, the delivery of ncRNA-based therapeutics faces significant hurdles, including stability, off-target effects, and the need for tissue-specific targeting.

The endogenous ncRNA expressions are highly dysregulated in BC. Specifically, tumor-suppressive ncRNAs are frequently downregulated, whereas oncogenic

Table 10.5 List of clinical studies of dietary bioactives on breast cancer regulation

Bioactive	Condition	Sample size	Phase	Current status	NCT number Reference
Genistein	Preventing breast or endometrial cancer in healthy postmenopausal women	30	I	Completed	NCT00099008, Pop et al. (2008)
	Women at high risk for breast cancer	126	II	Completed	NCT00290758, Khan et al. (2012)
	Combination therapy for stage IV breast cancer	19	II	Completed	NCT00244933
Quercetin	Chemoresistance in triple-negative breast cancer	10	II	Recruiting	NCT06355037
	Adjuvant therapy for early, metastatic breast cancer and triple-negative breast cancer	200	I	Unknown	NCT05680662
Sulforaphane	Breast cancer	34	II	Completed	NCT00982319
	Precancerous breast cancer	54	II	Completed	NCT00843167
	Breast cancer patients undergoing radiotherapy	30	II	Completed	NCT01740323
	Cardiac dysfunction In breast cancer patients under chemotherapy	70	N.A.	Recruiting	NCT03934905
	Breast cancer	60	II	Completed	NCT02970682
	Radiation-induced dermatitis in patients with breast cancer	12	N.A	Completed	NCT00894712
Curcumin	Preoperative breast cancer	30	N.A.	Unknown	NCT03847623
	Breast cancer chemoprevention	29	N.A	Completed	NCT01975363
	Radiation-induced dermatitis in patients with breast cancer	191	II	Completed	NCT02556632
	Breast cancer	41	N.A	Completed	NCT00949923
	Skin damage prevention in patients with breast cancer receiving adjuvant radiotherapy	68	II	Enrolling by invitation	NCT02580279
	Chemotherapy-treated breast cancer patients	30	II	Completed	NCT01740323
	Breast cancer primary tumors	22	I	Completed	NCT03980509
	Breast and ovarian cancer syndrome (HBOC)	34	N.A	Completed	NCT05306002

(continued)

Table 10.5 (continued)

Bioactive	Condition	Sample size	Phase	Current status	NCT number Reference
EGCG	Hormone receptor-negative stage I-III breast cancer	40	I	Completed	NCT00516243
	Breast cancer chemoprevention	1075	II	Completed	NCT00917735
	Preoperative breast cancer	32	II	Completed	NCT00676793
Diindolylmethane	Breast cancer	144	N.A	Completed	NCT01391689
	Breast cancer chemoprevention	60	III	Completed	NCT02525159
	Breast tissue density among female *BRCA* mutation carriers over 2 years	23	NA	Completed	NCT02197000
	BRCA1 mutation carriers	300	I	Recruiting	NCT01022333

ncRNAs are upregulated (Slack and Chinnaiyan 2019). Dietary bioactives have been shown to modulate this dysregulation in ncRNA expression. Therefore, dietary bioactives offer a potential strategy to harness endogenous ncRNAs for metabolic interventions, thereby enhancing treatment efficacy while minimizing systemic toxicity and off-target effects. This suggests a novel strategy for leveraging dietary interventions to refine metabolic targeting while minimizing toxicity and off-target effects. However, the bioavailability of these compounds and their direct impact on ncRNA profiles remain underexplored, with no clinical studies to date establishing a direct link between dietary bioactives and ncRNA regulation in cancer.

Future research should focus on elucidating the complex interactions between ncRNAs and dietary bioactives to develop precision nutrition strategies for breast cancer management. The identification of ncRNA signatures responsive to dietary interventions could pave the way for personalized dietary recommendations aimed at modulating cancer metabolism. Moreover, clinical trials assessing the synergistic potential of ncRNA-targeting therapies with conventional treatments, such as chemotherapy or immunotherapy, are crucial for their translational success. Moreover, Winkle et al. proposed that the effective use of RNA-based therapies necessitates an unprecedented interdisciplinary strategy that includes technological developments in molecular biology, immunology, pharmacology, chemistry, and nanotechnology (Winkle et al. 2021). Although RNA therapeutics have made great strides in each of these fields independently, their effective translation relies on additional transdisciplinary development to enhance delivery, tolerance, and selectivity. By addressing these clinical and translational challenges, ncRNA-based strategies hold significant promise in shaping the future of breast cancer treatment through dietary and molecular interventions.

10.8 Conclusion

Metabolic reprogramming of breast cancer is driven by complex regulatory networks in which ncRNAs play a central role by modulating key pathways such as glycolysis, lipid metabolism, and amino acid metabolism. Targeting dysregulated ncRNAs presents a promising strategy for metabolic intervention, offering potential avenues for therapeutic development. Additionally, dietary bioactives have emerged as potential modulators of ncRNA expression, highlighting their relevance in shaping metabolic responses and tumor progression. However, despite compelling preclinical data, the clinical translation of these findings remains challenging due to issues related to bioavailability, heterogeneity in breast cancer subtypes, and the complexity of ncRNA regulation. Addressing these challenges requires an interdisciplinary approach, integrating molecular biology, pharmacology, and nanotechnology to enhance the efficacy and specificity of ncRNA-based interventions. Future research should prioritize clinical validation of ncRNA-targeting dietary strategies, paving the way for precision nutrition in breast cancer therapy. By bridging the gap between diet, ncRNAs, and breast cancer cell metabolism, this emerging field holds significant promise for advancing personalized, targeted treatment strategies.

Ethical Approval This chapter does not contain any studies with human participants or animals performed by any of the authors.

References

Abd-Aziz N, Kamaruzman NI, Poh CL (2020) Development of MicroRNAs as potential therapeutics against cancer. J Oncol 2020:1–14. Available at: https://doi.org/10.1155/2020/8029721

Aden D et al (2025) Metabolic reprogramming in cancer: implications for immunosuppressive microenvironment. Immunology 174(1):30–72. Available at: https://doi.org/10.1111/imm.13871

Altman BJ, Stine ZE, Dang CV (2016) From Krebs to clinic: glutamine metabolism to cancer therapy. Nat Rev Cancer 16(10):619–634. Available at: https://doi.org/10.1038/nrc.2016.71

Alves-Filho JC, Pålsson-McDermott EM (2016) Pyruvate kinase M2: a potential target for regulating inflammation. Front Immunol 7:145. Available at: https://doi.org/10.3389/fimmu.2016.00145

Anand P et al (2007) Bioavailability of curcumin: problems and promises. Mol Pharm 4(6):807–818. Available at: https://doi.org/10.1021/mp700113r

Anastasiadou E, Jacob LS, Slack FJ (2018) Non-coding RNA networks in cancer. Nat Rev Cancer 18(1):5–18. Available at: https://doi.org/10.1038/nrc.2017.99

Atwell LL et al (2015) Sulforaphane bioavailability and chemopreventive activity in women scheduled for breast biopsy. Cancer Prev Res (Philadelphia, Pa) 8(12):1184–1191. Available at: https://doi.org/10.1158/1940-6207.CAPR-15-0119

Bao G et al (2020) Long noncoding RNA CERS6-AS1 functions AS a malignancy promoter in breast cancer by binding to IGF2BP3 to enhance the stability of CERS6 mRNA. Cancer Med 9(1):278–289. Available at: https://doi.org/10.1002/cam4.2675

Baselga-Escudero L et al (2014) Resveratrol and EGCG bind directly and distinctively to miR-33a and miR-122 and modulate divergently their levels in hepatic cells. Nucleic Acids Res 42(2):882–892. Available at: https://doi.org/10.1093/nar/gkt1011

Bray F et al (2024) Global cancer statistics 2022: GLOBOCAN estimates of incidence and mortality worldwide for 36 cancers in 185 countries. CA Cancer J Clin 74(3):229–263. Available at: https://doi.org/10.3322/caac.21834

Chen X et al (2008) Characterization of microRNAs in serum: a novel class of biomarkers for diagnosis of cancer and other diseases. Cell Res 18(10):997–1006. Available at: https://doi.org/10.1038/cr.2008.282

Chen YG et al (2017) Sensing self and foreign circular RNAs by intron identity. Mol Cell 67(2):228–238.e5. Available at: https://doi.org/10.1016/j.molcel.2017.05.022

Chen YG et al (2019) N6-Methyladenosine modification controls circular RNA immunity. Mol Cell 76(1):96–109.e9. Available at: https://doi.org/10.1016/j.molcel.2019.07.016

Chen X et al (2021) Long non-coding RNA H19 regulates glioma cell growth and metastasis via miR-200a-mediated CDK6 and ZEB1 expression. Front Oncol 11:757650. Available at: https://doi.org/10.3389/fonc.2021.757650

Chin AR et al (2016) Cross-kingdom inhibition of breast cancer growth by plant miR159. Cell Res 26(2):217–228. Available at: https://doi.org/10.1038/cr.2016.13

Chow H-HS et al (2010) Resveratrol modulates drug- and carcinogen-metabolizing enzymes in a healthy volunteer study. Cancer Prev Res 3(9):1168–1175. Available at: https://doi.org/10.1158/1940-6207.CAPR-09-0155

Ciska E, Verkerk R, Honke J (2009) Effect of boiling on the content of ascorbigen, indole-3-carbinol, indole-3-acetonitrile, and 3,3'-diindolylmethane in fermented cabbage. J Agric Food Chem 57(6):2334–2338. Available at: https://doi.org/10.1021/jf803477w

Dai W et al (2017) SOX4 promotes proliferative signals by regulating glycolysis through AKT activation in melanoma cells. J Invest Dermatol 137(11):2407–2416. Available at: https://doi.org/10.1016/j.jid.2017.06.026

David CJ et al (2010) HnRNP proteins controlled by c-Myc deregulate pyruvate kinase mRNA splicing in cancer. Nature 463(7279):364–368. Available at: https://doi.org/10.1038/nature08697

DeBerardinis RJ et al (2008) The biology of cancer: metabolic reprogramming fuels cell growth and proliferation. Cell Metab 7(1):11–20. Available at: https://doi.org/10.1016/j.cmet.2007.10.002

Del Pozo-Acebo L et al (2022) Therapeutic potential of broccoli-derived extracellular vesicles as nanocarriers of exogenous miRNAs. Pharmacol Res 185:106472. Available at: https://doi.org/10.1016/j.phrs.2022.106472

Delgir S et al (2021) The expression of miR-513c and miR-3163 was downregulated in tumor tissues compared with normal adjacent tissue of patients with breast cancer. BMC Med Genet 14(1):180. Available at: https://doi.org/10.1186/s12920-021-01029-3

Dickinson B et al (2013) Lack of detectable oral bioavailability of plant microRNAs after feeding in mice. Nat Biotechnol 31(11):965–967. Available at: https://doi.org/10.1038/nbt.2737

Dong G et al (2020) HAND2-AS1 works as a ceRNA of miR-3118 to suppress proliferation and migration in breast cancer by upregulating PHLPP2. BioMed Res Int 2020:8124570. Available at: https://doi.org/10.1155/2020/8124570

Du Y, Wei N, Hong J et al (2020a) Long non-coding RNASNHG17 promotes the progression of breast cancer by sponging miR-124-3p. Cancer Cell Int 20:40. Available at: https://doi.org/10.1186/s12935-020-1129-y

Du Y, Wei N, Ma R, Jiang S et al (2020b) A miR-210-3p regulon that controls the Warburg effect by modulating HIF-1α and p53 activity in triple-negative breast cancer. Cell Death Dis 11(9):731. Available at: https://doi.org/10.1038/s41419-020-02952-6

Du Y, Wei N, Ma R, Jiang S-H et al (2020c) Long noncoding RNA MIR210HG promotes the Warburg effect and tumor growth by enhancing HIF-1α translation in triple-negative breast cancer. Front Oncol 10:580176. Available at: https://doi.org/10.3389/fonc.2020.580176

Esau CC (2008) Inhibition of microRNA with antisense oligonucleotides. Methods (San Diego, Calif.) 44(1):55–60. Available at: https://doi.org/10.1016/j.ymeth.2007.11.001

Fang W et al (2022) microRNA-128-3p inhibits CD4+ regulatory T cells enrichment by targeting interleukin 16 in gastric cancer. Bioengineered 13(1):1025–1038. Available at: https://doi.org/10.1080/21655979.2021.2017566

Feng H et al (2020) LncRNA TTN-AS1 regulates miR-524-5p and RRM2 to promote breast cancer progression. Onco Targets Ther 13:4799–4811. Available at: https://doi.org/10.2147/OTT.S243482

Fong MY et al (2015) Breast-cancer-secreted miR-122 reprograms glucose metabolism in premetastatic niche to promote metastasis. Nat Cell Biol 17(2):183–194. Available at: https://doi.org/10.1038/ncb3094

Fouad YA, Aanei C (2017) Revisiting the hallmarks of cancer. Am J Cancer Res 7(5):1016–1036

Gao X et al (2019) PTENP1/miR-20a/PTEN axis contributes to breast cancer progression by regulating PTEN via PI3K/AKT pathway. J Exper Clin Cancer Res 38(1):256. Available at: https://doi.org/10.1186/s13046-019-1260-6

Ge X et al (2015) Overexpression of miR-206 suppresses glycolysis, proliferation and migration in breast cancer cells via PFKFB3 targeting. Biochem Biophys Res Commun 463(4):1115–1121. Available at: https://doi.org/10.1016/j.bbrc.2015.06.068

Guo X, Connick MC et al (2015) MicroRNA-16 modulates HuR regulation of cyclin E1 in breast cancer cells. Int J Mol Sci 16(4):7112–7132. Available at: https://doi.org/10.3390/IJMS16047112

Han X, Zhang S (2021) Role of long non-coding RNA LINC00641 in cancer. Front Oncol 11:829137. Available at: https://doi.org/10.3389/fonc.2021.829137

Hanieh H (2015) Aryl hydrocarbon receptor-microRNA-212/132 axis in human breast cancer suppresses metastasis by targeting SOX4. Mol Cancer 14:172. Available at: https://doi.org/10.1186/s12943-015-0443-9

Hansen TB et al (2013) Natural RNA circles function as efficient microRNA sponges. Nature 495(7441):384–388. Available at: https://doi.org/10.1038/nature11993

He Y et al (2019) Analysis of miRNA-mRNA network reveals miR-140-5p as a suppressor of breast cancer glycolysis via targeting GLUT1. Epigenomics 11(9):1021–1036. Available at: https://doi.org/10.2217/epi-2019-0072

Helms MW et al (2008) Squalene epoxidase, located on chromosome 8q24.1, is upregulated in 8q+ breast cancer and indicates poor clinical outcome in stage I and II disease. Br J Cancer 99(5):774–780. Available at: https://doi.org/10.1038/sj.bjc.6604556

Hsu PP, Sabatini DM (2008) Cancer cell metabolism: Warburg and beyond. Cell 134(5):703–707. Available at: https://doi.org/10.1016/j.cell.2008.08.021

Hu A et al (2021) Long non-coding RNA ROR recruits histone transmethylase MLL1 to upregulate TIMP3 expression and promote breast cancer progression. J Transl Med 19(1):95. Available at: https://doi.org/10.1186/s12967-020-02682-5

Huang X, Lyu J (2018) Tumor suppressor function of miR-483-3p on breast cancer via targeting of the cyclin E1 gene. Exp Ther Med 16(3):2615–2620. Available at: https://doi.org/10.3892/etm.2018.6504

Huang ZM et al (2019) MicroRNA-26a-5p inhibits breast cancer cell growth by suppressing RNF6 expression. Kaohsiung J Med Sci 35(8):467–473. Available at: https://doi.org/10.1002/KJM2.12085

Huang J et al (2022) Circular RNA KIF4A promotes liver metastasis of breast cancer by reprogramming glucose metabolism. J Oncol 2022:8035083. Available at: https://doi.org/10.1155/2022/8035083

Huang X et al (2023) EIF4A3-induced circZFAND6 promotes breast cancer proliferation and metastasis through the miR-647/FASN axis. Life Sci 324:121745. Available at: https://doi.org/10.1016/j.lfs.2023.121745

Jeanes A, Gottardi CJ, Yap AS (2008) Cadherins and cancer: how does cadherin dysfunction promote tumor progression? Oncogene 27(55):6920–6929. Available at: https://doi.org/10.1038/onc.2008.343

Ji W, Sun B, Su C (2017) Targeting MicroRNAs in cancer gene therapy. Genes 8(1) Available at: https://doi.org/10.3390/genes8010021

Jiang S et al (2012) A novel miR-155/miR-143 cascade controls glycolysis by regulating hexokinase 2 in breast cancer cells. EMBO J 31(8):1985–1998. Available at: https://doi.org/10.1038/emboj.2012.45

Jin T et al (2017) microRNA-200c/141 upregulates SerpinB2 to promote breast cancer cell metastasis and reduce patient survival. Oncotarget 8(20):32769–32782. Available at: https://doi.org/10.18632/oncotarget.15680

Kalluri R, Weinberg RA (2009) The basics of epithelial-mesenchymal transition. J Clin Invest 119(6):1420–1428. Available at: https://doi.org/10.1172/JCI39104

Kansara S et al (2023) The emerging regulatory roles of non-coding RNAs associated with glucose metabolism in breast cancer. Semin Cancer Biol 95:1–12. Available at: https://doi.org/10.1016/j.semcancer.2023.06.007

Khan SA et al (2012) Soy isoflavone supplementation for breast cancer risk reduction: a randomized phase II trial. Cancer Prev Res (Philadelphia, Pa.) 5(2):309–319. Available at: https://doi.org/10.1158/1940-6207.CAPR-11-0251

Kim S et al (2018) microRNA-155 positively regulates glucose metabolism via PIK3R1-FOXO3a-cMYC axis in breast cancer. Oncogene 37(22):2982–2991. Available at: https://doi.org/10.1038/s41388-018-0124-4

Kluiver J et al (2012) Rapid generation of microRNA sponges for microRNA inhibition. PLoS One 7(1):e29275. Available at: https://doi.org/10.1371/journal.pone.0029275

Koufaris C et al (2016) Systematic integration of molecular profiles identifies miR-22 as a regulator of lipid and folate metabolism in breast cancer cells. Oncogene 35(21):2766–2776. Available at: https://doi.org/10.1038/onc.2015.333

Koundouros N, Poulogiannis G (2020) Reprogramming of fatty acid metabolism in cancer. Br J Cancer 122(1):4–22. Available at: https://doi.org/10.1038/s41416-019-0650-z

Kristensen LS et al (2019) The biogenesis, biology and characterization of circular RNAs. Nat Rev Genet 20(11):675–691. Available at: https://doi.org/10.1038/s41576-019-0158-7

Krol J, Loedige I, Filipowicz W (2010) The widespread regulation of microRNA biogenesis, function and decay. Nat Rev Genet 11(9):597–610. Available at: https://doi.org/10.1038/nrg2843

Lewinska A et al (2017) Sulforaphane-induced cell cycle arrest and senescence are accompanied by DNA Hypomethylation and changes in microRNA profile in breast cancer cells. Theranostics 7(14):3461–3477. Available at: https://doi.org/10.7150/thno.20657

Li X et al (2007) Ubiquitin- and ATP-independent proteolytic turnover of p21 by the REGgamma-proteasome pathway. Mol Cell 26(6):831–842. Available at: https://doi.org/10.1016/j.molcel.2007.05.028

Li Y et al (2013) Epigenetic regulation of multiple tumor-related genes leads to suppression of breast tumorigenesis by dietary Genistein. PLoS One 8(1):e54369. Available at: https://doi.org/10.1371/journal.pone.0054369

Li M et al (2014) Identification, expression and bioactivity of hexokinase in amphioxus: insights into evolution of vertebrate hexokinase genes. Gene 535(2):318–326. Available at: https://doi.org/10.1016/j.gene.2013.10.068

Li H, Jiang X, Niu X (2017a) Long non-coding RNA reprogramming (ROR) promotes cell proliferation in colorectal cancer via affecting P53. Med Sci Monit 23:919–928. Available at: https://doi.org/10.12659/msm.903462

Li L et al (2017b) miR-30a-5p suppresses breast tumor growth and metastasis through inhibition of LDHA-mediated Warburg effect. Cancer Lett 400:89–98. Available at: https://doi.org/10.1016/j.canlet.2017.04.034

Li S et al (2018) Long noncoding RNA GAS5 suppresses triple negative breast cancer progression through inhibition of proliferation and invasion by competitively binding miR-196a-5p. Biomed Pharmacother 104:451–457. Available at: https://doi.org/10.1016/j.biopha.2018.05.056

Li M et al (2019a) Plant MIR156 regulates intestinal growth in mammals by targeting the Wnt/β-catenin pathway. Am J Physiol Cell Physiol 317(3):C434–C448. Available at: https://doi.org/10.1152/ajpcell.00030.2019

Li X-Y et al (2019b) The long noncoding RNA MIR210HG promotes tumor metastasis by acting as a ceRNA of miR-1226-3p to regulate mucin-1c expression in invasive breast cancer. Aging 11(15):5646–5665. Available at: https://doi.org/10.18632/aging.102149

Li J et al (2020a) Ferroptosis: past, present and future. Cell Death Dis 11(2):88. Available at: https://doi.org/10.1038/s41419-020-2298-2
Li Y et al (2020b) LINC00346 regulates glycolysis by modulation of glucose transporter 1 in breast cancer cells. Mol Cell Probes 54:101667. Available at: https://doi.org/10.1016/j.mcp.2020.101667
Li Y, Li H, Wang W, Yu X, Xu Q (2020c) LINC00346 regulates glycolysis by modulation ofglucose transporter 1 in breast cancer cells. Mol Cell Probes, 54, 101667.https://doi.org/10.3233/CBM-190721
Liao X-H et al (2017) STAT3 is required for MiR-17-5p-mediated sensitization to chemotherapy-induced apoptosis in breast cancer cells. Oncotarget 8(9):15763–15774. Available at: https://doi.org/10.18632/oncotarget.15000
Lin X et al (2021) Non-coding RNAs rewire cancer metabolism networks. Semin Cancer Biol 75:116–126. Available at: https://doi.org/10.1016/j.semcancer.2020.12.019
Liu C-Y et al (2015) Vimentin contributes to epithelial-mesenchymal transition cancer cell mechanics by mediating cytoskeletal organization and focal adhesion maturation. Oncotarget 6(18):15966–15983. Available at: https://doi.org/10.18632/oncotarget.3862
Liu H et al (2019) Long non-coding RNAs involved in cancer metabolic reprogramming. Cell Mol Life Sci 76(3):495–504. Available at: https://doi.org/10.1007/s00018-018-2946-1
Liu L, Zhang Y, Lu J (2020a) The roles of long noncoding RNAs in breast cancer metastasis. Cell Death Dis 11(9):749. Available at: https://doi.org/10.1038/s41419-020-02954-4
Liu X et al (2020b) LncRNA NEAT1 accelerates breast cancer progression through regulating miR-410-3p/ CCND1 axis. Cancer Biomark 29(2):277–290. Available at: https://doi.org/10.3233/CBM-190721
Liu X et al (2020c) Long noncoding RNA SOX21-AS1 regulates the progression of triple-negative breast cancer through regulation of miR-520a-5p/ORMDL3 axis. J Cell Biochem 121(11):4601–4611. Available at: https://doi.org/10.1002/jcb.29674
Liu X, Yao W, Xiong H, Li Q, Li Y (2020d) LncRNA NEAT1 accelerates breast cancerprogression through regulating miR-410-3p/ CCND1 axis. Cancer Biomarkers : Section A of Disease Markers,29(2), 277–290. https://doi.org/10.3233/CBM-190721
Lukey MJ, Katt WP, Cerione RA (2017) Targeting amino acid metabolism for cancer therapy. Drug Discov Today 22(5):796–804. Available at: https://doi.org/10.1016/j.drudis.2016.12.003
Luo Q et al (2013a) MiRNA-497 regulates cell growth and invasion by targeting cyclin E1 in breast cancer. Cancer Cell Int 13(1):1–8. Available at: https://doi.org/10.1186/1475-2867-13-95/FIGURES/7
Luo Q et al (2013b) MiRNA-497 regulates cell growth and invasion by targeting cyclin E1 in breast cancer. Cancer Cell Int 13(1):95. Available at: https://doi.org/10.1186/1475-2867-13-95
Meeran SM, Patel SN, Tollefsbol TO (2010) Sulforaphane causes epigenetic repression of hTERT expression in human breast cancer cell lines. PLoS One 5(7):e11457. Available at: https://doi.org/10.1371/journal.pone.0011457
Mierziak J et al (2021) Influence of the bioactive diet components on the gene expression regulation. Nutrients 13(11) Available at: https://doi.org/10.3390/nu13113673
Mitchell PS et al (2008) Circulating microRNAs as stable blood-based markers for cancer detection. Proc Natl Acad Sci U S A 105(30):10513–10518. Available at: https://doi.org/10.1073/pnas.0804549105
Mondal P et al (2022) Progress and promises of epigenetic drugs and epigenetic diets in cancer prevention and therapy: a clinical update. Semin Cancer Biol 83:503–522. Available at: https://doi.org/10.1016/j.semcancer.2020.12.006
Mota MSV et al (2018) Deficiency of tumor suppressor Merlin facilitates metabolic adaptation by co-operative engagement of SMAD-Hippo signaling in breast cancer. Carcinogenesis 39(9):1165–1175. Available at: https://doi.org/10.1093/carcin/bgy078
Natesh J, Penta D, Meeran SM (2022) Epigenetics in precision medicine of breast cancer. In: Epigenetics in precision medicine. Elsevier, pp 43–67. Available at: https://doi.org/10.1016/B978-0-12-823008-4.00004-4

O'Brien J et al (2018) Overview of MicroRNA biogenesis, mechanisms of actions, and circulation. Front Endocrinol 9:402. Available at: https://doi.org/10.3389/fendo.2018.00402
Otsuka K, Yamamoto Y, Ochiya T (2018) Regulatory role of resveratrol, a microRNA-controlling compound, in HNRNPA1 expression, which is associated with poor prognosis in breast cancer. Oncotarget 9(37):24718–24730. Available at: https://doi.org/10.18632/oncotarget.25339
Ozfiliz-Kilbas P et al (2021) In vitro investigations of miR-33a expression in estrogen receptor-targeting therapies in breast cancer cells. Cancer 13(21) Available at: https://doi.org/10.3390/cancers13215322
Pamudurti NR et al (2017) Translation of CircRNAs. Mol Cell 66(1):9–21.e7. Available at: https://doi.org/10.1016/j.molcel.2017.02.021
Patel N et al (2016) miR-15a/miR-16 induces mitochondrial dependent apoptosis in breast cancer cells by suppressing oncogene BMI1. Life Sci 164:60–70. Available at: https://doi.org/10.1016/j.lfs.2016.08.028
Pavlova NN, Thompson CB (2016) The emerging hallmarks of cancer metabolism. Cell Metab 23(1):27–47. Available at: https://doi.org/10.1016/j.cmet.2015.12.006
Peng X et al (2020) The role of long non-coding RNAs in thyroid cancer. Front Oncol 10:941. Available at: https://doi.org/10.3389/fonc.2020.00941
Penta D et al (2022) Diindolylmethane promotes metabolic crisis and enhances the efficacy of Centchroman in breast cancer: a 1H NMR-based approach. ACS Omega 7(47):43147–43160. Available at: https://doi.org/10.1021/acsomega.2c05832
Pop EA et al (2008) Effects of a high daily dose of soy isoflavones on DNA damage, apoptosis, and estrogenic outcomes in healthy postmenopausal women: a phase I clinical trial. Menopause (New York, N.Y.) 15(4 Pt 1):684–692. Available at: https://doi.org/10.1097/gme.0b013e318167b8f2
Pujalte-Martin M et al (2024) Targeting cancer and immune cell metabolism with the complex I inhibitors metformin and IACS-010759. Mol Oncol 18(7):1719–1738. Available at: https://doi.org/10.1002/1878-0261.13583
Qi C et al (2020) Circ_0072995 promotes cell carcinogenesis via up-regulating miR-149-5p-mediated SHMT2 in breast cancer. Cancer Manag Res 12:11169–11181. Available at: https://doi.org/10.2147/CMAR.S272274
Qiao K et al (2019) LINC00673 is activated by YY1 and promotes the proliferation of breast cancer cells via the miR-515-5p/MARK4/Hippo signaling pathway. J Exper Clin Cancer Res 38(1):418. Available at: https://doi.org/10.1186/s13046-019-1421-7
Qin Y et al (2021) A novel long non-coding RNA lnc030 maintains breast cancer stem cell stemness by stabilizing SQLE mRNA and increasing cholesterol synthesis. Adv Sci (Weinheim, Baden-Wurttemberg, Germany) 8(2):2002232. Available at: https://doi.org/10.1002/advs.202002232
Rezaie F, Mokhtari MJ, Kalani M (2021) Quercetin arrests in G2 phase, upregulates INXS LncRNA and downregulates UCA1 LncRNA in MCF-7 cells. Int J Mol Cell Med 10(3):208–216. Available at: https://doi.org/10.22088/IJMCM.BUMS.10.3.207
Riggio AI, Varley KE, Welm AL (2021) The lingering mysteries of metastatic recurrence in breast cancer. Br J Cancer 124(1):13–26. Available at: https://doi.org/10.1038/s41416-020-01161-4
Romero-Cordoba SL et al (2018) Loss of function of miR-342-3p results in MCT1 over-expression and contributes to oncogenic metabolic reprogramming in triple negative breast cancer. Sci Rep 8(1):12252. Available at: https://doi.org/10.1038/s41598-018-29708-9
Safi A et al (2023) The role of noncoding RNAs in metabolic reprogramming of cancer cells. Cell Mol Biol Lett 28(1):37. Available at: https://doi.org/10.1186/s11658-023-00447-8
Samavat H et al (2017) A randomized controlled trial of green tea extract supplementation and mammographic density in postmenopausal women at increased risk of breast cancer. Cancer Prev Res 10(12):710–718. Available at: https://doi.org/10.1158/1940-6207.CAPR-17-0187
Schiliro C, Firestein BL (2021) Mechanisms of metabolic reprogramming in cancer cells supporting enhanced growth and proliferation. Cells 10(5) Available at: https://doi.org/10.3390/cells10051056

Segal M, Slack FJ (2020) Challenges identifying efficacious miRNA therapeutics for cancer. Expert Opin Drug Discov 15(9):987–992. Available at: https://doi.org/10.1080/17460441.2020.1765770

Shankaraiah RC et al (2018) Non-coding RNAs in the reprogramming of glucose metabolism in cancer. Cancer Lett 419:167–174. Available at: https://doi.org/10.1016/j.canlet.2018.01.048

Sharma U et al (2021) LncRNA ZFAS1 inhibits triple-negative breast cancer by targeting STAT3. Biochimie 182:99–107. Available at: https://doi.org/10.1016/j.biochi.2020.12.026

Sheng XY, Wang C-H, Wang C-F, Xu H-Y (2020) Long-chain non-coding SOX21-AS1 promotes proliferation and migration of breast cancer cells through the PI3K/AKT signaling pathway. Cancer Manag Res 12:11005–11014. Available at: https://doi.org/10.2147/CMAR.S270464

Shi Y et al (2015) miR-7-5p suppresses cell proliferation and induces apoptosis of breast cancer cells mainly by targeting REGγ. Cancer Lett 358(1):27–36. Available at: https://doi.org/10.1016/j.canlet.2014.12.014

Shi Y et al (2020) Let-7a-5p inhibits triple-negative breast tumor growth and metastasis through GLUT12-mediated Warburg effect. Cancer Lett 495:53–65. Available at: https://doi.org/10.1016/j.canlet.2020.09.012

Shim H et al (1997) c-Myc transactivation of LDH-A: implications for tumor metabolism and growth. Proc Natl Acad Sci U S A 94(13):6658–6663. Available at: https://doi.org/10.1073/pnas.94.13.6658

Shukla K et al (2015a) MicroRNA-30c-2-3p negatively regulates NF-κB signaling and cell cycle progression through downregulation of TRADD and CCNE1 in breast cancer. Mol Oncol 9(6):1106–1119. Available at: https://doi.org/10.1016/J.MOLONC.2015.01.008

Shukla K et al (2015b) MicroRNA-30c-2-3p negatively regulates NF-κB signaling and cell cycle progression through downregulation of TRADD and CCNE1 in breast cancer. Mol Oncol 9(6):1106–1119. Available at: https://doi.org/10.1016/j.molonc.2015.01.008

Si H et al (2019) Long non-coding RNA H19 regulates cell growth and metastasis via miR-138 in breast cancer. Am J Transl Res 11(5):3213–3225

Singh R et al (2015) MicroRNA-195 inhibits proliferation, invasion and metastasis in breast cancer cells by targeting FASN, HMGCR, ACACA and CYP27B1. Sci Rep 5:17454. Available at: https://doi.org/10.1038/srep17454

Sinha S et al (2015) Epigenetic reactivation of p21CIP1/WAF1 and KLOTHO by a combination of bioactive dietary supplements is partially ERα-dependent in ERα-negative human breast cancer cells. Mol Cell Endocrinol 406:102–114. Available at: https://doi.org/10.1016/j.mce.2015.02.020

Slack FJ, Chinnaiyan AM (2019) The role of non-coding RNAs in oncology. Cell 179(5):1033–1055. Available at: https://doi.org/10.1016/j.cell.2019.10.017

Song Q et al (2020) Long non-coding RNA LINC00473 acts as a microRNA-29a-3p sponge to promote hepatocellular carcinoma development by activating Robo1-dependent PI3K/AKT/mTOR signaling pathway. Therap Adv Med Oncol 12:1758835920937890. Available at: https://doi.org/10.1177/1758835920937890

Springer M, Moco S (2019) Resveratrol and its human metabolites—effects on metabolic health and obesity. Nutrients 11(1):143. Available at: https://doi.org/10.3390/nu11010143

Statello L et al (2021) Gene regulation by long non-coding RNAs and its biological functions. Nat Rev Mol Cell Biol 22(2):96–118. Available at: https://doi.org/10.1038/s41580-020-00315-9

Stenvang J et al (2012) Inhibition of microRNA function by antimiR oligonucleotides. Silence 3(1):1. Available at: https://doi.org/10.1186/1758-907X-3-1

Su H et al (2022) LncRNA HOXC-AS3 promotes non-small-cell lung cancer growth and metastasis through upregulation of YBX1. Cell Death Dis 13(4):307. Available at: https://doi.org/10.1038/s41419-022-04723-x

Sun M-D et al (2018) Long noncoding RNA UCA1 promotes cell proliferation, migration and invasion of human leukemia cells via sponging miR-126. Eur Rev Med Pharmacol Sci 22(8):2233–2245. Available at: https://doi.org/10.26355/eurrev_201804_14809

Sun T et al (2020) LNC942 promoting METTL14-mediated m6A methylation in breast cancer cell proliferation and progression. Oncogene 39(31):5358–5372. Available at: https://doi.org/10.1038/s41388-020-1338-9

Sun H et al (2021a) p53 transcriptionally regulates SQLE to repress cholesterol synthesis and tumor growth. EMBO Rep 22(10):e52537. Available at: https://doi.org/10.15252/embr.202152537

Sun L et al (2021b) H19 promotes aerobic glycolysis, proliferation, and immune escape of gastric cancer cells through the microRNA-519d-3p/lactate dehydrogenase a axis. Cancer Sci 112(6):2245–2259. Available at: https://doi.org/10.1111/cas.14896

Sundaram GM (2019) Dietary non-coding RNAs from plants: fairy tale or treasure? Non-coding RNA Res 4(2):63–68. Available at: https://doi.org/10.1016/j.ncrna.2019.02.002

Tarrado-Castellarnau M, de Atauri P, Cascante M (2016) Oncogenic regulation of tumor metabolic reprogramming. Oncotarget 7(38):62726–62753. Available at: https://doi.org/10.18632/oncotarget.10911

Tian X et al (2011) E-cadherin/β-catenin complex and the epithelial barrier. J Biomed Biotechnol 2011:567305. Available at: https://doi.org/10.1155/2011/567305

Tian Y et al (2016) miR156a mimic represses the epithelial-mesenchymal transition of human nasopharyngeal cancer cells by targeting junctional adhesion molecule a. PLoS One 11(6):e0157686. Available at: https://doi.org/10.1371/journal.pone.0157686

Tuli HS et al (2019) Molecular mechanisms of action of Genistein in cancer: recent advances. Front Pharmacol 10. Available at: https://doi.org/10.3389/fphar.2019.01336

Vettore L, Westbrook RL, Tennant DA (2020) New aspects of amino acid metabolism in cancer. Br J Cancer 122(2):150–156. Available at: https://doi.org/10.1038/s41416-019-0620-5

Wang N et al (2019) Long non-coding RNA HULC promotes the development of breast cancer through regulating LYPD1 expression by sponging miR-6754-5p. Onco Targets Ther 12:10671–10679. Available at: https://doi.org/10.2147/OTT.S226040

Wang L, Zhang S, Wang X (2020a) The metabolic mechanisms of breast cancer metastasis. Front Oncol 10:602416. Available at: https://doi.org/10.3389/fonc.2020.602416

Wang F et al (2020b) Quercetin protects human oral keratinocytes from lipopolysaccharide-induced injury by downregulating microRNA-22. Hum Exp Toxicol 39(10):1310–1317. Available at: https://doi.org/10.1177/0960327120918291

Wang Y et al (2020c) Long noncoding RNA HOTAIR promotes breast cancer development by targeting ZEB1 via sponging miR-601. Cancer Cell Int 20:320. Available at: https://doi.org/10.1186/s12935-020-01410-9

Wang Y et al (2021a) Novel strategies to improve tumour therapy by targeting the proteins MCT1, MCT4 and LAT1. Eur J Med Chem 226:113806. Available at: https://doi.org/10.1016/j.ejmech.2021.113806

Wang J et al (2021b) Knockdown of circular RNA septin 9 inhibits the malignant progression of breast cancer by reducing the expression of solute carrier family 1 member 5 in a microRNA-149-5p-dependent manner. Bioengineered 12(2):10624–10637. Available at: https://doi.org/10.1080/21655979.2021.2000731

Wang B et al (2022a) A miR-34a-guided, tRNAiMet-derived, piR_019752-like fragment (tRiMetF31) suppresses migration and angiogenesis of breast cancer cells via targeting PFKFB3. Cell Death Discov 8(1):355. Available at: https://doi.org/10.1038/s41420-022-01054-w

Wang S et al (2022b) A novel circular RNA confers trastuzumab resistance in human epidermal growth factor receptor 2-positive breast cancer through regulating ferroptosis. Environ Toxicol 37(7):1597–1607. Available at: https://doi.org/10.1002/tox.23509

White E et al (2010) Role of autophagy in suppression of inflammation and cancer. Curr Opin Cell Biol 22(2):212–217. Available at: https://doi.org/10.1016/J.CEB.2009.12.008

Williams DE (2021) Indoles derived from Glucobrassicin: cancer chemoprevention by Indole-3-Carbinol and 3,3'-Diindolylmethane. Front Nutrit 8. Available at: https://doi.org/10.3389/fnut.2021.734334

Winkle M et al (2021) Noncoding RNA therapeutics – challenges and potential solutions. Nat Rev Drug Discov 20(8):629–651. Available at: https://doi.org/10.1038/s41573-021-00219-z

Wong RSY (2011) Apoptosis in cancer: from pathogenesis to treatment. J Exper Clin Cancer Res 30(1):87. Available at: https://doi.org/10.1186/1756-9966-30-87

Wu H et al (2021) CircARL8B contributes to the development of breast cancer via regulating miR-653-5p/HMGA2 Axis. Biochem Genet 59(6):1648–1665. Available at: https://doi.org/10.1007/s10528-021-10082-7

Xiang S et al (2024) Role of ferroptosis and ferroptosis-related long non'coding RNA in breast cancer. Cell Mol Biol Lett 29(1):40. Available at: https://doi.org/10.1186/s11658-024-00560-2

Xiao X et al (2016) The miR-34a-LDHA axis regulates glucose metabolism and tumor growth in breast cancer. Sci Rep 6:21735. Available at: https://doi.org/10.1038/srep21735

Xiao M et al (2018) MiR-128 regulation of glucose metabolism and cell proliferation in triple-negative breast cancer. Br J Surg 105(1):75–85. Available at: https://doi.org/10.1002/bjs.10646

Xiao Y et al (2023) Emerging therapies in cancer metabolism. Cell Metab 35(8):1283–1303. Available at: https://doi.org/10.1016/j.cmet.2023.07.006

Xie Y, Ye J, Luo H (2023) HOXC cluster antisense RNA 3, a novel long non-coding RNA as an oncological biomarker and therapeutic target in human malignancies. Onco Targets Ther 16:849–865. Available at: https://doi.org/10.2147/OTT.S425523

Xu X et al (2023) Metabolic reprogramming and epigenetic modifications in cancer: from the impacts and mechanisms to the treatment potential. Exp Mol Med 55(7):1357–1370. Available at: https://doi.org/10.1038/s12276-023-01020-1

Xu A et al (2024a) The therapeutic potential of circular RNA in triple-negative breast cancer. Cancer Drug Resist (Alhambra, Calif.) 7:13. Available at: https://doi.org/10.20517/cdr.2023.141

Xu S et al (2024b) Metabolism-regulating non-coding RNAs in breast cancer: roles, mechanisms and clinical applications. J Biomed Sci 31(1):25. Available at: https://doi.org/10.1186/s12929-024-01013-w

Yadav P et al (2021) SLC7A11/ xCT is a target of miR-5096 and its restoration partially rescues miR-5096-mediated ferroptosis and anti-tumor effects in human breast cancer cells. Cancer Lett 522:211–224. Available at: https://doi.org/10.1016/j.canlet.2021.09.033

Yan L et al (2020) Long noncoding RNA H19 acts as a miR-340-3p sponge to promote epithelial-mesenchymal transition by regulating YWHAZ expression in paclitaxel-resistant breast cancer cells. Environ Toxicol 35(9):1015–1028. Available at: https://doi.org/10.1002/tox.22938

Yang Z et al (2012) Bioavailability and pharmacokinetics of genistein: mechanistic studies on its ADME. Anti Cancer Agents Med Chem 12(10):1264–1280. Available at: https://doi.org/10.2174/187152012803833107

Yang J-R, Shi M-X, Zeng Y (2019a) LncRNA HAND2-AS1 inhibits proliferation and promotes apoptosis of chronic myeloid leukemia cells by sponging with micRNA-1275. Eur Rev Med Pharmacol Sci 23(5):2103–2111. Available at: https://doi.org/10.26355/eurrev_201903_17254

Yang Y, Xun N, Wu J-G (2019b) Long non-coding RNA FGF14-AS2 represses proliferation, migration, invasion, and induces apoptosis in breast cancer by sponging miR-205-5p. Eur Rev Med Pharmacol Sci 23(16):6971–6982. Available at: https://doi.org/10.26355/eurrev_201908_18737

Yao A et al (2019) PKM2 promotes glucose metabolism through a let-7a-5p/Stat3/hnRNP-A1 regulatory feedback loop in breast cancer cells. J Cell Biochem 120(4):6542–6554. Available at: https://doi.org/10.1002/jcb.27947

Yao J et al (2022) lnc-MICAL2-1 sponges miR-25 to regulate DKK3 expression and inhibits activation of the Wnt/β-catenin signaling pathway in breast cancer. Int J Mol Med 49(2) Available at: https://doi.org/10.3892/ijmm.2021.5078

Ye T et al (2020) MicroRNA-16-1-3p represses breast tumor growth and metastasis by inhibiting PGK1-mediated Warburg effect. Front Cell Develop Biol 8:615154. Available at: https://doi.org/10.3389/fcell.2020.615154

Yu B et al (2005) Methylation as a crucial step in plant microRNA biogenesis. Science (New York, N.Y.) 307(5711):932–935. Available at: https://doi.org/10.1126/science.1107130

Yu H, Luo H, Liu X (2022) Knockdown of circ_0102273 inhibits the proliferation, metastasis and glycolysis of breast cancer through miR-1236-3p/PFKFB3 axis. Anti-Cancer Drugs 33(4):323–334. Available at: https://doi.org/10.1097/CAD.0000000000001264

Yuan M et al (2022) Circ_0062558 promotes growth, migration, and glutamine metabolism in triple-negative breast cancer by targeting the miR-876-3p/SLC1A5 axis. Arch Gynecol Obstet 306(5):1643–1655. Available at: https://doi.org/10.1007/s00404-022-06481-9

Zhang L et al (2012) Exogenous plant MIR168a specifically targets mammalian LDLRAP1: evidence of cross-kingdom regulation by microRNA. Cell Res 22(1):107–126. Available at: https://doi.org/10.1038/cr.2011.158

Zhang Z et al (2016) Associations between cruciferous vegetable intake and selected biomarkers among women scheduled for breast biopsies. Public Health Nutr 19(7):1288–1295. Available at: https://doi.org/10.1017/S136898001500244X

Zhang G et al (2017) MicroRNA-200c and microRNA- 141 are regulated by a FOXP3-KAT2B axis and associated with tumor metastasis in breast cancer. Breast Cancer Res 19(1):73. Available at: https://doi.org/10.1186/s13058-017-0858-x

Zhang E et al (2018) A novel long noncoding RNA HOXC-AS3 mediates tumorigenesis of gastric cancer by binding to YBX1. Genome Biol 19(1):154. Available at: https://doi.org/10.1186/s13059-018-1523-0

Zhang Z et al (2019) PKM2, function and expression and regulation. Cell Biosci 9:52. Available at: https://doi.org/10.1186/s13578-019-0317-8

Zhang Q et al (2020) Quercetin stimulates osteogenic differentiation of bone marrow stromal cells through miRNA-206/connexin 43 pathway. Am J Transl Res 12(5):2062–2070

Zhang D et al (2021a) LINC01189-miR-586-ZEB1 feedback loop regulates breast cancer progression through Wnt/β-catenin signaling pathway. Mol Therap Nucleic Acids 25:455–467. Available at: https://doi.org/10.1016/j.omtn.2021.06.007

Zhang Y et al (2021b) Long non-coding RNA IGF2-AS represses breast cancer tumorigenesis by epigenetically regulating IGF2. Exper Biol Med (Maywood, N.J.) 246(4):371–379. Available at: https://doi.org/10.1177/1535370220966253

Zhao M, Zhang Z (2016) Glucose transporter regulation in cancer: a profile and the loops. Crit Rev Eukaryot Gene Expr 26(3):223–238. Available at: https://doi.org/10.1615/CritRevEukaryotGeneExpr.2016016531

Zhao P et al (2019) Long noncoding RNA DLX6-AS1 promotes breast cancer progression via miR-505-3p/RUNX2 axis. Eur J Pharmacol 865:172778. Available at: https://doi.org/10.1016/j.ejphar.2019.172778

Zheng Q-X et al (2021) TTN-AS1 as a potential diagnostic and prognostic biomarker for multiple cancers. Biomed Pharmacother 135:111169. Available at: https://doi.org/10.1016/j.biopha.2020.111169

Zhou Q et al (2020) Linc-ROR promotes the progression of breast cancer and decreases the sensitivity to rapamycin through miR-194-3p targeting MECP2. Mol Oncol 14(9):2231–2250. Available at: https://doi.org/10.1002/1878-0261.12700

Zhu D et al (2019a) MT1JP inhibits tumorigenesis and enhances cisplatin sensitivity of breast cancer cells through competitively binding to miR-24-3p. Am J Transl Res 11(1):245–256

Zhu M et al (2019b) MEG3 overexpression inhibits the tumorigenesis of breast cancer by downregulating miR-21 through the PI3K/Akt pathway. Arch Biochem Biophys 661:22–30. Available at: https://doi.org/10.1016/j.abb.2018.10.021

Zhu P et al (2021) tRNA-derived fragment tRFLys-CTT-010 promotes triple-negative breast cancer progression by regulating glucose metabolism via G6PC. Carcinogenesis 42(9):1196–1207. Available at: https://doi.org/10.1093/carcin/bgab058

Zhu W et al (2022) Low glucose-induced overexpression of HOXC-AS3 promotes metabolic reprogramming of breast cancer. Cancer Res 82(5):805–818. Available at: https://doi.org/10.1158/0008-5472.CAN-21-1179

Chapter 11
MitomiRs as Key Players in Metabolic Reprogramming of Blood Cell Development

Vijay Prasad Koppineedi, Mahesh Gutti, Bheriprasad Padhan, Amit Mishra, and Ravi Kumar Gutti

Contents

Abstract Mitochondria are key players in regulating cellular metabolism. Short noncoding RNAs, or microRNAs (miRNAs), have become significant modulators of gene expression and cellular functions. Recent research has demonstrated the presence and activity of mitochondrial miRNAs (MitomiRs) in various cell types, including blood cells. The role of MitomiRNAs in metabolic reprogramming

V. P. Koppineedi · B. Padhan · R. K. Gutti (✉)
Department of Biochemistry, School of Life Sciences, University of Hyderabad, Gachibowli, Hyderabad, India

M. Gutti
K.V. Gachibowli, GPRA Campus Rd Indira Nagar, Gachibowli, Hyderabad, Telangana, India

A. Mishra
Department of Bioscience & Bioengineering, Cellular and Molecular Neurobiology Unit, Indian Institute of Technology Jodhpur, Jodhpur, Rajasthan, India
e-mail: amit@iitj.ac.in

M. F. Ullah, A. Ahmad (eds.), *Non-coding RNAs*, Experientia Supplementum 115,
https://doi.org/10.1007/978-3-032-06948-1_11

throughout blood cell formation, including their biosynthesis, function, and possible therapeutic implications, was discussed in this chapter. The discovery of mitochondrial microRNAs (MitomiRs) transformed our understanding of gene regulation in these critical organelles. These MitomiRs were discovered as they were influencing mitochondrial gene expression, with considerable effects on various cellular functions like oxidative stress responses, production of energy, and apoptosis during blood cell development.

Keywords miRNAs · Hematopoietic stem cells · Metabolic reprogramming · Mitochondrial miRNA or mitomiRs · Metabolism · and Blood cell development

Abbreviations

3′UTR	3′ untranslated region
ACACB	Acetyl-CoA carboxylase beta
AGO2	Argonaut
ARL2	ARF like GTPase 2
ARL2	ARF like GTPase 2
ASS1	Arginosuccinate synthase 1
ATP5	ATP synthase peripheral stalk subunit F6
ATP5G1	ATP synthase membrane subunit c locus 1
ATP5L	ATP synthase membrane subunit g
BAG5	BAG cochaperone 5
BAX	BCL2 associated X, apoptosis regulator
BCL2	BCL2 apoptosis regulator
BCL2	BCL2 apoptosis regulator
BNIP3	BCL2 interacting protein 3
CIP2A	Cellular inhibitor of PP2A
COX1	Prostaglandin-endoperoxide synthase 1
COX10	Cytochrome c oxidase assembly factor heme A:farnesyltransferase COX10
COXIV	Cytochrome c oxidase subunit 4I1
CYC	Cytochrome c, somatic
DGCR8	DiGeorge syndrome chromosomal region 8
FASN	Fatty acid synthase
FIS1	Fission, mitochondrial 1
FOXA2	Fork head box A2
GLS	Glutaminase 2
GLUT1	Solute carrier family 2-member 1
H2AX	H2A.X variant histone
HSCs	Hematopoietic stem cells
LASS2	Ceramide synthase 2

MAP2K5	Mitogen-activated protein kinase 5
MED13	Mediator complex subunit 13
MFF	Mitochondrial fission factor
MFN1	Mitofusin 1
miRNAs	MicroRNAs
MitomiR's	Mitochondrial miRNA
mRNA	Messenger RNA
MTFR1	Mitochondrial fission regulator 1
PARP-2	Poly(ADP-ribose) polymerase 2
PDK1	3-phosphoinositide dependent protein kinase 1
PNPase	Polynucleotide phosphorylase
pri-miRNAs	Primary miRNAs
PRKAA1	Protein kinase AMP-activated catalytic subunit alpha 1
RISC	RNA induced silencing complex
RNaseIII	Ribonuclease III
ROS	Reactive oxygen species
SLC16A1	Solute carrier family 16-member 1
SLC2A1	Solute carrier family 2 member 1
SLC2A1	Solute carrier family 2-member 1
SLC2A4	Solute carrier family 2-member 4
SOD2	SOD2 overlapping transcript 1
SREBF1	Sterol regulatory element binding transcription factor 1
UCP2	Uncoupling protein 2
ZEB1	Zinc finger E-box binding homeobox 1

11.1 Introduction

11.1.1 Hematopoiesis and Metabolic Reprogramming

11.1.1.1 Overview of Hematopoiesis

Hematopoiesis is the process of differentiating hematopoietic stem cells (HSCs) into diverse adult blood cell lines, as shown in Fig. 11.1. Hematopoiesis is a constant process that produces the cells that are part of blood whenever it is required. Blood cells are of three types: erythrocytes (RBC), leukocytes (WBC) and thrombocytes. HSCs give rise to two different lineages, those are myeloid and lymphoid lineages. Myeloid lineage cells produce erythrocytes and megakaryocytes from a common megakaryocyte–erythrocyte progenitor and granulocytes (neutrophiles, basophiles, and eosinophiles), monocytes from granulocyte–monocyte progenitor. From lymphoid lineage, lymphocytes like T cell, B cells, and NK cells are produced as shown in the Fig. 11.1a.

Blood cells are produced in specific organs, which are also called as blood forming organs, such as bone marrow, instead of being produced in the bloodstream. In

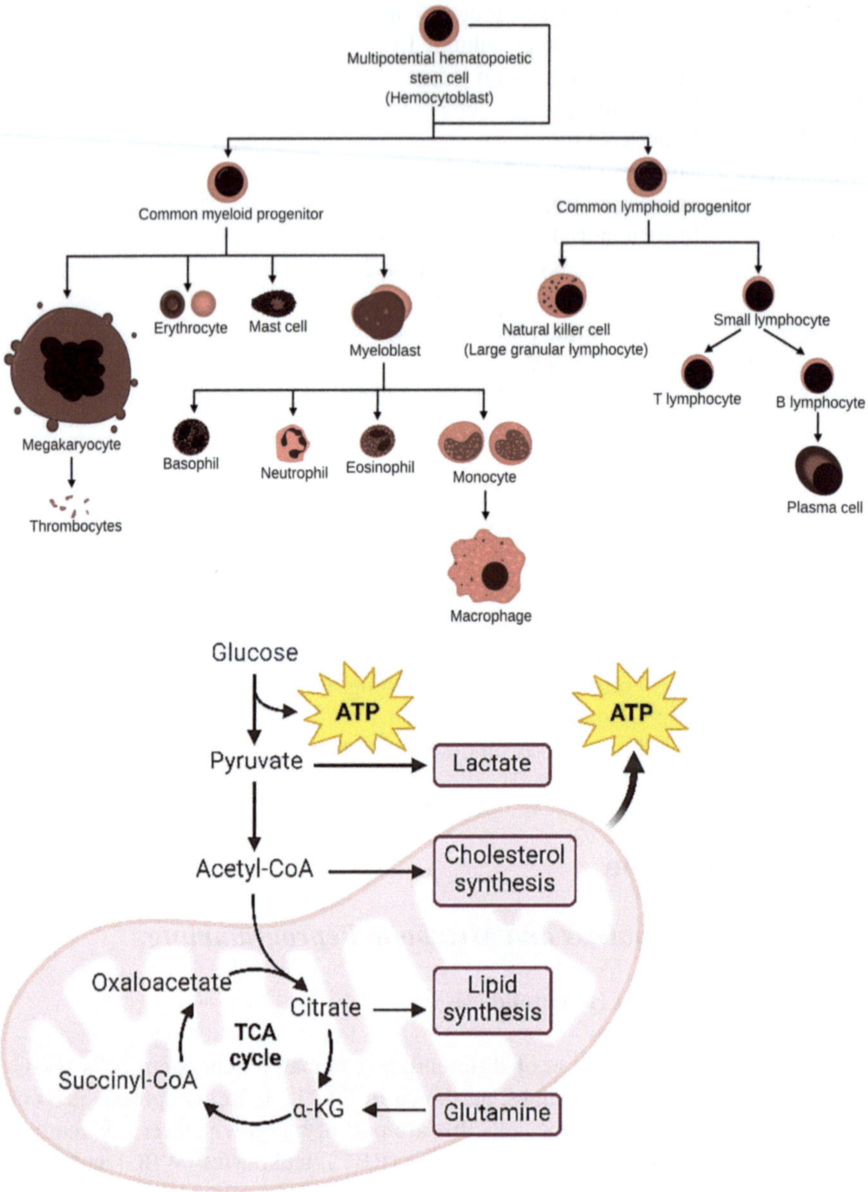

Fig. 11.1 (**a**) Hematopoiesis—Development of Various Blood Cells from Hematopoietic Stem Cell (HSC). (**b**) Significance of mitochondria

mature human, the bone marrow produces erythrocytes, leukocytes, and megakaryocytes, which will be fragmented to produce cellular fragments from their cytoplasmic extensions called platelets. Tissues, like spleen, thymus, and lymph nodes which are described as lymphatic tissues, produce lymphocytes and monocytes from the reticulo-endothelium of tissues in the liver, spleen, lymph nodes, and other organs.

Platelets, which are derived from megakaryocyte mother cell through fragmentation from cytoplasmic extensions and released into blood stream. The bone marrow of healthy individual produces roughly half a litre of erythrocytes per week. Every day, over 1% of body erythrocytes are produced, and the balance between red cell synthesis and aged red cell clearance in blood stream is maintained properly. The rate of blood cell formation varies from individual to individual, but a typical production might be around 10 billion white cells, 400 billion platelets and 200 billion red cells per day. During hematopoiesis, along with many cell organelles, mitochondria play a crucial role.

(a) *Mitochondria*

The cytosolic organelles known as mitochondria first came to light in the late 1800s, although their original name, "bioblasts," persisted. But it wasn't until efforts to isolate mitochondria were made 30–40 years later that the biochemical functions of mitochondria were revealed. Unlike other organelles, mitochondria were discovered to be maternally inherited, and they possess a double membrane structure and have their own DNA (Ernster and Schatz 1981). As most of the proteins present in mitochondria are encoded in the nucleus, they evolved alongside their host, despite having evolved from symbiotic bacteria. Nonetheless, mitochondria have a small, circular, self-replicating genome that encodes proteins required for the process of respiration. Now mitochondria are widely recognized as an important organelle that plays a crucial role in cellular energy metabolism (Fig. 11.1b) and are required for the oxidative phosphorylation process, which generates ATP (Ernster and Schatz 1981). Furthermore, mitochondrion is very essential for the synthesis of nucleotides, heme, lipids, and iron–sulphur clustal domains, among other macromolecules.

Because mitochondria are involved in so many different biological activities, they have gained significant attention as research subjects in many different fields. Along with its role of controlling metabolism (anabolism and catabolism), it is also involved in signaling, which is required for regulating various biological outcomes, like cell death, gene expression, differentiation, Ca^{2+}-homeostasis, and certain aging processes like chronic inflammation, cellular senescence, and age-dependent stem cell activity (Galluzzi et al. 2008). Furthermore, mitochondrion was recently demonstrated that the structure and proteome of the mitochondria alter significantly based on cell type, indicating a high degree of dynamic nature of the mitochondrial architecture and their ability to work in an adaptive manner to satisfy specific cell requirements (Calvo et al. 2016a, b). Altered mitochondrial activity is related with clinical illnesses such as metabolic disorders, cardiomyopathies, dementia, and cancer, which are closely linked to dysregulated cell death, in line with its multifunctional role (Calvo et al. 2016a, b).

(b) *miRNA*

A subclass of noncoding RNA molecules is called microRNAs (miRNAs). They are small, single-stranded molecules, as their name implies, ranging in length from 19 to 25 nucleotides. They have lately become important regulators of gene expression. Lin-4 is the original member of miRNA family; these were the first miRNA to be identified as regulatory molecules during a nematode (*Caenorhabditis elegans*) change from one larval stage to another stage. By translationally suppressing the lin-14 gene, which is implicated in the programs for larval development, it does this (Bartel 2004a, b). There are various methods, which miRNAs can be encoded, including independent transcription units, multi-cistron clusters, and the introns of pre-mRNAs. There are numerous methods by which miRNAs might repress genes of the target. mRNA transcripts are degraded and cleaved if the miRNA binding sequence and the starting point on the untranslated region of the mRNA are perfectly complementary. On the other hand, translational inhibition occurs at the initiation and elongation levels of mRNA due to improper base pairing (Bartel 2009; Stefani and Slack 2008). The fact that the inhibitory activity of miRNAs transcends the one-miRNA—one-mRNA paradigm is a crucial characteristic of these molecules. According to Bracken et al. (2016), a single miRNA has the ability to target several mRNA targets, and vice versa. Till date, hundreds of miRNAs have been found in numerous organisms and cloned.

As a result, understanding gene expression regulation through miRNAs is an important area of research. Many of these miRNAs play diverse functional roles in a wide range of physiological and pathological conditions (Lin and Gregory 2015). The majority of miRNAs are expressed ubiquitously; however, the presence of some miRs is more specific tissues than others. The pre-miRNA-142 possesses a distinctive stem-loop structure and encodes two miRNA species (miR-142-3p and miR-142-5p). miR-142-5p and -3p have been well conserved in organisms, such as rat, mice, and humans as well (Griffiths 2004; Griffiths-Jones et al. 2006, 2008; Kozomara and Griffiths-Jones 2019). miR-142 is predominantly expressed in hematopoietic organs. Its main function in hematopoietic tissue has been found to be lineage differentiation (Merkerova et al. 2021). Over time, a considerable body of literature has emerged emphasizing the importance of miR-142 in inflammatory and immunological responses as well. Despite study on miR-142's numerous functional impacts in different biological settings, its critical regulatory involvement in several aspects of immunity remains prevalent.

(c) *MitomiRs*

MiRNAs known as mitomiRs (mitochondrial miRNA) are mitochondrial miRNAs or nuclear that reside in mitochondria and regulate mitochondrial activity and its metabolic activity. According to accumulating data, mature miRNA is found in a variety of subcellular compartments, including mitochondria, endoplasmic reticulum, and exosomes. MitomiRs are classified into two types, the first class of mitomiRs suppress the expression of its target by binding to mRNA within cytoplasm to adjust function of mitochondria, and the second class of mitomiRs are carried into mitochondria in the form of either RISC complex or pre-RISC complex and target mRNAs, which were produced by mitochondrial genome. Most of the mitomiRs are known to be encoded by the nucleus; nevertheless, human mitochondrial DNA appears to contain mitomiR sequences (namely, miR-1974, miR-1977, and miR-1978) (Bandiera et al. 2011). Whether miRNAs are actually translated from the mitochondrial DNA is still unknown. MitomiRs govern multiple signaling pathways, which is an unknown field that has to be further investigated. Recent studies have highlighted the presence and function of mitochondrial miRNAs (MitomiRs) in various cell types, including blood cells.

11.1.1.2 Metabolic Reprogramming in Blood Cell Development

Different blood cell types have distinct metabolic needs, necessitating metabolic reprogramming to support their specialized functions. This involves a switch between different metabolic pathways, such as glycolysis and oxidative phosphorylation. Hematopoietic stem cells or HSCs have been programmed by metabolic changes, which can be able to modify HSC function.

MitomiRs regulate numerous mitochondrial metabolic processes and are involved in blood cell formation due to their potential to control gene expression. Mitochondrial malfunction caused by altered oxidative phosphorylation and a shift from oxidative phosphorylation to enhanced glycolysis (Warburg Phenomenon), lipid metabolism, reactive oxygen species (ROS) signaling, and apoptosis (Vyas et al. 2016). MitomiRs modulate and alter all of the aforementioned mitochondrial activities; therefore, studying mitomiRs may help to learn about hematopoiesis and to diagnose blood cell development-related disorders.

There are 37 genes encoded by the mitochondrial genome, 13 of which code for proteins, 22 for tRNA, and 2 for rRNA. Despite having their own genome and translation machinery, mitochondria rely heavily on cargo which was transported from the cytoplasm to function. Transport of cargo from the cytosol to the mitochondrion is facilitated by outer and inner membrane transporters. Before the first report surfaced more than 20 years ago, the import of this miRNA cargo to mitochondria was mostly unknown. It suggested that tRNALysCUU import into mitochondria in yeast is contingent upon the protein translocation mechanism of working mitochondria. (Tarassov et al. 1995). Earlier, it was thought that miRNAs might be transported into mitochondria through the same pathways as proteins. It was discovered that mitochondrial tRNAmet binds with the human AGO2 protein, and Maniataki et al. were the first to suggest that the AGO2 protein might be in charge of miRNA import into mitochondria. Additionally, Mukherjee et al. found that RIC, a multi-subunit RNA import complex, promotes RNA import into mitochondria in Leishmania via a mechanism independent of protein import (Mukherjee S. et al. 2007). There is an RNAi component in the mitochondrial sub-compartment, which was known to co-localization of AGO2 and AGO3 to mitochondria (Bandiera et al. 2011). And, scientists found that there is an interaction between protein AGO2 with another protein named poly nucleotide phosphorylase (PNPase), which was identified as a carrier of small RNAs to mitochondria (Shepherd and Wang 2017). However, there is no concrete proof that miRNAs are among the small RNAs that PNPase imports into organelles. It is evident that PNPase is an RNAi component in the mitochondrial sub-compartment because of its distribution in distinct mitochondrial RNA decay foci called degradosomes (Borowski et al. 2012). According to Danielle, Shepherd et al., there is a translocational increase of miR-378 to sub-compartment of mitochondrion, which controls bioenergetics, caused by overexpression of PNPase (Wang et al. 2017). Altogether, research outcomes of various scientists suggest that Ago and PNPase are crucial mediators that help mitochondrial miRNA enter the mitochondrion (Fig. 11.2), but the precise pathways of their mechanism and other elements of the mitochondrial miRNA import machinery are yet unclear.

11.2 Biogenesis and Localization of MitomiRs

11.2.1 miRNA Biogenesis

Single-stranded, noncoding RNAs, which were described as miRNAs have a length of 19–24 nucleotides and play a very important role as essential post-transcriptional regulators of gene expression. By attaching complementary sequences in the 3′ untranslated regions (3′UTR) of target mRNAs, they can modify the expression of genes by causing translation inhibition or mRNA degradation (Bartel 2004a, b). miRNA production is a multistep, intricate process that occurs in the cytoplasm and nucleus and involves multiple enzymatic processes (Fig. 11.2) (Lin and Gregory 2015). Primary-miRNA, or pre-miRNA, is typically transcribed from DNA to initiate miRNA biogenesis in the nucleus. A microprocessor complex, which is made up of DROSHA/DGCR8 (the DiGeorge syndrome chromosomal region 8) and ribonuclease III (RNaseIII) enzymes, subsequently processes pri-miRNA in order to produce a pre-miRNA (precursor miRNA). The nuclear export factor exportin 5 then carries this pre-miRNA into the cytoplasm in combination with RAN-GTP, which is a GTP-binding nuclear protein. After entering the cytoplasm, another RNase III enzyme called Dicer cleaves the pre-miRNA further, resulting in a mature, double-stranded miRNA (Srinivasan and Das 2015).

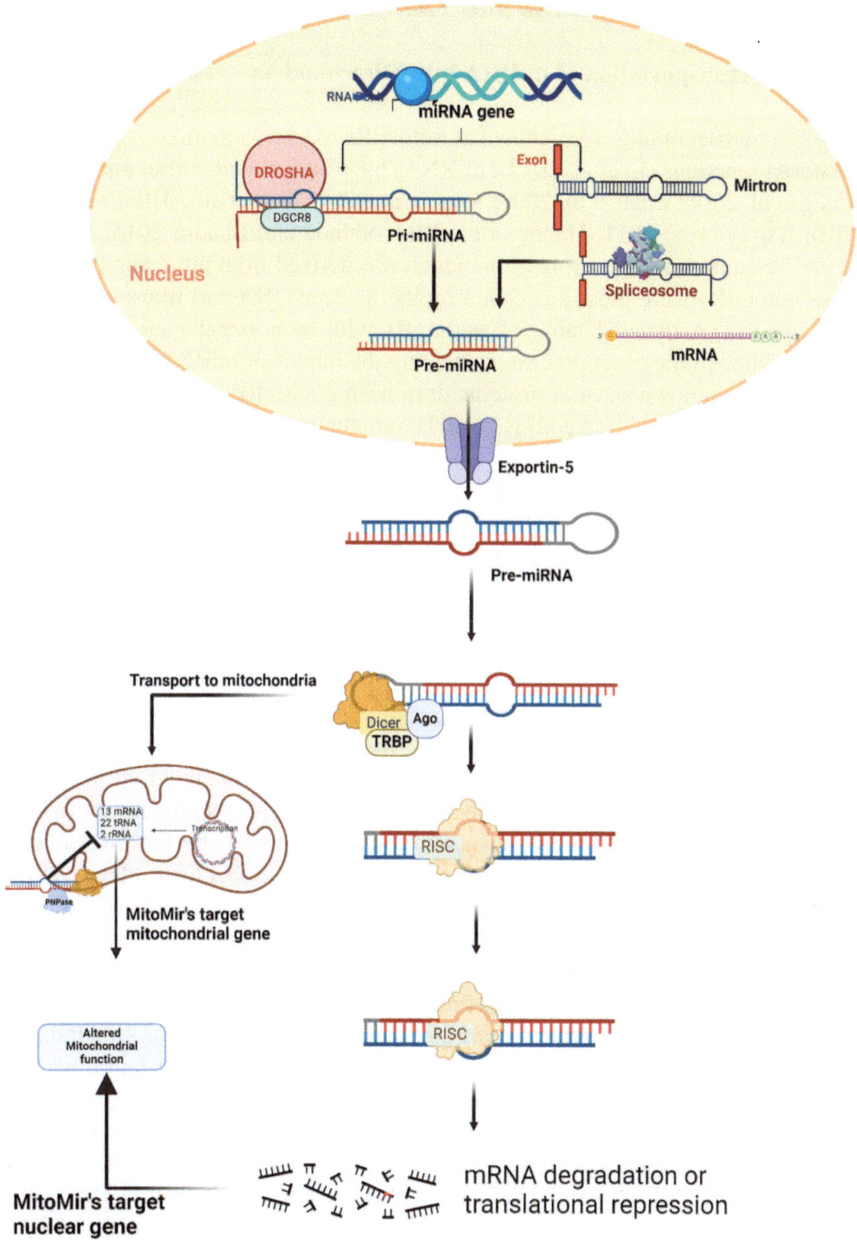

Fig. 11.2 Biogenesis of miRNA and transport into the mitochondria

11.2.2 Transport of miRs and Their Functions in Metabolism

11.2.2.1 Transportation of miRNA into Mitochondria

A specific subset of miRNAs, known as mitomiRs, is known to finely regulate mitochondrial functions (Li et al. 2012). miRNAs have been found within mitochondria (Lung et al. 2006; Chan et al. 2009; Kren et al. 2009; Bian et al. 2010; Favaro et al. 2010; Barrey et al. 2011; Huang et al. 2011; Shinde and Bhadra 2015; Fan et al. 2019). To control mRNA expression, which was derived from mitochondrial DNA, these mitomiRs were mostly encoded by the nuclear DNA and transported to the mitochondria. A detailed table of mitomiRs with their target genes were given below. Although the exact processes enabling the import of miRNA into mitochondria are yet unknown, several proteins have been connected to this intricate transport system. Argonaute2 (Ago2) is a crucial protein that is found in mitochondria of all different kinds of cells. Studies show that the mitochondrial translocation of mitomiRs depends critically on miRNA–Ago2 interactions (Latronico and Condorelli 2012; Wang et al. 2015). Zhang et al. (2014), for example, showed that the myogenesis-induced upregulation of miR-1 can efficiently infiltrate mitochondria through the formation of a complex with Ago2, which in turn enhances the translation of transcripts encoded by the mitochondria. The function of mitomiRs in the pathophysiology of the fragile X-associated tremor/ataxia syndrome (FXTAS) was examined by Gohel et al. in 2021. In HEK293 cells with increased CGG repeats, miR-320a was shown to bind with Ago2, and this complex was significantly more abundant in the mitochondrial matrix. Using the combined efforts of the sorting and assembly machinery (SAM50), translocase of the outer mitochondrial membrane 20 (TOM20), and translocase of the inner mitochondrial membrane (TIM), the miRNA-Ago2 complex can be transported into mitochondria by forming an association with the RNA-induced silencing complex (RISC) (Jusic et al., 2020). PNPase, a crucial component which was present in the inner mitochondrial membrane, which extends into the intermembrane gap, plays another important role. One essential element for miRNA mitochondrial import has been identified: PNPase. According to Wang et al. (2010), in murine hepatic cells, mutation of the PNPase gene (pnpt1) causes perturbations in mitochondrial architecture and function, partially due to RNA imports that control the transcription and translation of ETC proteins. Shepherd et al. (2017) went on to clarify that PNPase overexpression in HL-1 cardiomyocytes is associated with elevated levels of mitochondrial miRNA-378, highlighting its function in the transport of miRNAs. Another possible pathway for miRNA import is the voltage-dependent anion channel (VDAC). It is thought that VDAC, an outer membrane protein of mitochondria, which is highly conserved, helps short noncoding RNAs translocate into mitochondria. Although plant mitochondrial VDAC can bind to tRNA and mediate its import in vitro, as demonstrated by Salinas et al. (2006), its role in miRNA transport has not yet been fully explored, suggesting an exciting avenue for future research (Colombini 1980; Das et al. 2008; Bandiera et al. 2013; Macgregor-Das and Das 2018a).

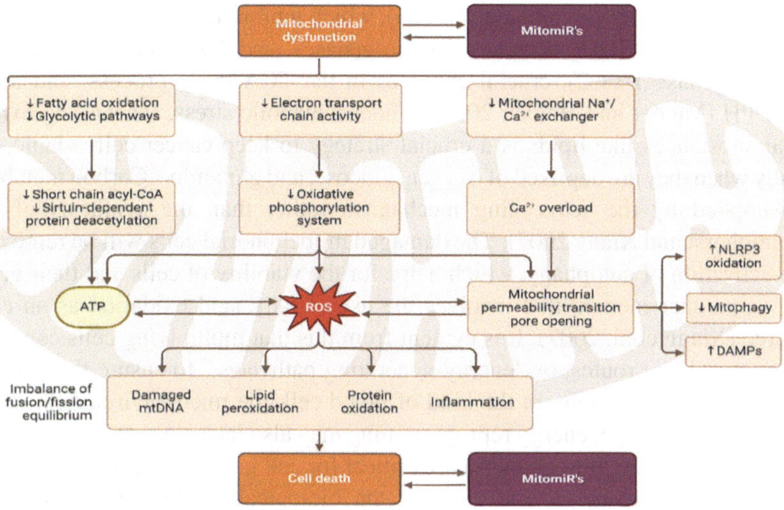

Fig. 11.3 MitomiRs as metabolic regulators

11.3 Metabolic Regulation of MitomiRs

MiRNAs control a variety of processes related to mitochondrial homeostasis and metabolism, including apoptosis, mitophagy, fusion, fission dynamics, mitochondrial membrane potential, metabolic reprogramming, and calcium transport (Porporato and Lin 2018).

During the development there is a metabolic switch in mitochondrial metabolism (Fig. 11.3). The metabolic "switch," referred to as the "Warburg Effect," causes dividing cells to rely on glycolysis instead of oxygen and properly functioning mitochondria to produce energy.

In order to enhance cell proliferation and anabolic processes, such as the biosynthesis of proteins, lipids, and nucleotides, this switch increases the concentration of CO2 and the absorption of glucose.

Multiplying blood cells absorb large amounts of glutamine, a nonessential amino acid which plays a crucial role in the metabolism of carbon. Several biosynthetic precursor molecules, electrons for OXPHOS (oxidative phosphorylation), and carbons for the tricarboxylic acid cycle (TCA) are all produced by glutamate metabolism (Palm et al. 2011). Glutamate is an important biomolecule for the denovo-synthesis of nucleotides, amino acids, and amino sugars, along with its role in nitrogen metabolism. (Vander and Heiden 2009).

The metabolism of nicotinamide adenine dinucleotide phosphate (NADPH), that provides electrons for anabolic processes and redox equilibrium, is one of the several metabolic pathways influenced by glutamine. The pentose phosphate and serine-dependent one-carbon pathways are two ways that NADPH can be produced. These pathways contribute to the synthesis of metabolites, including fumarate, succinate, and 2-hyroxyglutarate (2-HG), which inhibit the formation of FH (fumarate hydratase), SDH(succinate dehydrogenase), and IDH1 or IDH2 (isocitrate dehydrogenase 1 or 2), crucial enzymes in the TCA cycle process, and activate NADPH (Vander and Heiden 2009). Under metabolic stress, scavenging extracellular substances like lipids is a crucial strategy to keep cancer cells viable, especially when they are deprived of oxygen, glucose, and glutamine. Carbon metabolism is supplied by the scavenging mechanism rather than the synthesis of lipids (Kamphorst and Xiang 2007). The damaged mitochondrial cells will be removed by the activation of autophagy, which helps for the viability of cells and their growth, and is another method that involves the use of fatty acid oxidation as an energy source (Xiang et al. 2007). It is evident from this that multiplying cells can rely on many metabolic routes, or "energy-generating pathways," to ensure their survival. Furthermore, depending on the kind of blood cell, the microenvironment, and the developmental stage, energy reprogramming may also take place (Vyas et al. 2016). As a result, healthy mitochondria are critical for cell growth (Fig. 11.4).

production of energy and also for the biological production of metabolites required for cell growth, division, differentiation, regulation of reactive oxygen

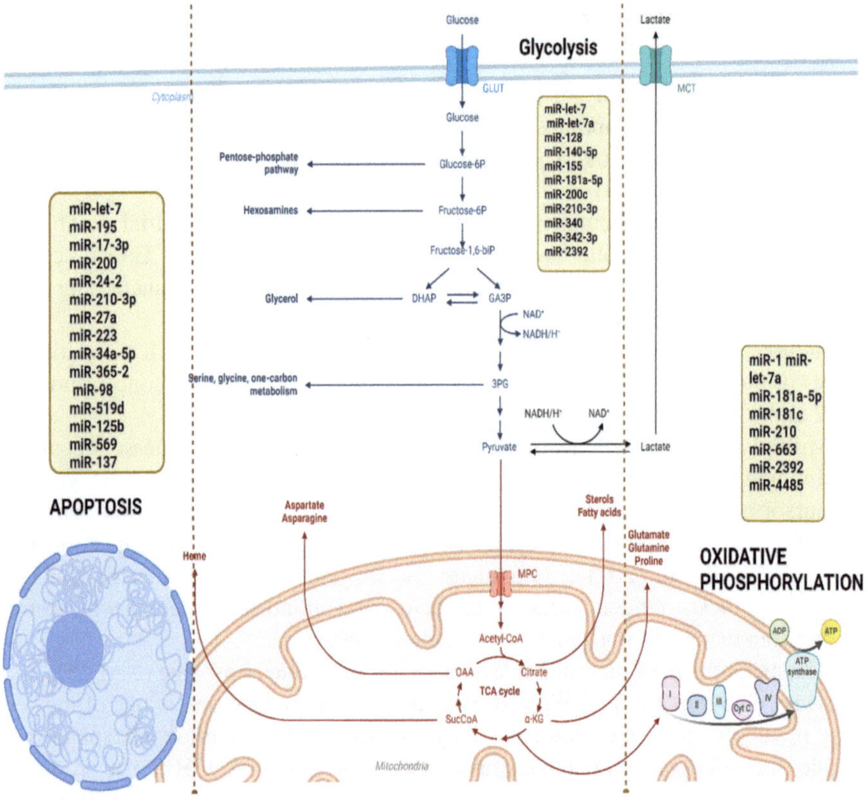

Fig. 11.4 Key mitomiRs regulating apoptosis, glycolysis, and oxidative phosphorylation

Table 11.1 List of MitomiRs with their gene targets and pathways affected

Target genes	miR	Metabolic pathway affected	Regulation	Reference
FASN	miR-33	Lipid metabolism	−vely	Bandiera et al. (2011)
MAP2k5	miR-143	Adipogenesis	−vely	Bandiera et al. (2013)
GLUT1	miR-138	Glucose metabolism	−vely	Macgregor-Das and Das (2018)
PDK1			−vely	
SLC2a4	miR-150		−vely	
SLC2A1			+vely	
SLC16A1	miR-29a	Oxidative metabolism	−vely	Shepherd et al. (2017)
	miR-29b			
	miR-124			
PRKAA1	miR-33	Lipid metabolism	−vely	Latronico and Condorelli (2012)
ACACB	miR-204	Lipid metabolism	−vely	Sripada et al. (2012)
ZEB1	miR-200	Lipid metabolism	+vely	Simão et al. (2021)
ZEB 2	miR-200	Lipid metabolism	+vely	Borralho et al. (2015)
Bcl2	miR-16	ATP production	−vely	Duarte et al. (2014)
ARL2	miR-195	ATP production	−vely	Geiger and Dalgaard (2017)
FOXA2	miR-29a-3p	Lipid metabolism	+vely	Sharma et al. (2019)
MFN1	miR-19b	Apoptosis	−vely	Rencelj et al. (2021)
COX1	miR-181c	Mitochondria oxidative metabolism	−vely	Kuthethur et al. (2022)
SREBF1	miR-33	Lipid metabolism	+vely	Fan et al. (2019)
GLS	miR-23b	Glutamine metabolism	−vely	McCarron et al. (2013)
H2aX	miR-24	Insulin signaling pathway	−vely	Youle and van der Bliek (2012)
SLC2A1	miR-1291	Mitochondrial	−vely	Wang et al. (2012)
ASS1		Metabolism		Duroux-Richard et al. (2016)

species (ROS) production and release, calcium ion transport, homeostasis, and cell death (Vyas et al. 2016). Various mitomiRs affecting different metabolic pathways were given in Table 11.1.

11.4 Mitochondrial Fission/Fusion and MitomiRs

A highly dynamic cellular organelle, the mitochondrion frequently goes through synchronised cycles of fission and fusion, also known as mitochondrial dynamics. In order to repair and eliminate damaged mitochondria, mitochondrial dynamics are crucial (Youle et al. 2012). Autophagy, Apoptosis, inflammation, and contractile failure are all significantly impacted by mitochondrial dynamics (Wang 2012; Duroux-Richard 2016; Wang et al. 2011). It has been demonstrated that MitomiRs control the fission and fusion of mitochondria. In various cell types, 42 MitomiRs have been identified. Among these, miR-146a, miR-34a, and miR-181a may target Bcl-2 to control mitochondrial dynamics. It has been shown that other MitomiRs directly target the fission/fusion proteins in the mitochondria.

Table 11.2 MitomiRs and their target genes which are associated with different diseases

S. no	Gene target	Disease	Related miR	References
1	DRP1, PPP3CA	Myocardial infarction	miR-499	Hata et al. (2015)
2	MFN-1	Cardiomyocyte apoptosis	miR-140	Yap et al. (2011)
3	PINK1, PRKN	Parkinson disease	miR-27 a/b	Trebinska-Stryjewska et al. (2019)
4	BAX	Type II diabetes	miR-27b	Sheng and Ni (2015)
5	FIS1	Myocardial infarction	miR-484	Liu et al. (2014)
6	MED13	Obesity and metabolic syndrome	MiR 378& miR378*	Hu et al. (2018)
7	COX10	Colon cancer	miR-210-5p	Shi et al. (2012)
8	COXIV	Neurological disorder	miR-338-5p	Wu et al. (2018)
9	MFF	Myocardial infarction	miR-761	Danza et al. (2016)
		Liver disease	miR-27	Weng et al. (2014)
10	UCP2	Kidney fibrosis	miR-30e	Sun et al. (2016)
11	BNIP3	Cardiovascular disease	miR-145	Xie et al. (2017)
12	MTFR1	Cardiac disease	miR-324-5p	Zhang et al. (2019)
13	LASS2	Bladder cancer	miR-98	Morita et al. (2016)
14	SOD2	Breast cancer	miR-509-5p	He et al.
15	CIP2A	Hepatocellular carcinoma	miR-548b-3p	Luan et al. (2018)
16	CYC	Cardiovascular and neurodegenerative disease	miR-34-a	Chen et al.
17	ATP5	Liver disease	miR-127-5p	Rottiers and Näär (2012)
				Boominathan et al. (2016)
18	ATP5L	Platelet disease	miR-570	Lees et al. (2018)
19	SDHD	Lung cancer	miR-210	Li et al. (2019)
				Yan et al. (2019)
20	ATP5G1	Metabolic disease	MiR-338-5p	Bian et al. (2010)
21	BAG5/ PINK1	Aging	miR-155	Ernster et al. (1981)
22	BAX	Cardiac fibrosis	miR-29b	Osellame et al. (2012)
				Calvo et al. (2016)
23	BCL2	Neurodegenerative disease	miR-181	Wang et al. (2009)
				Lee et al. (1993)
24	PGC1α	Metabolic disease	miR-130b-p	Wightman et al. (1993)
25	PARP-2	Metabolic disease	miR-149	Horvitz et al. (1980)

11.5 MitomiRs in Various Diseases

Abnormal miRNA expression and mitochondrial dysfunction have been documented in three major metabolic disorders: obesity, diabetes, and cardiovascular disease. In insulin-resistant type 2 diabetes (T2DM), a number of independent investigations have found altered gene expression, impaired biogenesis, changed mitochondrial structure, and decreased oxidative metabolism. Type-I diabetes has been shown to diminish mitochondrial ATP generation, and insulin withdrawal has been linked to decreased expression of oxidative phosphorylation-related muscle mitochondrial genes. By targeting respective subunits, a number of miRNAs have been shown to modify the activity of ETC complexes. For example, miR-181c and miR-338-5p modify the activity of mitochondrial complex IV by targeting distinct cytochrome oxidase subunits. Here are some MitomiRs which are associated with various diseases, were given in Table 11.2 (Purohit et al. 2021).

11.6 Conclusion

In summary, this chapter provides crucial evidences for the role of mitomiRNAs and their associated mechanism, functions, and impact on the metabolic reprogramming of blood cells during development and it is also shown that they are the key players which regulate mitochondrial dynamics and metabolic pathways, also mitomiRs were demonstrated as players in cell metabolic processes, mitophagy, apoptosis, fission and fusion, and redox signaling through their regulation of target genes of many signaling pathways. It is possible to ascertain mitomiRNAs' ability as a biological marker for therapy response and drug targets due to mounting evidence of their involvement in metabolic reprogramming and other related hallmarks of blood cell-related disorders.

Acknowledgments This work was supported by University of Hyderabad-Institution of Eminence Grant (No:UOH-IOE-RC3-21-006); Science and Engineering Research Board (SERB, EEQ/2022/00211); Indian Council of Medical Research (ICMR, No: EMDR/SG/12/2023/-1701); DBT-BUILDER, DST-FIST, ANRF-PAIR and Council of Scientific and Industrial Research (CSIR, No: 27(0343)/19/EMR-II) grants of Government of India.

Conflict of Interest All the authors declare that they have no conflict of interest.

Ethical Approval This chapter is a review of previously published accounts; as such, no animal or human studies were performed.

Data Availability All the data has been described within the manuscript.

References

Bandiera S, Rüberg S, Girard M, Cagnard N, Hanein S, Chrétien D, Munnich A, Lyonnet S, Henrion-Caude A (2011) Nuclear outsourcing of RNA interference components to human mitochondria. PLoS One 6:e20746

Bandiera S, Matégot R, Girard M, Demongeot J, Henrion-Caude A (2013) MitomiRs delineating the intracellular localization of microRNAs at mitochondria. Free Radic Biol Med 64:12–19

Bartel DP (2004a) MicroRNAs: genomics, biogenesis, mechanism, and function. Cell 116(2):281–97. https://doi.org/10.1016/s0092-8674(04)00045-5. PMID: 14744438

Bartel DP, Chen CZ (2004b) Micromanagers of gene expression: the potentially widespread influence of metazoanmicroRNAs. Nat Rev Genet, 5(5):396-400.https://doi.org/10.1038/nrg1328. PMID: 15143321

Bartel DP (2009) MicroRNAs: target recognition and regulatory functions. Cell, 136(2):215–33. https://doi.org/10.1016/j.cell.2009.01.002. PMID: 19167326; PMCID: PMC3794896.

Bian Z, Li L-M, Tang R, Hou D-X, Chen X, Zhang C-Y, Zen K (2010) Identification of mouse liver mitochondria-associated miRNAs and their potential biological functions. Cell Res 20(9):1076–1078

Boominathan A, Vanhoozer S, Basisty N, Powers K, Crampton AL, Wang X, Friedricks N, Schilling B, Brand MD, O'Connor MS (2016) Stable nuclear expression of ATP8 and ATP6 genes rescues a mtDNA complex V null mutant. Nucleic Acids Res 44(19):9342–9357

Borowski LS, Dziembowski A, Hejnowicz MS, Stepien PP, Szczesny RJ (2012) Human mitochondrial RNA decay mediated by PNPase–hSuv3 complex takes place in distinct foci. Nucleic Acids Res 41(2):1223–1240

Borowski M, Giovino-Doherty M, Ji L, Shi MJ, Smith KP, authors (2012) Laning J, editor. Basic pluripotent stem cellculture protocols. In: StemBook [Internet]. Cambridge (MA): Harvard Stem Cell Institute, 2008. https://doi.org/10.3892/ijo.2021.5285. PMID: 23658979

Borralho PM, Rodrigues CM, Steer CJ (2015) MicroRNAs in mitochondria: an unexplored niche. Adv Exp Med Biol 887:31–51

Calvo SE, Clauser KR, Mootha VK (2016a) MitoCarta2.0: an updated inventory of mammalian mitochondrial proteins. Nucleic Acids Res 44(D1):D1251–D1257

Calvo SE, Clauser KR, Mootha VK (2016b) MitoCarta2.0: an updated inventory of mammalian mitochondrial proteins. Nucleic Acids Res 44:D1251–D1257

Calvo SE, Clauser KR, Mootha VK (2016) MitoCarta2.0: an updated inventory of mammalian mitochondrial proteins. Nucleic Acids Res 44(D1):D1251–7s

Danza K, Silvestris N, Simone G, Signorile M, Saragoni L, Brunetti O, Monti M, Mazzotta A, De Summa S, Mangia A (2016) Role of miR-27a, miR-181a and miR-20b in gastric cancer hypoxia-induced chemoresistance. Cancer Biol Ther 17(4):400–406

Duarte FV, Palmeira CM, Rolo AP (2014) The role of microRNAs in mitochondria: small players acting wide. Genes 5:865–886

Duroux-Richard I, Roubert C, Ammari M, Presumey J, Grun JR, Haupl T, Grutzkau A, Lecellier CH, Boitez V, Codogno P et al (2016) miR-125b controls monocyte adaptation to inflammation through mitochondrial metabolism and dynamics. Blood 128:3125–3136

Ernster L, Schatz G (1981) Mitochondria: a historical review. J Cell Biol 91(3 Pt 2):227s–255s

Fan S, Tian T, Chen W, Lv X, Lei X, Zhang H, Sun S, Cai L, Pan G, He L (2019) Mitochondrial miRNA determines chemoresistance by reprogramming metabolism and regulating mitochondrial transcription. Cancer Res 79:1069

Galluzzi L, Joza N, Tasdemir E, Maiuri MC, Hengartner M et al (2008) No death without life: vital functions of apoptotic effectors. Cell Death Differ 15:1113–1123

Geiger J, Dalgaard LT (2017) Interplay of mitochondrial metabolism and microRNAs. Cell Mol Life Sci 74:631–646

Ha M, Kim VN (2014) Regulation of microRNA biogenesis. Nat Rev Mol Cell Biol 15(8):509–24. https://doi.org/10.1038/nrm3838. PMID: 25027649

Hata AN, Engelman JA, Faber AC (2015) The BCL2 family: key mediators of the apoptotic response to targeted anticancer therapeutics. Cancer Discov 5(5):475–487. https://doi.org/10.1158/2159-8290.CD-15-0011

Horvitz HR, Sulston JE (1980) Isolation and genetic characterization of cell-lineage mutants of the nematode Caenorhabditiselegans. Genetics, 96:435–54

Hu G, Zhao X, Wang J, Lv L, Wang C, Feng L, Shen L, Ren W (2018) miR-125b regulates the drug-resistance of breast cancer cells to doxorubicin by targeting HAX-1. Oncol Lett 15(2):1621–1629

Jagannathan R, Thapa D, Baseler WA, Shepherd DL, Croston TL, Nichols CE, Lewis SE, Hollander JM (2013) Translational regulation of the mitochondrial genome following redistribution of mitochondrial microRNA (MitomiR) in the diabetic heart. Federation of American Societies for Experimental Biology, Bethesda

Kozomara A, Birgaoanu M, Griffiths-Jones S (2019) miRBase: from microRNA sequences to function. Nucleic AcidsRes 47(D1):D155–D162. https://doi.org/10.1093/nar/gky1141. PMID: 30423142; PMCID: PMC6323917

Kuthethur R, Shukla V, Mallya S, Adiga D, Kabekkodu SP, Ramachandra L, Saxena PUP, Satyamoorthy K, Chakrabarty S (2022) Expression analysis and function of mitochondrial genome-encoded microRNAs. J Cell Sci 135:jcs258937

Latronico MV, Condorelli G (2012) The might of microRNA in mitochondria. Circ Res 110:1540–1542

Lee RC, Feinbaum RL, Ambros V (1993) The C. elegans heterochronic gene lin-4 encodes small RNAs with antisensecomplementarity to lin-14. Cell 75:843–54

Lees JG, Gardner DK, Harvey AJ (2018) Mitochondrial and glycolytic remodeling during nascent neural differentiation of human pluripotent stem cells. Development 145(20)

Li Z, Rich JN (2010) Diverse effects of hypoxia on tumor progression. New York: Springer. Hypoxia and hypoxia inducible factors in cancer stem cell maintenance; pp. 21–30

Li K, Zhang J, Ji C, Wang L (2016) MiR-144-3p and its target gene β-amyloid precursor protein regulate 1-methyl-4-phenyl-1, 2–3, 6-tetrahydropyridine-induced mitochondrial dysfunction. Mol Cells 39(7):543

Li H, Dai B, Fan J, Chen C, Nie X, Yin Z, Zhao Y, Zhang X, Wang DW (2019) The different roles of miRNA-92a-2-5p and let-7b-5p in MITOCHONDRIAL TRANSLATIon in db/db mice. ("The Different Roles of miRNA-92a-2-5p and let-7b-5p in Mitochondrial ..."). Mol Ther Nucleic Acids. 17:424–435. https://doi.org/10.1016/j.omtn.2019.06.013. (DOI) (PMC free article) (PubMed) (Google Scholar)

Li Y, Sun J, Wu R, Bai J, Hou Y, Zeng Y, Zhang Y, Wang X, Wang Z, Meng X (2020) Mitochondrial MPTP: a novel target of ethnomedicine for stroke treatment by apoptosis inhibition. Front Pharmacol 11:352

Lin MT, Beal MF (2006) Mitochondrial dysfunction and oxidative stress in neurodegenerative diseases. Nature 443(7113):787

Lin S, Gregory RI (2015) MicroRNA biogenesis pathways in cancer. Nat Rev Cancer 15(6):321–333

Liu B, Li J, Cairns MJ (2014) Identifying miRNAs, targets and functions. Brief Bioinform. 15(1):1–19. https://doi.org/10.1093/bib/bbs075. PMID: 23175680; PMCID: PMC3896928

Luan T, Fu S, Huang L, Zuo Y, Ding M, Li N, Chen J, Wang H, Wang J (2018) MicroRNA-98 promotes drug resistance and regulates mitochondrial dynamics by targeting LASS2 in bladder cancer cells. Exp Cell Res 373(1–2):188–197

Macgregor-Das AM, Das S (2018) A microRNA's journey to the center of the mitochondria. American Journal of Physiology-Heart and Circulatory Physiology 315:H206–H215

McCarron JG, Wilson C, Sandison ME, Olson ML, Girkin JM, Saunter C, Chalmers S (2013) From structure to function: mitochondrial morphology, motion and shaping in vascular smooth muscle. J Vasc Res 50:357–371

Merkerova MD, Krejcik Z (2021) Transposable elements and Piwi-interacting RNAs in hemato-oncology with a focus onmyelodysplastic syndrome (Review). Int J Oncol 59(6):105. https://doi.org/10.3892/ijo.2021.5285. PMID: 34779490

Morita K, Shirabe K, Taketomi A, Soejima Y, Yoshizumi T, Uchiyama H, Ikegami T, Yi Y, Sugimachi K, Harimoto N (2016) Relevance of micro RNA-18a and micro RNA-199a-5p to hepatocellular carcinoma recurrence after living donor liver transplantation. Liver Transpl 22(5):665–676

Mukherjee S, Basu S, Home P, Dhar G, Adhya S (2007) Necessary and sufficient factors for the import of transfer RNA into the kinetoplast mitochondrion. EMBO Rep 8(6):589–595

O'Connell RM, Rao DS, Chaudhuri AA, Baltimore D (2010) Physiological and pathological roles for microRNAs in theimmune system. Nat Rev Immunol 10(2):111–22. https://doi.org/10.1038/nri2708. PMID: 20098459

Osellame LD, Blacker TS, Duchen MR (2012) Cellular and molecular mechanisms of mitochondrial function. Best Pract Res Clin Endocrinol Metab 26(6):711–723

Purohit PK, Saini N (2021) Mitochondrial microRNA (MitomiRs) in cancer and complex mitochondrial diseases: current status and future perspectives. Cell Mol Life Sci 78(4):1405–1421

Purohit PK, Edwards R, Tokatlidis K, Saini N (2019) MiR-195 regulates mitochondrial function by targeting mitofusin-2 in breast cancer cells. RNA Biol 16:1–12

Rencelj A, Gvozdenovic N, Cemazar M (2021) MitomiRs: their roles in mitochondria and importance in cancer cell metabolism. Radiol Oncol 55:379–392

Rottiers V, Näär AM (2012) MicroRNAs in metabolism and metabolic disorders. Nat Rev Mol Cell Biol 13(4):239–250

Sayed D, Abdellatif M (2011) MicroRNAs in development and disease. Physiol Rev 91(3):827–87. https://doi.org/10.1152/physrev.00006.2010. PMID: 21742789

Sharma P, Bharat DN, Singh S (2019) Small regulatory molecules acting big in cancer: potential role of Mito-miRs in cancer. Curr Mol Med 19:621–631

Sheng C, Ni Q (2015) Expression of HAX1 and Ki-67 in breast cancer and its correlations with patient's clinicopathological characteristics and prognosis. Int J Clin Exp Med 8(11):20904–20910

Shepherd DL, Hathaway QA, Pinti MV, Nichols CE, Durr AJ, Sreekumar S, Hughes KM, Stine SM, Martinez I, Hollander JM (2017) Exploring the mitochondrial microRNA import pathway through polynucleotide phosphorylase (PNPase). J Mol Cell Cardiol 110:15–25

Shi L, Zhang S, Feng K, Wu F, Wan Y, Wang Z, Zhang J, Wang Y, Yan W, Fu Z (2012) MicroRNA-125b-2 confers human glioblastoma stem cells resistance to temozolomide through the mitochondrial pathway of apoptosis. Int J Oncol 40(1):119–129

Simão AL, Borralho PM, Steer CJ, Castro RE, Rodrigues CMP (2021) Isolation of mitochondria from liver and extraction of Total RNA and protein: analyses of miRNA and protein expression. Methods Mol Biol 2310:1–15

Sripada L, Tomar D, Singh R (2012) Mitochondria: one of the destinations of miRNAs. Mitochondrion 12:593–599

Srinivasan H, Das S (2015) Mitochondrial miRNA (MitomiR): a new player in cardiovascular health. Can J PhysiolPharmacol 93(10):855–61. https://doi.org/10.1139/cjpp-2014-0500. PMID: 26030833

Stefani G, Slack FJ (2008) Small non-coding RNAs in animal development. Nat Rev Mol Cell Biol 9(3):219–30. https://doi.org/10.1038/nrm2347. PMID: 18270516

Sun Y, He N, Dong Y, Jiang C (2016) MiR-24-BIM-Smac/DIABLO axis controls the sensitivity to doxorubicin treatment in osteosarcoma. Sci Rep 6(1):1–12

Tarassov I, Entelis N, Martin RP (1995) An intact protein translocating machinery is required for mitochondrial import of a yeast cytoplasmic tRNA. New York: Academic Press. J Mol Biol 245:315

Trebinska-Stryjewska A, Szafron L, Rembiszewska A, Wakula M, Tabor S, Sienkiewicz R, Owczarek J, Balcerak A, Felisiak-Golabek A, Grzybowska EA (2019) cytoplasmic HAX1 is an independent risk factor for breast cancer metastasis. J Oncol 2019:6375025

van Rooij E, Olson EN (2012) MicroRNA therapeutics for cardiovascular disease: opportunities and obstacles. Nat RevDrug Discov 11(11):860–72. https://doi.org/10.1038/nrd3864. PMID: 23080337; PMCID:PMC6813819

Vander Heiden MG, Cantley LC, Thompson CB (2009) Understanding the Warburg effect: the metabolic requirements of cell proliferation. Science 324(5930):1029–1033

Vyas S, Zaganjor E, Haigis MC (2016) Mitochondria and cancer. Cell 166(3):555–566

Wang C, Youle RJ (2009) The role of mitochondria in apoptosis*. Annu Rev Genet 43:95–118

Wang G, Chen H-W, Oktay Y, Zhang J, Allen EL, Smith GM, Fan KC, Hong JS, French SW, McCaffery JM (2010) PNPASE regulates RNA import into mitochondria. Cell 142(3):456–467

Wang K, Long B, Jiao JQ, Wang JX, Liu JP, Li Q, Li PF (2012) miR-484 regulates mitochondrial network through targeting Fis1. Nat Commun 3:781

Wang X, Song C, Zhou X, Han X, Li J, Wang Z, Shang H, Liu Y, Cao H (2017) Mitochondria associated microRNA expression profiling of heart failure. Biomed Res Int 2017:1–10

Weng H, Huang H, Dong B, Zhao P, Zhou H, Qu L (2014) Inhibition of miR-17 and miR-20a by oridonin triggers apoptosis and reverses chemoresistance by derepressing BIM-S. Can Res 74(16):4409–4419

Wightman B, Ha I, Ruvkun G (1993) Posttranscriptional regulation of the heterochronic gene lin-14 by lin-4 mediates temporalpattern formation in C. elegans. Cell 75:855–62

Wu G, Zhou W, Pan X, Sun Y, Xu H, Shi P, Li J, Gao L, Tian X (2018) miR-100 reverses cisplatin resistance in breast cancer by suppressing HAX-1. Cell Physiol Biochem 47(5):2077–2087

Xie Q, Wang S, Zhao Y, Zhang Z, Qin C, Yang X (2017) MiR-519d impedes cisplatin-resistance in breast cancer stem cells by down-regulating the expression of MCL-1. Oncotarget 8(13):22003

Yan K, An T, Zhai M, Huang Y, Wang Q, Wang Y, Zhang R, Wang T, Liu J, Zhang Y (2019) Mitochondrial miR-762 regulates apoptosis and myocardial infarction by impairing ND2. Cell Death Dis 10(7):1–16

Yap SV, Koontz JM, Kontrogianni-Konstantopoulos A (2011) HAX-1: a family of apoptotic regulators in health and disease. J Cell Physiol 226(11):2752–2761

Youle RJ, van der Bliek AM (2012) Mitochondrial fission, fusion, and stress. Science 337:1062–1062-5 https://doi.org/10.1126/science.1219855. PMID: 22936770; PMCID: PMC4762028

Zhang S, Liu C, Zhang X (2019) Mitochondrial damage mediated by miR-1 overexpression in cancer stem cells. Mol Therapy Nucleic Acids 18:938–953

Chapter 12
Noncoding RNA in Clinical Trials: Diagnostic and Therapeutical Prospects

Samar A. Zailaie, Amjad A. Aljagthmi, Farah M. Hmaid, and Consolato M. Sergi

Contents

Abstract The clinical potential of noncoding RNA has proven to have major advancements recently. Understanding their mechanisms and biological functions drives the use of ncRNAs in clinical applications. ncRNAs play crucial roles in regulating all cellular processes by influencing key proteins involved in the singling

S. A. Zailaie · A. A. Aljagthmi · F. M. Hmaid
Research Center, King Faisal Specialist Hospital & Research Center-Jeddah (KFSHRC-J), Jeddah, Saudi Arabia

C. M. Sergi (✉)
Children's Hospital of Eastern Ontario (CHEO), University of Ottawa, Ottawa, ON, Canada

Department of Laboratory Medicine and Pathology, University of Alberta, Edmonton, AB, Canada
e-mail: CSergi@cheo.on.ca

M. F. Ullah, A. Ahmad (eds.), *Non-coding RNAs*, Experientia Supplementum 115,
https://doi.org/10.1007/978-3-032-06948-1_12

pathways, both at gene transcriptional and translational levels. Additionally, ncRNAs can be detected intracellularly and extracellularly in different body fluids such as blood, urine, and cerebrospinal fluid, facilitating their noninvasive use in clinical settings. Several clinical trials have investigated ncRNAs' therapeutic and diagnostic use, leading to models that mimic the mechanistic functions of dysregulated ncRNAs. These models have provided new clinical tools for early disease diagnosis and therapy. While the FDA has approved some ncRNA-based diagnostic/therapeutic applications, other clinical trials were withdrawn or terminated. Challenges associated with ncRNAs include off-target effects due to a lack of specificity, decreased stability of specific ncRNAs in vivo, low cellular uptake, and limitations in delivery systems. Additionally, most exciting studies focus primarily on some types of ncRNA, such as microRNAs, highlighting the need for broader research into investigating other types of ncRNAs. Unlocking the full potential of ncRNAs will pave the way for a myriad of possibilities for developing novel strategies for early diagnosis, disease prognosis, and targeted therapies for different human diseases.

Keywords Noncoding RNAs · microRNA · lncRNA · siRNA · Clinical trials · circRNA · AoS · Circulating RNA · Serum-RNA · Exosomes · Gene therapy · Therapeutical approaches · Biomarkers · Early diagnosis

Abbreviations

AA	Advanced adenomas
AAV	adenovirus-associated viral vector
ADAR	Adenosine deaminases acting on RNA
ADPKD	Autosomal dominant polycystic kidney disease
ADQ	7-nitro-2-[4-(7-nitro-3-oxo-4, 9-dihydrofuro[3, 2-b] quinoxalin-2-yl) phenyl]-4, 9-dihydrofuro[3, 2-b]quinoxalin-3-one
AMI	Acute myocardial infarction
ASO	Antisense oligonucleotide
ATTRv	Hereditary amyloid transthyretin
ALAS1	Delta aminokevulinic acid
BC	breast cancer
BDS	bio-switchable delivery system
CAD	Coronary artery disease
CC	cholecystitis
CCC	Cholangiocellular carcinoma
circRNA	circular RNA
CLIP-ON	Circularization promoting oligonucleotide
CRC	Colorectal cancer
CTC	Circulating tumor cells

CVD	Cardiovascular disease
ESCC	esophageal squamous cell carcinoma
EMT	epithelial to mesenchymal transition
Ex-ncRNA	extracellular noncoding RNA
GalNAc	N-acetylgalactosamine conjugates
GBC	Gallbladder cancer
H3K27	histone3 lysine 27
HCC	Hepatocellular carcinoma
HCV	Hepatitis C virus
hnRNPs	heterogeneous nuclear riboproteins
HOTAIR	HOX antisense intergenic RNA
IS	Ischemic stroke
LC	Lung cancer
lncRNA	Long noncoding RNA
LNP	Lipid nanoparticle
MALAT1	Metastasis Associated Lung Adenocarcinoma Transcript 1
miRISC	miRNA induced silencing complex
MRG-201	Remlarsen (a miR-29 mimic therapy)
NAAT	Nucleic acid amplification test
NB	Neuroblastoma
NEAT2	Nuclear Enriched Abundant Transcript 2
NIHSS	national institutes of health stroke scale
PCA3	Prostate cancer antigen 3
PCA	Primary prostate cancer
Phi	Prostate health index
PH	primary hyperoxaluria
PH1	Primary hyperoxaluria type 1
PRC2	Polycomb repressive complex 2
PSA	prostate-specific antigen
RNAi	RNA interference
shRNA	short hairpin RNA
siRNA	Small interfering RNA
SMN2	survival of motor neuron 2
SNP	Single nucleotide polymorphism
snRNA	Small noncoding RNA
TDN	Tetrahedral DNA nanostructure
TLR	Toll-like receptor
TSTA	Two-step transcriptional amplification
TTR	transthyretin gene
UTR	Untranslated region
ZFN	zinc-finger nuclease
3STA	Three-step transcriptional amplification
4Kscore	Four-kallikrein panel

12.1 Introduction

Noncoding RNAs, including long noncoding RNA (lncRNA) and small noncoding RNA (snRNA), have been linked to human health and disease. The majority of the human DNA genome is transcribed as ncRNA, indicating its crucial roles in physiological homeostasis. The disturbances in ncRNA levels have been associated with different human diseases, particularly across all types of cancers. The clinical relevance of dysregulated ncRNA, especially microRNAs, has gained considerable attention due to their biological functions in promoting or inhibiting the functions of certain proteins. ncRNA can be expressed intracellularly, modulating diverse cellular processes such as chromatin remodeling, cell growth and apoptosis, and cellular migration and invasion, or packaged and released extracellularly to modulate other cells. In fact, the initial understanding of extracellular RNA was thought to be a byproduct of damaged or dying cells because of their low abundance and nonconserved sequences (Poliseno et al. 2024). However, their existence in many living organisms has led to the hypothesis that exRNA are functional and are involved in intercellular communication (Gruner and McManus 2021). ncRNAs are important in various biological processes, including gene expression, RNA maturation and processing, protein synthesis, and signaling pathways (Loganathan and Doss 2023). Eventually, those RNAs could regulate a broad range of cellular processes such as cell proliferation, DNA synthesis, and transcription (S. Kim et al. 2020). Soon after, it was recognized that exRNA is cell-specific and can be utilized as a biomarker or therapeutic target (Das et al. 2019).

The association of ncRNA with human diseases has established significant clinical potential of ncRNAs as diagnostic, prognostic, and therapeutic targets. ncRNA-based biomarkers and therapeutics have revolutionized over the years. The first RNA class, mRNA, was used as a prognostic biomarker for cancer patients (Martinez-Ledesma et al. 2015). Eventually, other RNA classes were found to be valuable, especially the extracellular RNA (exRNA). Several ncRNA-based therapies have been developed and approved for clinical use. For example, spinal cord muscular atrophy is treated with 18-mer antisense oligonucleotide (AOS) that targets the survival of motor neuron 2 (SMN2) pre-mRNA splicing (exon 7 inclusion). Additionally, acute hepatic porphyria was treated with 21 nt ds-siRNA that targets delta aminolevulinic acid synthase 1 (ALAS1) mRNA (Winkle et al. 2021).

In this chapter, we will present the current picture of ncRNA in clinical practice for different pathological applications of noncoding RNA, providing an overview of the utilization of ncRNA classes in the medical field. We will also discuss the limitations of ncRNA translation into clinical practice, as challenges in their clinical validation still exist.

12.2 ncRNA Characteristics Supporting their Relevance in Clinical Practice

The use of ncRNAs as biomarkers and therapeutic approaches has been evolving due to their biological properties. In addition to their presence within the intercellular space, ncRNAs are present in various body fluids, including serum, plasma, urine, saliva, ascites, cerebrospinal fluid, and amniotic fluid (Safdar et al. 2016). Extracellular noncoding RNAs (Ex-ncRNA), including small ncRNA, lncRNA, and circular RNA, have emerged as novel mediators and therapeutic targets, and they form the majority of extracellular transcripts (Jiapaer et al. 2023; Lin et al. 2021; Sato-Kuwabara et al. 2015; Wu et al. 2022a). ExncRNA is a heterogenous RNA population that circulates in the extracellular space as cell-free nucleic acids, bound to lipids or proteins or packed in membranous vesicles, protecting them from degradation (Sadik et al. 2018).

The distinct profile of exRNA has been established between healthy and disease groups, suggesting that exRNA are implicated in disease initiation and progression.

Many studies have been conducted to characterize exRNA, including their biogenesis, release mechanisms, packing, transportation, and degradation (Patton et al. 2015). Of note is that the chemical and physical properties of exRNA have been investigated. Despite the high activity of ribonuclease enzymes in the extracellular environment, researchers have found that exRNA is resistant to degradation. The existence of exRNA in biological fluids as chemically stable molecules made them favorable targets, especially in the field of liquid biopsy. It was scientifically evident that an RNA molecule is first packed in extracellular vesicles, such as exosomes, and released from a donor cell to a recipient cell nearby for RNA delivery (Tosar et al. 2021). Exosomal ncRNAs are highly clinically relevant and regulate biological processes related to physiological homeostasis and pathophysiology. Exosomes contain unique compositions, including ncRNAs, and undergo a highly selective sorting process that depends on several mechanisms, such as a sumoylated heterogeneous nuclear ribonucleoproteins (hnRNPs)-dependent mechanism, the miRNA-induced silencing complex (miRISC)-related pathway, and the 3′-end of the miRNA sequence-dependent pathway (Wei et al. 2021). Therefore, RNA-containing exosomes present unique signatures of their secreted cells and can be of high clinical relevance in diagnosis, prognosis, disease monitoring, and targeted therapy (Wei et al. 2021).

The fact that disease-specific ncRNAs can be differentially expressed and secreted in body fluids has enabled their use as noninvasive diagnosis and prognosis approaches. Liquid biopsies have become increasingly used in precision medicine, particularly in cancer screening, and have facilitated the accessibility of disease-derived information. Opposite to the standard tissue biopsy, liquid biopsy is considered a minimally invasive procedure for fluid collection, such as blood, followed by the isolation of specific analytes for testing. This technique was first initiated in the late 1980s, when circulating tumor cells (CTCs) were detected in cancer patients

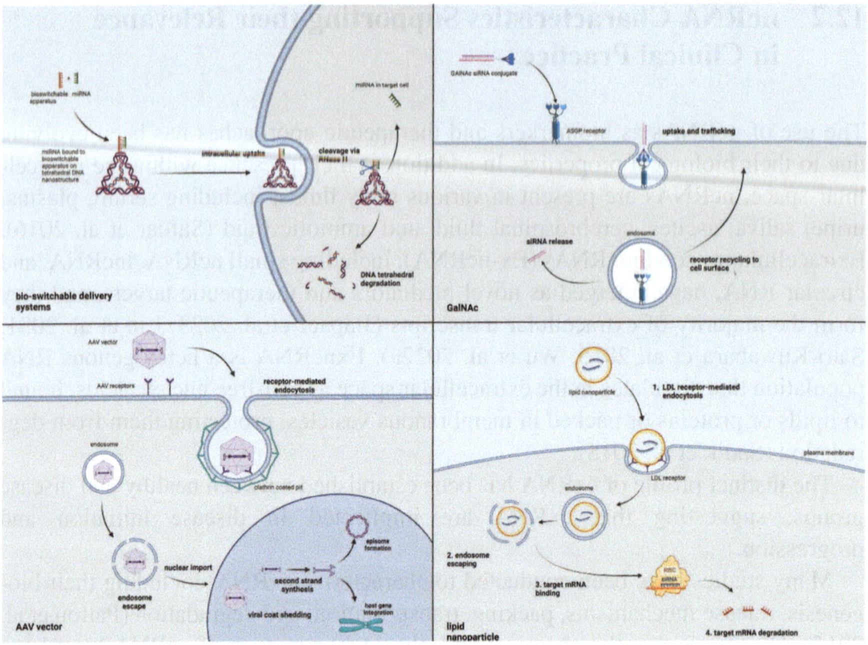

Fig. 12.1 The bio-switchable delivery systems

and manifested a similar molecular profile to tumor-isolated cells (Alba-Bernal et al. 2020). Since then, exploring other blood constituents as noninvasive biomarkers have become of great interest in medical and pharmaceutical research. Liquid biopsy as a source of multiple exRNA is an evolving field, and there is plenty of evidence that has linked many exRNA to diseases such as cancer and cardiovascular disorders. It is widely accepted that exRNA signatures are different from those of a diseased population to those of corresponding healthy individuals. Moreover, the subclasses of a given disease showed distinct RNA profiles, suggesting a valuable clinical application of exRNA. In Fig. 12.1, a cartoon showing from top left, top right, bottom right and bottom left, the bio-switchable delivery systems, the N-acetylgalactosamine conjugates (GalNAc), a lipid nanoparticle, and the adenovirus-associated viral vector (AAV vector).

12.3 The Clinical Relevance of lncRNA

LncRNA are classified according to their genomic location as intergenic, intronic, sense, and antisense lncRNA (L. Ma et al. 2013; Mattick et al. 2023). It was suggested that each class of lncRNA acts at different cellular locations, and their existence is less conserved among species relative to other ncRNA (Johnsson et al. 2014). Moreover, lncRNA was found to have many functional mechanisms, such as transcriptional, translational, post-transcriptional, and splicing regulators, making it a potential candidate as a biomarker or therapeutic agent. The clinical utility of lncRNA was first described as a breast cancer biomarker when HOX antisense intergenic RNA (HOTAIR) was detected in archived tissue-restricted from primary and metastatic breast carcinomas (Chisholm et al. 2012). This study has promoted comprehensive research to elucidate the application of lncRNA in cancer diagnosis and treatment. Following the valuable insight of lncRNA as a cancer biomarker, the noninvasive analysis of circulating lncRNA has been studied in other diseases such as cardiovascular, infectious, diabetes, and autoimmune diseases (Abdallah et al. 2022; Bär et al. 2016; Dong et al. 2021; Lodde et al. 2020; Niknam et al. 2023; Wang et al. 2022). High-throughput sequencing studies have focused on lncRNA profiling across different types of tissue, including body fluids. Many deregulated lncRNAs have been reported; however, only a few have a strong clinical application, including the HOTAIR, prostate cancer antigen 3 (*PCA3*), and metastasis-associated lung adenocarcinoma transcript 1 (MALAT1), which will be discussed in this section. In Table 12.1, the governmental ID and the title of the current clinical trials using lncRNA are shown.

12.3.1 HOX Antisense Intergenic RNA (HOTAIR)

HOTAIR lncRNA was among the first well-characterized lncRNA. It is transcribed from the antisense strand of HoxC gene, located between HoxC11 and HoxC12 on chromosome 12 (Hajjari and Salavaty 2015). It acts as a scaffold for the histone methyltransferase polycomb repressive complex 2 (PRC2), regulating chromatin remodeling and leading to transcriptional silencing (Bhan and Mandal 2015). Considering its important role in modulating gene expression, the clinical value of circulating HOTAIR was extensively explored in the diagnosis, prognosis, and treatment response of different diseases.

12.3.1.1 HOTAIR as a Biomarker

Increased expression of HOTAIR lncRNA is observed in breast cancer (BC), associated with disease metastasis and progression, and can be a powerful predictor (Gupta et al. 2010). The expression level of HOTAIR is elevated in tissue and plasma samples from breast cancer patients and is linked to lymph node metastasis and BC molecular subtypes, indicating a high potential for diagnostic accuracy (Zhang et al. 2016). The level of HOTAIR was also assessed in serum samples collected from BC patients, and it shows increased levels in BC patients compared to healthy individuals (Zhang et al. 2015). These studies provide valuable insight into HOTAIR diagnostic utility for BC utilizing different clinical samples, including noninvasive liquid biopsies to measure HOTAIR expression.

Table 12.1 Current clinical trials on noncoding RNAs (government ID and title)

1	NCT05995067: The potential association of long non-coding RNA (NEAT1 and MALAT1) in the process of epithelial-mesenchymal transition in Periodontitis.	59	NCT02756559: MiRNAs evaluating lymphocyte proliferation and apoptosis as well as prognosis of sepsis.	117	NCT03838991: Epigenetic regulation in fibrous dysplasia of bone: mirDYS study.
2	NCT04175691: Circulating non-coding RNA in acute ischemic stroke (AISRNA).	60	NCT01240369: Association between VEGF-C and miRNA and clinical non-small cell lung cancer and esophagus squamous cell carcinoma.	118	NCT04413916: MiRNA in kidney transplantation: Association with kidney graft function and disease process.
3	NCT06544005: Implication of long non-coding RNA HOTTIP haplotype on liver cancer metastasis.	61	NCT04767750: Role of LncRNA H19 in the regulation of IGF-1R expression	119	NCT05048381: miRNA in sweat, a novel digital biomarker for detection of active tuberculosis (TB).
4	NCT05928416: ALS diagnosis from a saliva sample: a non-coding RNA analysis approach.	62	NCT02353221: CPMC early undifferentiated arthritis registry.	120	NCT04661176: Evaluation of the sentinel PCC4 assay for diagnosis, prognosis and monitoring of prostate cancer in Puerto Rico.
5	NCT04230785: Circulating non-coding RNA in acute ischemic stroke with endovascular treatment (EVTRNA).	63	NCT04937855: The mechanism of lncRNA NEAT1 in alleviating acute respiratory distress syndrome through miR-27b regulated Nrf2 pathway.	121	NCT04009733: Epigenetic regulation of osteogenesis imperfecta severity: miROI study.
6	NCT06213493: Bace1as role in heart failure (BACE1AS-HF).	64	NCT06149351: Salivary biomarkers for concussion.	122	NCT04017520: Breast milk: Influence of the micro-transcriptome profile on atopy in children over time.
7	NCT04946266: Prospective validation of the prognostic value of long non-coding MFI2-AS1 RNA in localized clear cell kidney cancers.	65	NCT04176276: Determining serum and urinary levels of miRNA 192 and miRNA 25 in patients with and without type 2 diabetes.	123	NCT04029740: Exosomal microRNAs as a biomarker in panic disorder and in response to CBT.
8	NCT06334835: Long non-coding RNAs and their role on epigenome as diagnostic markers in childhood acute lymphoblastic leukemia of T cells.	66	NCT03377335: Dapagliflozin, cardio-metabolic risk factors and type-2 diabetes.	124	NCT03088839: Circulating biomarker for amyotrophic lateral sclerosis (ALS).

(continued)

Table 12.1 (continued)

9	NCT02641847: TA(E) C-GP versus A(E)C-T for the high risk TNBC patients and validation of the mRNA-lncRNA signature.	67	NCT03885297: 3 mg liraglutide for overweight or obesity.	125	NCT04100811: Identification of clinically insignificant or significant prostate cancer with the miR scientific sentinel platform.
10	NCT04788914: lncRNAs as a biomarker to assess the therapeutic impact of oral absorbent Â ± probiotics in CKD patients with PAD	68	NCT04765410: The impact of tissue micro-RNA profile from EUS-FNA in pancreatic adenocarcinoma.	126	NCT06427720: INC00511/ miR-185-3p axis and miR-301a-3p markers for breast cancer diagnosis.
11	NCT04792320: Oral absorbent and probiotics in CKD patients with PAD on gut microbiota, IncRNA, metabolome, and vascular function.	69	NCT05621343: New strategies for assessment of the persistence of viable bacilli in latent and active tuberculosis.	127	NCT05318898: Effect of dietary protein on the regulation of exosome microRNA expression in patients with insulin resistance.
12	NCT06387836: LINC01844 as a diagnostic biomarker for POCD in elderly patients.	70	NCT04427371: The effect and mechanism of lncRNA NBR2 regulating endothelial pyroptosis by targeting GSDMD in sepsis.	128	NCT05927467: Eurbio-Alport (RaDiCo cohort) (RaDiCo eurbio-alport).
13	NCT02751060: Evaluation of plasma non-coding RNAs as biomarkers in coronary heart disease.	71	NCT06313645: Vascular senescence and atherosclerotic plaque vulnerability.	129	NCT03991871: HARapan kiTa ECP (external counter pulsation) study HARTEC study.
14	NCT04542902: Non-coding RNAs analysis of eosinophil subtypes in asthma.	72	NCT06325137: Transcriptome analysis in idiopathic nephrotic syndrome: Steroid responsiveness.	130	NCT04641741: Effect of Mepolizumab on severe eosinophilic asthma
15	NCT06706687: A study of the behavioral variant of frontotemporal dementia and bipolar disorder: a neuroimaging and epigenetics integrated approach.	73	NCT01515605: Molecular biological and moleculargenetic monitoring of therapy after kidney transplantation.	131	NCT03999931: Study on the difference of plasma microRNA expression in patients with genetic susceptibility to mental disorders.
16	NCT05139914: Assessment of Dapagliflozin on vascular health in patients with type 2 diabetes.	74	NCT03076580: An integrative-"omics" study of cardiomyopathy patients for diagnosis and prognosis in China.	132	NCT03104075: Genomics and epigenomics of the elderly response to pneumococcal vaccines.

(continued)

Table 12.1 (continued)

17	NCT06320418: The epigenetic regulatory role of P-element induced wimpy testis (Piwi) interacting RNA-823 (piR-823) in ovarian cancer progression.	75	NCT03000764: RNA and heat shock protein biomarkers in radiation-induced fibrosis in breast cancer.	133	NCT05888636: Transcriptomics and epigenetics analysis in drug-resistance of multiple myeloma.
18	NCT05733026: Effect of genetic and epigenetic factors on the clinical response and toxicity to doxorubicin among Egyptian breast cancer patients.	76	nct04509271: Study on novel peripheral blood diagnostic biomarkers for MCI due to Alzheimer's disease.	134	NCT03102268: ncRNAs in exosomes of cholangiocarcinoma.
19	NCT06531902: LncRNA nicotinamide nucleotide transhydrogenase-antisense RNA1 NNT-AS1 in CRC.	77	NCT03905083: miRNA and myokines in patients with PAH.	135	NCT02062476: Epigenetic effects elicited by lactobacillus gg in children with cow's milk allergy.
20	NCT06649253: Childhood B-acute lymphoblastic leukaemia and role of CD9 gene regulation in relapse.	78	NCT02791217: Identification of hematological malignancies and therapy predication using microRNAs as a diagnostic tool.	136	NCT05544175: EUROMOdulation pain therapy in combination with intensive physiotherapy.
21	NCT02522026: Follow-up-study of COPD patients and healthy controls for evaluation of predictive non-coding RNA biomarkers	79	NCT05683769: microRNA in erosive hand osteoarthritis and psoriatic arthritis	137	NCT05529251: De-escalation study for stage IIa/IIb < 3 cm seminoma
22	NCT06427278: Implication of long non-coding RNA CCDC144NL-AS1/ micro RNA-143-3p axis expression level as novel signature in colorectal cancer.	80	NCT04860024: Epigenetics in ADHD	138	NCT04184700: Epigenetic effects in children with Cow's Milk allergy treated with different formulas
23	NCT04895124: Second follow-up study of COPD patients and healthy controls for evaluation of predictive non-coding RNA biomarkers.	81	NCT06442852: Study on the relationship between peripheral blood mirna and risk and severity of Alzheimer's disease.	139	NCT06203496: Monitoring of patients with diffuse gliomas using circulating miRNAs.

(continued)

Table 12.1 (continued)

24	NCT05906277: Biobanking and new biomolecular metrics.	82	NCT05735704: Early detection and screening of hematological malignancies—SANGUINE.	140	NCT02466035 epigenetic effects involved in children with cow's milk allergy (EPICMA).
25	NCT05398952: Post-COVID-19 outpatient care and biomarkers.	83	NCT04390269: Immunogenetics predictors with COVID-19.	141	NCT04175678: Project 1: Diet and exercise modulate the sperm epigenome in men.
26	NCT06131268: The role of Negr1 in modulating neuroplasticity in major depression (RONIN).	84	NCT02864329: Small RNA pathways in mammalian gametogenesis	142	NCT06065592: Exploring cancer-associated thromboembolism prognosis biomarkers and polymorphisms.
27	NCT04269746: Assessment of long noncoding RNA CCAT1 in colorectal cancer patients.	85	NCT04381351: Noncoding RNAs in COVID-19 and COVID-19 related kidney dysfunction.	143	NCT04021823: Neurobiological analyses within the FORESEE III study.
28	NCT03185767: Long noncoding RNA in complications of systemic lupus.	86	NCT03268135: Heart failure and aortic stenosis transcriptome.	144	NCT04594720: Circulating biomarkers to identify thyroid cancer.
29	NCT03738319: Non-coding RNA in the exosome of the epithelia ovarian cancer.	87	NCT05833360: Prospective study of OncRNA stratification of cancer by size and stage.	145	NCT06466850: Mesenchymal stem cells derived exosomes in osteoarthritis patients.
30	NCT06213493: Bace1as role in heart failure (BACE1AS-HF).	88	NCT06617585: The role of CircDENND4C in epithelial ovarian cancer.	146	NCT06536686: Silent myocardial ischemia in patients undergoing non-oncological abdominal surgeries.
31	NCT04175691: Circulating non-coding RNA in acute ischemic stroke (AISRNA).	89	NCT02419625:The epidemiology of hepatitis e virus infection in Israel and potential risk factors.	147	NCT06307249: Precision therapy for solid tumors: Synergistic CDK4/6 inhibition and anti-VEGF targeting LncRNA.
32	NCT03469544: Long noncoding RNA HOTAIR and midkine as biomarkers in thyroid cancer.	90	NCT05672602: Biomarkers of LONG COVID (LONG COVID_LCM).	148	NCT01141530: Deep sequencing of the breast cancer transcriptome.
33	NCT06357689: Association of SNPs in long intergenic noncoding RNA 00511 (LINC00511) with breast cancer among the egyptian population.	91	NCT02245633: Correlation between vitamin D levels to ADAMTS13, VWF and micro RNA expression in diabetic hemodialysis patients.	149	NCT02448056: MiRNA as a diagnostic and prognostic biomarker of hepatocellular carcinoma.

(continued)

Table 12.1 (continued)

34	NCT04230785: Circulating non-coding RNA in acute ischemic stroke with endovascular treatment (EVTRNA).	92	NCT05477667: Study of let-7a and miRNA-124 in cases of non-hodgkin's lymphoma and acute leukemia.	150	NCT02747784: Evaluation to assess cognitive training for the prevention of postoperative cognitive decline.
35	NCT03057171: A study on the gastrointestinal disease and helicobacter pylori controlled long non-coding RNA.	93	NCT04381156: MicroRNAs in renal recovery—Pilot study.	151	NCT03495921: A trial for participants with ewing's sarcoma treated with vigil in combination with irinotecan and temozolomide.
36	NCT06534242: Implication of genetic variations in long intergenic non-coding RNA 00511 (LINC00511) in colorectal cancer.	94	NCT03089242: MicroRNAs in acute kidney injury.	152	NCT03227510: MicroRNAs as diagnostic biomarkers in hepatocellular carcinoma among Somali patients.
37	NCT05943093: H19 in acute lymphoblastic leukemia.	95	NCT03962088: Timisnar—Biomarkers substudy (Timisnar-mirna).	153	NCT06726070: Investigation of the diagnostic role of miRNAs in PCa and BPH
38	NCT01187602: Short non-coding RNA biomarkers of predisposition to ovarian cancer.	96	NCT04288739: Immunophenotyping and xist gene in AML.	154	NCT05488418 clinical study of biomarkers of stress resilience: Role of ELK1 and GPR56.
39	NCT05840133: Study of long non-coding RNA SNHG15 as a novel biomarker in HBV associated HCC.	97	NCT05921812: Biochemical role of circulating microRNAs expression as diagnostic markers for non-hodgkin's lymphoma patients.	155	NCT05646901: Involvement of the gut microbiota-brain cross-talk in the loss of eating control.
40	NCT03830619: Serum exosomal long noncoding RNAs as potential biomarkers for lung cancer diagnosis.	98	NCT03457935: Detection of early idiopathic pulmonary fibrosis.	156	NCT06610851: Monitoring of patients with low-grade gliomas using circulating miRNA.
41	NCT06432413: Emerging role of NOTCH-related long non-coding RNA(s) as biomarkers in liquid biopsy from colorectal cancer patients	99	NCT05449847: Plasma MicroRNA for prediction of hepatocellular carcinoma	157	NCT01510054: Endometrial luteal phase characteristics and luteal phase support in controlled ovarian stimulation protocols with GnRH antagonists: Focusing on MicroRNA.

(continued)

Table 12.1 (continued)

42	NCT04788914: lncRNAs as a biomarker to assess the therapeutic impact of oral absorbent Ã,Â ± probiotics in CKD patients with PAD	100	NCT06058182: RNA as prognostic biomarkers in patients with acute coronary syndrome.	158	NCT04382950: Combination of recombinant bacterial ACE2 receptors—Like enzyme of B38-CAP and isotretinoin could be promising treatment for COVID-19 infection—And its inflammatory complications.
43	NCT06459895: Biochemical role of long non-coding RNA MALAT 1 and serum tumor necrosis factor alpha in bronchial asthma patients.	101	NCT05077007: Evaluation of renal damage after PCNL and ESWL using novel RNA based biomarkers.	159	NCT04950569: Effect of levosimendan on miRNAs regulation in the failing hearts.
44	NCT02602808: Study of peripheral blood non-coding RNAs as diagnosis and prognosis biomarker for acute pancreatitis.	102	NCT02521337: PreTerm RNA in maternal serum.	160	NCT04624581: FIbromyalgia and genetics subgroups (FIDGIS).
45	NCT06147700: Identifying molecular determinants of infertility in men (MODIFY).	103	NCT05739513: A specific miRNA encoded by SARS-CoV-2 as a diagnostic tool to predict disease severity in COVID-19 patients.	160	NCT06324084: Pathophysiology of osteoporosis: Role of hidden cortisol excess and its predictors in bone fragility.
46	NCT05088811: The role of long non-coding RNAs WRAP53 and UCA-1 as potential biomarkers in diagnosis of hepatocellular carcinoma.	104	NCT04584996: CIRcular and non-coding RNAs as clinically useful biomarkers in pancreaticobiliary cancers.	161	NCT05544396: Study on the probiotics regulating miRNA in *H. Pylori*-induced Wnt/catenin gastric carcinogenesis.
47	NCT04638556: Effect of circulating lncRNAs on type 2 diabetic peripheral neuropathy.	105	NCT04835454: Role of Piwi-protein interacting RNA, miRNA-194 and amino acids in patients with prostate cancer.	162	NCT05946915: miRNA as a growth plate marker after GH.
48	NCT04671498: Droplet-BC screening test for the detection of breast cancer, the DROPLET-BC study.	106	NCT06342401: Early onset colorectal cancer detection.	163	nct05846126: Digital multimodal rehabilitation for people with post-acute covid-19 syndrome.
49	NCT06532890: Optimal pediatric heart transplant immunosuppression with MicroRNAs (OPTIMA)	107	NCT03279770: Circulating non-coding rnas and cardiovascular outcomes in patients with dyslipidemia and atherosclerosis.	164	NCT06336174: Cognitive decline and underlying mechanisms in symptomatic intracranial artery stenosis patients: a cohort study.

(continued)

Table 12.1 (continued)

50	NCT05141383: Comparative study of diagnostic and prognosis biomarkers of prostate cancer in liquid biopsy.	108	NCT05514028: Evaluation of salivary RNAs in the presence of an adnexal mass of ovarian origin.	165	NCT02247453: Plasma microRNA profiling as first line screening test for lung cancer detection: a prospective study.
51	NCT03779022: miRNA and relevant biomarkers of bc patients undergoing neoadjuvant treatment.	109	NCT06081075: Newborn genomics programme.	166	NCT04285476: Predictive value of microRNA in thyroid cytologies of undetermined type.
52	NCT04525586: lncRNA-based nomogram to assist childhood asthma diagnosis.	110	NCT02304471: Circulating lncRNA and CV morbidities in CKD and ESRD.	167	NCT05346757: Validation of a microRNA-based fecal (miRFec) test for colorectal cancer screening (miRFec).
53	NCT06351722: Non-surgical periodontal treatment in smokers on SIRT-1, NLRP3 inflammasome and LncRNAs.	111	NCT05633342: Project CADENCE (CAncer detected early caN be CurEd).	168	NCT03601052: Efficacy, safety, and tolerability of Remlarsen (MRG-201) following intradermal injection in subjects with a history of keloids.
54	NCT06392997: Evaluation of salivary RNA in subjects scheduled for surgery for a gynecologic pathology.	112	NCT06736379: Intratumoral delivery of viral replicon (saRNA) particles expressing IL-12 in head and neck cancer.	169	NCT02603224: Safety, tolerability and pharmacokinetic study of MRG-201 in healthy volunteers.
55	NCT05708209: The long noncoding MALAT1 as a potential salivary diagnostic biomarker in oral squamous cell carcinoma through targeting mi RNA 124.	113	NCT03299335: Molecular profile of the evolution of inclusion body myositis.	170	NCT01829971: A multicenter phase I study of MRX34, MicroRNA miR-RX34 liposomal injection.
56	NCT05400057: The oncogenic potential of salivary miRNA-93 and miRNA-412-3P in oral lichen planus patients.	114	NCT03152630: Research on the early and prognosis diagnosis of vascular dementia.	171	NCT05521191: A study of RGLS8429 in patients with autosomal dominant polycystic kidney disease.
57	NCT02948855: Regulation of LncRNA for Breg in patients with thymoma and autoimmune diseases.	115	NCT04523389: Contents of circulating extracellular vesicles: Biomarkers in colorectal cancer patients.		
58	NCT02901821: Predicting concussion outcomes with salivary MiRNA.	116	NCT02780375: Feasibility study testing transcriptional responses as an indicator of individualized responses to radiation effects.		

HOTAIR can also be used as a diagnostic biomarker for GBC. The increased expression level of serum HOTAIR in GBC patients was reported, with a diagnostic value reaching 75% (Mishra et al. 2024). To further elucidate the clinical value of HOTAIR in GBC diagnosis, HOTAIR expression level was also evaluated in cholecystitis (CC), and significant differences were observed between serum CC and early stage GBC. The consistency in HOTAIR expression level indicates the reliability of its diagnostic value.

The expression profiling of HOTAIR was not limited to serum and plasma; feces were also used to assess the expression of several lncRNA in colon cancer patients as an early diagnostic tool (Gharib et al. 2021). The study reported a high expression of fecal HOTAIR in CRC patients compared to their corresponding healthy individuals, with approximately 90% sensitivity and specificity.

The development of cardiovascular disease (CVD) due to genetic or epigenetic change has suggested the involvement of genetic modulators such as ncRNA in the disease initiation and progression. Studies have shown differences in lncRNA expression among CVD patients, and that lncRNA is required to coordinate the normal physiology of the heart (Gangwar et al. 2018). HOTAIR lncRNA was the most studied lncRNA in CVD. Using microarray and real-time polymerase chain reaction-based profiling, the expression of HOTAIR was found to be upregulated in the plasma and blood of coronary artery disease (CAD) (Avazpour et al. 2018; Cai et al. 2016). On the contrary, the serum level of HOTAIR was reduced in patients diagnosed with Acute myocardial infarction (AMI) (Gao et al. 2017). Interestingly, not only was the expression profile of HOATAIR associated with CVD, but also single-nucleotide polymorphisms in the hosting gene increased the susceptibility to developing coronary artery disease (Kim et al. 2021).

In type 2 diabetes, elevated level of HOTAIR was associated with disease progression (PMID: 38618258) and diabetic complications (Shaker et al. 2019; Wang et al. 2021). Similar to the studies mentioned above, serum HOTAIR expression was also elevated in the serum of glioblastoma patients, esophageal squamous cell carcinoma patients (ESCC), and prostate cancer patients. The association between HOTAIR lncRNA and the prognosis of many cancers, including solid and liquid tumors, has also been investigated (Ma et al. 2019; Shen et al. 2018; Tan et al. 2017; Wang et al. 2017). Most studies have correlated the expression of HOTAIR with poor survival rate, cancer recurrence, and drug resistance (Miao et al. 2016; Zhou et al. 2014; Zhu et al. 2022).

12.3.1.2 HOTAIR as a Therapeutic Target

HOTAIR is a key epigenetic regulator, and lncRNA-based therapy would be considered a novel strategy for treating many pathologies. Aberrant HOTAIR expression is linked to drug resistance and poor prognosis in cancer patients, and targeting HOTAIR for therapeutic intervention has received significant attention. Several studies have reported different approaches to modulating the expression of HOTAIR; however, the transition from bench to clinic is still challenging. The antisense oligonucleotides (ASOs) approach, where a synthetic oligonucleotide binds to the target RNA and triggers its degradation, has been proposed for lncRNA knockdown, but only a few have been tested against HOTAIR (Liang and Peng 2022; Sarkar et al. 2024). Using small molecules was also proposed to interfere with lncRNA function. To our knowledge, ADQ (7-nitro-2-[4-(7-nitro-3-oxo-4, 9-dihydrofuro[3, 2-b] quinoxalin-2-yl)phenyl]-4, 9-dihydrofuro[3, 2-b]quinoxalin-3-one) was the only small molecule used to block the activity of HOTAIR (Ren et al. 2019). ADQ specifically interferes with HOTAIR and the chromatin-modifying protein EZH2 interaction, inhibiting HOTAIR transcriptional activity (Ren et al. 2019).

Most lncRNAs, including HOATIR, are poorly conserved among species, making it challenging to generate a pre-clinical model for lncRNA research. In this perspective, in late 2022, the generation of a transgenic mouse model was among the pioneering approaches to studying the HOATAIR effect on cancer metastases and providing valuable insights for cancer therapy (Q. Ma et al. 2022). This mouse model used a mutant form of HOATAIR lncRNA called HOTAIR-*sbid* that selectively inhibits HOTAIR function in cancer. HOTAIR silencing leads to methylation repression of H3 lysine 27 (H3K27) and subsequent inhibition of genes involved in epithelial-to-mesenchymal transition (EMT), a hallmark of cancer invasiveness and metastasis (Zhu et al. 2022). This model can be used in future research to study the modulation of HOTAIR expression by using small molecules or gene editing machinery as a potential therapeutic strategy.

12.3.2 Prostate Cancer Antigen 3 (PCA3)

PCA3 is an intronic, prostate-specific lncRNA that is highly expressed in primary prostate cancer (PCa) and used as a biomarker for this type of cancer (Bussemakers et al. 1999). The profiling of PCA3 in prostate cancer tissue relative to non-cancerous tissue back in 1999 has shed light on the possibility of using PCA3 as a biomarker for PC. Clinically, the diagnosis of PC is achieved by a blood-based test known as prostate-specific antigen (PSA) alone or combined with imaging and tissue biopsy (Descotes 2019). However, PSA can be expressed by normal prostate cells as well as benign prostate hyperplasia (Day et al. 2011). Additionally, other medical conditions, such as an inflamed prostate, can also result in high PSA levels. Alternative blood biomarkers have been introduced for PCa diagnosis, including prostate health index (Phi) and four-kallikrein panel (4Kscore). Both tests use the total, free, and intact blood-PSA measurement along with other parameters as a scoring system for PCa prediction (Filella and Foj 2016). However, all the aforementioned tests lack specificity and sensitivity, with a false positive rate that could reach 70% (Ilic et al. 2018; Tikkinen et al. 2018). To overcome the disadvantage of the available PCa diagnostic tests, several multi-centric clinical trials confirm the potential use of PCA3 as a specific and sensitive biomarker that could aid in PCa diagnosis and risk assessment. Using molecular biology-based techniques, PCA3 was detected in urine, giving more promise to patients for the noninvasiveness of PCa diagnosis.

Clinical diagnostic companies have adopted the establishment of urinary-based PCA3 tests, with the first RNA-based assay approved in 2012. PROGENSA PCA3 assay, developed by Gen-Probe company, utilized the nucleic acid amplification test (NAAT) technology to detect PCA3 in urine samples. The availability of PCA3-assay has eased the diagnosis of PCa and reduced the need for re-biopsy for men who had a previous negative result. On the other hand, the PCA3 assay, when combined with other PCa biomarkers, showed a high specificity in scoring the aggressiveness of prostate cancer (Gunelli et al. 2021).

Transcription expression-based nanotechnologies have been developed to understand prostate epithelial cell transformation and to develop better therapeutic strategies for PCA (Bhang et al. 2011). For example, molecular-genetic imaging, in which gene expression is used to image cell-specific pathological characteristics, involves an imaging reporter gene coupled with an imaging agent to measure gene expression in vivo (Bhang et al. 2011). The tissue-specific promoter can be utilized to detect gene expression in specific tissues; thereby, it can be used to detect the target cells that cannot be simply imaged with traditional techniques (Pouliot et al. 2009). For example, the two-step transcriptional amplification (TSTA) system successfully enhanced amplification driven by the PSA promoter. However, this system unsuccessfully detected primary PCa in vivo (Pouliot et al. 2011). PCA3 promoter has weak activity, and several approaches have been exploited to enhance promoter signals to target primary PC (PCA) cells (Neveu et al. 2016). However, using the three-step transcriptional amplification (3STA) technique has induced the PCA3 promoter activity and led to high amplification, allowing it to be used to detect PCA cells in vivo (Neveu et al. 2016).

12.3.3 Metastasis-Associated Lung Adenocarcinoma Transcript 1 (MALAT1)

MALAT1, also known as nuclear enriched abundant transcript 2 (NEAT2), is a lncRNA that was first detected in lung cancer tissue as a novel RNA associated with poor survival and increased probability of cancer metastasis (Gunelli et al. 2021). Later, MALAT1 was identified in other cancer types, such as breast, skin, cervical, and gastric cancers (Amodio et al. 2018; Arun et al. 2016; Guo et al. 2010; Okugawa et al. 2014). It is mainly retained in the nucleus, where it regulates the transcription of several genes involved in cancer metastasis, cell migration, cell cycle, and proliferation. MALAT1 functions by forming molecular scaffolds for ribonucleoprotein complexes. Like HOTAIR lncRNA, the functional role of MALAT1 has been extensively studied, giving its expression in a wide range of tissues. Also, unlike other lncRNAs, the MALAT1 gene is more conserved among eukaryotes, giving more opportunities for researchers to investigate the physiological and pathological role of MALAT1 using different model systems (Arun et al. 2020). The clinical potential of MALAT1 was rapidly recognized following its discovery and characterization, and more research studies have been conducted to evaluate its prognostic and diagnostic value.

In cancer, single-nucleotide polymorphisms (SNPs) in cancer-associated genes are considered a screening tool to identify people at risk of developing cancer or as a tool that helps physicians in identifying chemoresistant patients (Deng et al. 2017). Different polymorphisms have been identified in MALAT1 that could increase the likelihood of developing colon cancer, endometrial cancer, lung cancer, and thyroid cancer (G. Chen et al. 2020; Gao et al. 2022; Tong et al. 2022; Wen et al. 2019; Wu et al. 2019). This association has corroborated the clinical significance of using MALAT1 in cancer screening. Interestingly, examining the circulating level of MALAT1 as a cancer biomarker was not limited to peripheral blood. Urine-derived MALAT1 was also associated with bladder cancer (Yazarlou et al. 2018).

Like HOTAIR, the expression of MALAT1 is not restricted to cancer but is also related to cardiovascular disease. A multicenter study conducted by Li et al. reported a positive correlation between high MALAT1 expression and acute myocardial infarction (AMI) (Li et al. 2022b). This correlation corresponded to other markers used in AMI, such as C-reactive protein, cardiac troponin I, different cytokines, and lipoproteins. The study provided insight into lncRNA as a pathogenic factor in development of a cardiovascular condition.

The role of lncRNA in brain injuries was also investigated, given that some lncRNAs are enriched in the brain. The neuroprotection role of MALAT1 was elucidated recently when it was reported that several proteins could interact with MALAT1 and modulate its expression in the normal physiological state within the brain (Wu et al. 2022b; Yang et al. 2018). On the other hand, it was found that MALAT1 is expressed during neurite outgrowth, a critical signaling pathway in early neuronal development (Chen et al. 2016). Moreover, in vitro and in vivo studies reported that the expression of MALAT1 in brain hypoxic conditions was compromised, leading to a reduction in the overall angiogenesis pathway and neovascularization (Michalik et al. 2014). These observations lead to the hypothesis that MALAT1 might be involved in the pathogenesis of ischemic stroke (IS), one of the most cerebrovascular diseases. In 2020, the diagnostic potential of extracellular MALAT1 was examined between 120 ischemic patients and a healthy control group. The study showed that the expression of MALAT1 was reduced in the disease population and correlated with the National Institutes of Health Stroke Scale (NIHSS) (Ren et al. 2020). A cross-sectional study on IS patients also reported a lower expression level of MALAT1 in the IS group compared to healthy subjects (Bayat et al. 2023). Moreover, SNPs in the MALAT1 gene seem to play a role in the IS susceptibility or prevention (Wang et al. 2020).

All the above studies provide valuable insights about lncRNA to be translated into clinical use. No doubt, there are plenty of obstacles yet to be resolved, but a handful of scientific publications have provided solid evidence about their prospective use. Although modulating lncRNA expression by ASO or small molecules gave promising data for targeted therapy, the delivery mechanisms of those molecules remain challenging.

12.4 miRNA as Diagnostic and Therapeutical Tools

miRNA is among the first and most studied ex-ncRNA classes. The biological characteristics of miRNAs have been investigated and are valuable for clinical use. Many miRNAs have found their way into clinical trials as therapeutic targets, and biotech companies that develop miRNA-diagnostic signatures have adopted many. miRNAs are stable and abundant, facilitating their isolation and examination in different biological samples, including liquid biopsies and formalin-fixed tissues. Additionally, they mechanistically function by binding to the 3′-untranslated region (UTR) of the mRNA of the target gene, preventing translation or leading to transcript degradation. A single miRNA can target several mRNAs. Moreover, a single 3' UTR can be targeted by different miRNAs; thus, miRNAs can control the whole human genome. The fact that a miRNA can target several cellular pathways in health and disease is a distinctive trait. However, it comprises an obstacle to off-target effects in miRNA therapy. Despite several clinical trials being terminated, the miRNA-based diagnostic and therapeutic field is still evolving. As most healthcare organizations are shifting toward personalized medicine, miRNAs represent the industry's most promising approach. miRNA is involved in a wide range of human diseases, including inflammatory, neurological, hereditary, cancer, infectious, and cardiovascular, either enhancing the progression of a given disease or preventing the initiation of a pathological condition (Ha 2011). For instance, miR-10b was found to act as an oncogene in glioblastoma, whereas it is a tumor suppressor in gastric cancer (Gabriely et al. 2011; Li et al. 2015). The initial profiling of exRNA identified the miRNAs as a unique sncRNA profiled in circulation, and other classes of RNA have also been found using advanced technology.

MiRNAs are the most abundant sncRNA in biofluids relative to other noncoding RNA (Godoy et al. 2018). Moreover, several miRNAs have been characterized as tissue-specific, and large-scale transcriptomic data support that. For example, the liver-specific miRNA-122 is involved in regulating lipid metabolism, as miR-122-deficient mice showed reduced plasma cholesterol levels and lipid depositions (Esau et al. 2006). miRNA-122 accounts for 70% of RNA within the liver tissue and has been used to diagnose liver disease. The biotech company TAmiRNA introduced a plasma-based miRNA liver biomarker known as hepato-miRs. This test measures the expression of miR-122 with other miRNA (miR-192 and miR-151a) in postoperative patients with hepatocellular (HCC) or cholangiocellular (CCC) carcinoma, as well as in chronic liver disease and uses in-house analysis tools to assess the risk of liver failure. Other blood-based miRNA-analysis kits were also developed by the same company, such as osteomiR, which measures 19 serum miRNAs as biomarkers for bone quality. A thrombomiRs test was also developed to measure platelet-specific miRNAs and evaluate platelet function. ThyraMIR is another example of a miRNA-based diagnosis panel developed by Interpace Biosciences. This molecular test showed high sensitivity to detect thyroid cancer when combined with other well-known cancer genes. ID3EAL™ PanoramiR Knowledge Panel, sold by a Singapore-based company (Mirxes), is a miRNA-based diagnostic panel with the ability to detect more than one disease; however, due to the large number of miRNAs included in the panel (376 miRNAs), the specificity of the kit is questionable.

One of the most important objectives in patient care is the early detection of diseases, increasing treatment options and enhancing the overall prognosis. miRNAs are promising early diagnostic biomarkers. In this regard, an ongoing large prospective clinical trial (ClinicalTrials.gov ID NCT02247453) on lung cancer (LC) was initiated in 2013, aiming to establish a miRNA-surveillance panel for heavy smokers at risk of developing LC. This screening panel comprises 24 miRNAs that can classify high-risk patients and detect LC in its early stages (Sozzi et al. 2014). Another interesting clinical trial is examining the efficacy of miRNA in determining thyroid nodule malignancy with indeterminate cytology (Lithwick-Yanai et al. 2017) (clinical trial # NCT04285476). The study group has already developed the RosettaGX Reveal assay (Rosetta Genomics), which is commercially available for clinicians. Moreover, miR-421 and miR-27a-3p were identified by Duran-Sanchon and his group in fecal samples with a high sensitivity to distinguish colorectal cancer (CRC) from advanced adenomas (AAs) (Duran-Sanchon et al. 2020). The study group started a validation phase of the established diagnostic assay under the name of miRFec test (clinical trial # NCT05346757).

Dysregulated miRNA expression can alter normal physiological processes within the cells, leading to an overall cellular disturbance and the development of a pathological condition. MiRNA expression can be affected by several factors, such as their biogenesis pathways or interaction with the target mRNAs. Also, miRNAs expression can be altered at the transcriptional levels via different genetic factors. miRNA-therapeutics aim to restore normal miRNA expression by using a miRNA-mimic, a synthesized double-stranded RNA molecule mimicking the mature miRNA sequence, or a miRNA inhibitor, a synthesized antisense sequence of a miRNA that binds to it and represses its function (Esau 2008; Jin et al. 2015). However, the delivery mechanism of these artificial miRNAs remains a challenge. Intravenous injection, local administration route, inhalation, and oral delivery have all been investigated as miRNA delivery systems (Jiménez-Morales et al. 2022). Limitations related to miRNAs' stability and cellular uptake in vivo may pose significant challenges. However, these issues can be addressed by introducing chemical modifications to the synthetic miRNA structure or by encapsulating miRNAs within a carrier (Dasgupta and Chatterjee 2021; Fu et al. 2019; Seyhan 2024).

Despite the promising advancements in miRNA delivery systems, concerns remain regarding off-target effects, dosing, and immune reactions against exogenous miRNAs. Recently, Li and his group introduced a tetrahedral DNA nanostructure (TDN)-based delivery known as a bio-switchable delivery system (BDS) (Li et al. 2024). This approach uses a DNA fragment hybridized with a synthetic miRNA that can be easily released to the target tissue by a nuclease enzyme known as RNase H. This enzyme cleaves the RNA from the RNA–DNA hybrid and minimizes the toxic effect of the delivery system. Interestingly, a high degree of in vivo stability was observed using this system. The introduction of BDS is promising, yet it requires a careful design of miRNA oligos to ensure successful miRNA delivery. Up to date, no pre-clinical studies have evaluated the safety profile of BDS.

Despite the challenges in miRNA-based therapeutics, different pre-clinical and clinical studies have been introduced. Among the first miRNA-based therapy is MRX34, a miRNA mimic encapsulated in a liposomal vehicle aimed to be up taken by the liver cells and to restore the expression of miR-34. Initially, it was reported that miR-34 is a tumor suppressor gene that is functionally lost in many solid tumors such as hepatocellular carcinoma, melanoma and lung cancer (Li et al. 2014). Hence, introducing a double-stranded miRNA that could act as miR-34 was promising. Phase 1 clinical trial was initiated with an intravenous delivery of MRX34; however, the trial was terminated due to serious adverse events leading to mortality. (ClinicalTrials.gov identifier: NCT01829971 and NCT02862145). Additionally, miRNA therapy has been explored beyond cancer therapy, where many clinical trials have examined the efficiency of miRNA in treating skin conditions, genetic disorders, and cardiovascular disease (Diener et al. 2022). Yet, most clinical trials were stopped or terminated due to off-target toxicity.

MiR-29 family is among the most studied miRNA families, and its role in regulating the extracellular matrix is well investigated (Kriegel et al. 2012). miR-29 can modulate the mRNA of many collagen isoforms and act as an anti-fibrotic molecule, reducing scar formation and pathological tissue remodeling (Sassi et al. 2017). Giving it a bio-vital role, investigating the safety of miR-29 mimic, known as MRG-201 (Remlarsen) as a local treatment in wound healing, entered clinical trial in 2015. The result of phase 1 was encouraging, and the study group tolerated the drug well. Phase 2 clinical trial of MRG-201 was completed in 2020 as the first miRNA-based therapy successfully passed phase two (ClinicalTrials.gov ID NCT03601052 and NCT02603224). MiRagen Therapeutics, Inc., the developer of MRG-201, has not yet announced the progression of the drug into phase 3 and no data from phase 2 is reported. To our knowledge, the company rebranded as Viridian Therapeutics, Inc. (www.viridiantherapeutics.com) with no updates about miRNA-based therapy on their website yet.

In genetic disorders, Regulus Therapeutic started a pre-clinical evolution in early 2019 of anti-miRNA (antagomir/inhibitor) to slow the progression of cyst growth seen in the autosomal dominant polycystic kidney disease (ADPKD) (Lee et al. 2019). The pre-clinical safety assessment from the PKD mouse model revealed favorable pharmacokinetic properties of the drug, leading to the initiation of an interventional clinical trial sponsored by the company in 2022 with an estimated completion in 2025 (NCT05521191, company website Regulus Therapeutics (regulusrx.com)).

12.5 Small Interference RNA

Small interfering RNA (siRNA), also known as RNA interference (RNAi), is a naturally occurring double-stranded RNA that is found to be expressed in many organisms, including humans. It was initially discovered in plants as a defense mechanism against exogenous mRNA introduced to petunia flowers in 1990 (Sen and Blau

2006). The precise mechanism by siRNA induces gene silencing through the degradation of a target mRNA was elucidated years after its discovery, and since then, the concept of siRNA has been a remarkable milestone in gene therapy.

The siRNA-therapeutics are advancing rapidly compared to miRNAs. While a single miRNA can interact with many mRNAs, leading to translation inhibition, a siRNA can act in a sequence-specific manner, ending with the degradation of a single target mRNA (Hu et al. 2020; Williams et al. 1985). This advantage of siRNA-mRNA specific base pairing has granted a rapid growth in siRNA therapy with six FDA-approved drugs to date and many in clinical trials.

In 2018, Patisiran was the first siRNA therapy approved by the FDA to treat Hereditary amyloid transthyretin (ATTRv) amyloidosis. ATTRv is an autosomal dominant disease caused by a mutation in the transthyretin gene (TTR or TBPA), leading to TTR protein misfolding and subsequent amyloid deposition in multiple organs (Russo et al. 2020). Patisiran was formulated in a lipid nanoparticle (LNP) for hepatocyte-targeted delivery, the site for TTR synthesis. Once in the liver, the double-stranded siRNA is processed, and a complementary base pair with TTR mRNA is made, leading to TTR mRNA degradation. Thus, reducing the production of misfolded proteins and delaying the progression of the disease's onset. Due to this innovative treatment, Alnylam, the producing company of Patisiran (sold under the brand name ONPATTRO), became a leading sector in siRNA therapeutics for the treatment of rare genetic disorders. Currently, three other siRNA-based drugs are available for commercial use that are designed and produced by Alnyam, including GIVLAARI (givosiran) for acute hepatic porphyria, OXLUMO® (lumasiran) for primary hyperoxaluria type 1 (PH1) and AMVUTTRA® (vutrisiran) for the treatment of polyneuropathy seen in ATTRv patients. In addition, an injectable siRNA-based medicine for the treatment of hypercholesterolemia was developed by Novartis and sold under the name Leqvio (Inclisiran). In 2023, the sixth RNAi drug (Nedosiran (RIVFLOZA™), Dicerna Pharmaceuticals), has been approved for the treatment of Primary hyperoxaluria (PH).

Although progress in RNAi therapeutics is advancing, this class of medicine has shown limited use in other diseases originating outside the liver. The cellular delivery of siRNA has been riddled by the instability of siRNA in the bloodstream and the fact that siRNA can induce a rabid immune response by activating Toll-like receptors (TLRs), leading to drug clearance before reaching its destination (Lee et al. 2023). Scientists are actively testing alternative delivery systems for siRNA, including C16 lipophilic conjugates, ligand-based targeting, and N-acetylgalactosamine conjugates (GalNAc). All these delivery mechanisms were shown to increase siRNA stability, permeability, and uptake by the targeted tissue (Jadhav et al. 2024; Tang and Khvorova 2024).

12.6 Circular RNA

Circular RNA (circRNA) is a distinct class of ncRNA generated from the back-spliced transcript. It forms a covalent bond between 5′ and 3′ splice sites, leading to a closed-circular RNA with a 5′ cap and 3′ polyadenylated tails (Pisignano et al. 2023). The first circRNA was initially identified in humans in the late 1980s, but its function as an essential molecule in gene expression was not revealed until the middle of 2000s (Ebbesen et al. 2016). While the majority of circRNAs have not been functionally characterized, their role as regulator in the transcription and translation process has been acknowledged. One of the significant functions of circRNA is miRNA sponging, where circRNA binds to miRNA and regulates miRNA–mRNA activity (Hansen et al. 2013). A circRNA may have several binding sites for miRNA, creating competition between circRNA and the target mRNAs. This unique interaction between circRNA and miRNA makes the circRNA adaptable as a synthetic miRNA antagonist. In 2019, Lavenniah et al. successfully engineered synthetic circRNA known as circmiRs for in vivo miR-212/132 inhibition, a pathogenic miRNA family involved in cardiac hypertrophy (Lavenniah et al. 2020). Another group has shown a compelling example of miRNA sponging using synthetic circRNA in the hepatitis C viruses (HCV) life cycle (Jost et al. 2018). It was previously reported that the liver-specific miR-122 is required for HCV replication within the hepatocytes, leading to viral propagation (Schult et al. 2018). Thus, delivering circRNA that binds explicitly to miR-122 sequestered the HCV replication cycle and inhibited viral protein synthesis (Jost et al. 2018).

Moreover, elevated circRNA expression has been widely detected in many diseases, and using knockdown strategies for pathogenic circRNAs seems promising. Delivering small interfering RNA and short hairpin RNA (shRNA) molecules to target circRNA is the method of choice in mediating circRNA inhibition (He and Zhu 2023; Reddy et al. 2020). For example, circ-CUX1 is a circRNA found to play a key role in aerobic glycolysis seen in neuroblastoma (NB). It may contribute to tumor aggressiveness and poor outcomes (Li et al. 2019). In vitro transfection shRNA targeting circ-CUX1 suppresses aerobic glycolysis and subsequent NB progression, suggesting circ-CUX1 as a potential therapeutic target in cancer. Opposite to circRNA knockdown, overexpression of circular RNA by direct delivery is another possible approach in cancer therapy where circRNA is lost (Niu et al. 2023). Several techniques have been proposed for circRNA delivery, including LNP and exosomes. However, the recombinant adeno-associated viral (AAV) vector delivery system holds promise in circRNA-gene therapy (Meganck et al. 2018). Recently, circularization-promoting oligonucleotide (CLIP-ON), a chemically modified oligos, was proposed as a new strategy for the endogenous production of biologically relevant circRNA (Ni et al. 2024).

12.7 CircRNA Vaccine

Since the COVID-19 pandemic, mRNA vaccine has been at the forefront of vaccination due to its rapid and cost-effective development (Chaudhary et al. 2021). Nonetheless, as COVID-19 and other respiratory viruses continue to spread, several mutant types have been reported, and an effective vaccine is in high demand. The closed shape of the circRNA is a distinct feature that limits circRNA degradation by endonuclease enzymes, making it an exceptionally stable molecule. Furthermore, circRNA showed lower immunogenicity and longer antigen production than mRNA vaccine (Niu et al. 2023). Owing to this inherent feature, circRNAs represent a novel next-generation RNA-based vaccine. In 2022, a potent circular RNA vaccine was developed, and it showed steady antigen production and broad-spectrum neutralizing antibodies compared to the traditional linear RNA vaccine (Qu et al. 2022). Another innovative study has established an anti-tumor circRNA- LNP vaccine, which was tested in multiple tumor-mouse models. The results showed that circRNA vaccine triggered a robust innate and adaptive immune response, stable antigen presentation capability, and higher anti-tumor efficacy (Li et al. 2022a).

CircRNA-based vaccines are still in its early stages; nevertheless, they exhibit more advantages over mRNA vaccine. Giving their stability, cost efficiency, and easy production, it promises to become an alternative for mRNA-vaccine.

12.8 Ethical Considerations

The capacity to modify human nature through genomic alteration is not novel. Genome editing has significant technological and ethical obstacles that restrict its application in therapeutic environments. The fundamental concern is that genomic-DNA engineering may result in unforeseen detrimental genomic consequences that become apparent only when reversal is no longer feasible. Novel DNA base-editing technologies demonstrate enhanced specificity relative to traditional CRISPR-Cas and zinc-finger nuclease (ZFN) methods. Novel methods for altering genetic information have emerged in the last decade. This entails manipulating and directing endogenous RNA-editing enzymes to target a specific RNA transcript while preserving the integrity of the DNA. A-to-I RNA editing may be achieved by introducing an antisense oligonucleotide that binds to its target and forms the double-stranded RNA configuration necessary for editing by adenosine deaminases acting on RNA (ADARs). RNA editing is a promising technology with significant promise to transform genetic engineering. Ethical considerations about the accessibility of this revolutionary technology for economically disadvantaged individuals must be examined within the larger framework of inequities and inequality in the healthcare system. Nonetheless, the distinctive characteristics of RNA editing may alter the ethical dialogue surrounding genetic engineering in humans. Typically, genetic research concerning humans is problematic. Eugenics is the scientifically flawed belief that humans can be enhanced via the selective breeding of populations. Eugenicists believed in a prejudiced and inaccurate conception of Mendelian genetics that abstract human traits (e.g., intelligence and social behaviors) were inherited straightforwardly. They similarly claimed that complicated diseases and disorders were exclusively the result of genetic inheritance. The execution of eugenics policies has inflicted enormous damage, especially on underprivileged people. Eugenics is not an obscure movement. Beginning in the late 1800s, global leaders and intellectuals propagated eugenic ideologies and programs rooted in prevalent racist and xenophobic sentiments. RNA-editing technology could offer novel opportunities for genetic interventions in study, medicine, and human enhancement, but ethical considerations need to be ahead of any advancement to preserve our humanity.

We would like to further emphasize that RNA editing may function as a preliminary measure to evaluate the consequences of genome editing. This genetic engineering technique's ethical and social implications are extensive and substantial, and cannot be summarized in a short paragraph. However, it is never too late and prudent to contemplate today on this promising technology's future ethical and social ramifications that can have dramatic consequences in the future.

12.9 Conclusion

ncRNAs represent a subclass of RNA that forms most of the human genome and play crucial roles in numerous physiological functions. With the rapid development of RNA profiling techniques, a variety of ncRNAs have been identified in humans and are involved in pathological processes. Given their physiological importance and biological characteristics, many ncRNAs have been targeted as a clinical tool. In this chapter, we discussed the ncRNAs that have a long evolutionary history, yet other ncRNAs have been looked at from a clinical perspective. We presented several examples that illustrate the applicability of ncRNAs in patient care. These examples show that advancements in the ncRNA field have full potential for disease diagnosis and are effective for gene therapy for cancer and non-cancerous diseases. However, several challenges regarding using ncRNA in clinical practice must be overcome. Significant limitations involve the non-specificity, particularly in miRNAs-based therapies, low cellular internalization, and ineffective delivery systems. To enhance the use of ncRNAs as disease biomarkers and therapeutical tools, further investigations are needed to optimize their clinical applications.

Data Availability Statement All data are publicly available.

Acknowledgment for Funding Dr. C. Sergi's research has been funded by the generosity of the Children's Hospital of Eastern Ontario, Ottawa, Ontario, and the Stollery Children's Hospital Foundation and supporters of the Lois Hole Hospital for Women through the Women and Children's Health Research Institute (WCHRI, Grant ID #: 2096), Natural Science Foundation of Hubei Province for Hubei University of Technology (100-Talent Grant for Recruitment Program of

Foreign Experts Total Funding: Digital PCR and NGS-based diagnosis for infection and oncology, 2017–2022), Österreichische Krebshilfe Tyrol (Krebsgesellschaft Tirol, Austrian Tyrolean Cancer Research Institute, 2007 and 2009—"DMBTI and cholangiocellular carcinomas" and "Hsp70 and HSPBP1 in carcinomas of the pancreas"), Austrian Research Fund (Fonds zur Förderung der wissenschaftlichen Forschung, FWF, Grant ID L313-B13), Canadian Foundation for Women's Health ("Early Fetal Heart-RES0000928"), Cancer Research Society (von Willebrand factor gene expression in cancer cells), Canadian Institutes of Health Research (Omega-3 Fatty Acids for Treatment of Intestinal Failure Associated Liver Disease: A Translational Research Study, 2011–2014, CIHR 232514), and the Saudi Cultural Bureau, Ottawa, Canada. The funders had no role in study design, data collection and analysis, publication decision, or manuscript preparation.

Conflict of Interest The authors declare that they have no conflict of interest.

Compliance with Ethical Standards All authors declare full compliance with Good Laboratory Practice standards and guidelines, according to the Declaration of Helsinki. This study did not require institutional review approval as this chapter does not contain any studies with human participants performed by any of the authors.

References

Abdallah HY, Tawfik NZ, Soliman NH, Eldeen LAT (2022) The lncRNA PRINS-miRNA-mRNA Axis gene expression profile as a circulating biomarker panel in psoriasis. Mol Diagn Ther 26(4):451–465. https://doi.org/10.1007/s40291-022-00598-y

Alba-Bernal A, Lavado-Valenzuela R, Domínguez-Recio ME, Jiménez-Rodriguez B, Queipo-Ortuño MI, Alba E, Comino-Méndez I (2020) Challenges and achievements of liquid biopsy technologies employed in early breast cancer. EBioMedicine 62:103100. https://doi.org/10.1016/j.ebiom.2020.103100

Amodio N, Stamato MA, Juli G, Morelli E, Fulciniti M, Manzoni M et al (2018) Drugging the lncRNA MALAT1 via LNA gapmeR ASO inhibits gene expression of proteasome subunits and triggers anti-multiple myeloma activity. Leukemia 32(9):1948–1957. https://doi.org/10.1038/s41375-018-0067-3

Arun G, Diermeier S, Akerman M, Chang KC, Wilkinson JE, Hearn S et al (2016) Differentiation of mammary tumors and reduction in metastasis upon Malat1 lncRNA loss. Genes Dev 30(1):34–51. https://doi.org/10.1101/gad.270959.115

Arun G, Aggarwal D, Spector DL (2020) MALAT1 long non-coding RNA: functional implications. Noncoding RNA 6(2). https://doi.org/10.3390/ncrna6020022

Avazpour N, Hajjari M, Yazdankhah S, Sahni A, Foroughmand AM (2018) Circulating HOTAIR RNA is potentially up-regulated in coronary artery disease. Genome Inform 16(4):e25. https://doi.org/10.5808/GI.2018.16.4.e25

Bär C, Chatterjee S, Thum T (2016) Long noncoding RNAs in cardiovascular pathology, diagnosis, and therapy. Circulation 134(19):1484–1499

Bayat M, Tabrizi R, Saied Salehi M, Karimi N, Rahimi M, Hooshmandi E et al (2023) Association of Long non-Coding RNA Malat1 with serum levels of Interleukin-1 Beta and Vitamin D in patients with ischemic stroke. Galen Med J 12:1–10. https://doi.org/10.31661/gmj.v12i0.2457

Bhan A, Mandal SS (2015) LncRNA HOTAIR: a master regulator of chromatin dynamics and cancer. Biochim Biophys Acta 1856(1):151–164. https://doi.org/10.1016/j.bbcan.2015.07.001

Bhang HE, Gabrielson KL, Laterra J, Fisher PB, Pomper MG (2011) Tumor-specific imaging through progression elevated gene-3 promoter-driven gene expression. Nat Med 17(1):123–129. https://doi.org/10.1038/nm.2269

Bussemakers MJ, van Bokhoven A, Verhaegh GW, Smit FP, Karthaus HF, Schalken JA et al (1999) DD3: a new prostate-specific gene, highly overexpressed in prostate cancer. Cancer Res 59(23):5975–5979

Cai Y, Yang Y, Chen X, Wu G, Zhang X, Liu Y et al (2016) Circulating 'lncRNA OTTHUMT00000387022'from monocytes as a novel biomarker for coronary artery disease. Cardiovasc Res 112(3):714–724

Chaudhary N, Weissman D, Whitehead KA (2021) mRNA vaccines for infectious diseases: principles, delivery and clinical translation. Nat. Rev. Drug Discov 20(11):817–838

Chen L, Feng P, Zhu X, He S, Duan J, Zhou D (2016) Long non-coding RNA Malat1 promotes neurite outgrowth through activation of ERK/MAPK signalling pathway in N2a cells. J Cell Mol Med 20(11):2102–2110

Chen G, Zhang M, Liang Z, Chen S, Chen F, Zhu J et al (2020) Association of polymorphisms in MALAT1 with the risk of endometrial cancer in Southern Chinese women. J Clin Lab Anal 34(4):e23146. https://doi.org/10.1002/jcla.23146

Chisholm KM, Wan Y, Li R, Montgomery KD, Chang HY, West RB (2012) Detection of long non-coding RNA in archival tissue: correlation with polycomb protein expression in primary and metastatic breast carcinoma. PLoS One 7(10):e47998

Das S, Abdel-Mageed AB, Adamidi C, Adelson PD, Akat KM, Alsop E et al (2019) The extracellular RNA communication consortium: establishing foundational knowledge and technologies for extracellular RNA research. Cell 177(2):231–242

Dasgupta I, Chatterjee A (2021) Recent advances in miRNA delivery systems. Methods Protoc 4(1). https://doi.org/10.3390/mps4010010

Day JR, Jost M, Reynolds MA, Groskopf J, Rittenhouse H (2011) PCA3: from basic molecular science to the clinical lab. Cancer Lett 301(1):1–6

Deng N, Zhou H, Fan H, Yuan Y (2017) Single nucleotide polymorphisms and cancer susceptibility. Oncotarget 8(66):110635–110649. https://doi.org/10.18632/oncotarget.22372

Descotes JL (2019) Diagnosis of prostate cancer. Asian J Urol 6(2):129–136. https://doi.org/10.1016/j.ajur.2018.11.007

Diener C, Keller A, Meese E (2022) Emerging concepts of miRNA therapeutics: from cells to clinic. Trends Genet 38(6):613–626

Dong K, Shen J, He X, Hu G, Wang L, Osman I et al (2021) CARMN is an evolutionarily conserved smooth muscle cell-specific LncRNA that maintains contractile phenotype by binding Myocardin. Circulation 144(23):1856–1875. https://doi.org/10.1161/circulationaha.121.055949

Duran-Sanchon S, Moreno L, Augé JM, Serra-Burriel M, Cuatrecasas M, Moreira L et al (2020) Identification and validation of microRNA profiles in fecal samples for detection of colorectal cancer. Gastroenterology 158(4):947–957. e944

Ebbesen KK, Kjems J, Hansen TB (2016) Circular RNAs: identification, biogenesis and function. Biochim Biophys Acta 1859(1):163–168. https://doi.org/10.1016/j.bbagrm.2015.07.007

Esau CC (2008) Inhibition of microRNA with antisense oligonucleotides. Methods 44(1):55–60. https://doi.org/10.1016/j.ymeth.2007.11.001

Esau C, Davis S, Murray SF, Yu XX, Pandey SK, Pear M et al (2006) miR-122 regulation of lipid metabolism revealed by in vivo antisense targeting. Cell Metab 3(2):87–98. https://doi.org/10.1016/j.cmet.2006.01.005

Filella X, Foj L (2016) Prostate cancer detection and prognosis: from prostate specific antigen (PSA) to exosomal biomarkers. Int J Mol Sci 17(11):1784

Fu Y, Chen J, Huang Z (2019) Recent progress in microRNA-based delivery systems for the treatment of human disease. ExRNA 1(1):1–14

Gabriely G, Yi M, Narayan RS, Niers JM, Wurdinger T, Imitola J et al (2011) Human glioma growth is controlled by microRNA-10b. Cancer Res 71(10):3563–3572. https://doi.org/10.1158/0008-5472.Can-10-3568

Gangwar RS, Rajagopalan S, Natarajan R, Deiuliis JA (2018) Noncoding RNAs in cardiovascular disease: pathological relevance and emerging role as biomarkers and therapeutics. Am J Hypertens 31(2):150–165

Gao L, Liu Y, Guo S, Yao R, Wu L, Xiao L et al (2017) Circulating long noncoding RNA HOTAIR is an essential mediator of acute myocardial infarction. Cell Physiol Biochem 44(4):1497–1508

Gao XR, Wang XY, Li XY, Sun YQ, Zhang SL (2022) Clinical significance and function of MALAT1 gene expression and the rs619586 polymorphism in colorectal cancer. Biomed Environ Sci 35(8):768–771. https://doi.org/10.3967/bes2022.101

Gharib E, Nazemalhosseini-Mojarad E, Baghdar K, Nayeri Z, Sadeghi H, Rezasoltani S et al (2021) Identification of a stool long non-coding RNAs panel as a potential biomarker for early detection of colorectal cancer. J Clin Lab Anal. 35(2):e23601

Godoy PM, Bhakta NR, Barczak AJ, Cakmak H, Fisher S, MacKenzie TC et al (2018) Large differences in small RNA composition between human biofluids. Cell Rep 25(5):1346–1358. https://doi.org/10.1016/j.celrep.2018.10.014

Gruner HN, McManus MT (2021) Examining the evidence for extracellular RNA function in mammals. Nat Rev Genet 22(7):448–458

Gunelli R, Fragalà E, Fiori M (2021) PCA3 in prostate cancer. Methods Mol Biol 2292:105–113

Guo F, Liu Y, Li Y, Li G (2010) Inhibition of ADP-ribosylation factor-like 6 interacting protein 1 suppresses proliferation and reduces tumor cell invasion in CaSki human cervical cancer cells. Mol Biol Rep 37(8):3819–3825. https://doi.org/10.1007/s11033-010-0037-y

Gupta RA, Shah N, Wang KC, Kim J, Horlings HM, Wong DJ et al (2010) Long non-coding RNA HOTAIR reprograms chromatin state to promote cancer metastasis. Nature 464(7291):1071–1076. https://doi.org/10.1038/nature08975

Ha TY (2011) MicroRNAs in human diseases: from cancer to cardiovascular disease. Immune Netw 11(3):135–154. https://doi.org/10.4110/in.2011.11.3.135

Hajjari M, Salavaty A (2015) HOTAIR: an oncogenic long non-coding RNA in different cancers. Cancer Biol Med 12(1):1–9. https://doi.org/10.7497/j.issn.2095-3941.2015.0006

Hansen TB, Jensen TI, Clausen BH, Bramsen JB, Finsen B, Damgaard CK, Kjems J (2013) Natural RNA circles function as efficient microRNA sponges. Nature 495(7441):384–388. https://doi.org/10.1038/nature11993

He Z, Zhu Q (2023) A circular RNAs: Emerging roles and new insights in human cancers. Biomed Pharmacother 165:115217

Hu B, Zhong L, Weng Y, Peng L, Huang Y, Zhao Y, Liang XJ (2020) Therapeutic siRNA: state of the art. Signal Transduct Target Ther 5(1):101. https://doi.org/10.1038/s41392-020-0207-x

Ilic D, Djulbegovic M, Jung JH, Hwang EC, Zhou Q, Cleves A et al (2018) Prostate cancer screening with prostate-specific antigen (PSA) test: a systematic review and meta-analysis. BMJ 362:k3519. https://doi.org/10.1136/bmj.k3519

Jadhav V, Vaishnaw A, Fitzgerald K, Maier MA (2024) RNA interference in the era of nucleic acid therapeutics. Nat Biotechnol 42(3):394–405

Jiapaer Z, Li C, Yang X, Sun L, Chatterjee E, Zhang L et al (2023) Extracellular non-coding RNAs in cardiovascular diseases. Pharmaceutics 15(1). https://doi.org/10.3390/pharmaceutics15010155

Jiménez-Morales JM, Hernández-Cuenca YE, Reyes-Abrahantes A, Ruiz-García H, Barajas-Olmos F, García-Ortiz H et al (2022) MicroRNA delivery systems in glioma therapy and perspectives: a systematic review. J Control Release 349:712–730

Jin HY, Gonzalez-Martin A, Miletic AV, Lai M, Knight S, Sabouri-Ghomi M et al (2015) Transfection of microRNA mimics should be used with caution. Front Genet:6. https://doi.org/10.3389/fgene.2015.00340

Johnsson P, Lipovich L, Grandér D, Morris KV (2014) Evolutionary conservation of long non-coding RNAs; sequence, structure, function. Biochim Biophys Acta 1840(3):1063–1071

Jost I, Shalamova LA, Gerresheim GK, Niepmann M, Bindereif A, Rossbach O (2018) Functional sequestration of microRNA-122 from hepatitis C virus by circular RNA sponges. RNA Biol 15(8):1032–1039. https://doi.org/10.1080/15476286.2018.1435248

Kim S, Jeon OH, Jeon Y-J (2020) Extracellular RNA: emerging roles in cancer cell communication and biomarkers. Cancer Lett 495:33–40. https://doi.org/10.1016/j.canlet.2020.09.002

Kim IJ, Lee JY, Park HW, Park HS, Ko EJ, Sung JH, Kim NK (2021) Association between HOTAIR lncRNA polymorphisms and coronary artery disease susceptibility. J Pers Med 11(5). https://doi.org/10.3390/jpm11050375

Kriegel AJ, Liu Y, Fang Y, Ding X, Liang M (2012) The miR-29 family: genomics, cell biology, and relevance to renal and cardiovascular injury. Physiol Genomics 44(4):237–244. https://doi.org/10.1152/physiolgenomics.00141.2011

Lavenniah A, Luu TDA, Li YP, Lim TB, Jiang J, Ackers-Johnson M, Foo RS-Y (2020) Engineered circular RNA sponges act as miRNA inhibitors to attenuate pressure overload-induced cardiac hypertrophy. Mol Ther 28(6):1506–1517

Lee EC, Valencia T, Allerson C, Schairer A, Flaten A, Yheskel M et al (2019) Discovery and preclinical evaluation of anti-miR-17 oligonucleotide RGLS4326 for the treatment of polycystic kidney disease. Nat Commun 10(1):4148. https://doi.org/10.1038/s41467-019-11918-y

Lee JW, Shim MK, Kim H, Jang H, Lee Y, Kim SH (2023) RNAi therapies: Expanding applications for extrahepatic diseases and overcoming delivery challenges. Adv Drug Deliv Rev 201:115073

Li XJ, Ren ZJ, Tang JH (2014) MicroRNA-34a: a potential therapeutic target in human cancer. Cell Death Dis 5(7):e1327. https://doi.org/10.1038/cddis.2014.270

Li Z, Lei H, Luo M, Wang Y, Dong L, Ma Y et al (2015) DNA methylation downregulated mir-10b acts as a tumor suppressor in gastric cancer. Gastric Cancer 18(1):43–54. https://doi.org/10.1007/s10120-014-0340-8

Li H, Yang F, Hu A, Wang X, Fang E, Chen Y et al (2019) Therapeutic targeting of circ-CUX1/EWSR1/MAZ axis inhibits glycolysis and neuroblastoma progression. EMBO Mol Med 11(12):e10835. https://doi.org/10.15252/emmm.201910835

Li H, Peng K, Yang K, Ma W, Qi S, Yu X et al (2022a) Circular RNA cancer vaccines drive immunity in hard-to-treat malignancies. Theranostics 12(14):6422–6436. https://doi.org/10.7150/thno.77350

Li R, Jin J, Liu E, Zhang J (2022b) A novel circulating biomarker lnc-MALAT1 for acute myocardial infarction: its relationship with disease risk, features, cytokines, and major adverse cardiovascular events. J Clin Lab Anal 36(12):e24771. https://doi.org/10.1002/jcla.24771

Li S, Tian T, Zhang T, Lin Y, Cai X (2024) A bioswitchable delivery system for microRNA therapeutics based on a tetrahedral DNA nanostructure. Nat Protoc 20:336–362

Liang H, Peng J (2022) LncRNA HOTAIR promotes proliferation, invasion and migration in NSCLC cells via the CCL22 signaling pathway. PLoS One 17(2):e0263997

Lin Z, Lu F, Ma X, Xia X, Zou F, Jiang J (2021) Roles of circular RNAs in the pathogenesis of intervertebral disc degeneration (review). Exp Ther Med 22(5):1221. https://doi.org/10.3892/etm.2021.10655

Lithwick-Yanai G, Dromi N, Shtabsky A, Morgenstern S, Strenov Y, Feinmesser M et al (2017) Multicentre validation of a microRNA-based assay for diagnosing indeterminate thyroid nodules utilising fine needle aspirate smears. J Clin Pathol 70(6):500–507. https://doi.org/10.1136/jclinpath-2016-204089

Lodde V, Murgia G, Simula ER, Steri M, Floris M, Idda ML (2020) Long noncoding RNAs and circular RNAs in autoimmune diseases. Biomolecules 10(7). https://doi.org/10.3390/biom10071044

Loganathan T, Doss CG (2023) Non-coding RNAs in human health and disease: potential function as biomarkers and therapeutic targets. Funct Integr Genomics 23(1):33. https://doi.org/10.1007/s10142-022-00947-4

Ma L, Bajic VB, Zhang Z (2013) On the classification of long non-coding RNAs. RNA Biol 10(6):924–933

Ma Y, Hu M, Zhou L, Ling S, Li Y, Kong B, Huang P (2019) Long non-coding RNA HOTAIR promotes cancer cell energy metabolism in pancreatic adenocarcinoma by upregulating hexokinase-2. Oncol Lett 18(3):2212–2219

Ma Q, Yang L, Tolentino K, Wang G, Zhao Y, Litzenburger UM et al (2022) Inducible lncRNA transgenic mice reveal continual role of HOTAIR in promoting breast cancer metastasis. Elife 11:e79126

Martinez-Ledesma E, Verhaak RG, Treviño V (2015) Identification of a multi-cancer gene expression biomarker for cancer clinical outcomes using a network-based algorithm. Sci Rep 5(1):11966

Mattick JS, Amaral PP, Carninci P, Carpenter S, Chang HY, Chen L-L et al (2023) Long non-coding RNAs: definitions, functions, challenges and recommendations. Nat Rev Mol Cell Biol 24(6):430–447

Meganck RM, Borchardt EK, Castellanos Rivera RM, Scalabrino ML, Wilusz JE, Marzluff WF, Asokan A (2018) Tissue-dependent expression and translation of circular RNAs with recombinant AAV vectors in vivo. Mol Ther Nucleic Acids 13:89–98. https://doi.org/10.1016/j.omtn.2018.08.008

Miao Z, Ding J, Chen B, Yang Y, Chen Y (2016) HOTAIR overexpression correlated with worse survival in patients with solid tumors. Minerva Med 107(6):392–400

Michalik KM, You X, Manavski Y, Doddaballapur A, Zörnig M, Braun T et al (2014) Long noncoding RNA MALAT1 regulates endothelial cell function and vessel growth. Circ Res 114(9):1389–1397

Mishra S, Srivastava P, Pandey A, Agarwal A, Shukla S, Husain N (2024) Panel of serum long non-coding RNAs as potential non-invasive biomarkers for gallbladder carcinoma. Noncoding RNA Res 9(2):583–593

Neveu B, Jain P, Têtu B, Wu L, Fradet Y, Pouliot F (2016) A PCA3 gene-based transcriptional amplification system targeting primary prostate cancer. Oncotarget 7(2):1300–1310. https://doi.org/10.18632/oncotarget.6360

Ni L, Yamada T, Nakatani K (2024) Utility of oligonucleotide in upregulating circular RNA production in a cellular model. Sci Rep 14(1):8096

Niknam N, Nikooei S, Ghasemi H, Zadian SS, Goudarzi K, Ahmadi SM, Alipoor B (2023) Circulating levels of HOTAIR- lncRNA are associated with disease progression and clinical parameters in type 2 diabetes patients. Rep Biochem Mol Biol 12(3):448–457. https://doi.org/10.61186/rbmb.12.3.448

Niu D, Wu Y, Lian J (2023) Circular RNA vaccine in disease prevention and treatment. Signal Transduct. Target. Ther 8(1):341

Okugawa Y, Toiyama Y, Hur K, Toden S, Saigusa S, Tanaka K et al (2014) Metastasis-associated long non-coding RNA drives gastric cancer development and promotes peritoneal metastasis. Carcinogenesis 35(12):2731–2739. https://doi.org/10.1093/carcin/bgu200

Patton JG, Franklin JL, Weaver AM, Vickers K, Zhang B, Coffey RJ et al (2015) Biogenesis, delivery, and function of extracellular RNA. J Extracell Vesicles 4:27494. https://doi.org/10.3402/jev.v4.27494

Pisignano G, Michael DC, Visal TH, Pirlog R, Ladomery M, Calin GA (2023) Going circular: history, present, and future of circRNAs in cancer. Oncogene 42(38):2783–2800. https://doi.org/10.1038/s41388-023-02780-w

Poliseno L, Lanza M, Pandolfi PP (2024) Coding, or non-coding, that is the question. Cell Res 34(9):609–629. https://doi.org/10.1038/s41422-024-00975-8

Pouliot F, Johnson M, Wu L (2009) Non-invasive molecular imaging of prostate cancer lymph node metastasis. Trends Mol Med 15(6):254–262. https://doi.org/10.1016/j.molmed.2009.04.004

Pouliot F, Karanikolas BD, Johnson M, Sato M, Priceman SJ, Stout D et al (2011) In vivo imaging of intraprostatic-specific gene transcription by PET. J Nucl Med 52(5):784–791. https://doi.org/10.2967/jnumed.110.084582

Qu L, Yi Z, Shen Y, Lin L, Chen F, Xu Y et al (2022) Circular RNA vaccines against SARS-CoV-2 and emerging variants. Cell 185(10):1728–1744.e1716. https://doi.org/10.1016/j.cell.2022.03.044

Reddy PN, Patop IL, Aishwarya K, Reut A-F, Osnat B, Sebastian K (2020) An in vivo strategy for knockdown of circular RNAs. Cell Discov. 6:52

Ren Y, Wang YF, Zhang J, Wang QX, Han L, Mei M, Kang CS (2019) Targeted design and identification of AC1NOD4Q to block activity of HOTAIR by abrogating the scaffold interaction with EZH2. Clin Epigenetics 11(1):29. https://doi.org/10.1186/s13148-019-0624-2

Ren H, Wu F, Liu B, Song Z, Qu D (2020) Association of circulating long non-coding RNA MALAT1 in diagnosis, disease surveillance, and prognosis of acute ischemic stroke. Braz J Med Biol Res 53(12):e9174. https://doi.org/10.1590/1414-431x20209174

Russo M, Gentile L, Toscano A, Aguennouz M, Vita G, Mazzeo A (2020) Advances in treatment of ATTRv amyloidosis: state of the art and future prospects. Brain Sci 10(12). https://doi.org/10.3390/brainsci10120952

Sadik N, Cruz L, Gurtner A, Rodosthenous RS, Dusoswa SA, Ziegler O et al (2018) Extracellular RNAs: A new awareness of old perspectives. Methods Mol Biol 1740:1–15. https://doi.org/10.1007/978-1-4939-7652-2_1

Safdar A, Saleem A, Tarnopolsky MA (2016) The potential of endurance exercise-derived exosomes to treat metabolic diseases. Nat Rev Endocrinol 12(9):504–517. https://doi.org/10.1038/nrendo.2016.76

Sarkar S, Moitra P, Bera S, Bhattacharya SJAAM, Interfaces. (2024) Antisense oligonucleotide embedded context responsive nanoparticles derived from synthetic Ionizable lipids for lncRNA targeted therapy of. Breast Cancer 16(35):45871–45887

Sassi Y, Avramopoulos P, Ramanujam D, Grüter L, Werfel S, Giosele S et al (2017) Cardiac myocyte miR-29 promotes pathological remodeling of the heart by activating Wnt signaling. Nat Commun 8(1):1614. https://doi.org/10.1038/s41467-017-01737-4

Sato-Kuwabara Y, Melo SA, Soares FA, Calin GA (2015) The fusion of two worlds: non-coding RNAs and extracellular vesicles--diagnostic and therapeutic implications (Review). Int J Oncol 46(1):17–27. https://doi.org/10.3892/ijo.2014.2712

Schult P, Roth H, Adams RL, Mas C, Imbert L, Orlik C et al (2018) microRNA-122 amplifies hepatitis C virus translation by shaping the structure of the internal ribosomal entry site. Nat Commun 9(1):2613

Sen GL, Blau HM (2006) A brief history of RNAi: the silence of the genes. FASEB J 20(9):1293–1299. https://doi.org/10.1096/fj.06-6014rev

Seyhan AA (2024) Trials and tribulations of MicroRNA therapeutics. Int J Mol Sci 25(3):1469

Shaker OG, Abdelaleem OO, Mahmoud RH, Abdelghaffar NK, Ahmed TI, Said OM, Zaki, O. M. (2019) Diagnostic and prognostic role of serum miR-20b, miR-17-3p, HOTAIR, and MALAT1 in diabetic retinopathy. IUBMB Life 71(3):310–320

Shen J, Hodges TR, Song R, Gong Y, Calin GA, Heimberger AB, Zhao H (2018) Serum HOTAIR and GAS5 levels as predictors of survival in patients with glioblastoma. Mol Carcinog 57(1):137–141

Sozzi G, Boeri M, Rossi M, Verri C, Suatoni P, Bravi F et al (2014) Clinical utility of a plasma-based miRNA signature classifier within computed tomography lung cancer screening: a correlative MILD trial study. J Clin Oncol 32(8):768–773. https://doi.org/10.1200/jco.2013.50.4357

Tan SK, Pastori C, Penas C, Sarkaria J, Komotar RJ, Ayad N (2017) GENE-03 Serum long noncoding RNA HOTAIR as novel diagnostic and prognostic biomarker in glioblastoma multiforme. Mol Cancer *19*(Suppl 6):vi92

Tang Q, Khvorova A (2024) RNAi-based drug design: considerations and future directions. Nat. Rev. Drug Discov 23(5):341–364

Tikkinen KA, Dahm P, Lytvyn L, Heen AF, Vernooij RW, Siemieniuk RA et al (2018) Prostate cancer screening with prostate-specific antigen (PSA) test: a clinical practice guideline. *BMJ* 362:k358

Tong G, Tong W, He R, Cui Z, Li S, Zhou B, Yin Z (2022) MALAT1 polymorphisms and lung cancer susceptibility in a Chinese Northeast Han population. Int J Med Sci 19(8):1300–1306. https://doi.org/10.7150/ijms.73026

Tosar JP, Witwer K, Cayota A (2021) Revisiting extracellular RNA release, processing, and function. Trends Biochem Sci 46(6):438–445

Wang W, He X, Zheng Z, Ma X, Hu X, Wu D, Wang M (2017) Serum HOTAIR as a novel diagnostic biomarker for esophageal squamous cell carcinoma. Mol Cancer 16:1–5

Wang Y, Gu XX, Huang HT, Liu CH, Wei YS (2020) A genetic variant in the promoter of lncRNA MALAT1 is related to susceptibility of ischemic stroke. Lipids Health Dis 19(1):57. https://doi.org/10.1186/s12944-020-01236-4

Wang H, Xia Y, Zhang Y (2021) Diagnostic significance of serum lncRNA HOTAIR and its predictive value for the development of chronic complications in patients with type 2 diabetes mellitus. Diabetol Metab Syndr 13(1):97. https://doi.org/10.1186/s13098-021-00719-3

Wang R, Zhao J, Wei Q, Wang H, Zhao C, Hu C et al (2022) Potential of circulating lncRNA CASC2 as a biomarker in reflecting the inflammatory cytokines, multi-organ dysfunction, disease severity, and mortality in sepsis patients. J Clin Lab Anal 36(8):e24569. https://doi.org/10.1002/jcla.24569

Wei H, Chen Q, Lin L, Sha C, Li T, Liu Y et al (2021) Regulation of exosome production and cargo sorting. Int J Biol Sci 17(1):163–177. https://doi.org/10.7150/ijbs.53671

Wen J, Chen L, Tian H, Li J, Zhang M, Cao Q et al (2019) Effect of MALAT1 polymorphisms on papillary thyroid cancer in a Chinese population. J Cancer 10(23):5714–5721. https://doi.org/10.7150/jca.28887

Williams JG, North MJ, Mahbubani H (1985) A developmentally regulated cysteine proteinase in Dictyostelium discoideum. EMBO J 4(4):999–1006. https://doi.org/10.1002/j.1460-2075.1985.tb03730.x

Winkle M, El-Daly SM, Fabbri M, Calin GA (2021) Noncoding RNA therapeutics — challenges and potential solutions. Nat Rev Drug Discov 20(8):629–651. https://doi.org/10.1038/s41573-021-00219-z

Wu S, Sun H, Wang Y, Yang X, Meng Q, Yang H et al (2019) MALAT1 rs664589 polymorphism inhibits binding to miR-194-5p, contributing to colorectal cancer risk, growth, and metastasis. Cancer Res 79(20):5432–5441. https://doi.org/10.1158/0008-5472.Can-19-0773

Wu D, Tao T, Eshraghian EA, Lin P, Li Z, Zhu X (2022a) Extracellular RNA as a kind of communication molecule and emerging cancer biomarker. Front Oncol 12:960072. https://doi.org/10.3389/fonc.2022.960072

Wu N, Cheng CJ, Zhong JJ, He JC, Zhang ZS, Wang ZG et al (2022b) Essential role of MALAT1 in reducing traumatic brain injury. Neural Regen Res 17(8):1776–1784. https://doi.org/10.4103/1673-5374.332156

Yang H, Xi X, Zhao B, Su Z, Wang Z (2018) KLF4 protects brain microvascular endothelial cells from ischemic stroke induced apoptosis by transcriptionally activating MALAT1. Biochem Biophys Res Commun 495(3):2376–2382

Yazarlou F, Modarressi MH, Mowla SJ, Oskooei VK, Motevaseli E, Tooli LF et al (2018) Urinary exosomal expression of long non-coding RNAs as diagnostic marker in bladder cancer. Cancer Manag Res 10:6357–6365. https://doi.org/10.2147/cmar.S186108

Zhang L, Song X, Wang X, Xie Y, Wang Z, Xu Y (2015) Circulating DNA of HOTAIR in serum is a novel biomarker for breast cancer. Breast Cancer Res Treat 152:199–208

Zhang Y, Zhang K, Luo Z, Liu L, Wu L, Liu J (2016) Circulating long non-coding HOX transcript antisense intergenic ribonucleic acid in plasma as a potential biomarker for diagnosis of breast cancer. Thorac Cancer 7(6):627–632

Zhou X, Chen J, Tang W (2014) The molecular mechanism of HOTAIR in tumorigenesis, metastasis, and drug resistance. Acta Biochim Biophys Sin Shanghai 46(12):1011–1015. https://doi.org/10.1093/abbs/gmu104

Zhu C, Wang X, Wang Y, Wang K (2022) Functions and underlying mechanisms of lncRNA HOTAIR in cancer chemotherapy resistance. Cell Death Discov 8(1):383. https://doi.org/10.1038/s41420-022-01174-3

Chapter 13
Bioinformatics Approaches in Noncoding RNAs Research

Fariya Khan, Ajay Kumar, and Salman Akhtar

Contents

F. Khan
Department of Bioengineering, Integral University, Lucknow, India

A. Kumar
Department of Biotechnology, Faculty of Engineering and Technology, Rama University, Kanpur, Uttar Pradesh, India

S. Akhtar (✉)
Department of Bioengineering, Integral University, Lucknow, India

Novel Global Community Educational Foundation, Hebersham, NSW, Australia

M. F. Ullah, A. Ahmad (eds.), *Non-coding RNAs*, Experientia Supplementum 115,
https://doi.org/10.1007/978-3-032-06948-1_13

Abstract In recent years, noncoding RNAs have sparked significant interest in understanding the diverse roles of ncRNAs in cellular regulation and disease processes. Noncoding RNAs (ncRNAs), encompassing small noncoding RNAs (sncRNAs) and long noncoding RNAs (lncRNAs), are critical regulators of gene expression, epigenetic modifications, and various cellular processes within the human genome. The diverse nature ncRNAs, along with certain complex features, has made them difficult to study through traditional experimental methods. As a result, bioinformatics tools have expanded the possibilities for offering new insights through advanced computational strategies. This chapter explores the recent advancements in ncRNA databases, emphasizing their importance and the innovative in silico strategies that enable the prediction and analysis of biological interactions, particularly for miRNAs and lncRNAs. It offers an in-depth overview of the structural properties, classification, and functions of various types of noncoding RNAs, highlighting their crucial roles in cellular processes. Additionally, the chapter discusses the significant therapeutic potential of ncRNAs, focusing on their applications in treating cancer and other severe diseases. These insights are pivotal in advancing the development of targeted therapies and precision medicine.

Keywords Epigenetics · Transcriptome · miRNA · ncRNA · sncRNA · lncRNA · Bioinformatics · In silico · Genome

Abbreviations

ncRNAs	Noncoding RNAs
lncRNAs	Long noncoding RNAs
lincRNAs	Long intergenic ncRNAs
meRNAs	Multiexonic RNAs
circRNAs	Circular RNAs
miRNAs	MicroRNAs
CTLs	cytotoxic T lymphocytes
RNA-seq	RNA sequencing

13.1 Introduction

Over the past decade, recent findings have revealed that noncoding RNAs (ncRNAs) are extensively expressed and serve critical functions in the regulation of genes. A growing number of noncoding RNAs are being recognized for their involvement in gene expression and cellular regulation. However, many ncRNAs still await detailed functional analysis. Despite years of research, the field continues to progress, with each newly discovered ncRNA undergoing processes like biogenesis, characterization, and functional annotation. Both bioinformatics approaches and traditional biochemical techniques have been crucial in advancing our understanding of ncRNAs at various stages of their study (Statello et al. 2021). Depending on their localization and their specific interactions with DNA, RNA and proteins, ncRNAs are a wide spectrum of RNA molecules that, even though many of them still remain as oddities, are of great functional significance because they participate in both healthy and diseased manifestations of cells. Long noncoding RNAs (lncRNAs), which are RNA molecules exceeding the length of 200 nucleotides in size, usually share the appearance of coding RNAs. Deregulation of certain lncRNA usually causes the retraction of a number of diseases, including cancer and cardiovascular diseases, thus suggesting they can be targeted for the development of new medications (Amaral and Mattick 2008).

The growing body of evidence highlights noncoding RNAs (ncRNAs) as key regulators of gene expression in cancer development and progression. Their emerging role as potential diagnostic and prognostic biomarkers, especially in liquid biopsies and tissue-based diagnostics, holds significant promise (Gangotia et al. 2021). This chapter offers a comprehensive examination of the various types of ncRNAs, including microRNAs, long noncoding RNAs, and circular RNAs, discussing their functions and mechanisms of action. Additionally, we delve into the therapeutic potential of ncRNAs, such as anti-miRNA oligonucleotides, siRNAs, and other RNA-based cancer treatments. Despite this promise, substantial challenges remain in advancing ncRNA-based diagnostic technologies and therapies, such as issues with specificity, incomplete understanding of their mechanisms, and challenges related to delivery. Additionally, this chapter examines the latest advancements in noncoding RNA research technologies and the bioinformatic tools employed for their analysis. Finally, we discuss future directions for ncRNA research in cancer, focusing on the development of novel biomarkers, therapeutic targets, and treatment strategies. In summary, this review offers a comprehensive perspective on the role of ncRNAs in cancer and their potential clinical applications, emphasizing both the opportunities and challenges in this rapidly advancing field. Noncoding RNAs are involved in a wide range of cellular processes, including development, differentiation, and disease, with recent studies showing their potential as cancer biomarkers.

13.1.1 Classifications of Noncoding RNAs

Noncoding RNAs (ncRNAs) can be broadly categorized into two groups based on their regulatory roles. As a housekeeping ncRNA, it is highly and stably expressed in the majority of cells, thereby it becomes a crucial element of the cell function and maintenance (Cech and Steitz 2014). Contrary to this, the regulatory ncRNAs that act on the levels of transcriptional, epigenetic, and post-transcriptional are the ones that directly regulate the activities of a cell by expression of the respective genes (Ponjavic et al. 2007). All the noncoding RNAs are usually classified according to their lengths, with shorter ones being those that are less than 200 nucleotides and longer ones being those that are longer than 200 nucleotides (lncRNAs). Both

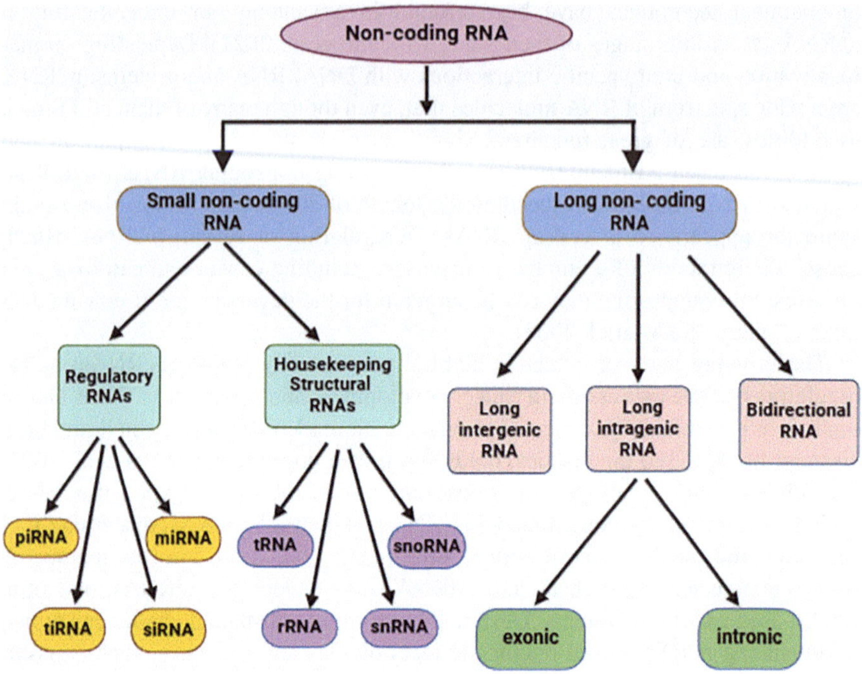

Fig. 13.1 Classification of noncoding RNA in various forms

lncRNAs and sncRNAs have various subtypes, most of which are important in the regulation of gene expression (Taft et al. 2009). These subtypes can be generated from different parts of the genome, and the mechanisms they use to get synthesized further show the complexity within them. Even though ncRNAs have the huge structural and functional diversity, their sequence length and the uncoding protein capability are still the genetic features of them to be differentiated by. LncRNAs can be further classified into subcategories according to their positions at the protein-coding genes (Fig. 13.1). Long intergenic ncRNAs (lincRNAs) are transcribed from both strands and lie at the center of the noncoding region between the protein-coding genes (Bhan and Mandal 2014). Long intronic ncRNAs are the products of the entire transcription of the introns of the protein-coding genes.

eRNAs are a newly discovered class of noncoding RNAs which are expressed from enhancer regions in the genome similarly to eRNAs, lncRNA show diverse characteristics, such as possibly, the potential to encode proteins, and also, the one that undergo bidirectional transcription in the construction of both sense and anti-sense transcripts. Enhanced regions very often express short, non-polyadenylated eRNAs (polyA-eRNAs), which are between 0.5 and 2 kb long (Lafontaine 2015). Differences are, however, seen with long (>4 kb), polyadenylated eRNAs (polyA+ eRNAs), which are mainly transcribed unidirectionally. Polyadenylated enhancers located outside the genes often give rise to eRNAs, which are \$polymζxed and mul-tiexonic types leading to the term meRNAs (Legnini et al. 2014). CircRNAs are a group of endogenous noncoding RNAs which form covalently closed loop structures, and their sizes can be very different, from less than 100 nucleotides to more than 10,000 nucleotides, even though most circRNAs in mammals and plants are usually only a few hundred nucleotides long (Jeck et al. 2013). Even though they have existed for a long time, recent technological developments in deep sequencing and bioinformatics have made circRNAs a hot topic.

From a regulatory perspective, noncoding RNAs (ncRNAs) can be classified into two major categories based on their roles and functional properties. The first category consists of housekeeping ncRNAs, which are typically short, ranging from 50 to 500 nucleotides in length. These ncRNAs are essential for basic cellular functions and are ubiquitously expressed across various cell types. They are constitutively involved in fundamental cellular processes such as protein synthesis, RNA processing, and maintenance of cellular homeostasis. Housekeeping ncRNAs include molecules like ribosomal RNAs (rRNAs) and transfer RNAs (tRNAs), which are crucial for the proper functioning of the translational machinery and the overall survival of the cell (Chen et al. 2018). Due to their vital and consistent roles, these ncRNAs are typically stable and expressed at relatively high levels across different conditions.

In contrast, regulatory ncRNAs play a more dynamic and complex role in coordinating gene expression. These molecules are not only involved in fine-tuning gene regulation but also in responding to various cellular signals and environmental changes. Regulatory ncRNAs function at multiple levels of gene expression regulation, including transcriptional, epigenetic, and post-transcriptional processes. For example, microRNAs (miRNAs) and long noncoding RNAs (lncRNAs) are pivotal in regulating gene expression by influencing chromatin structure, modulating transcription factors, and controlling mRNA stability and translation (Lagos-Quintana et al. 2001). By interacting with key regulatory elements of the genome, these ncRNAs serve as key modulators, ensuring that genes are expressed at appropriate times and levels to maintain cellular function and adapt to changing conditions. Their involvement in disease mechanisms, particularly in cancer and other complex disorders, highlights their critical role in cellular regulation and underscores their potential as therapeutic targets.

13.1.2 Functions of Noncoding RNAs

Numerous noncoding RNAs are key players in the control of cell regulation processes in both animals and plants. It is involved in a wide range of physiological phenomena such as regulation of gene expression, cytokine expression, translation control, inflammation, neuropathic pain, cholesterol biosynthesis and homeostasis, growth hormone and prolactin production, glucose metabolism, and RNA transport pathways (Atianand et al. 2016; Hu et al. 2016). Many RNA roles are transferred to proteins, but others remain because the study of the advantages given selectively by proteins and RNA during evolution came to the fore. To understand ncRNA function and mechanism, it may be useful to compare ncRNA function to that of

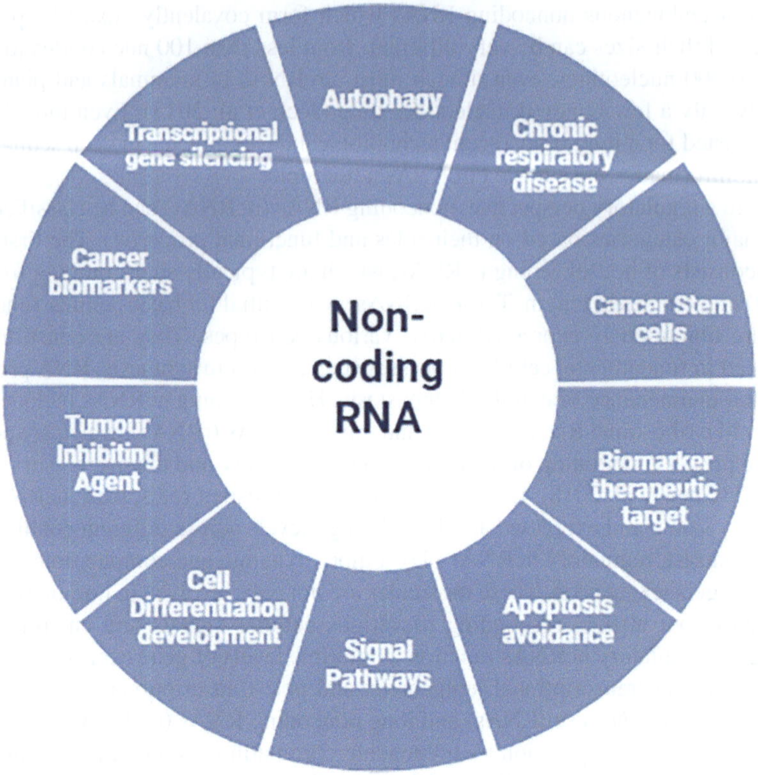

Fig. 13.2 Functions and therapeutic implications of noncoding RNA

proteins (Li et al. 2018a, b). Subsequent technological breakthroughs in ncRNAs have further subdivided ncRNAs according to function, with the new types possessing unique roles and applications, including the discovery of new biomarkers and the development of novel treatments (Fig. 13.2). Some of them function, as a rule of thumb, as powerful switches that can turn on/off transcription and translation of ncRNA as well as other ncRNAs. Thus, it can be inferred that ncRNAs are the originators of the complex human traits, operators of the normal physiological processes, and the catalysts of the development of the human diseases (Amaral et al. 2013). Long noncoding RNAs (lncRNAs) have become the focal point of attention due to their regulatory properties during viral infections. Recent literature depicted that lncRNAs are vital players in modulating infections with different levels of antiviral responses.

Revolutionary research has significantly advanced our understanding of how noncoding RNA (ncRNA) molecules are essential for gene coding, decoding, regulation, and expression. This insight into the regulatory roles of ncRNAs has revealed that distinct ncRNA networks function in various types of cancer (Slack and Chinnaiyan 2019). These findings have paved the way for the development of targeted cancer therapies and preventive strategies focusing on ncRNAs derived from the human genome. The innovation of ncRNA's involvement in cancer has offered a new perspective on the mechanisms behind cancer development and treatment, highlighting the broader influence of the genome. Aberrant ncRNA expression, coupled with the associated downstream signaling pathways, has been directly implicated in cancer initiation and progression. Several long noncoding RNAs (lncRNAs) exhibit dual roles as tumor suppressors and oncogenes that function as protective mechanisms against cancer development by inhibiting cell proliferation, preserving genomic stability, triggering apoptosis, or enhancing the expression of tumor suppressors (Johnsons et al. 2005). ncRNAs are also involved in maintaining chromosome stability, ensuring proper segregation, and regulating transcriptional gene silencing (Liu et al. 2017). Recent studies have shown that ncRNAs can promote immune evasion in tumors and cancer stem cells, exhibiting an oncogenic effect. They achieve this by inducing activation-induced cell death in cytotoxic T lymphocytes (CTLs) and TH1 cells, which helps the tumor escape immune surveillance (Fabbri et al. 2019, Khan et al. 2024a, 2024b).

13.1.3 1ncRNAs as Cancer Biomarkers

Noncoding RNAs (ncRNAs) are now known to be very significant molecular biomarkers, with their expression often being significantly altered in various cancers. Large-scale discovery studies have been performed in preclinical settings, and the utility of ncRNAs, including both small ncRNAs such as miRNAs and long ncRNAs such as lncRNAs, circRNAs, has been promising for diagnostic and prognostic purposes. Especially, circRNAs have recently drawn significant attention, due to their stable structure originating from the covalently closed ends preventing degradation by exoribonucleases (Hombach and Kretz, 2016; Khan et al. 2024a, 2024b). Some miRNAs, such as let-7 and miR-34a, are recognized as tumor suppressors. The let-7 family consists of several members. Notably, many of these biomarkers are detected in non-invasive bodily fluids, such as blood or urine, offering a less invasive alternative to obtaining a tumor biopsy. PCA3 stands as the first ncRNA to receive FDA approval for use as a cancer biomarker test. This marker, which is highly expressed in prostate cancer, can be easily detected through a non-invasive urine test (Tomlins et al. 2016).

The presence of specific ncRNAs can not only be used to identify cancer compared to normal tissues but also provide valuable insights into key clinical factors, including survival prognosis, metastasis, and treatment responses once cancer is diagnosed (Yan and Bu 2021).

13.1.3.1 ncRNAs as Biomarkers in Chronic Respiratory Disease

ncRNAs play a pivotal role in macrophage-mediated intercellular communication in lung cancer and inflammatory lung diseases, including asthma and pulmonary fibrosis. In these inflammatory conditions, ncRNAs facilitate the interactions between macrophages and lung tissue cells. Asthma is a persistent inflammatory disorder of the respiratory system, affecting an estimated 340 million individuals worldwide. It is commonly associated with symptoms such as difficulty breathing, wheezing, chest tightness, and coughing (Vos et al. 2012). Recent research has highlighted the significant impact of ncRNA-regulated macrophages in modulating macrophage polarization, airway inflammation and remodeling, and immune regulation, suggesting that targeting macrophages with ncRNAs may offer a promising therapeutic strategy for asthma (Zuyderduyn et al. 2008). The differentiation, activation, and functions of exosomal miRNAs that regulate the functions of different immune cells will influence a major mechanism shaping inflammatory conditions in asthma. It is notably elevated in the bronchoalveolar lavage fluid with regard to influencing the levels of expression for IL-13 and IL-5Ra and contributed to the lung inflammation as inflammatory damage that can worsen allergic inflammation of the airway (Zhao et al. 2019). Further investigations have indicated that exosomal miR-155 plays a role in promoting the release of double-stranded DNA (dsDNA) into the extracellular space, which in turn intensifies allergic airway inflammation and neutrophilic asthma phenotype (Kim et al. 2022). The release of dsDNA may be a useful therapeutic approach to mitigate severe asthma in order to reduce the levels of exosomal miRNA, such as delivery of miRNA inhibitors into specific cell types via engineered exosomes.

13.1.4 Significance of Bioinformatics Tools in ncRNA Research

Computational and bioinformatics approaches are crucial for advancing research on noncoding RNAs (ncRNAs), offering valuable tools to explore their intricate roles in cellular functions and diseases (Fig. 13.3). These methods are employed in several key areas, beginning with the identification, prediction and sequencing of novel ncRNAs (Yotsukura et al. 2017). By analyzing genomic data and integrating it with experimental techniques such as RNA sequencing (RNA-seq), bioinformatics allows for the discovery of previously uncharacterized ncRNA. Using bioinformatics and mathematical algorithms to analyze experimental data, and storing that data in databases with open-source tools, enables the prediction of molecular interactions between DNA, RNA, and proteins (Pinkney et al. 2020). This can help reduce both time and costs in research and development. In addition, bioinformatics tools predict miRNA target genes, which are crucial for understanding post-transcriptional regulation and the networks that miRNAs influence, particularly in diseases like cancer. Predicting the secondary and tertiary structures of ncRNAs is another key application, as these structures are critical to their function (Lin and Qiu 2025). Computational methods are used to predict these structures, providing insights into how ncRNAs interact with other molecules and perform their biological roles. Advances in bioinformatics have facilitated the exploration of the biological

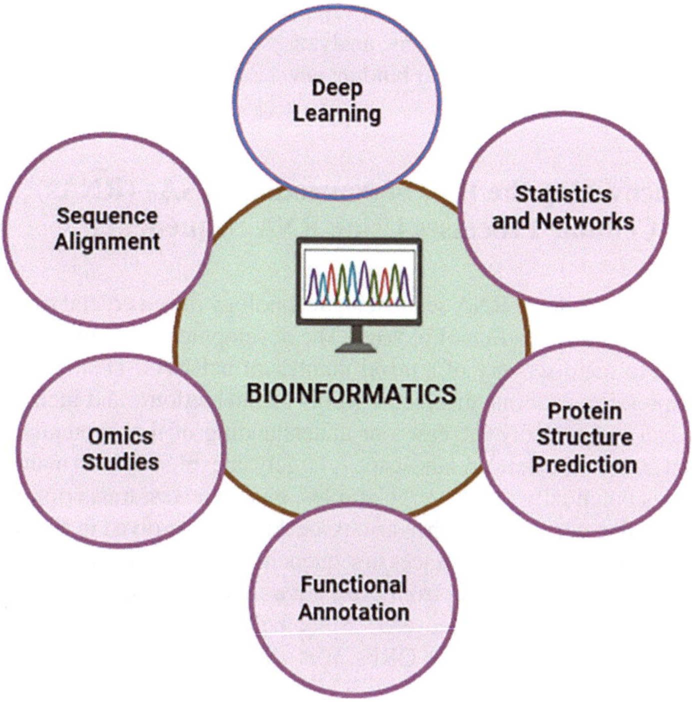

Fig. 13.3 Broad spectrum of bioinformatics analysis in noncoding RNA research

significance of ncRNA genes and miRNA target sites by analyzing their conservation and evolutionary dynamics (Rincon-Riveros et al. 2021). Modern computational technologies have enabled the prediction of ncRNA functions by examining their known interactions and relationships with other molecular entities. Furthermore, bioinformatics helps to construct integrated regulatory networks that include ncRNAs for gene regulation.

The creation of specialized databases and web servers for ncRNA research has made curated datasets, prediction tools, and analytical platforms more accessible, promoting collaboration and driving progress in the field. One major benefit of using multiple ncRNA databases and bioinformatics tools is their constant integration of experimentally obtained data, coupled with rigorous curation processes (Seifuddin and Pirooznia 2021). These computational approaches are transforming ncRNA research, facilitating the identification, functional annotation, and deeper insights into these molecules, thus advancing both fundamental scientific knowledge and clinical applications. The development of computational tools enables the prediction of the functional roles, structural features, and intermolecular interactions of RNA and has become a key resource in expanding our understanding of the mystery of the genome. These tools allow us to model and predict how various types of RNA, including miRNAs, lncRNAs, and piRNAs, interact with other molecules and contribute to different cellular functions. Therefore, computational methods offer valuable insights into the hidden regulatory mechanisms that govern gene expression and cellular activities by analyzing their RNA secondary structures, interaction networks, and potential binding sites.

13.2 Uncovering the Role of Noncoding RNAs (RNAs) in Cellular Processes Using RNA Sequencing

Overall, advancements in RNA sequencing technology play a crucial role in driving progress in our understanding of biology. The development of innovative technologies has led to the discovery of a novel number of ncRNAs. Their tissue-specific, distinct expression patterns, diverse networks of interactions, and their subcellular locations can undoubtedly increase our understanding of their potential functions (Schimmel 2018). The ncRNA landscape is largely unexplored, with many unknown RNA species potentially playing crucial roles, given the vast transcriptional capacity of mammalian genomes and the various mechanisms involved in ncRNA generation (Li et al. 2018a, b). With advances in sequencing technologies and bioinformatics analyses, novel ncRNAs arising from alternative splicing or the degradation of their precursor RNAs have been discovered. Since ncRNAs, particularly lncRNAs and circRNAs, can randomly contain ORFs, it is challenging to determine whether a transcript is truly noncoding. Therefore, it is important to measure the coding potential of newly identified ncRNAs since growing evidence indicates that some lncRNAs and circRNAs encode functional micropeptides (Sun and Chen 2020).

13.2.1 Several Tools for Sequence Alignment

In recent years, the importance of noncoding RNA sequences has been more apparent. Improved techniques in comparative analysis, especially multiple sequence alignments, have led to the discovery and study of new RNA families (Wright 2020). Alignment of RNA sequences is still a major challenge in computational biology despite many tools having been developed; there remains uncertainty over which are best suited for the task of RNA multiple sequence alignment(Fallmann et al. 2017). However, it is accepted that programs combining folding and alignment are more effective than those that do not consider structural considerations in the alignment process.

Rfam is a comprehensive ncRNA family database represented by multiple sequence alignments as well as covariance models used to detect homologs for sequence alignment. It provides an NCBI annotation of a set of nonredundant genomes (Griffiths-Jones et al. 2003). The text search of the database allows users to browse RNA families within certain species or taxonomic groups, compare RNAs across different genomes, and look at annotations of individual ncRNA sequences within a genome. The data provided by Rfam has been put to a wide variety of uses, from annotation of genomes to serving as training sets for the development of new algorithms (Kalvari et al. 2018a, b).

The DECIPHER algorithm for RNA sequences works very much like all the techniques that have previously been introduced to align a protein sequence (Wright 2015). Most multiple alignment programs attain their best accuracy at aligning just a few sequences and deteriorate rather quickly once several sequences are added into the mix. DECIPHER is an R package of many bioinformatics tools with capabilities such as primer design, probe design, etc. Remarkably, during the alignment process, DECIPHER offers a feature that helps maintain a consistent level of accuracy, even as the number of sequences increases (Wright et al. 2014).

13.2.2 Motif Discovery and Functional Regions

Motif discovery has been a central focus in bioinformatics research since the availability of genomic and protein sequence data. In order to establish the true biological significance of a motif, it is important to distinguish it from random occurrences (Sandve and Drablos 2006). This means that the motif must not occur at similar frequencies or show similar sequence conservation in any of the randomly selected or model-generated DNA sequences that are considered "biologically plausible." In other words, we can more accurately estimate the functional relevance of motifs by comparing them with motifs in biological sequences against random sequences. The function of ncRNAs is primarily sequence and structure-based, and often structural features are conserved evolutionarily. These structural motifs provide valuable means of identifying and classifying ncRNAs (Pavesi et al. 2004a, b). On the other hand, RNA structure analysis is not a very speedy process, leading to the need for the constant development of new algorithms for efficient and accurate predictions of structure. Such insight about the loading and intercellular transfer of miRNA calls for a systematic survey of short exosomal miRNAs toward finding motifs. Substantial issues continue to plague techniques, although extensive progress has been made to this point in terms of failing to identify important short motifs despite proper negative control data (Gao et al. 2020).

Motif discovery algorithms generally fall into two main categories: the enumeration approach and probabilistic techniques (Keith 2008). The enumeration approach focuses on identifying consensus sequences by systematically searching for motifs. This method involves enumerating words and computing their similarities, which is why it is often referred to as the "word enumeration approach."The word enumeration approach can be enhanced through the use of specialized data structures, such as suffix trees, or by utilizing parallel processing techniques (Pavesi et al. 2004a, b). Notable algorithms that rely on this approach include DREME, CisFinder, Weeder, FMotif, and MCES (Bailey 2011; Karaboga and Aslan 2016; Zhang et al. 2016). These algorithms are designed to efficiently identify motifs by leveraging advanced computational methods.

The second category of motif discovery approaches is the probabilistic method. This approach builds a probabilistic model, often referred to as a position-specific weight matrix (PSWM) or motif matrix, which defines the base distribution for each position in a transcription factor binding site (TFBS) (Yu et al. 2015). This model helps differentiate motifs from non-motifs and typically requires fewer search parameters. Some of the most widely used algorithms based on the probabilistic approach include MEME, STEME, EXTREME, AlignACE, and BioProspector (Sinha and Tompa 2003; Van Helden et al. 1998; Thomas-Chollier et al. 2012; Ma et al. 2012).

13.2.3 Comparative Genomics of ncRNAs

Comparative analysis is, with the ever-present availability of whole-genome sequences, becoming an indispensable tool in the evaluation of similarities and differences of the sequence among the organisms. In this respect, comparative genomics has been widely applied in finding genes, their regulatory elements, and gene structures as a whole (Sieglaff et al. 2009). For instances of comparative genomics, much valuable information could be extracted through its application regarding the study of ncRNAs by discovering patterns, functional roles, and even conserved features between species. Unlike protein-coding genes, ncRNAs do not encode proteins but contribute significantly to gene regulation, cellular processes, and maintaining biological homeostasis. By comparative analysis of ncRNA sequences in diverse organisms, scientists have amassed a great deal of knowledge regarding their conserved roles and evolutionary importance to biological systems. For instance, ncRNAs are usually associated with essential functions such as transcriptional regulation, RNA splicing, and epigenetic modifications (Peschansky and Wahlestedt 2013).

A remarkable strength of comparative genomics has been the identification of structured RNA motifs: regions of RNA that possess conserved secondary structures that are functionally significant. The method typically consists of clustering subsequences obtained from intergenic regions across species and making predictions of structural motifs. This type of approach, over the past decade, has revealed thousands of structured RNAs, mainly in bacterial species. These discoveries have significantly increased our knowledge about riboswitches, ribozymes, and other structured ncRNAs involved in crucial regulatory mechanisms within different cellular pathways.

Comparative genomics, however, has numerous limitations, such as computational complexity, which is one of the major drawbacks which limits the scope of analysis for a small fraction of intergenic regions in a single experiment to achieve a lower rate of false positives (Rivas et al. 2001). Comparative genomics often suffers from bias by focusing on subsets of intergenic regions that are associated with well-characterized classes of structured RNAs, such as ribozymes or specific riboswitch types. These RNAs are typically found in predictable genomic contexts and, therefore, do not allow a full exploration of novel or less-characterized ncRNA families (Axmann et al. 2005). Future improvements in computational power and algorithmic innovation will help overcome these limitations and allow for more comprehensive and unbiased exploration of ncRNAs. Integrating comparative genomics with experimental approaches and other genomic data types, such as epigenomic and transcriptomic datasets, will further enhance our understanding of ncRNAs and their evolutionary and functional significance in various biological systems. lncRNAs, the structural domains that determine functionality as well as define their interactome, are still not well understood. RNA structural analysis could enhance our comprehensive understanding of the modes of action of lncRNAs.

LncRNAs, typically over 200 nucleotides long, show cell-specific expression patterns and are involved in a variety of biological processes (Lennox et al. 2016). Their substantial length, sometimes reaching tens of kilobases, allows them to form intricate secondary and 3D structures that are still not fully understood. By examining the structure-function relationship of these noncoding RNAs, with an emphasis on their domain architecture, we can better categorize the large and rapidly growing group of ncRNAs. This approach will provide a solid foundation for more in-depth mechanistic research and, in the long run, open doors to clinical applications.

13.3 Identification and Functional Annotation of Noncoding RNA

Large-scale collaborative efforts have produced comprehensive catalogs documenting genomic elements and genetic variation across various cell types, including 3D genomic interactions, along with their molecular functions. A major portion of this data is stored in public databases, which play a crucial role as primary resources for human genomics research (Kircher and Ludwig 2022). Advances in technology now enable researchers to experimentally study the molecular consequences of genetic variations on a broad scale. Despite these advancements, our understanding of how noncoding variants influence phenotypic, cellular, and molecular levels remains limited. To deepen our understanding of the noncoding genome, it is critical to systematically catalog and annotate the relevant genomic elements based on their functional significance.

Functional maps are typically accessible through online platforms that offer graphical interfaces. While different molecular assays provide valuable insights into specific functional aspects, each generally captures only a single dimension of functionality, making it necessary to combine multiple approaches in order to obtain a more comprehensive functional annotation. These maps can be used to pinpoint the cell types most relevant to diseases by evaluating where noncoding variants are enriched in regulatory regions. Publicly accessible functional maps offer a wealth of opportunities to improve our understanding of noncoding risk variants (Cano-Gamez and Trynka 2020). However, a key challenge remains the need for consistent data processing across different studies. Due to the wide-ranging roles of noncoding elements play, a complete understanding of the regulatory principles and their complex interactions is still lacking.

Currently, data are available only for certain cell types, which restricts efforts when combining different datasets to predict variant effects. A primary objective in this field of research is to create effective techniques for engineering genomic changes to assess their functional impact within biological systems. There are various databases used for functional annotation that include BRAVO, which offers allele frequencies, functional annotations, and variant data from the TOPMed project; ClinVar is a repository from NCBI with clinically annotated variant effects; COSMIC is a Catalogue of Somatic Mutations in Cancer and their annotations; ENCODE is the Encyclopedia of DNA Elements which stores data on transcription factors, histone modifications, open chromatin, and gene expression; FANTOM is used for functional annotations for the mammalian genome, Ensembl Regulation is used for computational annotations of regulatory features and genome segmentation, and EnhancerAtlas provides experimentally derived enhancer annotations across nine species (Vu and Ernst 2022).

13.4 Structural Analysis of ncRNAs

Recent studies have highlighted the pivotal role of highly structured noncoding RNAs (ncRNAs) as key regulators of cellular processes, with their dysregulation being implicated in various human diseases (Esteller 2011). This recognition has sparked substantial interest in ncRNAs as promising targets for drug development and molecular recognition strategies. Over the past decade, research focus has shifted from the extensively studied small structural ncRNAs, like transfer RNAs (tRNAs) and ribosomal RNAs (rRNAs), to a group of small RNAs that play a key role in post-transcriptional regulation. This includes microRNAs (miRNAs), whose precursors (pre-miRNAs) adopt a characteristic hairpin structure, as well as small interfering RNAs (siRNAs) and piwi-interacting RNAs (piRNAs) (Bonnet et al. 2004). Different databases and their mode of prediction are mentioned in the Table 13.1.

Despite this, the classification of ncRNAs, including long noncoding RNAs (lncRNAs), into functional families remains challenging and imprecise. Structural characterization of lncRNAs in vivo is particularly difficult due to their large size and inherent heterogeneity. The structural features of lncRNAs provide valuable insights into their functions, especially regarding the relationship between structural architecture and functional roles. However, lncRNAs often exhibit limited evolutionary conservation, with small conserved regions interspersed within largely unconstrained sequences. This irregular conservation complicates the prediction of their functions based solely on structural modeling. To address these challenges, a comprehensive approach that combines structural determination with genetic manipulation is crucial for identifying functional domains within lncRNAs

Table 13.1 Different bioinformatics databases for the prediction of different ncRNA

Database	Type of ncRNA	Description
Rfam	miRNAs, snoRNAs, and lncRNAs	Database for RNA and ncRNA.
siRNAdb	siRNAs	Database for small interfering RNAs (siRNAs).
miRbase	miRNAs	Database for miRNA sequences and their annotations.
LncTarD	lncRNA	Database for long ncRNA interactions.
piRBase	piRNAs	Database for piwi-interacting RNA sequences and their functions.
LncBase	miRNAs and lncRNAs	Database for prediction of lncRNA interactions with microRNAs.
TarBase	miRNAs	Database for experimentally confirmed microRNA targets.
RNAcentral	lncRNAs, miRNAs, piRNAs, siRNAs	Database for prediction of noncoding RNA sequences
piRNAdb	piRNAs	Database for annotation and functionality of piRNAs
starBase	lncRNAsand miRNAs	For miRNAand lncRNAs interactions
PLncDB	lncRNAs	Database for prediction of plant-based long noncoding RNAs
miRPathDB 2.0	miRNAs and miRBase	Database for pathway interactions
deepBase	lncRNAs, miRNAs, piRNAs, and siRNAs	Database for prediction of expression and interaction of small RNAs and long RNAs
sRNAtoolbox	miRNAs, siRNAs, and piRNAs	Database for analyzing small RNA, miRNA, and piRNA sequencing data
DeePathology	mRNA and miRNA	Database for the prediction of deep neural networks
RPI-SAN	ncRNA	Database for the prediction of neural networks
lncRNASNP	lncRNAs	A database of SNPs in lncRNAs and their potential effects on gene expression
LncRNAPred	lncRNAs	A tool that predicts lncRNAs based on their sequence features
VIRmiRNA	miRNAs and RNAs	Database for prediction of interactions of microRNAs with host genes
LncRNAdb	lncRNAs	Database for prediction of experimentally validated long noncoding RNAs
lncPro	lncRNAs	Database for prediction of function, sequence, and structure of lncRNAs

(Fig. 13.4). Such a strategy enables a more nuanced understanding of their roles in cellular mechanisms (Panwar et al. 2014). Advancements in lncRNA structural analysis could lead to the development of domain mapping as a robust tool for predicting the functions and mechanisms of novel or poorly understood lncRNAs. By enhancing our understanding of lncRNA functionality, this approach holds significant potential to drive therapeutic innovation and broaden the applications of lncRNAs in medicine (Alhajlah et al. 2025).

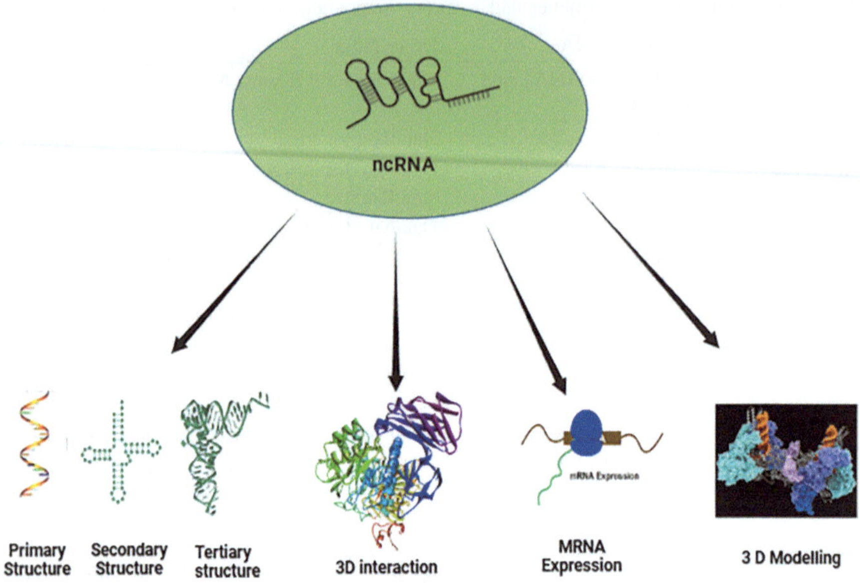

Fig. 13.4 Various modes of structural determination of ncRNA

13.4.1 Secondary Structure Prediction Tools

Accurate predictions of RNA secondary structures play a critical role in uncovering the functional roles of noncoding RNAs (ncRNAs). Unlike protein-coding RNAs, the functionality of ncRNAs is more closely tied to their secondary structures than to their primary sequences. As such, understanding the structural characteristics of ncRNAs is essential for elucidating their biological functions. However, datasets of ncRNAs, often provided in FASTA format, typically contain only primary sequence information, omitting the secondary structure details that are crucial for functional analysis. Since the secondary structures of RNAs are formed by sets of hydrogen-bonded base pairs, they provide vital clues for identifying the functional roles of RNAs and for predicting whether an ncRNA sequence belongs to a specific class (Sato et al. 2021).

Instead of experimental assays, which can be labor-intensive and time-consuming, computational methods are frequently employed to predict RNA secondary structures. Classical approaches rely on dynamic programming and thermodynamic calculations to identify energetically favorable base-pairing patterns. Over the past 4 years, advances in machine learning (ML), particularly deep learning (DL), have revolutionized biomolecular structure prediction. These ML-based methods have demonstrated remarkable improvements in predicting the structures of short RNA transcripts, often surpassing the performance of traditional techniques.

Numerous computational tools have been developed over the years in predicting RNA secondary structures, each with varying strengths and applications. Widely used tools include Mfold, RNAstructure, and the Vienna RNA package. Many of these methods enhance prediction accuracy by incorporating sequence homology, as seen in programs like RNAalifold, Construct, and CM-finder (Mathews et al. 2007). However, applying these tools to long noncoding RNAs (lncRNAs) remains challenging due to the lack of sequence diversity and the poor evolutionary conservation of lncRNAs. Other notable tools include CONTRAfold that leverages machine learning techniques to predict RNA secondary structures and stands out as one of the most versatile ML-based approaches (Do et al. 2006). It often outperforms deep learning models when applied to previously unseen structural families. EternaFold, introduced with multitask learning objectives, surpasses the capabilities of the CONTRAfold tool (Zhang and Yang 2021). By predicting multiple related outputs simultaneously, multitask learning enhances its versatility and prevents the model from overfitting or becoming too narrowly focused on a specific target.

A promising trend in RNA secondary structure prediction involves integrating experimental data with computational algorithms. For example, the RNAstructure package utilizes SHAPE (Selective 2'-Hydroxyl Acylation analyzed by Primer Extension) probing data as pseudo-free energy constraints within its energy minimization framework. By combining experimental probing data with computational approaches, the limitations of purely in silico methods and more accurate and biologically relevant structural predictions. This integrated strategy is particularly valuable for studying complex RNA molecules, such as lncRNAs, whose structures are critical for understanding their roles in cellular processes. As computational tools and experimental techniques continue to evolve, the field is poised to achieve unprecedented accuracy in RNA structure prediction, paving the way for deeper insights into the functional roles of ncRNAs and their potential applications in medicine and biotechnology.

13.4.2 Tertiary Structure Modeling Approaches

The uncovering of noncoding RNAs (ncRNAs) is advancing at a remarkable pace, accompanied by the rapid development of computational tools for predicting their three-dimensional (3D) structures. ncRNAs have emerged as key regulators in several tumor-associated processes, influencing transcriptional and post-transcriptional and signal transduction. Similar to proteins, ncRNAs adopt specific 3D structures that are fundamental to their function and therefore, these structures are essential for unraveling their biological roles (Singh and Gupta 2022). However, despite their importance, significant challenges remain in experimentally determining the 3D structures of ncRNAs, with progress in this area still limited.

Chemical probing-based techniques have been instrumental in studying RNA structures for several decades, offering detailed insights into the native conformations and flexibility of RNA molecules (Weeks 2010). These methods allow for high-throughput analysis and provide critical data for understanding RNA's three-dimensional organization (Lorenz et al. 2016). However, evident disparity exists between the number of experimentally resolved RNA structures and protein structures in the Protein Data Bank. While the noncoding human transcriptome vastly surpasses the protein-coding transcriptome by nearly 30-fold yet the number of experimentally resolved RNA structures remains disproportionately low, representing less than 1% of all determined protein structures to date (Cordero and Das 2015). This shortage underscores a major challenge, including that conventional computational approaches for protein structure prediction, like homology modeling, are not easily applicable to ncRNAs due to the limited availability of suitable structural templates.

Therefore, various computational tools can be utilized to predict the three-dimensional structures of ncRNA. The advancements in protein structure prediction methods have significantly influenced the field of RNA tertiary structure prediction. This has led to the development of dedicated RNA folding algorithms such as trRosettaRNA (Wang et al. 2023), RoseTTAFoldNA (Baek et al. 2024), DRfold (Li et al. 2023), and RhoFold (Chen et al. 2022). NuFold is an end-to-end deep neural network designed to predict RNA structures directly from input sequences (Kagaya et al. 2025). It leverages a nucleobase center representation, which allows for the flexible modeling of ribose ring conformations. Detailed analysis of the NuFold pipeline components revealed that its performance significantly improved by incorporating metagenome sequences for multiple sequence alignments and increasing the number of recycling steps. Another notable tool, AlphaFold 3, represents a significant advancement in RNA structure prediction, offering the ability to generate 3D RNA structures directly from primary sequences (Abramson et al. 2024). AlphaFold 3 integrates a sophisticated multiple sequence alignment (MSA) module, leveraging extensive genetic databases such as RNACentral, Rfam, and the nucleotide collection. This module identifies homologous sequences and structural templates, which AlphaFold 3 then uses to create refined vector-matrix representations of residues and their interactions, capturing intricate long-range interactions and structural details inferred from the MSA.

13.5 Integrative Multi-omics Approaches

Multi-Omics research is reshaping biological sciences by integrating data from proteomics, genomics, transcriptomics, metabolomics, and beyond to create a more complete picture of biological systems (Heo et al. 2021). This approach has been particularly impactful in cancer research, where large-scale multi-omics datasets are helping to uncover key molecular markers that influence responses to targeted therapies. By connecting these diverse data layers, researchers can better understand disease mechanisms and refine precision medicine strategies for more effective treatments (Chauvel et al. 2020).

Gaining deeper insights into the molecular traits of non-responsive or drug-resistant tumors can enhance the ability to predict therapy outcomes more accurately, ultimately improving treatment efficacy or overcoming resistance mechanisms. In recent years, advances in modeling techniques and high-throughput screening have transformed cancer research (Iadevaia et al. 2010). By integrating large-scale omics data, the progression of cancer can be understood at the tissue level, gaining valuable insights and how tumors respond to drugs and angiogenesis (Huang et al. 2013). Integrating transcriptomics and metabolomics data offers a holistic approach to understanding cellular metabolism and its regulation. This multi-omics integration allows for a more detailed and dynamic view of cellular processes. Similarly, the integration of proteomics and transcriptomics data has become essential in contemporary systems biology research. High-throughput technologies enable swift, balanced profiling of key molecular features, such as somatic mutations, and gene and protein expression. To make sense of this complex data, computational algorithms for multi-view clustering help identify meaningful patterns across different datasets. These innovations have been particularly impactful in defining clinical subtypes of cancer and other complex diseases, paving the way for more precise diagnostics and treatments.

13.5.1 Combining Transcriptomics, Genomics, and Proteomics

Therefore, integrating multi-omics data has become a crucial aspect of bioinformatics research, enabling a more holistic understanding of complex biological systems. Several strategies are now employed to combine transcriptomics, genomics, proteomics, and metabolomics data, each offering unique insights (Fig. 13.5). These approaches include co-expression analysis, which explores the relationships between gene expression levels; metabolite-gene networks, which link metabolic pathways to gene activity; and constraint-based models, which predict the flow of metabolites through biochemical networks. These strategies provide comprehensive views of cellular processes, facilitating deeper perceptions of the biomarker discovery, disease mechanisms, and probable therapeutic targets.

13.5.1.1 Genomics

The rapid advancements in DNA sequencing and genomic research have significantly transformed our understanding of genetic variation. The transition from studying individual genes to comparing entire genomes across populations has been made possible by increasingly efficient and high-speed data generation (Manolio et al. 2008). Standardized nomenclature has been established for both coding and noncoding genomic regions, as well as for different types of genetic variants, ensuring uniformity in scientific communication. Cutting-edge sequencing technologies, including Sanger sequencing (Sanger et al. 1977), DNA microarrays (Bumgarner 2013), and next-generation sequencing (NGS) (Van Dijk et al. 2014), have paved the way for comprehensive genomic analysis. The study of genomes heavily depends on reference sequences and insights into the distribution of common variants. To analyze genome-wide variant

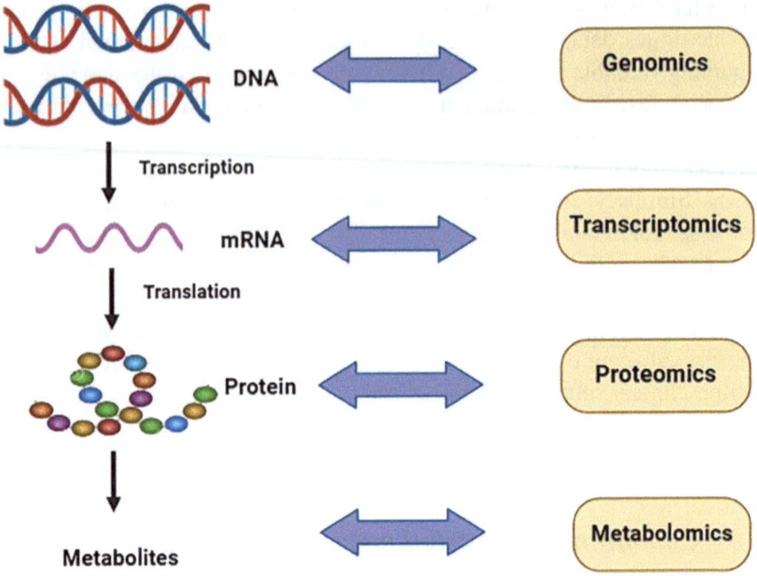

Fig. 13.5 Omics studies, an integrated comprehensive review of genomics, transcriptomics, proteomics, and metabolomics

data, numerous bioinformatics tools such as PLINK (Purcell et al. 2007), SNPTEST (Marchini et al. 2007), and Bioconductor (Gentleman et al. 2004) are widely used.

Additionally, several databases serve as crucial repositories for genomic information. Platforms like NCBI, Ensembl, UCSC Genome Browser, and the European Variation Archive provide access to vast genomic datasets (Manzoni et al. 2018). Within NCBI, dbSNP offers detailed information on single-nucleotide polymorphisms (SNPs), including their genomic location, functional implications, and prevalence across populations. Furthermore, ClinVar and the online Mendelian inheritance in man (OMIM) database play a vital role in linking genetic variants to specific traits and diseases, offering valuable insights into human genetic disorders.

13.5.1.2 Transcriptomics

Technologies with high-throughput capabilities, like microarrays and RNA sequencing, are opening up a new avenue in the generation of large-scale gene expression data from diverse biological systems (Larsson et al. 2020). In the last few years, machine learning has emerged as a critical tool for interpreting this vast amount of data. In fact, now it is possible to study complex patterns and relationships through the application of machine learning that would be rather difficult to trace using traditional methods. These computational tools have increasingly been employed to predict types of cancer as well as the identities of various cells based on gene expression data derived from microarrays, bulk RNA-Seq, and single-cell transcriptomics (Liu et al. 2021). The entire analysis usually commences with dataset selection, accessed through well-established repositories such as SRA, GEO (Barrett et al. 2013), and TCGA (Weinstein et al. 2013).

13.5.1.3 Proteomics

Proteomics is the comprehensive study of the proteome, employing a range of techniques including structural proteomics and protein–protein interaction analysis to gain insights into protein functions and relationships (Kelchtermans et al. 2014). The progression of high-throughput technologies, such as protein microarrays and mass spectrometry, has transformed the capacity to detect and measure thousands of proteins in a single experiment. Despite this progress, the vast amount and complexity of proteomics data present significant challenges for traditional approaches. To overcome this, Machine learning algorithms offer a robust alternative, enabling more efficient analysis and interpretation of these complex datasets (Sonsare and Gunavathi 2019). By harnessing the power of machine learning, researchers can uncover deeper insights into protein functions, interactions, and the molecular mechanisms behind diseases, ultimately advancing our understanding of biological processes (Khan et al. 2025). Tools such as the R package ReactomeGSA feature a function called "plot_correlations" that creates a comparative scatter plot between transcriptomics and proteomics data to predict similarity at the pathway level (Griss et al. 2020). GECKO models leverage the process of constraining metabolic fluxes in any GEM that includes enzymatic data, which helps reduce variability in constraint-based modeling outcomes and enhances the accuracy of predictions (Domenzain et al. 2022).

13.5.2 Insights from Integrative Analysis of ncRNA Data

Experimental research has played a fundamental role in uncovering the functions of ncRNAs in biological processes related to human health (Fritah et al. 2020). However, advancements in computational biology have provided a complementary approach, enabling the prediction of ncRNA interactions with other molecules, which can subsequently be confirmed through experimental analysis. This dynamic interplay between experimental and computational methods fosters continuous refinement—experimental findings enhance predictive models, while computational insights generate hypotheses for further validation—ultimately deepening our understanding of ncRNA functions (Dai et al. 2019). The use of ncRNA databases and bioinformatics tools provides a significant advantage, as they are continuously updated with newly acquired experimental data and undergo rigorous curation, thus enabling novel interactions. Additionally, studying ncRNA interactions with other molecules often involves constructing biological networks. One such approach is heterogeneous graph inference for MiRNA-disease association (HGIMDA), which employs a graph reconstruction method to predict potential miRNA-disease associations (Chen et al. 2016).

Several computational tools leverage advanced machine learning techniques to analyze ncRNA interactions. DeepTarget utilizes neural network-based auto-encoding and sequence–sequence interactions to predict miRNA–mRNA interactions (Lee et al. 2016). For ncRNA–protein interaction prediction, RPI-SAN identifies potential interaction pairs, while RPITER employs convolutional neural networks (CNN) and stacked auto-encoders (SAE) for more accurate predictions (Yi et al. 2018). Additionally, GCLMI integrates graph convolution and auto-encoding methods to model lncRNA–miRNA interactions (Huang et al. 2019). Meanwhile, eADAGE is designed for biological pathway enrichment by extracting meaningful patterns from expression data (Tan et al. 2017). Additionally, ncRNAs are thought to influence a large range of genomic interactions; therefore, research on enhancer RNAs (eRNAs) and certain lncRNAs has shown that these molecules facilitate promoter-enhancer interactions, ultimately regulating the expression of protein-coding genes (Licatalosi et al. 2008). Given the complexity of ncRNA interactions across multiple biological pathways, understanding their functions within intricate molecular networks remains a challenge. While the growth of knowledge in this field has been gradual, advancements in computational prediction models and novel tools based on network analysis are expected to enhance the efficiency of ncRNA research.

13.6 Future Pathways and Challenges in ncRNA Research

The rapid advancements in ncRNA research underscore their critical role which are essential for drug development and therapeutic pharmacology. RNA-based therapies are poised to be the next wave of treatment options, offering a novel approach that contrasts with traditional drug therapies, such as small molecules and protein-based treatments, which primarily focus on targeting proteins (Yu et al. 2016). These RNA-based treatments provide an innovative avenue for therapeutic intervention, expanding the possibilities in clinical medicine. Noncoding RNAs (ncRNAs) play crucial roles in cancer, acting both as oncogenes and tumor suppressors. A growing body of evidence shows that many ncRNAs are dysregulated in cancerous cells, contributing to the initiation and progression of the disease (Gao et al. 2020). These ncRNAs exert their influence by regulating a wide range of cellular processes critical for cancer development, including cell proliferation, apoptosis (programmed cell death), invasion, and metastasis. For instance, some ncRNAs promote tumorigenesis by enhancing cell division and preventing cell death, while others may function to suppress tumor growth by inhibiting these same processes. The dysregulation of these ncRNAs disrupts normal cellular homeostasis and facilitates the aggressive behavior of cancer cells, making them promising candidates for targeted therapeutic strategies in cancer treatment.

The expansion of RNA-based approaches presents distinct challenges due to the fundamental differences, particularly in terms of pharmacology and toxicity, between RNA drugs and traditional small-molecule or protein-based drugs. In order to expand the scope of druggable targets to include RNAs and the genome, challenges such as instability and difficulty in crossing cellular membranes remain significant hurdles. However, a new frontier is emerging with the use of recombinant or bioengineered RNA molecules, produced and folded in living cells, offering promising new opportunities for both research and therapeutic applications (Ho and Yu 2016). Although computational methods have advanced significantly in protein research, their application to noncoding RNAs (ncRNAs) is still emerging. Several challenges hinder progress in this area. Firstly, ncRNAs comprise a diverse range of molecular classes, each with distinct characteristics and regulatory mechanisms. Additionally, many ncRNA genes are less conserved than protein-coding genes and often have low expression levels without obvious knockout phenotypes, making functional studies more difficult. Furthermore, our understanding of ncRNAs remains incomplete, leading to a lack of sufficient training data for the effective application of machine learning or statistical models. An additional challenge is the generally low quality of ncRNA gene annotations, which further restricts the accuracy and reliability of computational analyses (Cheng et al. 2014). However, as technology advances and more data become available, these challenges are expected to be addressed. With the rapid development of computational tools and growing ncRNA databases, progress in this area is expected to accelerate (Fig. 13.6). One key issue when using circRNA quantification tools is the inconsistency between different tools. While each tool may demonstrate high precision in detecting circRNAs, their ability to accurately identify all true positives can vary significantly, making tool selection a critical factor in ensuring reliable results.

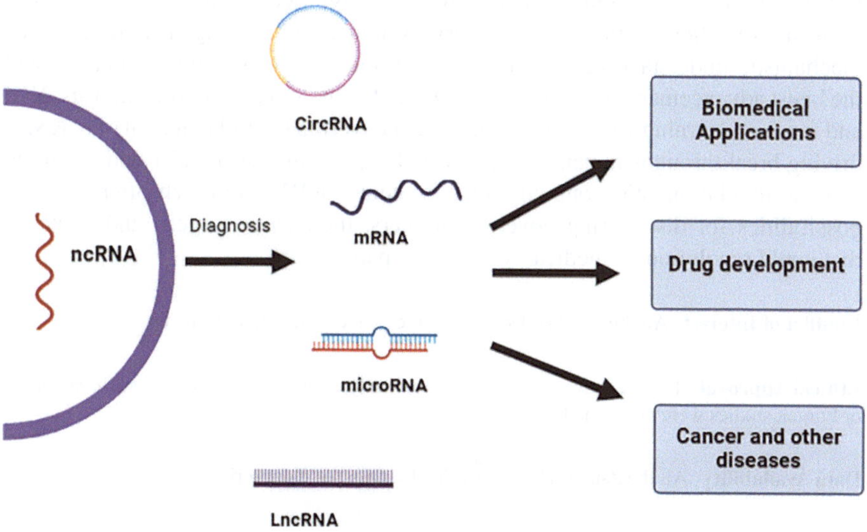

Fig. 13.6 Future prospects of ncRNA in therapeutic applications

13.7 Conclusion

In conclusion, bioinformatics has become a foundation in advancing our understanding of noncoding RNAs, particularly in the era of gene regulation and their involvement in a wide range of diseases, including cancer. As scientific research has increasingly focused on ncRNAs, especially long noncoding RNAs (lncRNAs), it has become clear that these molecules play critical roles in regulating cellular processes. The recognition of their importance has grown, and a wealth of evidence now connects the dysfunction of ncRNAs to various pathologies, offering new insights into disease mechanisms. Their growing significance in areas such as gene expression, gene regulation, functional annotation, chromatin remodeling, and cellular signaling pathways is highlighted by the emerging understanding of how their dysregulation can contribute to disease progression, including cancer, neurological disorders, respiratory diseases and cardiovascular diseases. Despite significant advances, a large proportion of ncRNAs remains poorly characterized, with many still lacking a clear understanding of their biogenesis, functional roles, and the specific mechanisms by which they exert their regulatory effects. Bioinformatics tools are essential in addressing these challenges, allowing researchers to annotate ncRNAs accurately, predict their interactions with other biomolecules, and map the complex networks of gene regulation that these molecules control. These tools provide a computational framework that supports the identification of novel ncRNAs, the prediction of their secondary and tertiary structures, sequencing and the modelling of their functional networks. The therapeutic potential of ncRNAs is particularly promising, as they can serve as diagnostic and prognostic biomarkers, offering valuable insights into disease onset and progression. Furthermore, their potential as therapeutic targets is immense, particularly in cancer treatment, where they could be harnessed for personalized medicine. However, significant hurdles remain, such as ensuring specificity, improving delivery systems, and gaining a more complete mechanistic understanding. As the field continues to evolve, bioinformatics, with the latest advancement combining with Omics technologies, network analysis, NGS and machine learning will remain central in uncovering the full potential of ncRNAs, driving breakthroughs in cancer therapy and expanding our broader understanding of gene regulation. With ongoing advancements, ncRNA research offers exciting possibilities for discovering novel biomarkers, therapeutic targets, and strategies that could revolutionize medical science and patient care.

Conflict of Interest All the authors declare that they have no conflict of interest.

Ethical Approval This chapter is a review of previously published accounts, as such, no animal or human studies were performed.

Data Availability All the data has been described within the manuscript.

References

Abramson J, Adler J, Dunger J, Evans R, Green T, Pritzel A, Ronneberger O, Willmore L, Ballard AJ, Bambrick J et al (2024) Accurate structure prediction of biomolecular interactions with AlphaFold 3. Nature 630:493–500

Alhajlah S, Jasim SA, Altalbawy FMA, Bansal P, Kaur H, Mohammed JS, Fenjan MN, Edan RT, Sharma MK, Zwamel AH (2025) Exploring the role of exosomal lncRNA in cancer immunopathogenesis: unraveling the immune response and EMT pathways. Exp Cell Res 445(1):114401

Amaral PP, Mattick JS (2008) Noncoding RNA in development. Mamm Genome 19(7):454–492

Amaral PP, Dinger ME, Mattick JS (2013) Non-coding RNAs in homeostasis, disease, and stress responses: an evolutionary perspective. Brief Funct Genomics 12(3):254–278

Atianand MK et al (2016) A long noncoding RNA lincRNA-EPS acts as a transcriptional brake to restrain inflammation. Cell 165:1672–1685

Axmann IM, Kensche P, Vogel J, Kohl S, Herzel H, Hess WR (2005) Identification of cyanobacterial non-coding RNAs by comparative genome analysis. Genome Biol 6(9):R73

Baek M, McHugh R, Anishchenko I, Jiang H, Baker D, DiMaio F (2024) Accurate prediction of protein-nucleic acid complexes using RoseTTAFoldNA. Nat Methods 21(1):117–121

Bailey TL (2011) DREME: motif discovery in transcription factor ChIP-seq data. Bioinformatics 27(12):1653–1659

Barrett T, Wilhite SE, Ledoux P, Evangelista C, Kim IF, Tomashevsky M, Marshall KA, Phillippy KH, Sherman PM, Holko M et al (2013) NCBI GEO: archive for functional genomics data sets—update. Nucleic Acids Res 41:D991–D995

Bhan A, Mandal SS (2014) Long noncoding RNAs: emerging stars in gene regulation, epigenetics, and human disease. ChemMedChem 9:1932–1956

Bonnet E, Wuyts J, Rouzé P, Van de Peer Y (2004) Evidence that microRNA precursors, unlike other non-coding RNAs, have lower folding free energies than random sequences. Bioinformatics 20:2911–2917

Bumgarner R (2013) Overview of DNA microarrays: types, applications, and their future. In: Current protocols in molecular biology. Chapter 22: unit 22.21

Cano-Gamez E, Trynka G (2020) From GWAS to function: using functional genomics to identify the mechanisms underlying complex diseases. Front Genet 11:424

Cech TR, Steitz JA (2014) The noncoding RNA revolution—trashing old rules to forge new ones. Cell 157:77–94

Chauvel C, Novoloaca A, Veyre P, Reynier F, Becker J (2020) Evaluation of integrative clustering methods for the analysis of multi-omics data. Brief Bioinform 21:541–552

Chen X, Clarence Yan C, Zhang X, You Z-H, Huang Y-A, Yan G-Y (2016) HGIMDA: heterogeneous graph inference for miRNA-disease association prediction. Oncotarget 7:65257–65269

Chen Q, Meng X, Liao Q, Chen M (2018) Versatile interactions and bioinformatics analysis of noncoding RNAs. Brief Bioinform 19(6):1110–1121

Chen J, Hu Z, Sun S, Tan Q, Wang Y, Yu Q, Zong L, Hong L, Xiao J, Shen T, King I, Li Y (2022) Interpretable RNA foundation model from unannotated data for highly accurate RNA structure and function predictions. arXiv. https://doi.org/10.1101/2022.08.06.503062

Cheng C, Moore J, Greene C (2014) Applications of bioinformatics to non-coding RNAs in the era of next-generation sequencing. Pac Symp Biocomput:412–416

Cordero P, Das R (2015) Rich RNA structure landscapes revealed by mutate-and-map analysis. PLoS Comput Biol 11(11):e1004473

Dai Q, Guo M, Duan X, Teng Z, Fu Y (2019) Construction of complex features for computational predicting ncRNA-protein interaction. Front Genet 10:18

Do CB, Woods DA, Batzoglou S (2006) CONTRAfold: RNA secondary structure prediction without physics-based models. Bioinformatics 22(14):e90–e98

Domenzain I, Sánchez B, Anton M, Kerkhoven EJ, Millán-Oropeza A, Henry C, Siewers V, Morrissey JP, Sonnenschein N, Nielsen J (2022) Reconstruction of a catalogue of genome-scale metabolic models with enzymatic constraints using GECKO 2.0. Nat Commun 13:3766

Esteller M (2011) Non-coding RNAs in human disease. Nat Rev Genet 12(12):861–874

Fabbri M, Girnita L, Varani G, Calin GA (2019) Decrypting noncoding RNA interactions, structures, and functional networks. Genome Res 29(9):1377–1388

Fallmann J, Will S, Engelhardt J, Grüning B, Backofen R, Stadler PF (2017) Recent advances in RNA folding. J Biotechnol 261:97–104

Fritah S, Muller A, Jiang W, Mitra R, Sarmini M, Dieterle M, Golebiewska A, Ye T, Van Dyck E, Herold-Mende C et al (2020) Temozolomide-induced RNA Interactome uncovers novel LncRNA regulatory loops in glioblastoma. Cancer 12:2583

Gangotia D, Gupta A, Mani I (2021) Role of bioinformatics in biological sciences. In: Singh V, Kumar A (eds) Advances in bioinformatics. Springer, Singapore, pp 3–22. https://doi.org/10.1007/978-981-33-6191-1_3

Gao N, Li Y, Li J, Gao Z, Yang Z, Li Y, Liu H, Fan T (2020) Long non-coding RNAs: the regulatory mechanisms, research strategies, and future directions in cancers. Front Oncol 10:598817

Gentleman RC, Carey VJ, Bates DM et al (2004) Bioconductor: open software development for computational biology and bioinformatics. Genome Biol 5:R80

Griffiths-Jones S, Bateman A, Marshall M, Khanna A, Eddy SR (2003) Rfam: an RNA family database. Nucleic Acids Res. 1;31(1):439-41. https://doi.org/10.1093/nar/gkg006. PMID: 12520045; PMCID: PMC165453

Griss J, Viteri G, Sidiropoulos K, Nguyen V, Fabregat A, Hermjakob H (2020) ReactomeGSA—efficient multi-omics comparative pathway analysis. Mol Cell Proteomics 19:2115–2124

Heo YJ, Hwa C, Lee GH, Park JM, An JY (2021) Integrative multi-omics approaches in cancer research: from biological networks to clinical subtypes. Mol Cells 44(7):433–443

Ho PY, Yu AM (2016) Bioengineering of noncoding RNAs for research agents and therapeutics. Wiley Interdiscip Rev RNA 7:186–197

Hombach S, Kretz M (2016) Non-coding RNAs: classification, biology and functioning. Adv Exp Med Biol, 937, 3–17. https://doi.org/10.1007/978-3-319-42059-2_1

Hu G et al (2016) LincRNA-Cox2 promotes late inflammatory gene transcription in macrophages through modulating SWI/SNF-mediated chromatin remodeling. J Immunol 196:2799–2808

Huang SS, Clarke DC, Gosline SJ, Labadorf A, Chouinard CR, Gordon W, Lauffenburger DA, Fraenkel E (2013) Linking proteomic and transcriptional data through the interactome and epigenome reveals a map of oncogene-induced signaling. PLoS Comput Biol 9:e1002887

Huang Y-A, Huang Z-A, You Z-H, Zhu Z, Huang W-Z, Guo J-X, Yu C-Q (2019) Predicting lncRNA-miRNA interaction via graph convolution auto-encoder. Front Genet 10:758

Iadevaia S, Lu Y, Morales FC, Mills GB, Ram PT (2010) Identification of optimal drug combinations targeting cellular networks: integrating phospho-proteomics and computational network analysis. Cancer Res 70:6704–6714

Jeck WR, Sorrentino JA, Wang K, Slevin MK, Burd CE, Liu J et al (2013) Circular RNAs are abundant, conserved, and associated with ALU repeats. RNA 19:141–157

Johnsons SM, Grosshans H, Shingara J, Byrom M, Jarvis R, Cheng A, Labourier E, Reinert KL, Brown D, Slack FJ (2005) RAS is regulated by the let-7 microRNA family. Cell 120:635–647

Kagaya Y, Zhang Z, Ibtehaz N et al (2025) NuFold: end-to-end approach for RNA tertiary structure prediction with flexible nucleobase center representation. Nat Commun 16:881

Kalvari I, Argasinska J, Quinones-Olvera N, Nawrocki EP, Rivas E, Eddy SR, Bateman A, Finn RD, Petrov AI (2018a) Rfam 13.0: shifting to a genome-centric resource for non-coding RNA families. Nucleic Acids Res 46(D1):D335–D342

Kalvari I, Nawrocki EP, Argasinska J, Quinones-Olvera N, Finn RD, Bateman A, Petrov AI (2018b) Non-coding RNA analysis using the Rfam database. Curr Protoc Bioinformatics 62(1):e51

Karaboga D, Aslan S (2016) A discrete artificial bee colony algorithm for detecting transcription factor binding sites in DNA sequences. Genet Mol Res 15(2):1–11

Keith JM (2008) Bioinformatics. Volume I. Data, sequence analysis and evolution (methods in molecular biology). Humana press, New York, p 562

Kelchtermans P, Bittremieux W, De Grave K, Degroeve S, Ramon J, Laukens K, Valkenborg D, Barsnes H, Martens L (2014) Machine learning applications in proteomics research: how the past can boost the future. Proteomics 14:353–366

Khan F, Khan MV, Kumar A, Akhtar S (2024a) Recent advances in the development of alpha-glucosidase and alpha-amylase inhibitors in type 2 diabetes management: Insights from in silico to in vitro studies. Curr Drug Targets. 25(12):782-795. https://doi.org/10.2174/1389450125666231120153151

Khan F, Shah AA, Kumar A, Akhtar S (2024b) In silico investigation against inhibitors of alpha-amylase using structure-based screening, molecular docking, and molecular simulations studies. Cell Biochem Biophys. 82(3):2873–2888. https://doi.org/10.1007/s12013-023-01169-2

Khan F, Ahmad S, Osama K et al (2025) Deoxynojirimycin derivatives as potent α-glucosidase inhibitors: in silico ADMET evaluation, molecular dynamics and in vitro validation studies. Mol Divers. https://doi.org/10.1007/s11030-025-11307-2

Kim JY, Stevens P, Karpurapu M et al (2022) Targeting ETosis by miR-155 inhibition mitigates mixed granulocytic asthmatic lung inflammation. Front Immunol 13:943554

Kircher M, Ludwig KU (2022) Systematic assays and resources for the functional annotation of non-coding variants. Med Genet 34(4):275–286

Lafontaine DL (2015) Noncoding RNAs in eukaryotic ribosome biogenesis and function. Nat Struct Mol Biol 22:11–19

Lagos-Quintana M, Rauhut R, Lendeckel W, Tuschl T (2001) Identification of novel genes coding for small expressed RNAs. Science 294:853–858

Larsson I, Uhlén M, Zhang C, Mardinoglu A (2020) Genome-scale metabolic modeling of glioblastoma reveals promising targets for drug development. Front Genet 11:459644

Lee B, Baek J, Park S, Yoon S (2016) deepTarget: end-to-end learning framework for microRNA target prediction using deep recurrent neural networks. In: Proceedings of the 7th ACM international conference on bioinformatics, computational biology, and health informatics, Seattle, pp 434–442

Legnini I, Morlando M, Mangiavacchi A, Fatica A, Bozzoni I (2014) A feedforward regulatory loop between HuR and the long noncoding RNA linc-MD1 controls early phases of myogenesis. Mol Cell 53:506–514

Lennox KA, and Behlke MA (2016) Cellular localization of long non-coding RNAs affects silencing by RNAi more than by antisense oligonucleotides. Nucleic Acids Res 44:863–877. https://doi.org/10.1093/nar/gkv1206

Li X, Yang L, Chen LL (2018a) The biogenesis, functions, and challenges of circular RNAs. Mol Cell 71:428–442

Li H, Zheng L, Jiang A, Mo Y, Gong Q (2018b) Identification of the biological affection of long noncoding RNA BC200 in Alzheimer's disease. Neuroreport 29(13):1061–1067

Li Y, Zhang C, Feng C et al (2023) Integrating end-to-end learning with deep geometrical potentials for ab initio RNA structure prediction. Nat Commun 14:5745

Licatalosi DD, Mele A, Fak JJ, Ule J, Kayikci M, Chi SW et al (2008) HITS-CLIP yields genome-wide insights into brain alternative RNA processing. Nature 456:464–469

Lin S, Qiu P (2025) Predicting microRNA target genes using pan-cancer correlation patterns. BMC Genomics 26:77

Liu C, Zhang YH, Deng Q, Li Y, Huang T, Zhou S, Cai YD (2017) Cancer-Related Triplets of mRNA-lncRNA-miRNA Revealed by Integrative Network in Uterine Corpus Endometrial Carcinoma. BioMed Res Int 3859582. https://doi.org/10.1155/2017/3859582

Liu J, Fan Z, Zhao W, Zhou X (2021) Machine intelligence in single-cell data analysis: advances and new challenges. Front Genet 12:655536

Lorenz R, Wolfinger MT, Tanzer A, Hofacker IL (2016) Predicting RNA secondary structures from sequence and probing data. Methods 103:86–98

Ma X, Kulkarni A, Zhang Z, Xuan Z, Serfling R, Zhang MQ (2012) A highly efficient and effective motif discovery method for ChIP-seq/ChIP-chip data using positional information. Nucleic Acids Res 40(7):e50

Manolio TA, Brooks LD, Collins FS (2008) A HapMap harvest of insights into the genetics of common disease. J Clin Invest 118:1590–1605

Manzoni C, Kia DA, Vandrovcova J, Hardy J, Wood NW, Lewis PA, Ferrari R (2018) Genome, transcriptome and proteome: the rise of omics data and their integration in biomedical sciences. Brief Bioinform 19(2):286–302

Marchini J, Howie B, Myers S et al (2007) A new multipoint method for genome-wide association studies by imputation of genotypes. Nat Genet 39:906–913

Mathews DH, Turner DH, Zuker M (2007) RNA secondary structure prediction. In: Current protocols in nucleic acid chemistry., Chapter 11, unit 11.2. https://doi.org/10.1002/0471142700.nc1102s28

Panwar B, Arora A, Raghava GP (2014) Prediction and classification of ncRNAs using structural information. BMC Genomics 15:127

Pavesi G, Mauri G, Pesole G (2004a) In silico representation and discovery of transcription factor binding sites. Brief Bioinform 5:217–236

Pavesi G, Mereghetti P, Mauri G, Pesole G (2004b) Weeder web: discovery of transcription factor binding sites in a set of sequences from co-regulated genes. Nucleic Acids Res 32. (Web Server issue), W199–W203:W199

Peschansky VJ, Wahlestedt C (2013) Non-coding RNAs as direct and indirect modulators of epigenetic regulation. Epigenetics 9(1):3–12. https://doi.org/10.4161/epi.27473

Pinkney HR, Wright BM, Diermeier SD (2020) The lncRNA toolkit: databases and in silico tools for lncRNA analysis. Non-Coding RNA 6:49

Ponjavic J, Ponting CP, Lunter G (2007) Functionality or transcriptional noise? Evidence for selection within long noncoding RNAs. Genome Res 17:556–565

Purcell S, Neale B, Todd-Brown K et al (2007) PLINK: a tool set for whole-genome association and population-based linkage analyses. Am J Hum Genet 81:559–575

Rincon-Riveros A, Morales D, Rodríguez JA, Villegas VE, López-Kleine L (2021) Bioinformatic tools for the analysis and prediction of ncRNA interactions. Int J Mol Sci 22(21):11397

Rivas E, Klein RJ, Jones TA, Eddy SR (2001) Computational identification of noncoding RNAs in *E. Coli* by comparative genomics. Curr Biol 11:1369–1373

Sandve GK, Drablos F (2006) A survey of motif discovery methods in an integrated framework. Biol Direct 1:11

Sanger F, Nicklen S, Coulson AR (1977) DNA sequencing with chain-terminating inhibitors. Proc Natl Acad Sci USA 74:5463–5467

Sato K, Akiyama M, Sakakibara Y (2021) RNA secondary structure prediction using deep learning with thermodynamic integration. Nat Commun 12:941

Schimmel P (2018) The emerging complexity of the tRNA world: mammalian tRNAs beyond protein synthesis. Nat Rev Mol Cell Biol 19:45–58

Seifuddin F, Pirooznia M (2021) Bioinformatics approaches for functional prediction of long noncoding RNAs. Methods Mol Biol 2254:1–13

Sieglaff DH, Dunn WA, Xie XS, Megy K, Marinotti O, James AA (2009) Comparative genomics allows the discovery of cis-regulatory elements in mosquitoes. Proc Natl Acad Sci USA 106:3053–3058

Singh KP, Gupta S (2022) 3D modeling of non-coding RNA interactions. In: Schmitz U, Wolkenhauer O, Vera-González J (eds) Systems biology of MicroRNAs in cancer. Advances in experimental medicine and biology, 1385. Springer, Cham. https://doi.org/10.1007/978-3-031-08356-3_11

Sinha S, Tompa M (2003) YMF: a program for discovery of novel transcription factor binding sites by statistical overrepresentation. Nucleic Acids Res 31(13):3586–3588

Slack FJ, Chinnaiyan AM (2019) The role of non-coding RNAs in oncology. Cell 179(5):1033–1055

Sonsare PM, Gunavathi C (2019) Investigation of machine learning techniques on proteomics: a comprehensive survey. Prog Biophys Mol Biol 149:54–69
Statello L, Guo CJ, Chen LL et al (2021) Gene regulation by long non-coding RNAs and its biological functions. Nat Rev Mol Cell Biol 22:96–118
Sun YM, Chen YQ (2020) Principles and innovative technologies for decrypting noncoding RNAs: from discovery and functional prediction to clinical application. J Hematol Oncol 13:109
Taft RJ, Glazov EA, Lassmann T, Hayashizaki Y, Carninci P, Mattick JS (2009) Small RNAs derived from snoRNAs. RNA 15:1233–1240
Tan J, Doing G, Lewis KA, Price CE, Chen KM, Cady KC, Perchuk B, Laub MT, Hogan DA, Greene CS (2017) Unsupervised extraction of stable expression signatures from public compendia with an ensemble of neural networks. Cell Syst 5:63–71.e6
Thomas-Chollier M, Herrmann C, Defrance M, Sand O, Thieffry D, van Helden J (2012) RSAT peak-motifs: motif analysis in full-size ChIP-seq datasets. Nucleic Acids Res 40(4):e31
Tomlins SA, Day JR, Lonigro RJ, Hovelson DH, Siddiqui J, Kunju LP, Dunn RL, Meyer S, Hodge P, Groskopf J et al (2016) Urine TmPrSs2:ERG plus PCA3 for individualized prostate cancer risk assessment. Eur Urol 70:45–53
Van Dijk EL, Auger H, Jaszczyszyn Y et al (2014) Ten years of next-generation sequencing technology. Trends Genet 30:418–426
Van Helden J, André B, Collado-Vides J (1998) Extracting regulatory sites from the upstream region of yeast genes by computational analysis of oligonucleotide frequencies. J Mol Biol 281(5):827–842
Vos T, Flaxman AD, Naghavi M, Lozano R, Michaud C, Ezzati M, Shibuya K, Salomon JA, Abdalla S, Aboyans V et al (2012) Years lived with disability (YLDs) for 1160 sequelae of 289 diseases and injuries 1990-2010: a systematic analysis for the global burden of disease study 2010. Lancet 380:2163–2196
Vu H, Ernst J (2022) Universal annotation of the human genome through integration of over a thousand epigenomic datasets. Genome Biol 23:9
Wang W, Feng C, Han R et al (2023) trRosettaRNA: automated prediction of RNA 3D structure with transformer network. Nat Commun 14:7266
Weeks KM (2010) Advances in RNA structure analysis by chemical probing. Curr Opin Struct Biol 20(3):295–304
Weinstein JN, Collisson EA, Mills GB, Shaw KRM, Ozenberger BA, Ellrott K, Sander C, Stuart JM, Chang K, Creighton CJ et al (2013) The cancer genome atlas pan-cancer analysis project. Nat Genet 45:1113–1120
Wright ES (2015) DECIPHER: harnessing local sequence context to improve protein multiple sequence alignment. BMC Bioinformatics 16:322
Wright ES (2020) RNAconTest: comparing tools for noncoding RNA multiple sequence alignment based on structural consistency. RNA 26(5):531–540
Wright ES, Yilmaz LS, Corcoran AM, Okten HE, Noguera DR (2014) Automated design of probes for rRNA-targeted fluorescence in situ hybridization reveals the advantages of using dual probes for accurate identification. Appl Environ Microbiol 80:5124–5133
Yan H, Bu P (2021) Non-coding RNA in cancer. Essays Biochem 65(4):625–639
Yi H-C, You Z, Huang D-S, Li X, Jiang T-H, Li L-P (2018) A deep learning framework for robust and accurate prediction of ncRNA-protein interactions using evolutionary information. Mol Ther Nucleic Acids 11:337–344
Yotsukura S, Duverle D, Hancock T, Natsume-Kitatani Y, Mamitsuka H (2017) Computational recognition for long non-coding RNA (lncRNA): software and databases. Brief Bioinform 18:9–27
Yu Q, Huo H, Chen X, Guo H, Vitter JS, Huan J (2015) An efficient algorithm for discovering motifs in large DNA data sets. IEEE Trans Nanobioscience 14(5):535–544
Yu AM, Tian Y, Tu MJ, Ho PY, Jilek JL (2016) MicroRNA pharmacoepigenetics: posttranscriptional regulation mechanisms behind variable drug disposition and strategy to develop more effective therapy. Drug Metab Dispos 44:308–319

Zhang Y, Yang Q (2021) A survey on multi-task learning. IEEE Trans Knowl Data Eng 34(12):5586–5609

Zhang Y, Wang P, Yan M (2016) An entropy-based position projection algorithm for motif discovery. BioMed Res Int 2016:9127474

Zhao M, Li YP, Geng XR et al (2019) Expression level of MiRNA-126 in serum exosomes of allergic asthma patients and lung tissues of asthmatic mice. Curr Drug Metab 20(10):799–803

Zuyderduyn S, Sukkar MB, Fust A, Dhaliwal S, Burgess JK (2008) Treating asthma means treating airway smooth muscle cells. Eur Respir J 32:265–274

Chapter 14
Challenges and Perspectives in ncRNA Research

Aamir Ahmad

Contents

Abstract The research on noncoding RNAs has come a long way since the realization that these RNA molecules hold a lot of promise, particularly in terms of diagnosis, prognosis, and perhaps treatment of several human diseases. A number of noncoding RNAs, ranging from linear microRNAs and long-noncoding RNAs to circular RNAs, and several others are now under investigation. A logical way forward is the testing of these noncoding RNAs to bring them from bench to diagnostic laboratories, clinics, and bedside. However, the transition has not been very smooth, and even challenging at times, because of the unique nature of these biological molecules and the expertise and fine technologies needed to evaluate them, along with the refinement of methodologies to deliver them as therapeutics. This chapter specifically focuses on the associated challenges, such as expression/detection, target

A. Ahmad (✉)
Dermatology Institute and Translational Research Institute, Academic Health System, Hamad Medical Corporation, Doha, Qatar

Department of Bioengineering, Integral University, Lucknow, Uttar Pradesh, India

Centre of Medical and Bio-allied Health Sciences Research, Ajman University, Ajman, UAE

Department of Biosciences and Bioinformatics, Xi'an Jiaotong-Liverpool University, Suzhou, China
e-mail: aahmad9@hamad.qa

M. F. Ullah, A. Ahmad (eds.), *Non-coding RNAs*, Experientia Supplementum 115,
https://doi.org/10.1007/978-3-032-06948-1_14

validation, immunogenicity, and delivery, that are hampering the utilization of noncoding RNAs in clinics.

Keywords Noncoding RNAs · miRNA · lncRNA · circRNA · Immune response · Clinical trial · Delivery

Abbreviations

circRNAs	Circular RNAs
lncRNAs	Long noncoding RNAs
miRNAs	microRNAs
piRNAs	Piwi-interacting RNAs
siRNAs	Small interfering RNAs
snoRNAs	Small nucleolar RNAs

14.1 Introduction

Before the current realization of an important role of noncoding RNAs in human health and diseases, they were considered a cellular "junk" and their story is truly that of a junk turning into treasure (Ahmad 2016). Noncoding RNAs, for a long time, were considered transcriptional noise, which was primarily due to their perceived lower levels of expression and lower identification/detection in common molecular screenings (Mattick et al. 2023). Improvement in detection methodologies has played a big role in the "visibility" of noncoding RNAs, and now they are believed to be ubiquitous (Benacka et al. 2023; George et al. 2024). There is abundant literature now available on noncoding RNAs, covering almost every disease (Nemeth et al. 2024; Alsayed et al. 2023; Loganathan and Doss 2023), and, more importantly, there is continued interest in the physiological activities and functions of these RNA molecules (Silva et al. 2024; Liao et al. 2025). As the name suggests, noncoding RNAs do not code for any protein products, albeit that definition is being challenged with evidence of some peptides being coded by noncoding RNAs (Tan et al. 2025; Xiao et al. 2024; Tornesello et al. 2025).

Among all the different classes of noncoding RNAs, microRNAs (miRNAs) are perhaps the most widely studied noncoding RNAs, followed by long noncoding RNAs (lncRNAs) (Chen and Kim 2024; Martino et al. 2025). miRNAs are small noncoding RNAs with an average length of 22 nucleotides (Adeeb et al. 2023; O'brien et al. 2018). They modulate the expression of their target genes by binding to their 3′ untranslated regions. This also defines the versatility of the mode of action of miRNAs because even though miRNAs repress the expression and function of their target genes, the function of their target genes signifies miRNA action. For example, in the context of cancer, suppression of an oncogene (a gene that positively correlates with cancer) would mean a tumor suppressive action of miRNA, whereas suppression of a tumor suppressor gene (a gene that suppresses cancer) would mean a tumor-inducing or oncogenic function of the miRNA. Similarly, a metastasis-promoting miRNA will be the one that suppresses metastasis-inhibiting genes, while a metastasis suppressor miRNA is the one that inhibits metastasis-promoting genes. LncRNAs are relatively longer RNA molecules—more than 200 nucleotides long (Mattick et al. 2023). They can influence cellular processes through different mechanisms and one of the important mode of action of lncRNAs is their sponging of miRNAs, thereby interfering with the activity of miRNA and inhibiting it (Ma et al. 2023). Circular RNAs (circRNAs), the circular noncoding RNAs, are not linear and open-ended as miRNAs and lncRNAs; however, similar to lncRNAs, they can also sponge and inhibit miRNAs, thus inhibiting miRNA functions (Ma et al. 2023). Additionally, circRNAs can affect transcription and serve as protein decoys, scaffolds and recruiters (Zhou et al. 2020). miRNAs, lncRNAs, and circRNAs are some of the better described noncoding RNAs (Nemeth et al. 2024; Gareev et al. 2023), but there are several other noncoding RNAs as well, such as small interfering RNAs (siRNAs), Piwi-interacting RNAs (piRNAs), and small nucleolar RNAs (snoRNAs). There is so much scientific data, mostly from in vitro investigations, but some also from in vivo and clinical trials, on the broad utility of investigating the different noncoding RNAs. Readers can easily access all this data, and this is not the goal of this write-up. This chapter focuses on the various challenges that researchers working on noncoding RNAs face.

14.2 Endogenous Expression Levels

The endogenous expression levels and the detection thereof are very important for the characterization of noncoding RNAs and the establishment of their role in diseases. The detection techniques have advanced over the years, but still need sophisticated instruments. Beyond techniques and methodologies, a bigger challenge is the endogenous expression pattern of noncoding RNAs. For example, it has been estimated that the top five expressed miRNAs can represent as much as ~50% of the entire miRNA pool (O'brien et al. 2018; De Rie et al. 2017) leading to an extreme overrepresentation of upregulated miRNAs and severe depletion of thousands of other miRNAs. Also, almost a quarter of all miRNAs are typically expressed at very low levels, irrespective of cell type. On a similar note, lncRNAs (Wang et al. 2024) and circRNAs (Tao et al. 2025) have also been reported to be expressed mostly at low expression levels endogenously, except for the ones that are highly upregulated in individual disease conditions. Thus, whereas expression of a few abundant noncoding RNAs is possible using existing techniques, the current research is missing out on transcripts that are not very abundant. It can not be ruled out that the noncoding RNAs with very low expression have a role to play in health and disease—its just that our methodologies to detect noncoding RNAs need to improve.

14.3 Disconnect Between Target Prediction and Verification

Bioinformatics-based target prediction is an important tool used by researchers in the field of noncoding RNAs to list the target genes that will be regulated by their individual noncoding RNA of interest. For miRNAs, researchers utilize these tools to list target genes, while for lncRNAs, predictions can be made for sponged miRNAs and the targeted genes. To begin with, the predictions list dozens, if not hundreds, of targets, which makes it impractical to screen the entire list in laboratory settings. Researchers resort to either picking up the top few targets or targets based on reported functionality that makes more sense in the context of disease being investigated. This already introduces an element of bias in the research, and the innovation is compromised. Additionally, some web-based tools rely entirely or heavily on published reports, which again compromises the innovation and pushes researchers to pursue what is perhaps already shown in other diseases/models. Another layer of complexity in target identification and validation comes from the fact that physiological systems are very different from laboratory-based investigations, which check the interactions between individual entities. Here, the interactions have to endure in the presence of multiple by standers and factors which may positively or negatively influence the interactions. It is because of these reasons that many noncoding RNAs-targets predicted and verified in vitro do not play, exactly as predicted, in viable physiological systems. Another reality often overlooked is the redundancy of targets, i.e., though it's appreciated that a noncoding RNA can target and modulate multiple gene and noncoding targets, it's not always remembered that a single gene or noncoding RNA can be targeted by multiple noncoding RNAs. What this means is that a single gene target might be targeted by multiple miRNAs and a single miRNA might be targeted by multiple lncRNAs and circRNAs, and therefore modulating a single miRNA or lncRNA or circRNA might not be adequate to see expected physiological changes as this might involve simultaneous targeting of multiple noncoding RNAs in order to make sure that the desired regulation of target is being achieved.

14.4 Adverse Immunogenic Effects of Noncoding RNAs

One possible drawback using noncoding RNAs as therapeutics is their immune response-eliciting capabilities. RNA molecules can act as immune-stimulatory factors that are recognized by the innate immune system (Uehata and Takeuchi 2020). Such induction of innate immunity by mRNA vaccines, for example, in the case of COVID-19, is a good example (Verbeke et al. 2022). The viral RNA, on the other hand, has evolved to escape immune surveillance. SARS-CoV-2 mRNA interacts with the host factors in an attempt to suppress immunogenic response (Zhao et al. 2024). While the pathogenic viral RNAs have evolved to survive, the individual noncoding RNAs are not adequately equipped to circumvent the host immune system. Activation of innate immunity is a necessary step in body's defence against a wide variety of external threats, including pathogens and RNA molecules. As would be expected, innate immunity is tightly regulated because aberrant and unnecessary activation can be harmful and even fatal (Cui et al. 2014). Thus, activation of innate immunity by noncoding RNAs can result in unwanted side effects and toxicities (Segal and Slack 2020). Systemic administration of noncoding RNAs can eventually drive both innate and adaptive immune responses, thereby amplifying the toxicity. While delivery of noncoding RNAs can elicit an immune response, we also need to recognize that noncoding RNAs are known to interact with and/or influence the immune system even as part of their normal physiological action. These effects range from a role of miRNAs (Gutierrez-Vazquez et al. 2017; Ahmad 2022; Das and Rao 2022; Fernando et al. 2012), lncRNAs (Ghahramani Almanghadim et al. 2025; Chini et al. 2023; Zhang et al. 2024; Brazao et al. 2016) as well as circRNAs (Amaya et al. 2023; Son et al. 2024; Qadir et al. 2023; Zurawska et al. 2021) in the activation of T cells to their role in macrophage activation, inflammation to B cell development and activation. In view of such an intricate relationship between noncoding RNAs and the immune system, it is important to keep these consequences in mind when evaluating the manipulation of noncoding RNA expression as a therapy. In addition, the packaging systems used to deliver noncoding RNAs can elicit their own immune responses, which also need to be considered (Liu et al. 2017).

14.5 Missed Promises at Clinical Trials

For any potential diagnostic and/or prognostic biomarkers, the interest of academicians is not enough. The end goal is to see the putative biomarkers being utilized in clinics and diagnostic laboratories for the benefit of patients to either diagnose the disease or predict its prognosis. This can not materialize without the conduct of optimum clinical trials that document the proof supporting such activity of the biomarkers in question. Despite the big promises, the results from clinical trials that tested noncoding RNAs have been inconclusive at best (Winkle et al. 2021). In fact, there are considerable differences between different classes of noncoding RNAs as well, when it comes to testing and results from clinical trials. For example, whereas miRNAs have been tested in fewer than twenty clinical trials, siRNAs have been tested in more than sixty clinical trials (Momin et al. 2021; Zhang et al. 2021). These numbers are a clear indication of a trend towards evaluation of siRNAs in clinical trials. It also turns out that none of the miRNAs have moved on to Phase III clinical trials, whereas at least some of the Phase III studies are testing siRNAs (Zhang et al. 2021). It has been speculated that perhaps the large number of targets for miRNAs might be a hindrance to their efficacy in clinical trials (Momin et al. 2021). The pleiotropic nature of anticancer agents is considered a highly desirable property (Ullah et al. 2016; Farooqi et al. 2018, 2020; Ahmad et al. 2012, 2015) because the pleiotropic nature allows therapies to simultaneously block multiple oncogenic signalling pathways thus limiting the ability of cancer cells to find alternate pathways for survival. However, simultaneous targeting of multiple gene targets by miRNAs might result in unintended side effects. This might explain the differences observed in clinical trials on siRNAs vs. the clinical trials on miRNAs. The miRNAs targets range from 30 to 250, whereas siRNAs are much more specific and their targets range only between 1 and 3 (Momin et al. 2021, Zhang et al. 2021). The more specific targeting by siRNAs might be the crucial factor determining their relative success in clinical trials, at least in comparison to miRNAs. A relatively higher number of early clinical trials (50%) have either been terminated or suspended when testing miRNAs, as compared to such number being 35% for siRNAs. To be fair, miRNAs are endogenously produced while the siRNAs being tested in clinical trials are designed and generated in labs, thus leaving much room in case of siRNAs for tweaking and designing a more efficient product. Also, as a testimony to the wide variety of diseases where noncoding RNAs can be effective, they have been tested in clinical trials for numerous and unrelated diseases and disorders (Zhang et al. 2021; Weng et al. 2019).

14.6 Challenges Related to Targeted Delivery and Bioavailability

Targeted delivery is one way to ensure optimum efficacy of any therapeutic. This is also true for noncoding RNAs. Targeted delivery becomes even more important because of the realization that often the ‘overexpressed’ molecular targets in diseases, such as on the surface of cancer cells, are also expressed on normal or noncancerous cells (Adashek et al. 2020). In such cases, a total destruction of every cell that harbours the target gene will result in the ablation of normal and healthy cells as well. This challenge can be circumvented if the therapeutic is either not administered systemically or it is activated only after reaching its intended target. If the therapy reached only its intended target/organ, this can significantly overcome the associated toxicity and side-effects. The current methodologies for targeted therapy involving noncoding RNAs are still in infancy and this represents one major area that needs particular attention and refinement.

The bioavailability and persistence of RNA molecules in biological fluids is challenging, given that many of these RNA molecules, particularly the likes of miRNAs, are single stranded and quite labile. This can render them unstable in circulation unless they are protected and stabilized through chemical modifications (Corey 2007; Chen et al. 2015). However, any modifications can themselves compromise the specificity and activity of noncoding RNAs, which is an important factor to be considered while designing and testing novel delivery mechanisms. As an example, native miRNAs are quite prone to degradation through the action of endogenous ribonucleases (Zhang et al. 2012) and specific chemical modifications (Segal et al. 2020; Lennox and Behlke 2011) can help noncoding RNAs escape ribonuclease action. These challenges are not limited to the cases which will involve testing of active noncoding RNAs for disease therapy. In several diseases, noncoding RNAs are up or downregulated (Lekka and Hall 2018; Moldogazieva et al. 2023; Benacka et al. 2023). This gives an opportunity to exploit noncoding RNAs as a therapeutic target where downregulation of an upregulated noncoding RNA or, conversely, upregulation of downregulated noncoding RNA can modulate or halt disease onset and/or progression. Pre-miRNA and anti-miRNA oligonucleotides have been used in laboratory-based in vitro experiments for a long time to increase or decrease, respectively, the levels of miRNAs and special formulations have been shown to be effective in vivo in modulating levels of miRNAs in animal models (Ahmad et al. 2011; Baigude and Rana 2014). However, it is well known that oligonucleotides, irrespective of whether they are single- or double-stranded, exhibit pharmacokinetic and pharmacodynamic properties that are significantly suboptimal (Ratti et al. 2020). A major reason for this inefficiency is the relatively large molecular size (>6000 Da) and, additionally, the negatively charged backbone of the oligonucleotides (Kole et al. 2012). Both, the noncoding RNAs as well as the targeting oligonucleotides face similar challenges—in vivo stability, precise targeting to intended cells/tissues, cell permeability (Zhao et al. 2019) and the associated nonspecific effects.

The normal trafficking of noncoding RNAs, especially miRNAs, starts in endosomes, and this is true for miRNAs that are delivered endogenously through exosomes, for example, or even those that would be delivered as part of an exogenous therapy. The movement from early endosomes to late endosomes and then onwards to lysosomes makes the intra-compartmental pH progressively acidic (Hu et al. 2015) and challenging to the stability and integrity of noncoding RNAs, and this is one of the focus for researchers working on optimizing the delivery of noncoding RNAs (Segal and Slack 2020).

14.7 Conclusions and Perspectives

It is irrefutable that noncoding RNAs have made a place for themselves when it comes to their promised utility as diagnostic, prognostic and therapeutic targets in almost all human diseases. The reported as well as emerging data are clearly very supportive of this. The promise, however, has not been effectively delivered as noncoding RNAs have not yet made it to the clinics. In this chapter, we discussed several challenges that still need to be overcome in order for the noncoding RNAs to fulfil their promises. One clear challenge is the relatively slow pace of advances in our capabilities. Detection (of particularly less abundant noncoding RNAs) as well as delivery of noncoding RNAs remains a barrier to the confirmation of noncoding RNAs in diagnosis/prognosis and therapy, respectively. While the ability of noncoding RNAs, such as that of miRNAs, to regulate multiple targets is currently being viewed as a negative attribute, even being blamed for the poor showing of noncoding RNAs in clinical trials, it is possible that future advancements in methodologies will exploit this very property, similar to what oncology has witnessed in the case of pleiotropic anticancer therapeutics. The transition of noncoding RNAs from bench to clinics and bedside has surely been slow, but continued sincere efforts are needed as we cannot afford to undo the progress made so far.

Conflict of Interest There exists no conflict of interest with regard to the content and writing of the manuscript.

Ethical Approval This chapter is a review of previously published accounts, as such, no animal or human studies were performed.

Data Availability All the data has been described within the manuscript.

References

Adashek JJ, Kato S, Lippman SM, Kurzrock R (2020) The paradox of cancer genes in non-malignant conditions: implications for precision medicine. Genome Med 12:16

Adeeb M, Therachiyil L, Moton S, Buddenkotte J, Alam MA, Uddin S, Steinhoff M, Ahmad A (2023) Non-coding RNAs in the epigenetic landscape of cutaneous T-cell lymphoma. Int Rev Cell Mol Biol 380:149–171

Ahmad A (2016) Non-coding RNAs: a tale of junk turning into treasure. Noncoding RNA Res 1:1–2

Ahmad A (2022) Epigenetic regulation of immunosuppressive tumor-associated macrophages through dysregulated microRNAs. Semin Cell Dev Biol 124:26–33

Ahmad A, Aboukameel A, Kong D, Wang Z, Sethi S, Chen W, Sarkar FH, Raz A (2011) Phosphoglucose isomerase/autocrine motility factor mediates epithelial-mesenchymal transition regulated by miR-200 in breast cancer cells. Cancer Res 71:3400–3409

Ahmad A, Sakr WA, Rahman KM (2012) Novel targets for detection of cancer and their modulation by chemopreventive natural compounds. Front Biosci (Elite Ed) 4:410–425

Ahmad A, Ginnebaugh KR, Li Y, Padhye SB, Sarkar FH (2015) Molecular targets of naturopathy in cancer research: bridge to modern medicine. Nutrients 7:321–334

Alsayed R, Sheikhan K, Alam MA, Buddenkotte J, Steinhoff M, Uddin S, Ahmad A (2023) Epigenetic programing of cancer stemness by transcription factors-non-coding RNAs interactions. Semin Cancer Biol 92:74–83

Amaya L, Grigoryan L, Li Z, Lee A, Wender PA, Pulendran B, Chang HY (2023) Circular RNA vaccine induces potent T cell responses. Proc Natl Acad Sci USA 120:e2302191120

Baigude H, Rana TM (2014) Strategies to antagonize miRNA functions in vitro and in vivo. Nanomedicine (Lond) 9:2545–2555

Benacka R, Szaboova D, Gulasova Z, Hertelyova Z, Radonak J (2023) Non-coding RNAs in human cancer and other diseases: overview of the diagnostic potential. Int J Mol Sci 24:16213

Brazao TF, Johnson JS, Muller J, Heger A, Ponting CP, Tybulewicz VL (2016) Long noncoding RNAs in B-cell development and activation. Blood 128:e10–e19

Chen LL, Kim VN (2024) Small and long non-coding RNAs: past, present, and future. Cell 187:6451–6485

Chen Y, Gao DY, Huang L (2015) In vivo delivery of miRNAs for cancer therapy: challenges and strategies. Adv Drug Deliv Rev 81:128–141

Chini A, Guha P, Malladi VS, Guo Z, Mandal SS (2023) Novel long non-coding RNAs associated with inflammation and macrophage activation in human. Sci Rep 13:4036

Corey DR (2007) Chemical modification: the key to clinical application of RNA interference? J Clin Invest 117:3615–3622
Cui J, Chen Y, Wang HY, Wang RF (2014) Mechanisms and pathways of innate immune activation and regulation in health and cancer. Hum Vaccin Immunother 10:3270–3285
Das K, Rao LVM (2022) The role of microRNAs in inflammation. Int J Mol Sci 23:15479
De Rie D, Abugessaisa I, Alam T, Arner E, Arner P, Ashoor H, Astrom G, Babina M et al (2017) An integrated expression atlas of miRNAs and their promoters in human and mouse. Nat Biotechnol 35:872–878
Farooqi AA, Khalid S, Ahmad A (2018) Regulation of cell signaling pathways and miRNAs by resveratrol in different cancers. Int J Mol Sci 19:652
Farooqi AA, Pinheiro M, Granja A, Farabegoli F, Reis S, Attar R, Sabitaliyevich UY, Xu B, Ahmad A (2020) EGCG mediated targeting of deregulated signaling pathways and non-coding RNAs in different cancers: focus on JAK/STAT, Wnt/β-catenin, TGF/SMAD, NOTCH, SHH/GLI, and TRAIL mediated signaling pathways. Cancers (Basel) 12:951
Fernando TR, Rodriguez-Malave NI, Rao DS (2012) MicroRNAs in B cell development and malignancy. J Hematol Oncol 5:7
Gareev I, Ahmad A, Wang J, Beilerli A, Ilyasova T, Sufianov A, Beylerli O (2023) Gastric juice non-coding RNAs as potential biomarkers for gastric cancer. Front Physiol 14:1179582
George TP, Subramanian S, Supriya MH (2024) A brief review of noncoding RNA. Egypt J Med Hum Genet 25:98
Ghahramani Almanghadim H, Karimi B, Valizadeh S, Ghaedi K (2025) Biological functions and affected signaling pathways by long non-coding RNAs in the immune system. Noncoding RNA Res 10:70–90
Gutierrez-Vazquez C, Rodriguez-Galan A, Fernandez-Alfara M, Mittelbrunn M, Sanchez-Cabo F, Martinez-Herrera DJ, Ramirez-Huesca M, Pascual-Montano A, Sanchez-Madrid F (2017) miRNA profiling during antigen-dependent T cell activation: a role for miR-132-3p. Sci Rep 7:3508
Hu YB, Dammer EB, Ren RJ, Wang G (2015) The endosomal-lysosomal system: from acidification and cargo sorting to neurodegeneration. Transl Neurodegener 4:18
Kole R, Krainer AR, Altman S (2012) RNA therapeutics: beyond RNA interference and antisense oligonucleotides. Nat Rev Drug Discov 11:125–140
Lekka E, Hall J (2018) Noncoding RNAs in disease. FEBS Lett 592:2884–2900
Lennox KA, Behlke MA (2011) Chemical modification and design of anti-miRNA oligonucleotides. Gene Ther 18:1111–1120
Liao Z, Liu W, Wang L, Xie W, Yao C, Huang Q, Jiang X, Ren C (2025) The role of non-coding RNA regulates stem cell programmed death in disease therapy. Noncoding RNA Res 13:57–70
Liu Y, Hardie J, Zhang X, Rotello VM (2017) Effects of engineered nanoparticles on the innate immune system. Semin Immunol 34:25–32
Loganathan T, Doss CG (2023) Non-coding RNAs in human health and disease: potential function as biomarkers and therapeutic targets. Funct Integr Genomics 23:33
Ma B, Wang S, Wu W, Shan P, Chen Y, Meng J, Xing L, Yun J, Hao L, Wang X, Li S, Guo Y (2023) Mechanisms of circRNA/lncRNA-miRNA interactions and applications in disease and drug research. Biomed Pharmacother 162:114672
Martino MTD, Tagliaferri P, Tassone P (2025) MicroRNA in cancer therapy: breakthroughs and challenges in early clinical applications. J Exp Clin Cancer Res 44:126
Mattick JS, Amaral PP, Carninci P, Carpenter S, Chang HY, Chen LL, Chen R, Dean C, Dinger ME, Fitzgerald KA, Gingeras TR, Guttman M, Hirose T, Huarte M, Johnson R, Kanduri C, Kapranov P, Lawrence JB, Lee JT, Mendell JT, Mercer TR, Moore KJ, Nakagawa S, Rinn JL, Spector DL, Ulitsky I, Wan Y, Wilusz JE, Wu M (2023) Long non-coding RNAs: definitions, functions, challenges and recommendations. Nat Rev Mol Cell Biol 24:430–447
Moldogazieva NT, Zavadskiy SP, Astakhov DV, Sologova SS, Margaryan AG, Safrygina AA, Smolyarchuk EA (2023) Differentially expressed non-coding RNAs and their regulatory networks in liver cancer. Heliyon 9:e19223

Momin MY, Gaddam RR, Kravitz M, Gupta A, Vikram A (2021) The challenges and opportunities in the development of MicroRNA therapeutics: a multidisciplinary viewpoint. Cells 10:3097

Nemeth K, Bayraktar R, Ferracin M, Calin GA (2024) Non-coding RNAs in disease: from mechanisms to therapeutics. Nat Rev Genet 25:211–232

O'brien J, Hayder H, Zayed Y, Peng C (2018) Overview of MicroRNA biogenesis, mechanisms of actions, and circulation. Front Endocrinol (Lausanne) 9:402

Qadir J, Wen SY, Yuan H, Yang BB (2023) CircRNAs regulate the crosstalk between inflammation and tumorigenesis: the bilateral association and molecular mechanisms. Mol Ther 31:1514–1532

Ratti M, Lampis A, Ghidini M, Salati M, Mirchev MB, Valeri N, Hahne JC (2020) MicroRNAs (miRNAs) and long non-coding RNAs (lncRNAs) as new tools for cancer therapy: first steps from bench to bedside. Target Oncol 15:261–278

Segal M, Slack FJ (2020) Challenges identifying efficacious miRNA therapeutics for cancer. Expert Opin Drug Discov 15:987–992

Segal M, Biscans A, Gilles ME, Anastasiadou E, De Luca R, Lim J, Khvorova A, Slack FJ (2020) Hydrophobically modified let-7b miRNA enhances biodistribution to NSCLC and downregulates HMGA2 in vivo. Mol Ther Nucleic Acids 19:267–277

Silva RG, Amaral PP, Franco GR, Goes-Neto A (2024) Exploring the hidden hot world of long noncoding RNAs in thermophilic fungus using a robust computational pipeline. Sci Rep 14:19797

Son CJ, Carnino JM, Lee H, Jin Y (2024) Emerging roles of circular RNA in macrophage activation and inflammatory lung responses. Cells 13:1407

Tan S, Yang W, Ren Z, Peng Q, Xu X, Jiang X, Wu Z, Oyang L, Luo X, Lin J, Xia L, Peng M, Wu N, Tang Y, Han Y, Liao Q, Zhou Y (2025) Noncoding RNA-encoded peptides in cancer: biological functions, posttranslational modifications and therapeutic potential. J Hematol Oncol 18:20

Tao X, Zhai SN, Liu CX, Huang Y, Wei J, Guo YL, Liu XQ, Li X, Yang L, Chen LL (2025) Degradation of circular RNA by the ribonuclease DIS3. Mol Cell 85:1674–1685. e8

Tornesello AL, Cerasuolo A, Starita N, Amiranda S, Cimmino TP, Bonelli P, Tuccillo FM, Buonaguro FM, Buonaguro L, Tornesello ML (2025) Emerging role of endogenous peptides encoded by non-coding RNAs in cancer biology. Noncoding RNA Res 10:231–241

Uehata T, Takeuchi O (2020) RNA recognition and immunity-innate immune sensing and its posttranscriptional regulation mechanisms. Cells 9:1701

Ullah MF, Bhat SH, Husain E, Abu-Duhier F, Hadi SM, Sarkar FH, Ahmad A (2016) Pharmacological intervention through dietary nutraceuticals in gastrointestinal neoplasia. Crit Rev Food Sci Nutr 56:1501–1518

Verbeke R, Hogan MJ, Lore K, Pardi N (2022) Innate immune mechanisms of mRNA vaccines. Immunity 55:1993–2005

Wang G, Lee-Yow Y, Chang HY (2024) Approaches to probe and perturb long noncoding RNA functions in diseases. Curr Opin Genet Dev 85:102158

Weng Y, Xiao H, Zhang J, Liang XJ, Huang Y (2019) RNAi therapeutic and its innovative biotechnological evolution. Biotechnol Adv 37:801–825

Winkle M, El-Daly SM, Fabbri M, Calin GA (2021) Noncoding RNA therapeutics — challenges and potential solutions. Nat Rev Drug Discov 20:629–651

Xiao Y, Ren Y, Hu W, Paliouras AR, Zhang W, Zhong L, Yang K, Su L, Wang P, Li Y, Ma M, Shi L (2024) Long non-coding RNA-encoded micropeptides: functions, mechanisms and implications. Cell Death Discov 10:450

Zhang Z, Qin YW, Brewer G, Jing Q (2012) MicroRNA degradation and turnover: regulating the regulators. Wiley Interdiscip Rev RNA 3:593–600

Zhang S, Cheng Z, Wang Y, Han T (2021) The risks of miRNA therapeutics: in a drug target perspective. Drug Des Devel Ther 15:721–733

Zhang Y, Liu H, Niu M, Wang Y, Xu R, Guo Y, Zhang C (2024) Roles of long noncoding RNAs in human inflammatory diseases. Cell Death Discov 10:235

Zhao C, Sun X, Li L (2019) Biogenesis and function of extracellular miRNAs. ExRNA 1:38

Zhao H, Cai Z, Rao J, Wu D, Ji L, Ye R, Wang D, Chen J, Cao C, Hu N, Shu T, Zhu P, Wang J, Zhou X, Xue Y (2024) SARS-CoV-2 RNA stabilizes host mRNAs to elicit immunopathogenesis. Mol Cell 84:490–505. e9

Zhou WY, Cai ZR, Liu J, Wang DS, Ju HQ, Xu RH (2020) Circular RNA: metabolism, functions and interactions with proteins. Mol Cancer 19:172

Zurawska AE, Mycko MP, Selmaj I, Raine CS, Selmaj KW (2021) Multiple sclerosis: circRNA profile defined reveals links to B-cell function. Neurol Neuroimmunol Neuroinflamm 8:e1041